# Classical Methods in Ordinary Differential Equations

## With Applications to Boundary Value Problems

# Classical Methods in Ordinary Differential Equations

## With Applications to Boundary Value Problems

Stuart P. Hastings
J. Bryce McLeod

Graduate Studies in Mathematics
Volume 129

American Mathematical Society
Providence, Rhode Island

2010 *Mathematics Subject Classification.* Primary 34B15, 34B16, 34C28, 34C37, 34E05, 35A24, 35C07, 37C29, 37D45.

---

For additional information and updates on this book, visit
**www.ams.org/bookpages/gsm-129**

---

**Library of Congress Cataloging-in-Publication Data**

Hastings, Stuart P., 1937–
Classical methods in ordinary differential equations : with applications to boundary value problems / Stuart P. Hastings, J. Bryce McLeod.
p. cm. — (Graduate studies in mathematics ; v. 129)
Includes bibliographical references and index.
ISBN 978-0-8218-4694-0 (alk. paper)
1. Boundary value problems. 2. Differential equations, Nonlinear. I. McLeod, J. Bryce, 1929– II. Title.

QA379.H377 2012
515′.352—dc23

2011029730

---



∞ The paper used in this book is acid-free and falls within the guidelines
established to ensure permanence and durability.
Visit the AMS home page at `http://www.ams.org/`

10 9 8 7 6 5 4 3 2 1 17 16 15 14 13 12

To our wives, Eileen and Eunice, whose support has made our careers possible, and to our mentors, Norman Levinson and Edward Titchmarsh, whose emphasis on classical mathematics can be seen throughout the book.

# Contents

# Preface

Mathematicians are sometimes categorized as "theory builders" or "problem solvers". The authors of this book belong firmly in the problem solver class and find most pleasure in delving into the details of a particular differential equation, usually one arising from science or engineering, with the aim of understanding how the solutions behave and determining existence and uniqueness of solutions with particular properties. On the other hand, no such classification is hard and fast, and usually our goal is to determine what properties of the equation are important in generating the desired behavior. For example, in what ranges of the parameters do we see this type of solution or that, and sometimes, how broad a class of equations can we discuss without losing the essential behavior. This is a step toward building a theory, but we have not usually been inclined to pursue this goal very far.

We are, of course, delighted if others are able to put our results in a broader context. This has been done for some examples in this text, and we have tried to point the reader towards these new theories. However it is our belief that usually, to derive our particular results, many of the details which we study are still important and need attention specific to the problem at hand. Exceptions, or borderline cases, are discussed, and we have tried to assess fairly the strengths of various approaches.

These problems arise in a variety of areas in science and engineering. Often the mathematical models in these fields consist of nonlinear partial differential equations, and the analysis of these equations leads to a system of nonlinear ordinary differential equations, for example by seeking a steady state, or by a similarity substitution. In other cases the original model is a system of ode's (ordinary differential equations). Knowledge of the behavior of the solutions to these ode systems can be vital to understanding the solutions of related pde's (partial differential equations), if any, and the corresponding physical phenomena. Thus our interests come into play and are, we hope, helpful to the modeler who originally obtained the equations.

The emphasis in this book is on mathematical techniques, rather than results or applications. We choose a variety of applied problems to illustrate these techniques, but we often do not give much discussion of the background of these problems. However we are careful to cite references where this background may be found. We also do not aim for great generality in our results. Instead, for ease of exposition, we usually discuss the simplest examples which illustrate the methods of interest. Again, we give citations where the reader will find more comprehensive discussions.

We wish to emphasize our belief that many of the important problems in differential equations arise from applications. There may be more general theories to be developed; indeed we hope this is the case. But we think that the inspiration for these theories will often come from particular models of new phenomena, discovered either by scientific research or by numerical experiments. We hope that the techniques we discuss in this book will be among those that are useful in analyzing the new phenomena on which the future development of the theory may depend.

The book is written under the assumption that the reader has had a basic course in ordinary differential equations which includes the following topics:

(1) The Picard theorem on existence and uniqueness of solutions to an initial value problem of the form

$$\mathbf{x}' = \mathbf{f}(t, \mathbf{x}),$$
$$\mathbf{x}(t_0) = \mathbf{x}_0$$

when the vector-valued function $\mathbf{f}$ is continuous and satisfies a local Lipshitz condition in $\mathbf{x}$.

(2) The continuous dependence of solutions on initial conditions and parameters.

(3) The general theory of linear systems of ode's with variable coefficients.

(4) Sturm-Liouville problems and Green's functions.

(5) An introduction to qualitative theory and phase plane analysis.

(6) Stability theory of equilibrium points for nonlinear autonomous systems, including the concepts of stable and unstable manifolds.

The material listed above is usually included in a standard graduate course in ode's, and also in some more advanced undergraduate courses.

Much of the material may be difficult for those with only a basic undergraduate course.

Some sections of the book use more advanced material, particularly nonlinear functional analysis and some topics in the calculus of variations. We attempt to outline some of the required background, but frankly, a student with only an ode prerequisite will find this material challenging. Such a student will have to consult the cited basic literature for a better understanding. However, in almost every case there is a classical approach to the same results offered later in the chapter, and these sections can be read independently.

In Chapter 1 we describe what we mean by "classical methods" for ode's and give some simple illustrative examples. Chapter 2 gives an introduction to the so-called "shooting" method for proving the existence of solutions to boundary value problems for ode's. Detailed examples of the shooting method will appear in a number of other chapters.

Chapters 3–18 are the heart of the book. Each chapter introduces one or more techniques, perhaps classical, perhaps modern, in a relatively simple setting that still includes the essential points. These are mostly examples which we have worked on, and often they have also been studied by other authors with alternative approaches. When this is the case, we discuss some of these alternative methods as well. Usually each approach has its own strong points, which we try to bring out in our discussion. For example, one approach may give a simpler proof while another may yield more information. Which type of proof, modern or classical, has which advantage varies from one problem to another. In some cases, the alternative method may not give the simplest proofs or most complete results for the ode problem at hand but has the advantage that it can be extended to cover related problems in partial differential equations. We do not attempt to cover these extensions, however.

Chapter 3 begins with an example where the shooting method appears not to work but where a proof using real analysis in infinite dimensions (Helly's theorem) can be replaced by a simple compactness argument in two dimensions. In the second part of this chapter we contrast two different shooting techniques for proving existence of certain important solutions to the second Painlevé transcendent, a second order nonlinear equation which arises in studying the Korteweg-de Vries equation for water waves.

In Chapter 4 we show how the Brouwer fixed point theorem can be used to prove the existence of periodic solutions to some autonomous systems. In Chapter 5 we describe three different approaches to a boundary value problem for a linear system.

In Chapter 6 we consider the existence of traveling wave solutions of the FitzHugh-Nagumo equations from neurobiology. Comparison is made with the technique of geometric perturbation theory.

In Chapter 7 we give elementary and rigorous proofs of the validity of matched asymptotic expansions for two example problems, in one case comparing our methods with those from geometric perturbation theory. Chapter 8 is something of a change of pace and is independent of the other sections. It explores the use of complex function theory techniques by extending a nonlinear ode into the complex plane. One point of interest is that the result established is the nonexistence of solutions to a simple looking third order boundary value problem.

In Chapter 9 we return to the question of periodic solutions. The well-known Falkner-Skan equation from fluid mechanics is rescaled, turning the question of existence of periodic solutions into a singularly perturbed problem. In Chapter 10 we study a problem in Poiseuille flow, comparing an elementary method with use of degree theory in a Sobolev space. Chapter 11 deals with buckling of a tapered rod. Classical methods are contrasted with the use of calculus of variations and bifurcation theory in a Hilbert space.

In Chapter 12 we give an extended discussion of uniqueness and multiplicity problems. We illustrate some techniques for proving that a boundary value problem has only one solution, and in addition we discuss some examples where the solution is not unique and the goal is to determine just how many solutions there are.

Chapter 13 gives an application of two-dimensional shooting to a problem from boundary layer theory in fluid mechanics. In Chapter 14 we give classical ode approaches to some important results of A. Lazer and coauthors, as well as short proofs of related pde theorems. In Chapter 15 we show how shooting techniques can lead to results about “chaos”. Comparison is made with the technique of Melnikov in the same setting of a forced pendulum equation.

This idea is carried forward in Chapter 16, where we discuss solutions with “spike” behavior and also a type of “chaos”. In Chapter 17 we outline a very recent approach of X. Chen and Sadhu to obtaining asymptotic expansions of solutions with boundary layers and spikes for a class of equations with quadratic nonlinear terms. The last of the core chapters is Chapter 18, in which families of spikes and transition layer solutions are found for another class of inhomogeneous reaction-diffusion equations. Three different proofs of a central result are discussed.

Finally, in Chapter 19, we describe three important unsolved problems in our area, problems which have challenged us and other researchers for a number of years and which we hope the reader will find attractive. It would be gratifying to see these problems solved by someone who learned of them from this book.

An experienced reader will now detect that important techniques have been neglected. Undoubtedly many will feel that their favorite method is omitted, or at least under-appreciated. Our main defense is that we have written most extensively about what we know best. Also, many of the omitted topics have been the subject of their own specialized monographs, which we have tried to cite appropriately. It is undoubtedly true that many other techniques have importance in a wide variety of problems, which we have neither the space nor the background to discuss in detail. Topics which are under-represented include Lin's method and others from the important and influential school of Hale, Chow, and Mallet-Paret, applications of the Moser twist theorem, use of bifurcation and degree theory (including center manifolds), comparison methods and ideas from the theory of competitive and cooperative systems (developed particularly by M. Hirsch and H. Smith), many topics generally related to chaos and to be found in the landmark monograph of Guckenheimer and Holmes, and others perhaps even farther from the realm of the classical techniques which are our focus. Our prejudice is that for the particular kinds of problems we study here, problems which appear frequently in applications, the methods illustrated are often effective and efficient. This is not meant to suggest that they would be best in all of the vast array of problems in ode's which are found in modern applied analysis.

Finally, we are delighted to thank the people who have assisted us with various parts of the book. We are indebted to Ina Mette, AMS acquisitions editor, whose emailed question "Have you thought of writing a book?" started the project off and whose steady encouragement helped keep it going. Other AMS staff, including Marcia Almeida, Barbara Beeton, our production editor, Arlene O'Sean, and others have been especially helpful as well.

Our colleague William Troy has been particularly helpful, providing useful advice and some clever proofs (as pointed out later). Thanks also to Matt Stoffregen, who while an undergraduate has gone over much of the material and done many of the problems, and to Susmita Sadhu, whose doctoral dissertation influenced several parts of the book. Others who have contributed useful discussions and proofs, and in some cases read entire chapters, include Professors Shangbing Ai, Xinfu Chen, David Kinderlehrer, Chunqing Lu, Patrick Rabier, Jon Rubin, Marshall Slemrod, Charles Stuart, Shin-Hwa Wang, and Yieh-Hei Wan. We are grateful to all for their help.

Chapter 1

# Introduction

## 1.1. What are classical methods?

By "classical" methods in ordinary differential equations we mean finite dimensional methods, largely from what is often called "classical analysis", as opposed to the infinite dimensional methods of functional analysis. We believe that ode's are in essence finite dimensional, as contrasted with pde's, integral equations, delay equations and various combinations of these, all of which may require infinite dimensional functional analysis to be understood. When we say that an ode is finite dimensional, we mean that the initial value problem is finite dimensional. In other words, initial conditions are points in $R^n$ rather than in some function space.

However, many difficult ode problems encountered in modern applied analysis are boundary value problems. It is quite common to cast these in infinite dimensional settings, where the solutions are elements of a Banach space. For example, a solution could be sought in the space $\mathcal{C}^2([a,b])$ of real-valued functions on an interval $[a,b]$ with continuous second derivatives, using an appropriate norm. One result of this approach is the need to discuss the smoothness of the solutions, making sure that they are in the desired space. Yet, smoothness should not be an issue for a system of the form $\frac{d\mathbf{x}}{dt} = f(t,x)$ if $f$ is smooth. Every solution is automatically at least as smooth as $f$, and to belabor this point in the notation is to waste time and perhaps to confuse those new to the area.

The reason for putting a boundary value problem into an infinite dimensional space is usually to allow the use of one form or another of degree theory, or perhaps an infinite dimensional fixed point theorem, to prove the existence of solutions. Indeed, we have sometimes found this approach useful ourselves. We do not claim that all boundary value problems in ode's can be handled with classical analysis. However when they can, often more information is obtained than in the abstract approaches. In particular, the methods used are more likely to be constructive in some sense and so can form the basis of a numerical method. They are also sometimes more "global", for instance by giving estimates of the size of a "small parameter".

In this text we introduce both classical and nonclassical approaches to a number of problems, offering the reader exposure to some of the newer methods which may be useful in other settings. We include as well some examples where modern techniques are the most efficient, or give the most information, or, indeed, are the only way we know to do the problem.

Not all fixed point theorems are infinite dimensional. The obvious example is the Brouwer fixed point theorem [**173**], which in its simplest form tells us that a continuous function $f$ mapping an interval $[a,b]$ into itself has a fixed point, i.e. a point $x \in [a,b]$ such that $f(x) = x$. This theorem definitely has a role in our subject. Consider, for example, the differential equation

$$x' = -x^3 + \sin t. \tag{1.1}$$

The periodicity of $\sin t$ raises the possibility that a solution $x(t)$ may also be periodic, with period $2\pi$. This is a boundary value problem, because a solution is periodic with period $2\pi$ if and only if it satisfies the boundary condition $x(0) = x(2\pi)$. (See Exercise 1.4.)

We will show that (1.1) has a periodic solution by considering, for each $\alpha \in [-2,2]$, the unique solution to this ode satisfying the initial condition $x(0) = \alpha$. Denote this solution by $x(t,\alpha)$. Since $x' < 0$ when $x > 1$ and $x' > 0$ when $x < -1$, all solutions beginning in $[-2,2]$ remain in this interval for all $t \geq 0$. Hence, $x(2\pi,\alpha)$ is defined for every $\alpha \in [-2,2]$ and also lies in $[-2,2]$. Applying the Brouwer fixed point theorem to the function $f(\alpha) = x(2\pi,\alpha)$, we conclude that there is some $\alpha^*$ such that $x(2\pi,\alpha^*) = \alpha^*$. Exercise 1.4 then tells us that the corresponding solution $x(t,\alpha^*)$ is $2\pi$-periodic.

In Chapter 4 we give an example where the Brouwer fixed point theorem is used in higher dimensions to obtain periodic solutions. Unfortunately, however, this theorem is not as useful as we might hope, for two reasons. One is that as soon as we move to higher dimensions, meaning systems of two or more ode's, it is usually difficult to verify the hypotheses of the theorem. The other is that even in one dimension, it is not hard to come up with a boundary value problem which has solutions but where this theorem does not apply.

Such a problem occurs if we make one simple change to the example just above. Instead of (1.1), consider the equation

$$x' = x^3 + \sin t, \tag{1.2}$$

and again ask if there are periodic solutions. If $x > 1$ then $x' > 0$. Therefore the solution $x(t, 2)$ with $x = 2$ at $t = 0$ is increasing for $t > 0$, and so the map $\alpha \to x(2\pi, \alpha)$, if it is defined at all,[1] does not map $[-2, 2]$ into itself.

We will show later that the question just posed, namely existence of periodic solutions of (1.2), can be answered with a powerful method which we will use frequently in this book, called the "shooting method". We will defer this discussion until Chapter 2, where we give an introduction to the shooting method designed mainly for those who have not studied the technique before. A reader with experience in the topic may wish to skim Chapter 2 and move on quickly to more interesting examples in later chapters.[2]

The contraction mapping theorem is another basic fixed point theorem that has many applications in ode's. We think of this as a borderline case between classical and modern analysis. Consider, for example, the existence and uniqueness theorem for initial value problems due to Picard (which we assume is familiar to the reader).

**Theorem 1.1.** *Suppose that a continuous real-valued function $f(t, x)$ is Lipschitz continuous in $x$ in some open neighborhood $\Omega$ of a point $(t_0, \alpha)$ in $R^2$. Then there is a unique solution $x(t, \alpha)$ of the initial value problem*

$$\begin{aligned} x' &= f(t, x), \\ x(t_0) &= \alpha \end{aligned} \tag{1.3}$$

*on some interval $[\tau_0, \tau_1]$ with $\tau_0 < t_0 < \tau_1$.*

The classical proof of this theorem, as given for example in the texts of Coddington and Levinson [**41**], or Hartman [**83**], is by the method of successive approximations. The problem (1.3) is rewritten as an integral equation

$$x(t) = \alpha + \int_0^t f(s, x(s))\, ds. \tag{1.4}$$

[1] It is not hard to show that if $\alpha > 1$, then the solution $x(t, \alpha)$ exhibits "finite time blowup", meaning that it exists for $t \geq 0$ only on some bounded interval $[0, T)$, with $\lim_{t \to T^-} x(t, \alpha) = \infty$. A term on the right-hand side such as $x^3$, which grows faster than linearly as $x \to \infty$, is the clue that this may happen. It is not immediately clear for any particular $\alpha$ if $T \leq 2\pi$ or $T > 2\pi$, but in fact, for $\alpha = 2$, $T < 2\pi$, and so the map described is not defined at $\alpha = 2$. See Exercise 1.1.

[2] However, the last few exercises in Chapter 2 may be of interest even to the more experienced reader.

A sequence of approximate solutions is defined iteratively by the equations

$$(1.5)\qquad \begin{aligned} x^{0}(t) &= \alpha, \\ x^{i+1}(t) &= \alpha + \textstyle\int_0^t f\left(s, x^{i}(s)\right)\, ds, \end{aligned}$$

and it is shown that there is an interval $[\tau_0, \tau_1]$, with $\tau_0 < t_0 < \tau_1$, such that each $x^i$ is defined in $[\tau_0, \tau_1]$ and moreover, $\lim_{i\to\infty} x^i(t)$ exists uniformly on $[\tau_0, \tau_1]$. It follows from standard advanced calculus arguments that the limiting function $x^\infty(t)$ is the desired solution to the initial value problem (1.3) and that this solution is unique. To prove all this, certain estimates are needed, which we omit. See [**41**] or many other texts for the details. Of course the proof easily extends to a system of $n$ equations.

The point of discussing this here is to observe that some other authors give a proof which uses Banach space terminology.[3] The integral equation (1.4) is used to define a mapping on the Banach space $\mathcal{C}([\tau_0, \tau_1])$ of continuous functions in an interval $[\tau_0, \tau_1]$. This requires estimates in order to choose an appropriate interval. Further estimates show that the mapping is a contraction and therefore has a unique fixed point. This fixed point is easily seen to be a solution of (1.3) on $[\tau_0, \tau_1]$ . Uniqueness is also a consequence of the contraction property.

The estimates required in these two proofs are exactly the same. Hence the proofs are of about the same length. Which proof is better? We believe that there is no significant difference, and the terminology used is a matter of taste. In this text we will consider either version a part of classical analysis, and in fact, Picard's existence proof [**192**] predates any of the functional analytic terminology developed by Banach and his successors. We often prefer the older terminology because we feel it brings the reader into closer contact with the technical manipulations that are necessary in any proof. We value this above the sense of generality that comes from more abstract notation but which we believe is sometimes of less practical use than it may seem at first glance. However, on occasion we will use the contraction mapping terminology, recognizing both its efficiency of notation in more routine applications of the method when the spaces involved are standard and its flexibility when a clever choice of norm makes a problem more tractable.

Another important fixed point theorem often used in our subject is that of Schauder (see Section 9.2.2 of [**62**]). We mention this theorem several times in this book, giving some interesting applications. However, it involves more sophisticated Banach space properties than the contraction mapping theorem, so that we do not consider it classical, and in every case that is solely an ode problem we give an alternative proof.

[3]See for example [**82**] or [**34**] for definitions and details.

The fuzzy line between classical and modern methods becomes even harder to define when we turn to some of the other problems considered in this book, where methods of dynamical systems have been used which are, at least sometimes, entirely finite dimensional. This occurs, for example, in Chapters 6 and 7. Our reasons for discussing problems of this sort vary. For example, in Chapter 6 we study an example where a newer approach, still finite dimensional, yields an elegant picture of the basic phenomenon, while our analysis of the details of the flow leads to some extra information. In Chapter 4, on the other hand, the proof we present of existence of a periodic solution is mainly of interest because of its simplicity, and a more sophisticated analysis yields more information.

We also include in our term "classical methods" the use of standard complex variable techniques. These are crucial in Chapter 8. However most of the other examples in this book remain in the real domain.

## 1.2. Exercises

**Remark about the exercises:** *Some exercises are called "routine", or "straightforward", or "interesting", etc. These comments are mainly for the more experienced readers of this book. We recommend that a student learning this material for the first time try every exercise, as we hope that even the "routine" ones will be instructive. We also hope that exercises with no comment are at least moderately interesting and moderately challenging.*

**Exercise 1.1.** Referring to footnote 1, obtain the best rigorous estimate you can for the value of $T$ at which the solution $x(t,2)$ of (1.2) blows up.

**Exercise 1.2.** One reason for expecting that (1.1) has periodic solutions is that the similar looking linear equation $x' = -x + \sin t$ has periodic solutions. (a) Find all periodic solutions of this linear equation. (b) Is the $2\pi$-periodic solution of (1.1) unique?

**Exercise 1.3.** (Needed later) For $n \geq 2$, consider the differential equation

$$\left(r^{n-1}u'\right)' + r^{n-1}f(u) = 0$$

where $f$ is continuously differentiable on $[0,\infty)$. We seek a solution $u = u(r)$ on an interval $(0, r_1)$ such that $\frac{d^2u}{dr^2}$ exists and is continuous on $[0, r_1)$ and $u$ satisfies the initial conditions

$$u(0) = u_0,$$
$$u'(0) = 0.$$

Note that the ode is singular at $r = 0$ and so Theorem 1.1 does not apply. Nevertheless, use a successive approximation scheme to prove that for every $u_0 > 0$ there is a unique solution to this initial value problem on some interval $[0, r_1)$ and that this solution depends continuously on $u_0$. In fact, $\frac{\partial u}{\partial u_0}$ is continuous. Hint: Consider the system of integral equations

$$u(r) = u_0 + \int_0^r v(\rho)\, d\rho,$$
$$v(r) = -\frac{1}{r^{n-1}} \int_0^r \rho^{n-1} f(u(\rho))\, d\rho.$$

For the $C^1$ dependence on $u_0$, follow [**41**]. For a proof using modern terminology, see [**181**, Proposition 2.35].

**Exercise 1.4.** (Routine; needed later) Suppose that $\phi$ is a solution to an equation of the form

$$x' = f(x) + \sin t$$

on $[0, 2\pi]$ and that $\phi(0) = \phi(2\pi)$. Give a condition on $f$ which ensures that $\phi$ is periodic of period $2\pi$. Hint: Consider $\psi(t) = \phi(t + 2\pi)$.

**Exercise 1.5.** Suppose that $n \geq 1$ and $f : R^n \to R^1$ is Lipschitz continuous in each bounded set. Let $\phi$ be a solution of the $n^{\text{th}}$ order scalar equation

$$x^{(n)} = f\left(x, x', x'', ..., x^{(n-1)}\right)$$

which exists on an interval $[0, T]$. Suppose that for some increasing sequence $\{t_k\}_{1 \leq k < \infty} \subset [0, T)$ which converges to $T$, $\phi(t_k)\, \phi'(t_k) = 0$ for each $k$. Show that $\phi$ is constant. Hint: Use Rolle's theorem. Conclude that if $\phi$ is not constant, then neither $\phi$ nor $\phi'$ has infinitely many zeros in $[0, T]$. Hence, for example, the function $t^7 \sin \frac{1}{t}$ (0 at the origin), though having three continuous derivatives at 0, cannot be a solution of an equation of the type considered above on an interval containing 0.

Chapter 2

# An introduction to shooting methods

## 2.1. Introduction

This chapter starts with some simple examples and continues through a variety of types of shooting, presented in considerable detail. Those who have some familiarity with the method may wish to start with Section 2.4, or merely take a look at some of the exercises, particularly the last three.

What has come to be called the shooting method has its origins in a more sophisticated technique, due mainly to Ważewski [**250**]. This method makes use of a topological lemma which, in $R^n$, is related to Brouwer's fixed point theorem. An exposition of Ważewski's method, and some applications to differential equations, may be found in the classic ode text by Hartman [**83**]. We will discuss this method further in the final section of this chapter.

While Ważewski's result has seen continued application over the years, examples where its hypotheses can be verified are relatively difficult to find. Shooting may be thought of as including Ważewski's method, but also simpler topological arguments involving only connectedness. These will form the bulk of the applications of shooting in this book. We will see that a wide range of existence problems can be handled with this technique. In the last two sections of the chapter, and in Chapter 13, we discuss topological principles similar to Ważewski's which may be viewed as extending the range of the shooting method.

## 2.2. A first order example

We begin by returning to the second example from Chapter 1, namely the equation

$$x' = x^3 + \sin t. \tag{2.1}$$

As before, our goal is to prove that there is a solution with period $2\pi$, and in Exercise 1.4 you were asked to show that it is sufficient to find a solution such that $x(0) = x(2\pi)$.

For each real number $\alpha$, let $x(t, \alpha)$ be the unique solution of (2.1) satisfying

$$x(0) = \alpha.$$

In the shooting method we consider $\alpha$ as an unknown and try to show that the equation

$$x(2\pi, \alpha) - \alpha = 0$$

has at least one solution.

We apply the intermediate value theorem from basic analysis. Define a function $F : R^1 \to R^1$ by letting

$$F(\alpha) = x(2\pi, \alpha) - \alpha.$$

We wish to show that the equation $F(\alpha) = 0$ has a solution.

To use the intermediate value theorem we look for numbers $\alpha_1$ and $\alpha_2$ such that $F(\alpha_1) > 0$ and $F(\alpha_2) < 0$. We also have to verify that $F$ is defined and continuous on an interval that contains $\alpha_1$ and $\alpha_2$. Unfortunately, in this problem the second step is not as straightforward as we would like, because of possible "blowup". As was pointed out in footnote 1.1 of Chapter 1, the solution $x(t, \alpha)$ might not exist on the entire interval $0 \le t \le 2\pi$. For the moment, let us ignore this complication. In Chapter 1, when discussing equation (2.1), we pointed out that if $x > 1$ then $x' > 0$. So if $\alpha > 1$ then the solution increases for as long as it exists.[1] For example, if $x(t, 2)$ exists on $[0, 2\pi]$ then $F(2)$ is defined, and $F(2) = x(2\pi, 2) - 2 > 0$. Similarly, $x' < 0$ if $x < -1$, and so if $F(-2)$ is defined then it is negative. Therefore, if $F$ is continuous on $[-2, 2]$ then there is an $\alpha^* \in [-2, 2]$ with $F(\alpha^*) = 0$. The corresponding solution $x(t, \alpha^*)$ satisfies $x(0) = x(2\pi)$ and so is periodic with period $2\pi$.

This simple argument is the essence of the shooting method. The reader should not proceed without understanding it. It is helpful to sketch two possible solutions of (2.1) on the same graph, one starting above 1, and hence increasing, and one starting below $-1$, and hence decreasing. Then

[1] We will often use "as long as" arguments. To be more precise, if $u$ exists on an interval $[0, T]$ then $u$ is increasing in this interval.

ask yourself what must be true about at least one of the solutions in between, assuming that they all exist on the interval $[0, 2\pi]$.

To complete the existence proof, we must deal with the question of blow-up. It is usually an easy issue to dispose of, by truncating the equation outside the region where the solutions you want can lie. Let $I = [-2, 2]$. We will assume that $x(0) \in I$. The idea is that once a solution leaves $I$ it is monotonic, because $x' \neq 0$, and so $x(t)$ does not return to $I$. Such a solution is not periodic, and so we do not care how it behaves. We are only interested in solutions such that $x(t) \in [-2, 2]$ for $0 \leq t \leq 2\pi$. Therefore, it is not necessary to assume that $x$ satisfies the same differential equation when $|x| > 2$. Replace (2.1) with a different ode, by setting

$$f(t, y) = \begin{cases} y^3 + \sin t & \text{if } -2 \leq y \leq 2, \\ 8 + \sin t & \text{if } y \geq 2, \\ -8 + \sin t & \text{if } y \leq -2, \end{cases}$$

and considering the initial value problem

$$\begin{aligned} y' &= f(t, y), \\ y(0) &= \alpha. \end{aligned} \tag{2.2}$$

Since $|f(t, y)| \leq 9$ for all $(t, y)$, the solution $y(t, \alpha)$ exists on $[0, 2\pi]$ for every $\alpha$. Hence the function

$$G(\alpha) = y(2\pi, \alpha) - \alpha$$

is defined for all $\alpha$. Since the function $f(t, y)$ is Lipschitz continuous in $y$, the solution $y(t, \alpha)$ is a continuous function of $\alpha$ and thus $G$ is continuous. As before, $y' > 0$ if $y > 1$ and $y' < 0$ if $y < -1$, and so $G(2) > 0$, $G(-2) < 0$. By the intermediate value theorem there is an $\alpha^* \in [-2, 2]$ such that $G(\alpha^*) = 0$.

The corresponding solution $y(t, \alpha^*)$ remains in $[-2, 2]$ on $[0, 2\pi]$, for otherwise $G(\alpha^*) \neq 0$. Hence, from the definition of $f$, $y(t, \alpha^*) = x(t, \alpha^*)$ and $F(\alpha^*) = 0$. We have therefore found an initial condition $\alpha^*$ such that $x(2\pi) = x(0)$, and the solution $x$ is $2\pi$-periodic, as desired.

In this argument, the interval $[-2, 2]$ was chosen for simplicity. It is clear that a periodic solution must lie in the smaller interval $[-1, 1]$, because every solution which enters the region $|x| > 1$ is monotonic thereafter. We may wish to know if a periodic solution $x$ can satisfy $|x(t_0)| = 1$ for some $t_0$. The technique for answering this arises fairly frequently in shooting methods. We must analyze higher derivatives.

For example, if $x(t_0) = 1$ then $x'(t_0) = 0$, because $x(t) \leq 1$ for all $t$. From (2.1) we see that $\sin t_0 = -1$. Differentiate (2.1) to obtain

$$x''(t_0) = 3x(t_0)^2 x'(t_0) + \cos t_0 = \cos t_0 = 0,$$
$$x'''(t_0) = -\sin t_0 = 1.$$

Therefore $x'''(t_0) > 0$, which implies that the solution enters the region $x > 1$ and so is not periodic. A similar argument applies if $x(t_0) = -1$, and so every periodic solution lies in the open interval $-1 < x < 1$.

We are not aware of any other method for demonstrating the existence of a periodic solution to (2.1). We remark also that shooting usually does not provide any information about the uniqueness of the desired solution. Uniqueness is often more difficult to prove than existence. For (2.1), however, a simple comparison method shows that there is only one periodic solution. In particular, there is no solution of period $2k\pi$, with $k > 1$, except for the solution (with **least** period $2\pi$) found above. This is implied by a more general result which you are asked to prove in Exercise 2.2 below.

**2.2.1. An alternative formulation of shooting.** In this subsection we show that shooting is essentially a topological method, by looking again at the existence of periodic solutions of (2.1). We will describe our approach to this problem somewhat abstractly. Our goal is to prove the existence of solutions with a certain property, say "property $P$". In this example, an initial choice might be that a solution has property $P$ if it is periodic. We identify a parameter which we are free to choose within some nonempty set $\Omega$. In the example the parameter is $\alpha$, and $\Omega = R^1$. There must be a topology defined on $\Omega$, and in all of our examples, $\Omega$ is connected in this topology.

We don't know if there are any points $\alpha$ in $\Omega$ such that $x(t, \alpha)$ has property $P$, but if not, then obviously there are values of $\alpha$ such that $x(t, \alpha)$ does not have this property. Let

$$\Lambda = \{\alpha \in \Omega \mid x(t, \alpha) \text{ does not have property } P\}.$$

In cases where there is only one "shooting parameter" (i.e. $\Omega \subset R^1$), we usually show that $\Lambda \neq \Omega$ by partitioning $\Lambda$ into two nonempty, disjoint, open subsets $A$ and $B$, sometimes called "bad sets". The connectedness of $\Omega$ then implies the desired existence result. More complicated problems, as in Section 2.6 below, require more sophisticated topological principles than connectedness.

For (2.1) we again consider the truncated initial value problem (2.2), so that all solutions exist on $[0, 2\pi]$. The property of "being periodic" is too vague, because a division of $\Lambda$ into two disjoint nonempty open sets is not obvious. Instead, we will say that a solution $y(t, \alpha)$ has property

$P$ if $y(2\pi,\alpha)=\alpha$. We saw earlier that this version of property $P$ implies periodicity with period $2\pi$, and Exercise 2.2 implies that if there is a solution with property $P$, then it is the only periodic solution.

Let $\Omega = R^1$, and define the bad sets $A$ and $B$ as follows:

$$A = \{\alpha \mid y(2\pi,\alpha) > \alpha\},$$
$$B = \{\alpha \mid y(2\pi,\alpha) < \alpha\}.$$

A solution $y(t,\alpha)$ has property $P$ if and only if $\alpha \notin \Lambda = A \cup B$. The sets $A$ and $B$ are obviously disjoint, simply from their definitions. Our argument above shows that $2 \in A$ and $-2 \in B$, so they are each nonempty. To show that $A$ is open, suppose that $\alpha_1 \in A$. Then $y(2\pi,\alpha_1) > \alpha_1$. Since $y(2\pi,\alpha)$ is a continuous function of $\alpha$, there is a $\delta > 0$ such that if $|\alpha-\alpha_1| < \delta$ then $y(2\pi,\alpha) > \alpha$. Hence $A$ is open, and similarly $B$ is open. This proves that $A \cup B$ is disconnected, and so there is some $\alpha^* \in R^1$ which is not in $A \cup B$. Then $y(t,\alpha^*)$ is a periodic solution of (2.1).

### 2.2.2. A problem on $[0,\infty)$.

We now look at a generalization of (2.1), by considering

$$x' = x^3 + f(t), \tag{2.3}$$

where we assume only that $f$ is continuous and bounded on $[0,\infty)$. Without assuming periodicity of $f$ we cannot expect there to be a periodic solution. Instead, we ask if there is a solution which is bounded on $[0,\infty)$. To search for such a solution we again consider an initial value problem, now consisting of (2.3) and the initial condition

$$x(0) = \alpha, \tag{2.4}$$

with solution $x(t,\alpha)$. We look for an $\alpha^*$ such that $x(t,\alpha^*)$ is bounded on $[0,\infty)$.

We use the topological formulation of shooting which was introduced in the previous subsection. The point here is to show that the choice of property $P$ is not always obvious. For this problem, one's first guess might be to say that a solution has property $P$ if it exists on $[0,\infty)$ and is bounded.

Proceeding with this suggestion, let $M$ be an upper bound for $|f|$ over $[0,\infty)$, and set $m = M^{1/3}$. From (2.3), if $x(t) > m$ then $x'(t) > 0$, while if $x(t) < -m$ then $x'(t) < 0$. It follows that if $x(t_1) > m$ for some $t_1 \geq 0$, then $x(t) > m$ for all $t \geq t_1$ as long as the solution exists, and a similar remark applies if $x(t_2) < -m$ for some $t_2 \geq 0$.[2] If a solution enters $x > m$ then it is increasing from there on, with $x' \geq x^3 - m^3$, and it easily follows

[2]One further explanation of "as long as": There is no $\tau > t_1$ such that the solution exists on $[t_1,\tau]$, with $x(t_1) > m$ and $x(\tau) \leq m$. If there were such a $\tau$, then there would have to be a $t \in (t_1,\tau)$ with $x(t) > m$ and $x'(t) < 0$, which is impossible from (2.3) and the definition of $m$.

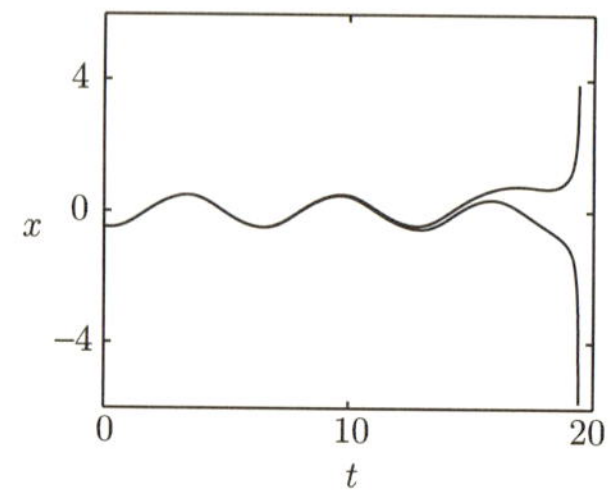


**Figure 2.1**

that $x$ is unbounded on its maximal interval of existence. Similarly, every solution entering $x < -m$ is unbounded.

These considerations suggest certain choices for the sets $A$ and $B$. We note that because of the $x^3$ term on the right of (2.3), unbounded solutions blow up in finite time.[3] To allow for this we let

(2.5)
$$A = \{\alpha \mid x \text{ exists on some interval } [0,\omega) \text{ and } \lim_{t\to\omega^-} x(t) = \infty\},$$
$$B = \{\alpha \mid x \text{ exists on some interval } [0,\omega) \text{ and } \lim_{t\to\omega^-} x(t) = -\infty\},$$

where in each case, $x = x(t,\alpha)$ and $\omega$ denotes either a positive number or $\infty$. We show solutions corresponding to each set in Figure 2.1.

We now consider the three properties required of $A$ and $B$, namely that they are nonempty, disjoint, and open subsets of $R^1$. The easiest is disjointedness, for the definitions in (2.5) themselves rule out the existence of a point $\alpha$ which is in both $A$ and $B$.

We saw above that if $\alpha > m$ then $\alpha \in A$, while if $\alpha < -m$ then $\alpha \in B$, so that $A$ and $B$ are nonempty. But the third property required of $A$ and $B$ is not so straightforward. It is not immediately clear that either is an open set.

Suppose, for example, that $\alpha_1 \in A$, so that $\lim_{t\to\omega^-} x(t,\alpha_1) = \infty$ for some $\omega \le \infty$. Must this be true for all $\alpha$ in some open neighborhood of $\alpha_1$? The property of "tending to infinity" is not intrinsically an "open property". As an example, for each $\alpha$ let $\phi(\cdot,\alpha)$ be the function defined by

$$\phi(t,\alpha) = \frac{t}{1+\alpha^2 t} \quad \text{for } t \ge 0. \tag{2.6}$$

The function $\phi(\cdot,0)$ is unbounded, but for $\alpha \ne 0$, $\phi(t,\alpha)$ is a bounded function of $t$. See Exercise 2.1 for related examples.

For this reason, our initial choice for property $P$ and the resulting definitions of the sets $A$ and $B$ are not optimal in this problem. Recall that we chose a fixed $m > 0$ such that $m^3 \ge \sup_{t\ge0} |f(t)|$. Recognizing that a

[3] See Exercise 1.1.

solution which takes values only in $[-m, m]$ is bounded, we will now say that $x(t, \alpha)$ has property $P$ if $|x(t, \alpha)| \leq m$ for all $t \geq 0$. We then set

$$A = \{\alpha | x(t, \alpha) > m \text{ for some } t \geq 0\},$$
$$B = \{\alpha | x(t, \alpha) < -m \text{ for some } t \geq 0\}.$$

If $\alpha \notin A \cup B$ then the solution $x(t, \alpha)$ can be extended to $[0, \infty)$ and is bounded on that interval.

From the definitions of $A$ and $B$ it is clear that if $\alpha > m$ then $\alpha \in A$, while if $\alpha < -m$ then $\alpha \in B$, so that both sets are nonempty. They are disjoint because, as remarked earlier, once a solution enters the region $|x(t)| > m$, it remains in this region and so does not cross from $x > m$ to $x < -m$ or vice versa.

The final property needed is that $A$ and $B$ are open subsets of $\Omega$. Suppose that $\alpha_0 \in A$. Then there is a $t_1$ with $x(t_1, \alpha_0) > m$. Since $x(t_1, \alpha)$ is a continuous function of $\alpha$, there is an open neighborhood $\mathcal{O}$ of $\alpha_0$ such that if $\alpha \in \mathcal{O}$ then $x(t_1, \alpha) > m$. Therefore $A$ is open, and similarly, $B$ is open. Hence $A \cup B$ does not equal the connected set $\Omega = (-\infty, \infty)$, and so there is an $\alpha^*$ such that $x(t, \alpha^*)$ is bounded on $[0, \infty)$.

We will not go into such detail about these properties again. Usually the openness of the bad sets, such as $A$ and $B$, follows from continuity of a solution to an initial value problem with respect to initial conditions or parameters. We have proved

**Theorem 2.1.** *If $f$ is continuous and bounded on $[0, \infty)$, with $|f(t)| \leq m$, then the equation (2.3) has a solution $x$ which exists on this interval and satisfies $-m \leq x(t) \leq m$ for all $t \geq 0$.*

In Exercise 2.2 you are asked to show that the bounded solution is unique.

## 2.3. Some second order examples

### 2.3.1. A linear problem.

We start with the linear second order scalar equation

$$x'' + q(t)x = f(t), \tag{2.7}$$

and we seek a solution satisfying the boundary conditions

$$\begin{aligned} x(0) &= 0, \\ x(1) &= 0. \end{aligned} \tag{2.8}$$

The theory of linear boundary value problems, especially the Fredholm alternative, tells us that there is a solution to (2.7)–(2.8) if and only if $f$

satisfies the orthogonality condition $\int_0^1 f(t)\,\phi(t)\;dt = 0$ for every solution $\phi(t)$ to the homogeneous equation

$$x'' + q(t)\,x = 0 \tag{2.9}$$

which satisfies (2.8).[4] If the only solution to (2.9) and (2.8) is $\phi = 0$, then (2.7)–(2.8) has a unique solution, which can be found by constructing a Green's function $G(t, \tau)$. We then have the formula

$$x(t) = \int_0^1 G(t,\tau)\,f(\tau)\;d\tau.$$

Looking at the problem another way, let $x_p$ denote the solution to (2.7) such that $x(0) = x'(0) = 0$, and let $x_h$ denote the solution to the corresponding homogeneous equation (2.9) such that $x(0) = 0$, $x'(0) = 1$. Then the general solution to (2.7) with $x(0) = 0$ is

$$x(t,\alpha) = x_p(t) + \alpha x_h(t). \tag{2.10}$$

The boundary value problem (2.7)–(2.8) has a solution if and only if there is an $\alpha$ such that $x(1,\alpha) = 0$. If $x_h(1) \neq 0$ then $x_p(1) + \alpha x_h(1) = 0$ can be solved for a unique $\alpha$.

To put this in the language of shooting, observe that $x(t,\alpha)$ is the unique solution of the **initial value problem** consisting of (2.7) and the initial conditions

$$x(0) = 0, \qquad x'(0) = \alpha. \tag{2.11}$$

For example, if $x_h(1) > 0$ then from (2.10), $x(1,\alpha) > 0$ if $\alpha$ is large and positive and $x(1,\alpha) < 0$ if $\alpha$ is large and negative. The intermediate value theorem can then be applied as before.

Shooting is not a particularly efficient method for this linear problem. To see its true power we need to consider nonlinear equations.

**2.3.2. A nonlinear problem.** Our next example is a forced pendulum equation in which the dependent variable $x$ is the angle that a swinging pendulum makes with the downward vertical and the applied force is a continuous function $f$. With appropriate units the equation becomes[5]

$$x'' + \sin x = f(t). \tag{2.12}$$

We again look for a solution satisfying the boundary conditions

$$x(0) = x(1) = 0. \tag{2.13}$$

---

[4]See [**201**] or [**82**, p. 146].

[5]This is not the usual linearized pendulum equation seen in elementary ode books. The equation is valid even with large oscillations. However, a first derivative term accounting for friction is omitted.

Once again consider an initial value problem. As in the linear example above, the natural set of initial conditions to choose is

$$x(0) = 0, \quad x'(0) = \alpha, \tag{2.14}$$

and we again denote the unique solution by $x(t,\alpha)$.

Because $f$ is continuous and $\sin x$ is bounded, $x(t,\alpha)$ is defined for $0 \le t \le 1$ and every $\alpha$. We wish to find an $\alpha^*$ such that $x(1,\alpha^*) = 0$. Then $x(t,\alpha^*)$ will be a solution to (2.12)–(2.13).

To use shooting, we look for $\alpha_1$ and $\alpha_2$ such that $x(1,\alpha_1) > 0$ and $x(1,\alpha_2) < 0$ and apply the intermediate value theorem to the continuous map $\alpha \to x(1,\alpha)$. To show that $\alpha_1$ and $\alpha_2$ exist, let

$$M = \max_{0 \le t \le 1} |f(t)|.$$

**Lemma 2.2.** *If $\alpha > \frac{M+1}{2}$ then $x(1,\alpha) > 0$. If $\alpha < -\frac{M+1}{2}$ then $x(1,\alpha) < 0$.*

**Proof.** Suppose that $\alpha > \frac{M+1}{2}$. Then on $[0,1]$,

$$\begin{aligned} x'' &= -\sin x + f(t) \ge -\ (M+1), \\ x' &\ge \alpha - (M+1)t > \frac{M+1}{2} - (M+1)t, \\ x &> \frac{M+1}{2}t - (M+1)\frac{t^2}{2}, \end{aligned}$$

and $x(1,\alpha) > 0$. A similar argument applies if $\alpha < -\frac{M+1}{2}$. □

This lemma and the continuity of $x(1,\alpha)$ with respect to $\alpha$ lead immediately to the existence of at least one $\alpha^*$ such that $x(1,\alpha^*) = 0$, giving the desired solution of (2.12)–(2.13).

We can also phrase this in the topological language of Section 2.2.1, letting

$$\begin{aligned} A &= \{\alpha | x(1,\alpha) > 0\}, \\ B &= \{\alpha | x(1,\alpha) < 0\}. \end{aligned}$$

As in Section 2.2.1, the continuity of $x(1,\alpha)$ as a function of $\alpha$ implies that $A$ and $B$ are open sets. Further, they are disjoint by their definitions. Lemma 2.2 implies that $A$ and $B$ are each nonempty. Since $R^1$ is connected, $A \cup B \ne R^1$, and hence there is an $\alpha^*$ in $R^1$ which is not in $A \cup B$. From the definitions of $A$ and $B$ it is seen that $x(1,\alpha^*) = 0$.

**2.3.3. Airy's equation on $[0, \infty)$.** Airy's equation in its simplest form is

$$y'' - xy = 0. \tag{2.15}$$

It was introduced by Airy, a distinguished nineteenth-century British scientist and mathematician, during his studies of optics [**6**], [**2**]. Mathematically it is important as the simplest example of a "turning point", because the nature of the solutions changes dramatically at $x = 0$. Using the Sturm oscillation theorem, it is easy to see that as $x$ decreases from zero all solutions oscillate more and more frequently, while no solution has more than one critical point in the region $x > 0$. The solutions of this equation are linear combinations of what are called Airy functions, which are among the "special functions" of mathematical physics. We will see one context where they are important in Chapter 3.

Here we will be concerned with a relatively simple problem, chosen to illustrate an important feature of most shooting arguments which has not been encountered in our previous examples. We look for a solution of (2.15) satisfying the conditions

$$\begin{gathered} y(0) = 1, \\ \lim_{x\to\infty} y(x) = 0. \end{gathered} \tag{2.16}$$

To find such a solution we consider, as usual, an initial value problem, letting $y(x) = y(x, \alpha)$ denote the solution of (2.15) such that $y(0) = 1$, $y'(0) = \alpha$. Further, let

$$\begin{aligned} A &= \{\alpha < 0 |\ y(x) < 0 \text{ for some } x > 0\}, \\ B &= \left\{\alpha < 0 |\ y'(x) > 0 \text{ for some } x > 0\right\}. \end{aligned}$$

We hope it is obvious that these sets are open, by the sort of argument used near the end of the previous subsection, since solutions depend continuously on $\alpha$. To show that $A$ and $B$ are disjoint, suppose that there is an $\alpha \in A \cap B$. Since $\alpha < 0$ and $y'(x) > 0$ for some $x > 0$, there is an $x_1 > 0$ with $y'(x_1) = 0$. There is also an $x_2 > 0$ with $y(x_2) = 0$. Choose the smallest possible $x_1$ and $x_2$, so that $y > 0$ on $[0, x_2)$ and $y' < 0$ on $[0, x_1)$.

We now point out the feature of most interest to us in this example. Suppose that $x_1 = x_2$. Then

$$y(x_1) = y'(x_1) = 0. \tag{2.17}$$

However the uniqueness theorem for initial value problems then implies that $y$ is the zero solution of (2.15), which is a contradiction because we chose $y(0) = 1$. Hence, no nontrivial solution of (2.15) is tangent to the $x$-axis. Such a "nontangency condition" appears in most shooting arguments, and often establishing nontangency is the key step in the proof.

We have shown that either $x_2 < x_1$ or $x_2 > x_1$. Suppose that $x_2 < x_1$. Then $y(x_2) = 0$ and $y'(x_2) < 0$. Therefore $y < 0$ and $y' < 0$ in some interval $(x_2, x_2 + \varepsilon)$. However if $y < 0$ and $x > 0$, then $y'' < 0$. Therefore $y'(x) < 0$ for all $x > x_2$ where the solution exists, and so $x_1$ is not defined. A similar contradiction is obtained if $x_1 < x_2$, because if $y > 0$ and $x > 0$, then $y'' > 0$, and so if $y'$ vanishes while $y > 0$, then $y$ is increasing from there on and cannot become negative. Hence, $x_2$ is not defined, completing the proof that $A$ and $B$ are disjoint.

Finally, we show that these sets are nonempty. Suppose that $\alpha < -3$. Then on some interval $[0, x_0)$, $y' < -2$. Suppose that there is an $x_1 \in (0,1)$ with $y' < -2$ on $(0, x_1)$ and $y'(x_1) = -2$. In the interval $(0, x_1)$, $y < 1$ and so $y'' < 1$. But then, $y'(x_1) < -3 + x_1 < -2$, a contradiction. Therefore, $y' < -2$ on $[0,1]$. Since $y(0) = 1$, we conclude that $y(1) < 0$. Hence $\alpha \in A$, so that $A$ is nonempty.[6]

On the other hand, if $y(0) = 1$ and $y'(0) = 0$, then $y''(0) = 0$ and $y'''(0) = y(0) > 0$ and so for some $x_1 > 0$, $y'(x_1) > 0$. Hence, $y'(x_1) > 0$ for sufficiently small $|\alpha|$. Therefore, small negative numbers are in $B$, so $B$ is nonempty.

We have shown that $A$ and $B$ are open, nonempty, and disjoint, and so there is an $\alpha < 0$ which is not in $A \cup B$. For such an $\alpha$, $y > 0$ and $y' < 0$ on $[0, \infty)$. Therefore $L = \lim_{x \to \infty} y(x)$ exists and is nonnegative. If $L > 0$ then $y''$ becomes large, which clearly causes $y'$ to become positive. Hence $L = 0$.

## 2.4. Heteroclinic orbits and the FitzHugh-Nagumo equations

### 2.4.1. Heteroclinic orbits.

We begin with

**Definition 2.3.** Suppose that an autonomous system

$$X' = F(X) \tag{2.18}$$

has at least two distinct equilibrium points, say $X_1$ and $X_2$. (Thus, $F(X_1) = F(X_2) = 0$.) Suppose that there is a solution such that $\lim_{t \to -\infty} X(t) = X_1$ and $\lim_{t \to \infty} X(t) = X_2$. Then the trajectory of this solution is called a "heteroclinic" orbit.

Note that if $\Gamma$ is a heteroclinic orbit then it is an invariant set for (2.18), meaning that if $X(0) \in \Gamma$ then $X(t) \in \Gamma$ for all $t$ where the solution is defined. The existence of compact invariant sets is a fundamental problem in dynamical systems.

[6] It is not necessary to choose the best possible estimates in the argument of this paragraph. Something weaker than $\alpha < -3$ would work, but why work harder than we have to?

For a basic example, consider the second order scalar equation

$$x'' + x - x^3 = 0 \tag{2.19}$$

and the equivalent first order system

$$\begin{aligned} x' &= y, \\ y' &= x^3 - x. \end{aligned} \tag{2.20}$$

This system is easy to analyze because it has a conserved quantity, namely the function $E(x, y) = y^2 + x^2 - \frac{1}{2}x^4$. If $(x, y)$ is a solution of (2.20) then

$$\frac{d}{dt}E(x(t), y(t)) = 0.$$

The curve $E(x, y) = \frac{1}{2}$ in the $(x, y)$-plane contains the points $(1, 0)$ and $(-1, 0)$, and by drawing the phase plane we see that there are two heteroclinic orbits, one going from $(-1, 0)$ to $(1, 0)$ in the upper half-plane and the other going from $(1, 0)$ to $(-1, 0)$ in the lower half-plane. No shooting is required to obtain these two solutions, and they can even be found exactly. The functions $x_1(t) = \tanh\left(\frac{t}{\sqrt{2}}\right)$ and $x_2(t) = -\tanh\left(\frac{t}{\sqrt{2}}\right)$ are solutions of (2.20) which behave as described. Since the equation is autonomous, $x_1(t - \tau)$ and $x_2(t - \tau)$ also give heteroclinic orbits, for every constant $\tau$.

Hence, to illustrate the use of shooting we need a slightly more difficult problem. A much studied example is the "reduced" form of the FitzHugh-Nagumo equations of neurobiology. These were formulated as a simplified version of the Hodgkin-Huxley equations, the most important quantitative model of nerve impulses [**69**], [**103**], [**59**].

**2.4.2. Traveling waves and the FitzHugh-Nagumo equations.** The form of the FitzHugh-Nagumo equation of interest here is obtained by looking for a "traveling wave" solution of a parabolic system of partial differential equations. To explain this terminology, consider a system of partial differential equations in one space dimension of the form

$$\frac{\partial V}{\partial t} = D\frac{\partial^2 V}{\partial x^2} + F(V), \tag{2.21}$$

where $V$ and $F(V)$ are in $R^n$ and $D$ is an $n \times n$ positive semidefinite symmetric matrix. Such systems are called "reaction-diffusion" equations. In the Hodgkin-Huxley and FitzHugh-Nagumo models $D$ is diagonal with some zero entries on the diagonal. We look for a solution of the form

$$V(x, t) = U(x + ct)$$

for some function $U : R^1 \to R^n$ and constant $c$. By considering the graphs of a component of the functions $\Phi_t(x) = U(x + ct)$ for different values of $t$, we see that $U$ represents a wave moving with speed $c$ along the $x$-axis. If

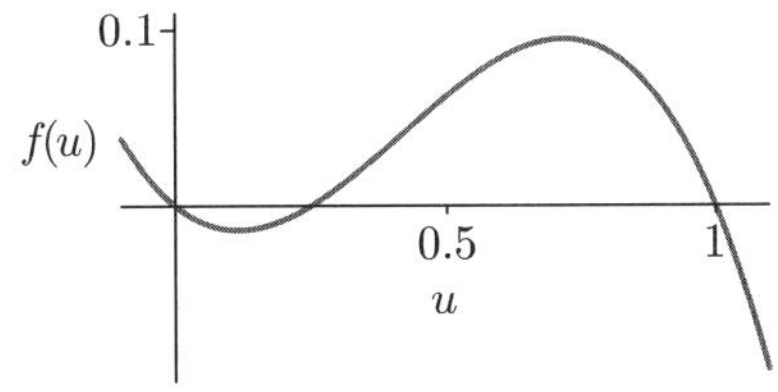


**Figure 2.2**

$c > 0$ the wave moves to the left. Substituting this into (2.21) results in the system of ode's

$$cU' = DU'' + F(U). \tag{2.22}$$

One type of traveling wave for (2.21) is a solution to (2.22) with a trajectory that forms a heteroclinic orbit between two equilibrium points. In some problems of this type we expect to find such a wave for only one or a few values of $c$, while in others, such waves exist when $c$ is in an appropriate interval [**12**].

The FitzHugh-Nagumo equations are

$$\begin{aligned}\frac{\partial u}{\partial t} &= \frac{\partial^2 u}{\partial x^2} + u(1-u)(u-a) - w,\\ \frac{\partial w}{\partial t} &= \varepsilon(u - \gamma w).\end{aligned} \tag{2.23}$$

The parameters $a$, $\varepsilon$, and $\gamma$ are positive, with $a \in \left(0, \frac{1}{2}\right)$. Making the traveling wave substitution $s = x + ct$ and considering $u$ and $w$ as functions of the single variable $s$ results in the system

$$\begin{aligned}u'' - cu' + f(u) - w &= 0,\\ w' &= \varepsilon(u - \gamma w),\end{aligned}$$

where $f(u) = u(1-u)(u-a)$. (See Figure 2.2.)

This system is difficult to study, and so at this stage we will set $\varepsilon = 0$. Then $w$ is constant, and we will take $w = 0$. In Chapter 6 we will consider the full system, with $\varepsilon > 0$ but small.

The reduced FitzHugh-Nagumo equation for a traveling wave is therefore the second order scalar ode

$$u'' - cu' + f(u) = 0. \tag{2.24}$$

Writing this as a system gives

$$\begin{aligned}u' &= v,\\ v' &= cv - f(u).\end{aligned} \tag{2.25}$$

The equilibrium points are $(0,0)$, $(a,0)$, and $(1,0)$. It turns out, as we discuss below, that a physically realistic heteroclinic orbit will connect $(0,0)$ and

$(1,0)$. To determine the direction of this orbit we consider the "energy"

$$E(u,v) = \frac{1}{2}v^2 + F(u),$$

where $F(u) = \int_0^u f(z)\ dz$. If $(u,v)$ is a solution of (2.25) and $W(s) = E(u(s), v(s))$, then

$$\frac{dW}{ds} = \dot{E}(u,v) := \frac{\partial E}{\partial u}u' + \frac{\partial E}{\partial v}v' = cv^2.$$

We note that $E(0,0) = 0$, $E(a,0) = F(a) = \int_0^a f(z)\,dz < 0$, and $E(1,0) = F(1) = \int_0^1 f(z)\,dz$. Because $0 < a < \frac{1}{2}$, $F(1) > 0$. If $c > 0$ then $W$ is increasing, so that the solution goes from a lower value of $E$ to a higher value. We reject a solution tending to $(a,0)$ as $s \to \infty$ because this is an unstable equilibrium point for the ode system satisfied by spatially homogeneous solutions of (2.23). (See below as well on this point.) Therefore it is reasonable to look for a heteroclinic orbit such that

$$\lim_{s\to-\infty}(u(s), v(s)) = (0,0), \tag{2.26}$$

$$\lim_{s\to\infty}(u(s), v(s)) = (1,0). \tag{2.27}$$

Our main goal is to show that for some $c > 0$ there is a solution of (2.25) satisfying (2.26) and (2.27).[7] At this point we continue our somewhat informal development of the theory and defer a formal statement of the main results until the end of the section.

To develop a further understanding of the solutions of (2.25), we linearize this system around the equilibria. The linearized system at an equilibrium $(u_0, v_0)$ is

$$\begin{aligned} y' &= z, \\ z' &= cz - f'(u_0)\,y. \end{aligned}$$

Around $(0,0)$ we get

$$Y' = MY \tag{2.28}$$

where $Y = \begin{pmatrix} y \\ z \end{pmatrix}$ and

$$M = \begin{pmatrix} 0 & 1 \\ a & c \end{pmatrix}.$$

The eigenvalues of $M$ are $\lambda_1 = \frac{1}{2}\left(c + \sqrt{c^2+4a}\right)$ and $\lambda_2 = \frac{1}{2}\left(c - \sqrt{c^2+4a}\right)$, with corresponding eigenvectors $\mathbf{v}_1 = \begin{pmatrix} 1 \\ \lambda_1 \end{pmatrix}$ and $\mathbf{v}_2 = \begin{pmatrix} 1 \\ \lambda_2 \end{pmatrix}$. Since

[7] In fact, the cubic form of $f$ allows the heteroclinic wave speed $c$ to be found exactly, as $c^* = \frac{1-2a}{\sqrt{2}}$, because at that speed there is an exact solution to (2.24). See Exercise 2.9. However, our goal here is to give a proof which can be extended to a class of "cubic like" functions $f(u)$.

$\lambda_1 > 0$ and $\lambda_2 < 0$, $(0,0)$ is a saddle point. The functions $e^{\lambda_1 s}\mathbf{v}_1$ and $e^{\lambda_2 s}\mathbf{v}_2$ are linearly independent solutions of the linearized system (2.28). Plotting the phase curves in the plane, we see that one is along the line of the vector $\mathbf{v}_1$, in the first quadrant, and tends to $(0,0)$ as $s \to -\infty$, while a second is in the fourth quadrant, $y > 0$, $z < 0$, and tends to $(0,0)$ as $s \to +\infty$. These are parts of the unstable manifold $\mathcal{U}_L$ and stable manifold $\mathcal{S}_L$ of the linearized system, with other branches contained in $y < 0$.

We now use the stable manifold theorem, which you can find, for example, in [**41**, Chapter 13, Section 4]. This result and the linearization of (2.25) around $(0,0)$ show that in the phase plane of the nonlinear system (2.25) there is a smooth curve $\mathcal{U}_c$, called the unstable manifold, such that if $(u,v)$ is a solution of (2.25) with $(u(0), v(0))$ on $\mathcal{U}_c$, then $(u(s), v(s)) \in \mathcal{U}_c$ for all $s \le 0$ and $(u,v)$ satisfies (2.26). A further consequence of the stable manifold theorem is that $\mathcal{U}_c$ is tangent to the eigenvector $\mathbf{v}_1$ at $(0,0)$. Also, if $(u,v)$ is a solution on $\mathcal{U}_c$ then $\lim_{s\to-\infty} \frac{u(s)}{e^{\lambda_1 s}}$ exists and is nonzero.

The linearized matrix at the point $(1,0)$ is

$$\begin{pmatrix} 0 & 1 \\ -f'(1) & c \end{pmatrix},$$

and since $f'(1)$ is negative, this is also a saddle point, with a local phase plane similar to the one around $(0,0)$. Our main interest near $(1,0)$ is in the stable manifold $\mathcal{S}_{c,1}$ at this equilibrium point.

The linearized matrix at $(a,0)$ has eigenvalues with positive real parts, if $c > 0$, which suggests the possibility of a heteroclinic orbit starting at $(a,0)$ and ending at one of the other two equilibria. In fact, such orbits exist for all $c > 0$. However, these orbits are not stable physically[8] and so are of less interest.

Linearization does not tell us the structure of the invariant manifolds $\mathcal{U}_c$ and $\mathcal{S}_c$ away from the equilibrium point. Here are the definitions of the **global** stable and unstable manifolds at some equilibrium point $P$:

$$\mathcal{S}_P = \left\{(u(0), v(0)) \mid \lim_{s\to\infty} (u(s), v(s)) = P\right\},$$

$$\mathcal{U}_P = \left\{(u(0), v(0)) \mid \lim_{s\to-\infty} (u(s), v(s)) = P\right\}.$$

The stable manifold theorem is a local theorem. It discusses these manifolds in a neighborhood of the equilibrium point and says nothing about their nature outside such a neighborhood. It is a much more difficult task to

[8] In other words, they are not stable solutions of the first equation in (2.23) when $w = 0$. See, for example, [**209**] for a discussion of the stability of traveling wave solutions of reaction-diffusion equations.

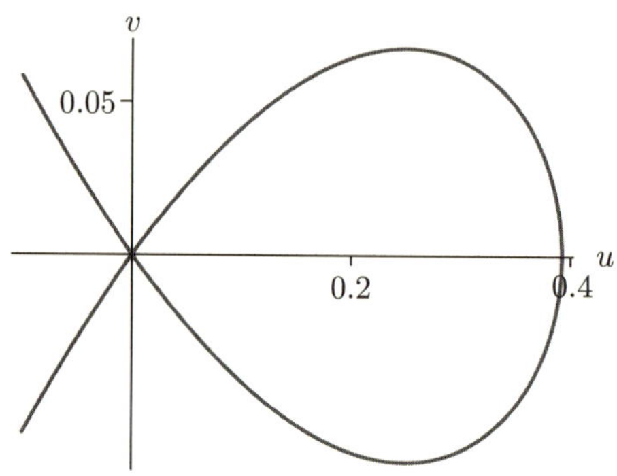


**Figure 2.3**

prove that the global stable and unstable manifolds, as defined above, are smooth. We will not pursue this here, but see [**215**] for further information.

We look for a solution $(u, v)$ connecting the equilibria $(0,0)$ and $(1,0)$. Let $\mathcal{U}_c^+$ denote the branch of $\mathcal{U}_c$ which tends to $(0,0)$ from the first quadrant, and let $\mathcal{S}_{c,1}^-$ denote the branch of $\mathcal{S}_{c,1}$ which tends to $(1,0)$ from the region $u < 1$, $v > 0$. Our goal is to find a $c$ such that $\mathcal{U}_c^+ = \mathcal{S}_{c,1}^-$.

Again we consider the function $E(u, v) = \frac{1}{2}v^2 + F(u)$ where $F(u) = \int_0^u f(z)\,dz$, and set $W(s) = E(u(s), v(s))$. As we pointed out above, $W'(s) = cv(s)^2$. The case $c = 0$ can be analyzed completely, since phase curves are the curves along which $E(u, v)$ is constant. Of most interest is the curve $E = 0$. Since $0 < a < \frac{1}{2}$, and so $F(1) > 0$, this level curve includes a closed curve through $(0,0)$, as shown in Figure 2.3.

A solution lying on this curve satisfies

$$\lim_{x \to -\infty} (u(s), v(s)) = (0,0),$$
$$\lim_{x \to \infty} (u(s), v(s)) = (0,0).$$

The trajectory of such a solution in the phase plane is called a "homoclinic orbit".

The local stable and unstable manifolds are defined just in a small neighborhood of $(0,0)$ and do not intersect except at $(0,0)$. It is only when we look at the global picture obtained by following the unstable manifold at $(0,0)$, $\mathcal{U}_0^+$, forward in $s$ that we see it returning as part of $\mathcal{S}_0^+$. In fact, $\mathcal{U}_0^+ = \mathcal{S}_0^+$. We cannot give such a precise analysis except when $c = 0$.

We have seen that when $c > 0$, $W$ is an increasing function. Also, if the solution is on $\mathcal{U}_c^+$ then $\lim_{s \to -\infty} W(s) = 0$. Hence $W(s) > 0$ and the solution lies outside the loop $E(u, v) = 0$.

The quantity $a$ is a physical parameter, called the "threshold", while $c$ was introduced in the traveling wave substitution and represents the wave speed, which must be determined. It is a consequence of the stable manifold theorem that the unstable manifold at $(0,0)$, $\mathcal{U}_c^+$, varies continuously with $c$. To understand this clearly we need to discuss $\mathcal{U}_c^+$ in more detail.

For $c = 0$, this curve is precisely defined as the set of points in the half-plane $u > 0$ where $E(u,v) = 0$. For $c \geq 0$, $\mathcal{U}_c^+$ is a smooth curve which is tangent to the eigenvector $\mathbf{v}_1$ of $\begin{pmatrix} 0 & 1 \\ a & c \end{pmatrix}$ corresponding to the positive eigenvalue $\lambda_1(c)$. This eigenvector has slope $\lambda_1(c) = \frac{c}{2} + \frac{1}{2}\sqrt{c^2+4a}$ , so that the slope of $\mathcal{U}_c^+$ near $(0,0)$ increases as $c$ increases. As $c \to \infty$, this slope tends to $\infty$.

Since the system (2.25) is autonomous, the trajectory of a solution starting on $\mathcal{U}_c^+$ is the same, no matter where on this curve it starts. Solutions on $\mathcal{U}_c^+$ differ from each other only by a shift of the independent variable, say $s$. We saw above that for each solution $(u,v)$ lying on $\mathcal{U}_c^+$, $\lim_{s\to-\infty} \frac{u(s)}{e^{\lambda_1 s}}$ exists and is nonzero. For every $s_0$,

$$\lim_{s\to-\infty} \frac{u(s-s_0)}{e^{\lambda_1 s}} = \lim_{r\to-\infty} \frac{u(r)}{e^{\lambda_1(r+s_0)}} = e^{-\lambda_1 s_0} \lim_{s\to-\infty} \frac{u(s)}{e^{\lambda_1 s}}.$$

Hence there is a unique $s_0$ such that

$$\lim_{s\to-\infty} \frac{u(s-s_0)}{e^{\lambda_1 s}} = 1.$$

Consequently there is a unique solution $(u,v) = (u_c, v_c)$ on $\mathcal{U}_c^+$ such that

$$\lim_{s\to-\infty} \frac{u(s)}{e^{\lambda_1 s}} = 1. \tag{2.29}$$

The statement that $\mathcal{U}_c^+$ is continuous in $c$ means that the function $(c,s) \to (u_c(s), v_c(s))$ is a continuous function at every point where it is defined. If the solution $(u_c, v_c)$ exists on $(-\infty, T]$, then $(u_c(s), v_c(s))$ is uniformly continuous in $c$ for $-\infty < s \leq T$, so that if $c$ changes by a small amount, the entire segment up to $(u_c(T), v_c(T))$ of $\mathcal{U}_c^+$ also changes by only a small amount. But $(u_c(s), v_c(s))$ is not uniformly continuous in $c$ for $-\infty < s < \infty$.

What we call "shooting from infinity", or in this case minus infinity, is to choose $c$, which determines the asymptotic behavior of $u_c(s)$ as $s \to -\infty$, to be the shooting parameter. Contrast this with our previous examples of shooting, where we used the value of some component of the solution at $s = 0$.[9]

[9]In Section 3.3 we discuss another instance of shooting from infinity.

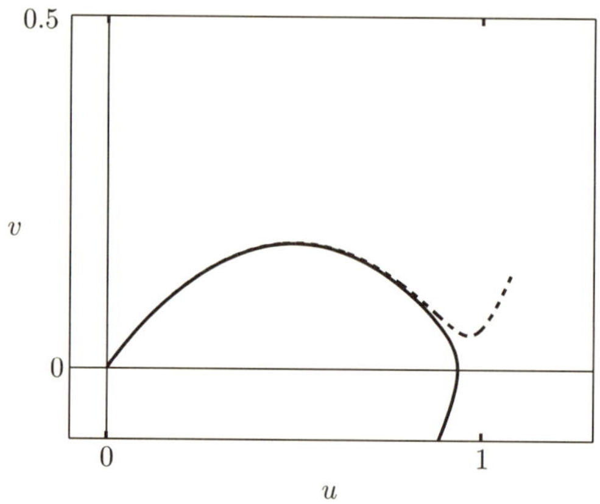


**Figure 2.4.** The solid curve is $(u_c, v_c)$ for a $c \in A$, while the dotted curve is for a $c \in B$.

We can now define the "bad sets" for this problem. Let

$$A = \{c > 0 \mid \text{for some } s_0,\ v_c(s_0) < 0 \text{ and } u_c < 1 \text{ on } (-\infty, s_0]\},$$
$$B = \{c > 0 \mid \text{for some } s_1,\ u_c(s_1) > 1 \text{ and } v_c > 0 \text{ on } (-\infty, s_1]\}.$$

(See Figure 2.4.)

We have the usual properties of $A$ and $B$ to verify, namely that they are open, nonempty, and disjoint. Disjointedness follows from the definitions of these sets, and openness follows from our earlier remark that $(u_c, v_c)$ depends continuously on $c$. To prove that $A$ and $B$ are nonempty, we will use the energy function $E$ and

$$W_c(s) = E(u_c(s), v_c(s)).$$

Since $(u_c(s), v_c(s)) \to (0,0)$ as $s \to -\infty$, $\lim_{s\to-\infty} W_c(s) = 0$.

To show that $A$ is nonempty we note that the solution $(u_0, v_0)$ enters the region $v < 0$ at a point $(u_0(\tau), 0)$, with $a < u_0(\tau) < 1$. At such a point, $v_0' < 0$, so that $v_0 < 0$ immediately to the right of $\tau$. Since $v_c(\tau)$ is a continuous function of $c$, $(u_c, v_c)$ also enters the region $v < 0$, near $(u_0(\tau), 0)$, for sufficiently small $|c|$. Furthermore, $u_c < 1$ at least until after $v_c < 0$. Therefore, a small number $c > 0$ lies in $A$.

To show that $B$ is nonempty, recall from earlier in this subsection that for $c > 0$, $\mathcal{U}_c^+$ lies outside $\mathcal{U}_0^+$. We wish to show first that the solution $(u_c, v_c)$ crosses the line $u = a$. To prove this it is convenient to consider $v$ as a function of $u$ on the trajectory of this solution. This assumption is valid for as long as $v_c (= u_c')$ is positive. In Exercise 2.7 you are asked to give a careful proof of this fact.

You are also asked in Exercise 2.7 to show that for as long as $v_c$ is positive, the slope of the tangent line to the curve $(u_c, v_c)$ is given by

$$\frac{dv_c}{du_c} = \frac{cv_c - f(u_c)}{v_c}. \tag{2.30}$$

Since $(u_c(s), v_c(s))$ is outside $\mathcal{U}_0^+$, $v_c(s) > 0$ as long as $u_c(s) \le a$. The slope (2.30) is bounded for (say) $v_c > 1$, so that the solution cannot blow up without crossing $u_c = a$. Hence, the solution does cross this line.

First, however, $(u_c, v_c)$ crosses $u = \frac{1}{2}a$. During the interval where $\frac{1}{2}a \le u \le a$, $f(u_c) \le 0$, and so $\frac{dv}{du} \ge c$. Hence, $c$ can be chosen so large that when $u_c(s) = a$, $\frac{1}{2}v_c^2(s) + F(a) > F(1) + 1$. Note that $F(1) = \max_{u \ge 0} F(u) > 0$. Since $W_c$ is nondecreasing, there is no subsequent $s$ where $\frac{1}{2}v_c^2 + F(u_c) = F(1) + 1$. In other words, after $u_c = a$,

$$\frac{1}{2}v_c^2 > F(1) + 1 - F(u_c) \ge 1.$$

Therefore $v_c$ is positive and bounded away from zero, and the solution crosses $u = 1$. Hence $B$ is nonempty, and so there is a $c^* > 0$ which is not in $A \cup B$.

Assume that $(u_{c^*}, v_{c^*})$ exists on the maximal interval $(-\infty, \omega)$, where $\omega \le \infty$. Then $u'_{c^*} = v_{c^*} > 0$ and $0 < u_{c^*} < 1$ on this interval. As before, $v_{c^*}$ can be considered as a function of $u_{c^*}$, and (2.30) then gives a bound on $v_{c^*}$ as well as $u_{c^*}$. Hence, $\omega = \infty$ and $\lim_{s \to \infty} u_{c^*}(s)$ exists.

Since $u'_{c^*} = v_{c^*}$, it is tempting to conclude immediately that $\lim_{s \to \infty} v_{c^*} = 0$. However a little care must be exercised here, and we refer the reader to Exercise 2.5 below for this detail.

The middle equilibrium point, $(a, 0)$, lies inside the loop $E = 0$, while for $c > 0$, $\mathcal{U}_c^+$ lies outside this loop. Hence the solution $(u_{c^*}, v_{c^*})$ does not tend to this point. But $\lim_{s \to \infty} (u_{c^*}(s), v_{c^*}(s))$ exists and is an equilibrium point. The only possibility is that the solution tends to $(1, 0)$ as $s \to \infty$, giving the desired heteroclinic orbit.

In Chapter 6 we will need to know a bit more about the solutions on $\mathcal{U}_c$ when $c$ is in $A$ or $B$.

**Proposition 2.4.** *If $c \in A$ then solutions on $\mathcal{U}_c^+$ enter the region $u < 0, v < 0$, and both variables tend to $-\infty$. If $c \in B$ then solutions on $\mathcal{U}_c^+$ enter the region $u > 1, v > 0$ and both variables tend to $\infty$.*

**Proof.** The proof is easy when we realize that the orbits lie outside the $E = 0$ curve and that $v' > 0$ when $v > 0, u > 1$. □

Finally, in Exercise 2.8 you are asked to prove that $c^*$ is unique.

**2.4.3. Summary of the results.** Since we will need the results in this section in Chapter 6, we summarize them here.

**Proposition 2.5.** (A) *For every $c > 0$, the system* (2.25) *has three equilibrium points, at* $(0,0)$, $(0,a)$, *and* $(1,0)$. *The points* $(0,0)$ *and* $(1,0)$ *are saddle points, while* $(a,0)$ *is either an unstable node or an unstable spiral point. The unstable manifolds at* $(0,0)$ *and* $(1,0)$ *have positive slope at the equilibrium point, while the stable manifolds have negative slope.*

(B) *When $c = 0$ and $0 < a < \frac{1}{2}$, there is a nonconstant solution $p_0 = (u,v)$ of* (2.25) *such that $p_0(\pm\infty) = (0,0)$. The trajectory of this solution lies on the curve $E(u,v) = 0$, where $E(u,v) = \frac{1}{2}v^2 + F(u)$.*

(C) *For each $c > 0$, let $\mathcal{U}_c^+$ denote the branch of the unstable manifold at* $(0,0)$ *which points into the region $u > 0, v > 0$. A solution on $\mathcal{U}_c^+$ exists at least until $u = a$, with $v > 0$ up to that point. Let $p_c = (u_c, v_c)$ be the unique solution of* (2.25) *such that $p_c(0) \in \mathcal{U}_c^+$, $v_c(s) > 0$ if $s \leq 0$, and $u_c(0) = a$. Then there is a unique $c^*(a) > 0$ such that $p_{c^*}(s)$ exists on $(-\infty, \infty)$ and $p_{c^*}(s) \to (1,0)$ as $s \to \infty$.*

(D) *If $c \in (0, c^*(a))$ then $v_c(s) = 0$ exactly once, at a point $s = s_1$ with $a < u_c(s_1) < 1$. In this case the solution exists on a semi-infinite interval $(-\infty, \omega_c)$, with $\omega_c < \infty$, and both $u_c(s)$ and $v_c(s)$ approach $-\infty$ as $s \to \omega_c$. Further, $v_c'(s_1) < 0$, and $u_c(s_1)$ is a continuous function of $c$ which tends monotonically to 1 as $c \to c^*(a)^-$.*

(E) *If $c > c^*(a)$ then again the solution $p_c$ exists on a semi-infinite interval $(-\infty, \omega_c)$. In this case, $v_c > 0$ and both $u_c$ and $v_c$ tend to $\infty$ as $s \to \omega_c$.*

**2.4.4. Results of Fife and McLeod and of Xinfu Chen.** When $\varepsilon = 0$ and $w = 0$, (2.23) becomes

$$u_t = u_{xx} + u(1-u)(u-a).$$

Much has been proved about this equation beyond the existence and uniqueness of traveling waves. In [**66**] P. Fife and J. B. McLeod showed that the traveling wave is asymptotically stable, and a region of attraction was described. In a second paper the same authors used a different approach, in which $p = u_x$ is the dependent variable and $u$ the independent variable [**67**]. This is reasonable because for the traveling wave here, $u_x > 0$. They obtained the same results as in their first paper, and their existence proof for the traveling wave is relatively short.

Subsequently, X. Chen extended the result to a wide range of evolution equations [**31**]. His main hypothesis was that a certain comparison principle between solutions was satisfied. Unfortunately, the needed comparison principle does not hold for the full system (2.21). We consider the existence of traveling waves for this system in Chapter 6.

## 2.5. Shooting when there are oscillations: A third order problem

We now discuss a third order equation of a particularly simple looking form. A natural boundary value problem for this equation is to prove the existence of a heteroclinic orbit connecting its two equilibrium points. The feature which adds interest is that the relevant eigenvalues for the linearized equations around each of these equilibrium points are complex. Hence the solutions must exhibit oscillations. Existence of oscillating solutions to a boundary value problem is often difficult to prove, with shooting or other methods. Proving uniqueness is even more problematic and is discussed in Chapter 12. The problem we discuss here has been studied several times before, so we were surprised to find that, contrary to our usual experience, existence follows by a relatively simple shooting argument. We thank our colleague W. C. Troy, who made a key observation enabling shooting to work.

The equation to be considered is

$$u''' = u^2 - 1. \tag{2.31}$$

To our knowledge it was first studied by Kopell and Howard in [**122**]. They gave a geometric proof of existence for the same problem we will study. However their proof was (it has now turned out, thirty years later!) much more complicated than necessary and relied on some four decimal point numerical calculations. Subsequently Conley considered the equation using what is now called the "Conley index", a sophisticated perturbation technique which is far from classical, and showed that there is a nonconstant bounded solution. As we will see below, this easily implies that there is a heteroclinic orbit. His result and proof extend to the equation $u^{(n)} = u^2 - 1$ for all integers $n$, but for $n > 3$ the existence of a heteroclinic orbit has not been proved, as far as we know, and is not true if $n$ is even. This problem is also discussed using the method of Conley index in [**223**] and [**225**].

More recently the problem arose again in a model of "syrup dripping from a spoon" [**17**], [**255**]. In this setting the desired solution is the "inner solution" of a so-called "matching" problem from singular perturbation theory. We will not discuss this topic now, but Chapter 7 is devoted to problems in matched asymptotic expansions. The problem also arose very recently in work of Bona and Weissler on conservation laws [**21**]. They gave independent existence and uniqueness proofs for the problem below of the same type as the ones in this chapter and in Chapter 12.

The equilibrium solutions of (2.31) are clearly $u = \pm 1$. We seek a solution such that

$$u(-\infty) = -1, \qquad u(\infty) = 1. \tag{2.32}$$

The linearization of (2.31) around $u = 1$ is

$$v''' = 2v,$$

with solutions $v = e^{\lambda t}$, where $\lambda$ is one of the three roots of 2 in the complex plane. The roots with negative real parts are both complex, and so the boundary conditions (2.32) force us to consider solutions which oscillate around $u = 1$ as $t \to \infty$. It is seen in the same way that the solutions also oscillate around $u = -1$ as $t \to -\infty$.

**2.5.1. Existence.** We now outline two slightly different shooting proofs of the existence of solutions to (2.31)–(2.32). The first is a repetition of a technique introduced in Section 2.2, in which we define bad sets $A$ and $B$ as disjoint nonempty open subsets of a connected set of initial conditions. The second proof uses the implicit function theorem at one stage.

For both proofs we consider, for each $\alpha$, the unique solution $u = u(t, \alpha)$ of (2.31) such that

$$u(0) = 0, \qquad u'(0) = \alpha, \qquad u''(0) = 0. \tag{2.33}$$

It is easy to see that the function $-u(-t)$ satisfies the same differential equation and initial conditions, so that (2.33) implies that the solution $u$ is an odd function. If $u(\infty) = 1$ then $u(-\infty) = -1$.

For a solution $u$ to (2.31) and (2.33) let

$$W(t) = u'u'' + u - \frac{1}{3}u^3.$$

Then

$$W' = u''^2,$$

and so $W$ is strictly increasing for every nonconstant solution. Further, $W(0) = 0$. The function $W$ enables us to prove two important lemmas for use in our existence proof.

**Lemma 2.6.** *If $u(t_1, \alpha) = \sqrt{3}$ for some first $t_1 > 0$, then $u(t, \alpha) > \sqrt{3}$ for as long after $t_1$ as the solution exists.*

**Proof.** The definition of $t_1$ implies that $u'(t_1) \geq 0$. If $u'(t_1) = 0$ then $W(t_1) = 0$, a contradiction since $W(0) = 0$ and $W$ is increasing. Hence $u'(t_1) > 0$. If $u(t, \alpha)$ has a subsequent maximum, say at $t_2 > t_1$, then $W(t_2) = u(t_2) - \frac{1}{3}u(t_2)^3 \leq 0$. Again, this is impossible, proving the lemma. □

**Lemma 2.7.** *If* $u(t,\alpha) > 0$ *on some interval* $(0,t_1)$ *and* $u(t_1,\alpha) = 0$, *then* $u'(t_1,\alpha) < 0$.

**Proof.** The only other possibility is that $u'(t_1,\alpha) = 0$. But then, $W(t_1) = 0$, again a contradiction. □

Next we observe that $u'''(0) = -1$, and so if $\alpha = 0$ then $u(t) < 0$ for sufficiently small positive values of $t$. By continuity of solutions with respect to $\alpha$, if $\alpha$ is positive but sufficiently small then $u = u(t,\alpha)$ first increases and then decreases and has a first zero $t_\alpha > 0$. By Lemma 2.7, $u'(t_\alpha,\alpha) < 0$.

On the other hand, suppose that $\alpha$ is large and positive. Since $u'''$ is bounded for as long as $u$ is bounded, it is easy to show that for sufficiently large $\alpha$, $u(t,\alpha)$ increases monotonically to beyond $u = \sqrt{3}$. By Lemma 2.6, this solution has no positive zero.

We now proceed in the usual way to define two sets $A$ and $B$ :

(2.34)
$$A = \{\alpha > 0 \mid u(t_1) = 0 \text{ for some } t_1 > 0\},$$
$$B = \left\{\alpha > 0 \mid u(t_2) = \sqrt{3} \text{ for some } t_2 > 0, \text{ and } u > 0 \text{ on } (0,t_2]\right\}.$$

We have shown above that $A$ contains an interval $(0,\alpha_1]$ and $B$ contains an interval $(\alpha_2,\infty)$, so that each of these sets is nonempty. In Exercise 2.6 you are asked to show that, as in several other examples in this chapter, $A$ and $B$ are also open and disjoint, and so there is an $\alpha^*$ such that $0 < u(t,\alpha^*) < \sqrt{3}$ on $(0,\infty)$.

We now outline a second proof that such an $\alpha^*$ exists. Our proof that the set $A$ is nonempty shows that for some positive $\alpha_1$, $u(t,\alpha_1) = 0$ has a solution $t_{\alpha_1} > 0$, with $u'(t_{\alpha_1},\alpha_1) < 0$ and $u(t,\alpha_1) > 0$ for $0 < t < t_{\alpha_1}$. The implicit function theorem then implies that the equation

$$u(t,\alpha) = 0 \tag{2.35}$$

has a unique solution $t = t_\alpha$ near to $t_{\alpha_1}$, for $\alpha$ in some neighborhood of $\alpha_1$. Using Lemma 2.7 we can show that the function $\alpha \to t_\alpha$ can be continued as a solution of (2.35) to a maximal interval $[\alpha_1,\alpha^*)$, where we know that $\alpha^* < \infty$ because the set $A$ is bounded above. It can further be shown that $\alpha^*$ has the same property as before, namely, $0 < u(t,\alpha^*) < \sqrt{3}$ on $(0,\infty)$.

Having found $\alpha^*$ by either of these methods, we claim that for the solution $u_*(t) = u(t,\alpha^*)$, $W$ is bounded. To prove this, recall that $W$ is positive and increasing, and so if $W$ is unbounded, then there is a $t_3 > 0$ with $W(t_3) > \frac{2}{3}$. But since $0 < u_* \le \sqrt{3}$, it follows that $u_*(t_3) - \frac{u_*(t_3)^3}{3} \le \frac{2}{3}$, and so $u_*'(t_3)\,u_*''(t_3) > 0$. Further, $u_*'u_*'' > 0$ for $t > t_3$, for if $u_*'u_*'' = 0$ at some $t > t_3$ then $W(t) \le \frac{2}{3}$, a contradiction. However, $u_*'u_*'' = \left(\frac{1}{2}u_*'^2\right)'$, and if this is positive on $(t_3, \infty)$ then $|u_*'|$ is increasing on this interval, which means that $u_*$ is unbounded, a contradiction. Hence $W$ is bounded.

Also, $W$ is increasing, and so $\lim_{t\to\infty} W(t)$ exists. Since $W' = u''^2$, it follows that $\int_0^\infty u_*''(s)^2\, ds < \infty$, and the boundedness of $u_*'''$ implies that $\lim_{t\to\infty} u_*''(t) = 0$. Since $u_*$ is bounded, Exercise 2.5(d) shows that $\lim_{t\to\infty} u_*'(t) = 0$, and then, using the definition of $W(s)$, we see that $\lim_{t\to\infty}\left(u_* - \frac{u_*^3}{3}\right)$ exists. By considering the graph of $u - \frac{u^3}{3}$ we conclude that $\lim_{t\to\infty} u_*(t)$ exists. Hence $(u_*, u_*', u_*'')$ tends to an equilibrium point of the system corresponding to (2.31). Also, $u_* > 0$ for $t > 0$, and so the only possible limit of $u_*$ is $u_*(\infty) = 1$. Since $u_*$ is odd, we have shown the existence of a solution to (2.31) and (2.32).

## 2.6. Boundedness on $(-\infty, \infty)$ and two-parameter shooting

This section relies on a more sophisticated topological property than connectedness. It may be considered somewhat specialized and perhaps be omitted in a first reading. We acknowledge that the concept of a classical method becomes hazy here.

Recall that in Section 2.2.2 we showed that if $f$ is continuous and bounded on $[0, \infty)$ then the first order equation $x' = x^3 + f(t)$ has a solution which is bounded on this interval. We now wish to consider the existence of solutions which are bounded on the entire real line $R^1$. However, it is not hard to see that this is unlikely for this first order equation, because in some sense we are imposing two boundary conditions, boundedness at $\infty$ and at $-\infty$.

Hence, to illustrate the use of shooting to prove the existence of solutions which are bounded on $R^1$ we will consider a second order equation

$$x'' = x^3 + f(t), \tag{2.36}$$

where $f$ is assumed to be continuous and bounded on $R^1$. Further, consider the initial value problem consisting of (2.36) and initial conditions

$$\begin{aligned} x(0) &= \alpha, \\ x'(0) &= \beta. \end{aligned} \tag{2.37}$$

We then have a two-parameter shooting problem. For each $(\alpha, \beta) \in R^2$ there is a unique solution $x(t, \alpha, \beta)$, and we wish to show that there is a pair $(\alpha^*, \beta^*)$ such that $x(t, \alpha^*, \beta^*)$ exists and is bounded on $R^1$.

Choose $M$ so that $|f(t)| < M$ for all $t$ and let $m = M^{1/3}$. We then consider the shooting set

$$\Omega = \{(\alpha, \beta) \mid -m - 1 < \alpha < m + 1\}. \tag{2.38}$$

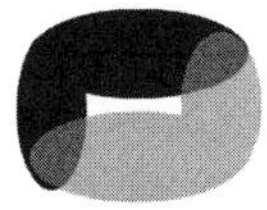

**Figure 2.5.** The intersection of the two sets is disconnected.

It turns out that this set is easier to work with than $R^2$ in this example. It also has a slight advantage over the more natural choice in which $m$ is used in the inequalities instead of $m+1$. It is important to realize that subtle differences in the choices of $\Omega$ and the bad sets can significantly affect a shooting method proof.[10]

In this and some other two-parameter problems, connectedness of $\Omega$ is not a sufficiently discriminating topological property. Various two-dimensional topological results might be employed. Here we will use the following lemma:

**Lemma 2.8.** *Suppose that $\Lambda_1$ and $\Lambda_2$ are two open subsets of $R^2$, with components (maximal connected subsets) $\tilde{\Lambda}_1 \subset \Lambda_1$ and $\tilde{\Lambda}_2 \subset \Lambda_2$ such that $\tilde{\Lambda}_1 \cap \tilde{\Lambda}_2$ is disconnected. Then $\Lambda_1 \cup \Lambda_2 \neq R^2$.*

This result may be found in a useful book on the topology of the plane by M. H. A. Newman [**180**]. It also applies to every subset of $R^2$ which is homeomorphic to the entire plane, and we will apply it to the set $\Omega$ without further comment. Figure 2.5 illustrates the lemma.

In Chapter 13, we will have occasion to use a similar result on the topology of the plane for another two-parameter shooting problem. However in many cases where it appears at first glance that such a two-dimensional theorem is required, it turns out that one-parameter shooting is applicable. An example is given in Section 3.3.

Observe in Lemma 2.8 that the sets $\Lambda_1$ and $\Lambda_2$ do not have to be connected. In the problem under discussion, the $\Lambda_1$ and $\Lambda_2$ to be defined will be shown to be connected, but this is not always possible.

We now need to identify the bad sets, $\Lambda_1$ and $\Lambda_2$, for our particular equation, namely (2.36), with initial conditions (2.37). The situation is a little different from our earlier work, on the first order equation (2.3), since $x > m = M^{1/3}$ does not imply that $x' > 0$, but instead that $x'' > 0$. Still, if $x(0) < m+1$ and for some $t_1 > 0$, $x(t_1) > m+1$, then there is a $t_0 \in (0, t_1)$ with $x(t_0) > m+1$ and $x'(t_0) > 0$. Further, $x'' > 0$ and $x' > 0$ for as long as the solution exists beyond $t_0$.

Once again we define a "property $P$", in such a way that if $(\alpha, \beta)$ has property $P$ then $x(t, \alpha, \beta)$ exists and is bounded on $(-\infty, \infty)$. Based on

[10] See Exercise 2.11.

what we found in the previous section, we do not define property $P$ simply as boundedness of $x(t,\alpha,\beta)$. Instead, we will say that $(\alpha,\beta)$ has property $P$ if $|x(t,\alpha,\beta)| \leq m+1$ for all $t$.

We can immediately identify **four** bad sets, which is to say four ways in which the solution to (2.36)–(2.37) can fail to have property $P$. These are

$$\begin{aligned}
\Gamma_1 &= \{(\alpha,\beta) \in \Omega \mid \text{for some } t>0,\ x(t,\alpha,\beta) > m+1\},\\
\Gamma_2 &= \{(\alpha,\beta) \in \Omega \mid \text{for some } t>0,\ x(t,\alpha,\beta) < -m-1\},\\
\Gamma_3 &= \{(\alpha,\beta) \in \Omega \mid \text{for some } t<0,\ x(t,\alpha,\beta) > m+1\},\\
\Gamma_4 &= \{(\alpha,\beta) \in \Omega \mid \text{for some } t<0,\ x(t,\alpha,\beta) < -m-1\}.
\end{aligned}$$

These sets are clearly open in $R^2$, because $x(t,\alpha,\beta)$ is a continuous function of $(\alpha,\beta)$ for every $t$ where it is defined. They are also each nonempty. For example, if $m<\alpha<m+1$ then for every $\beta \geq 0$, $(\alpha,\beta) \in \Gamma_1$. However, we do not have an appropriate topological lemma involving four sets.[11] Therefore we combine the $\Gamma_i$ to form two new sets, as follows:

$$\begin{aligned}
\Lambda_1 &= \Gamma_1 \cup \Gamma_3,\\
\Lambda_2 &= \Gamma_2 \cup \Gamma_4.
\end{aligned}$$

It is easy to see that if $(\alpha,\beta)$ is in $\Omega$ but not in $\Lambda_1 \cup \Lambda_2$, then $|x(t,\alpha,\beta)| \leq m+1$ on $(-\infty,\infty)$. Hence we wish to show that $\Lambda_1$ and $\Lambda_2$ satisfy the hypotheses of Lemma 2.8.

**Lemma 2.9.** *The sets $\Lambda_1$ and $\Lambda_2$ are connected.*

**Proof.** We first observe that every point in the strip $m<\alpha<m+1$ lies in $\Lambda_1$, which is true because $x''>0$ in this strip, and so every solution starting there crosses $x=m+1$ either as $t$ increases or as $t$ decreases or both. In particular, if $m<\alpha<m+1$ then $(\alpha,0) \in \Gamma_1 \cap \Gamma_3$.

Now suppose that $(\alpha_1,\beta_1) \in \Lambda_1$. We show that if $\alpha_2>\alpha_1$ then $(\alpha_2,\beta_1) \in \Lambda_1$. This follows from a simple comparison principle. Suppose, for example, that $x(t_1,\alpha_1,\beta_1) > m+1$ for some $t_1>0$. If $\alpha_2>\alpha_1$ then $x(t,\alpha_2,\beta_1) > x(t,\alpha_1,\beta_1)$ on some interval $[0,T)$. Furthermore, on every such interval, $x''(t,\alpha_2,\beta_1) > x''(t,\alpha_1,\beta_1)$, and so $x'(t,\alpha_2,\beta_1) > x'(t,\alpha_1,\beta_1)$ on $(0,T]$. It follows that $x(t,\alpha_2,\beta_1) > x(t,\alpha_1,\beta_1)$ for as long as they both exist, and therefore, $x(t,\alpha_2,\beta_1) > m+1$ for some $t \in (0,t_1)$. A similar argument applies if $t_1<0$.

We therefore see that every point in $\Lambda_1$ can be connected by a straight line within $\Lambda_1$ to the region $m<\alpha<m+1$, and this region lies entirely in $\Lambda_1$. Therefore $\Lambda_1$ is connected, and similarly, $\Lambda_2$ is connected. ☐

[11] One can construct such a lemma but it is simpler to work with Lemma 2.8.

Therefore, in applying Lemma 2.8 we take $\tilde{\Lambda}_1 = \Lambda_1, \tilde{\Lambda}_2 = \Lambda_2$.

We now show that $\Lambda_1 \cap \Lambda_2$ is disconnected. Clearly,

$$\Lambda_1 \cap \Lambda_2 = (\Gamma_1 \cap \Gamma_2) \cup (\Gamma_1 \cap \Gamma_4) \cup (\Gamma_3 \cap \Gamma_2) \cup (\Gamma_3 \cap \Gamma_4).$$

**Lemma 2.10.** *$\Gamma_1 \cap \Gamma_2$ and $\Gamma_3 \cap \Gamma_4$ are empty.*

**Proof.** Suppose that $(\alpha, \beta) \in \Gamma_1$, and let $x = x(\cdot, \alpha, \beta)$. Then $-m-1 < x(0) < m+1$ and there is a $t_1 > 0$ with $x(t_1) > m+1$. Since $x$ has no minimum below $x = -m$, there is no $t \in (0, t_1)$ with $x(t) < -m-1$, and further, as we saw earlier, $x$ increases after $t_1$ for as long as it exists. Hence $(\alpha, \beta) \notin \Gamma_2$. A similar proof shows that $\Gamma_3 \cap \Gamma_4$ is also empty. □

**Lemma 2.11.** *$\Gamma_1 \cap \Gamma_4$ and $\Gamma_2 \cap \Gamma_3$ are nonempty and disjoint.*

**Proof.** To show that these sets are nonempty, consider a point $\left(m + \frac{1}{2}, \beta\right)$. For each $\beta \leq 0$ this point lies in $\Gamma_3$. In Exercise 2.12 you are asked to show that if $-\beta$ is sufficiently large then $\left(m + \frac{1}{2}, \beta\right)$ also lies in $\Gamma_2$. This conclusion follows because $|x''|$ is bounded in the region $|x| \leq m+1$. A similar argument shows that for sufficiently large $\beta$, $\left(-m - \frac{1}{2}, \beta\right) \in \Gamma_1 \cap \Gamma_4$.

Disjointness of $\Gamma_1 \cap \Gamma_4$ and $\Gamma_2 \cap \Gamma_3$ follows from Lemma 2.10. □

We now apply Lemma 2.8 to show that there is a point $(\alpha^*, \beta^*) \in \Omega$ which is not in $\Lambda_1$ or $\Lambda_2$, and the corresponding solution $x(t, \alpha^*, \beta^*)$ is bounded on $(-\infty, \infty)$. While our results above only imply the bound $|x| \leq m+1$, it is easy to see that if $|x(t)| \geq m$ for some $t$, then $x$ is unbounded in one direction or the other. We have proved the following result.

**Theorem 2.12.** *If $f$ is continuous on $(-\infty, \infty)$, with $|f(t)| < m^3$ for all $t$, then the equation* (2.36) *has a solution $x$ which exists on this interval and satisfies $|x(t)| < m$ for all $t$.*

## 2.7. Ważewski's principle, Conley index, and an $n$-dimensional lemma

The shooting method is essentially topological, as is seen most clearly in the last section. Some authors have used the term shooting only for proofs that depend on connectedness. (See [**44**, Section 2.5].) We use it more broadly, for each argument in which a boundary value problem is shown to have a solution by considering the topology of the space of initial conditions. The term is also used for numerical procedures based on varying the initial conditions to find an approximate solution.

As we said at the beginning of this chapter, the shooting method can be traced back at least to a theorem of Ważewski [**250**] in 1947 which is similar in spirit to Lemma 2.8. In the classic textbook by Hartman [**83**] the

Ważewski principle is used to obtain some general results about the asymptotic behavior of perturbed linear systems. This text also has references to earlier uses of similar principles. A nice concrete application of Ważewski's principle to a fourth order system was given in [**53**]. More recent surveys of Ważewski's principle and the important extensions made by C. Conley are in [**223**], [**225**], and [**172**].

In Chapter 13 we will use a variant of Lemma 2.8 to establish existence for a problem from fluid mechanics. Also we will give an $n$-dimensional version of this topological result. Again, this is similar to Ważewski's theorem. Unfortunately we have found only one boundary value problem to which this higher dimensional result can be applied [**88**]. We would be happy if readers discover some others.

## 2.8. Exercises

**Exercise 2.1.** (a) Find a family of functions $\phi_\alpha$ such that $\phi_\alpha(t)$ is continuous in $(\alpha, t)$ except at $(0,1)$, $\lim_{t\to 1^-}\phi_\alpha(t)$ exists for each $\alpha \neq 0$, and $\lim_{t\to 1^-}\phi_0(t) = \infty$. (b) Find an initial value problem $x' = f(t,x)$, $x(0) = \alpha$, with a unique solution $x_\alpha$ for each $\alpha$, such that

$$\Omega = \{\alpha | x_\alpha \text{ is unbounded on } [0,\omega) \text{ for some } \omega < \infty\}$$

is not an open set. Hint: Try an equation of the form $x' = m(t)x^2$. (c) Prove that for an autonomous differential equation $x' = f(x)$ with $f'$ continuous on $R^1$, the set $\Omega$ as defined in (b) is open. (d) Find an autonomous system $x' = f(x,y)$, $y' = g(x,y)$ such that if $x(\cdot,\alpha,\beta)$ is the solution $(x,y)$ such that $x(0) = \alpha$, $y(0) = \beta$, then $x(\cdot,\alpha,\beta)$ is unbounded on $[0,\infty)$ if $\alpha = \beta \neq 0$, and bounded on $[0,\infty)$ if $\alpha \neq \beta$ or $\alpha = \beta = 0$. Hint: You might look for an example where $\alpha = \beta$ implies that $x = y$.

**Exercise 2.2.** (Easy) Prove that the solution found in Theorem 2.1 is unique.

**Exercise 2.3.** (Routine application of shooting) For each $\alpha$ let $u(t,\alpha)$ satisfy

$$u'' = u^3 - u + \cos t,$$
$$u(0) = \alpha, \ u'(0) = 0.$$

Prove that there is an $\alpha^*$ such that $u'(\pi, \alpha^*) = 0$.

**Exercise 2.4.** (Routine) For the solution obtained in Exercise 2.3, use the symmetry properties of the cosine function to show that the corresponding solution $u(t,\alpha)$ is periodic of period $2\pi$. (Hint: Consider the functions $v(t) = u(\pi + t)$ and $w(t) = u(\pi - t)$.) Conclude that this equation has both a periodic solution and unbounded solutions. Is this possible for a linear inhomogeneous equation? (See the last section of Chapter 14.)

**Exercise 2.5.** (Standard, but important to remember) (a) Give an example of a function $q$ such that $q''$ is continuous on $[0, \infty)$ and $\lim_{s\to\infty} q(s)$ exists, but $\lim_{s\to\infty} q'(s)$ does not exist. (b) Show that for every such example, $q''$ is unbounded. (c) For the solution $(u_{c^*}, v_{c^*})$ discussed in Section 2.4, show that $v''_{c^*}$ is bounded and therefore that $\lim_{s\to\infty} v_{c^*} = 0$. (d) Show that if $\lim_{s\to\infty} q''(s) = 0$ and $q$ is bounded, then $\lim_{s\to\infty} q'(s) = 0$.

**Exercise 2.6.** (Routine; needed in this chapter) Show that the sets $A$ and $B$ defined in (2.34) are open and disjoint. Hint: Use Lemmas 2.6 and 2.7.

**Exercise 2.7.** (Needed later) Consider a solution $(u, v)$ of (2.25) such that in some interval $I$, $v > 0$. Show that there is a unique function $\eta(u)$ defined on $u(I)$ such that for $s \in I$, $v(s) = \eta(u(s))$, and in $u(I)$,

$$\frac{d\eta(u)}{du} = \frac{c\eta(u) - f(u)}{\eta(u)}.$$

(In the future, we will use the notation $\frac{dv}{du} = \frac{cv - f(u)}{v}$ for this equation.) Hint: Which is it, the implicit function theorem or the inverse function theorem?

**Exercise 2.8.** By using the function $E$, prove that the wave speed $c^*$ of the heteroclinic orbit found for (2.25) is unique.

**Exercise 2.9.** Find the exact solution referred to in footnote 7. Hint: Look for a polynomial solution of the first order equation in Exercise 2.7.

**Exercise 2.10.** (Thanks to Prof. G. B. Ermentrout for this example.) Suppose that $f(u) = u(1-u)(u-a)$, where $a \in \left(0, \frac{1}{2}\right)$, and consider, for each $c > 0$, the system

$$\begin{aligned} cu' &= f(u) + v - u, \\ v' &= v - u. \end{aligned}$$

Show that for some $c > 0$ there is a heteroclinic orbit connecting $(0,0)$ and $(1,1)$. Hint: Watch out for the equilibrium point $(a, a)$, which might "get in the way".

**Exercise 2.11.** (a) Investigate the difficulties which crop up if we let $\Omega = R^2$ in the problem of finding a bounded solution to (2.36). Which step in the shooting proof becomes more difficult? (b) Suppose that in the definition of $\Omega$, $m + 1$ is replaced by $m$. What must change in the rest of the proof? (c) In the definition of $\Omega$ in (2.38), the assumption was that $|f(t)| < M = m^3$ for all $t$. Would the proof be any harder if we only assumed that $|f(t)| \le M$ for all $t$? (d) What other reasonable choices might be made for $\Omega$, $\Lambda_1$, and $\Lambda_2$? Can you find one that is better than what we used?

**Exercise 2.12.** (Routine) Complete the proof of Lemma 2.11 by showing that if $\alpha = m + \frac{1}{2}$ and $-\beta$ is sufficiently large, then $(\alpha, \beta) \in \Gamma_2$.

**Exercise 2.13.** (Important) Show that if $\beta > 0$ then the boundary value problem

$$f''' + ff'' + \beta\left(1 - f'^2\right) = 0,$$
$$f(0) = f'(0) = 0, \; f'(\infty) = 1$$

has a solution such that $f' > 0$ on $(0, \infty)$. Hint: Suppose that $f''(0) = \alpha$. Vary $\alpha$. Base your shooting sets on the behaviors of $f'$ and $f''$. You may wish to use a numerical ode solver to examine $f'$ for different values of $\alpha$. [12]

**Exercise 2.14.** Show that the boundary value problem

$$y'' + \frac{2}{r}y' + yy' = 0,$$
$$y(1) = 0, \; y(\infty) = 1$$

has at least one solution. Hint: Show that every solution $y$ of this ode with $y(1) = 0$, $y'(1) = c > 0$ satisfies the equation

$$y(r) = c\int_1^r \frac{1}{s^2} e^{-\int_1^s y(t)dt} ds.$$

Conclude that $\lim_{r\to\infty} y(r)$ exists. Show that this limit is a continuous function of $c$. Show that for some value of $c$, $y(\infty) = 1$. (See Chapter 7.)

**Exercise 2.15.** (Interesting) Consider a boundary value problem

$$u'' = f\left(t, u, u'\right), \quad u(0) = 0, \quad u(1) = 0.$$

Assume that $f$ is smooth and that there is an $R > 0$ such that $f(t, R, 0) > 0$ and $f(t, -R, 0) < 0$, for $t \in [0, 1]$. Assume also that there is a function $\psi : [0, \infty) \to (0, \infty)$ such that (i) $|f(t, u, u')| \le \psi(u')$ for all $t \in [0, 1]$ and (ii) $\int_0^\infty \frac{vdv}{\psi(v)} > R$. (a) Use shooting to prove that the given boundary value problem has a solution. (b) Extend to boundary conditions of the form $-\alpha u(0) + \beta u'(0) = 0$, $\gamma u(1) + \delta u'(1) = 0$, with $\alpha, \beta, \gamma, \delta$ all positive, and to periodic boundary conditions $u(0) = u(1)$, $u'(0) = u'(1)$. (c) Extend to $f$ which is continuous by approximating such an $f$ with smooth functions. This result is due to Bernstein and Nagumo, and a proof using Ważewski's method is given in [**225**].

[12] The differential equation in this problem, known as the Falkner-Skan equation, is important in the theory of boundary layers in fluid mechanics. The first proof of the result was given in 1942 by H. Weyl [**253**] and used a fixed point theorem in function space. See also Chapter XIV, part II of [**83**], where uniqueness of the solution is also proved. It is an indication of the strength of the shooting method that once the theory has been described, this problem can be given as an exercise, of average difficulty. Contrast this with the approach outlined in [**34**], where it is stated that the problem is "very difficult". Further results on the Falkner-Skan equation are given in Chapter 9.

*Chapter 3*

# Some boundary value problems for the Painlevé transcendents

## 3.1. Introduction

The equations studied in this chapter represent two out of a set of six families of second order nonlinear equations discovered by Painlevé and others over 100 years ago [**186**], [**106**]. The original work on these equations seems to have been purely for their mathematical interest, but in recent decades it has been found that these equations are of fundamental importance for a wide class of problems involving wave motion in physics. This realization has generated a large literature, for surveys of which see [**70**] or [**38**].[1]

Painlevé found these equations in the course of a study of second order equations of the form

$$y'' = F(y, y', z),$$

where the variables $y$ and $z$ are complex. The function $F$ is assumed to be rational in $y$ and $y'$, and analytic in $z$. Solutions to an initial value problem will then be locally analytic but may have singularities. Painlevé sought equations for which the locations of all essential singularities or branch points of the solutions are independent of the initial conditions and depend only

[1]The last section of this chapter includes more information about what makes the Painlevé transcendents interesting.

on the equation. This property is satisfied by linear equations and by six families of nonlinear equations. We will study examples of the first and second of these six families. For these Painlevé showed that solutions are meromorphic and have only poles as singularities. Here we consider only real solutions.

## 3.2. A boundary value problem for Painlevé I

The first two of Painlevé's equations are especially amenable to detailed analysis. In this section we will concentrate on a problem considered originally by Joshi and Kitaev [**114**], and then by Slemrod [**218**], for the first and simplest of these equations, universally referred to as Painlevé I. One form of this equation is

$$y'' = y^2 - t. \tag{3.1}$$

A previous study by P. Holmes and D. Spence in [**102**] divided the solutions of (3.1) into three sets, namely (i) those which are asymptotic to $\sqrt{t}$ as $t \to \infty$, (ii) those which are asymptotic to $-\sqrt{t}$ as $t \to \infty$, and (iii) those which blow up at a finite value of $t$. It was stated that the solutions in class (ii) all oscillate around $-\sqrt{t}$ as $t \to \infty$. But in [**114**] Joshi and Kitaev showed that there is a unique solution which remains below $-\sqrt{t}$ on $[0, \infty)$ and is in the Holmes-Spence class (ii). Their proof was complicated, and depended on series expansions of the solutions. It was simplified considerably by Slemrod in [**218**].[2] His proof is mostly in the spirit of this text, except for the use of Helly's selection theorem from real analysis [**121**], which is not usually required in ode problems.

In this section we discuss three approaches to proving existence of this solution, one by shooting, one modeled on the proof in [**218**] but at a more elementary level, and one which is perhaps the epitome of classical analysis, because it is based on asymptotic analysis of the solution. We show that the shooting method appears to run into a "tangency" problem which is difficult, perhaps impossible, to resolve and also show how to simplify the final steps of the proof in [**218**] so as to make it truly "elementary" in classical ode style.

**3.2.1. Shooting appears not to work.** Shooting does not work for every boundary value problem. For example, in Section 14.6 we discuss a problem due to Littlewood concerning the boundedness of solutions to a class of second order ode's. Unfortunately we have not been able to apply our methods to this class of equations. Painlevé I presents a much simpler example. The reader who is interested only in approaches which succeed,

[2]Slemrod was motivated to consider this type of solution by a problem in plasma physics where only the monotone solution could be stable.

not in those which fail, may want to skip to the next subsection. However we hope that a discussion of the reason that shooting does not work in this case (or, at least, we have not been able to make it work) may add to the reader's understanding of the method.

The typical shooting approach is to consider (3.1) with initial conditions

$$y(0) = \alpha, \qquad y'(0) = \beta \tag{3.2}$$

and attempt to show that there is a point $(\alpha, \beta)$ such that $y$ has the desired properties. The goal is to find a solution with $y' \leq 0$ and $y'' \geq 0$ on $[0, \infty)$, and it is easy to see that this is only possible if $\alpha < 0$ and $\beta < 0$. Denote this quadrant in the $(\alpha, \beta)$-plane by $\Omega$.

The so-called "shooting sets" are chosen to reflect the different ways in which a solution may fail to satisfy the required conditions. A natural initial guess is to let

$$\begin{aligned} A &= \left\{(\alpha, \beta) \in \Omega \mid y' \text{ becomes positive before } y'' \text{ becomes negative}\right\}, \\ B &= \left\{(\alpha, \beta) \in \Omega \mid y'' \text{ becomes negative before } y' \text{ becomes positive}\right\}. \end{aligned}$$

For example, $(\alpha, \beta) \in A$ if there is a $t_0 > 0$ such that $y'(t_0) > 0$ and $y'' \geq 0$ on $[0, t_0]$.

Suppose that $(\alpha^*, \beta^*)$ is a point in $\Omega$ which is not in $A \cup B$. If $y$ is the corresponding solution, then either

(i) $y' \leq 0, y'' \geq 0$ on $[0, \infty)$ or

(ii) for some $t_0 > 0$, $y'(t_0) = 0$, $y''(t_0) = 0$, and $y' \leq 0$, $y'' \geq 0$ on $[0, t_0)$.

If case (i) holds then we have found the desired solution. But unfortunately, we cannot rule out case (ii). In fact, it is easily seen in numerical computations that case (ii) is likely to occur. So we look for another shooting procedure. An important tool in such a search is the use of a simple numerical ode package to test out various ideas. The ode program XPP, developed by our colleague G. B. Ermentrout, has been extremely helpful to us over the years [**58**]. Often, if a numerical shooting procedure can be found using this program, then an existence theorem can be proved.

A numerical procedure which seems to work is as follows: Start with an arbitrary initial condition $(\alpha_0, \beta_0) \in \Omega$ and integrate forward until the solution graph crosses $y = -\sqrt{t}$, say at $t = T(\alpha_0, \beta_0)$. Most likely one finds that by moving $\alpha_0$ up or down slightly, $T(\alpha, \beta_0)$ can be increased, so that the point where $y$ crosses $-\sqrt{t}$ is further out. Repeatedly adjusting $\alpha$ in one direction or another, one finds an optimal $\alpha$, where $T(\alpha, \beta_0)$ is a maximum. Let $\alpha_1$ be this optimal $\alpha$.

Next, adjust $\beta$, looking to increase $T(\alpha_1, \beta)$. Sometimes $\beta$ must be increased, sometimes decreased, but again it is possible to find an optimal $\beta = \beta_1$, so that $T(\alpha_1, \beta_1)$ is a maximum. Now go back and adjust $\alpha$ again.

After a number of iterations one sees apparent convergence toward the desired monotonic solution. And there is some theory to support this. To describe this theory, we consider the partial derivatives $v = \frac{\partial y}{\partial \alpha}$ and $w = \frac{\partial y}{\partial \beta}$. Both $v$ and $w$ satisfy the equation of first variation[3] for (3.1), which is

$$Y'' = 2yY. \tag{3.3}$$

In fact, $v$ and $w$ are linearly independent solutions of this linear equation, as is seen from their initial conditions:

$$\begin{aligned} v(0) &= 1, \qquad & v'(0) &= 0, \\ w(0) &= 0, \qquad & w'(0) &= 1. \end{aligned}$$

Now consider the equation which locally defines the crossing point $T$:

$$y(T(\alpha, \beta), \alpha, \beta) = -\sqrt{T(\alpha, \beta)}. \tag{3.4}$$

Using the implicit function theorem we find that

$$\frac{\partial T}{\partial \alpha} = -\frac{\frac{\partial y}{\partial \alpha}}{y'(T) + \frac{1}{2\sqrt{T}}}, \qquad \frac{\partial T}{\partial \beta} = -\frac{\frac{\partial y}{\partial \beta}}{y'(T) + \frac{1}{2\sqrt{T}}}. \tag{3.5}$$

In order to increase $T$ by varying $\alpha$ or $\beta$, we need either $v = \frac{\partial y}{\partial \alpha}$ or $w = \frac{\partial y}{\partial \beta}$ to be nonzero at $t = T$. But as we observed, these are linearly independent solutions of the linear equation (3.3), and so they don't both vanish at the same time. Thus, our iteration procedure ought to work.

But to be sure, we need to check that the hypotheses of the implicit function theorem are satisfied where we use (3.4) to define $T$ as a function of $(\alpha, \beta)$. Clearly the method fails if the denominators in (3.5) are zero, which happens if the solution $y$ is tangent to the graph of $-\sqrt{t}$ at $t = T$. So we attempt to prove that this tangency does not occur if $y(0) < 0$, $y'(0) < 0$.

In fact, a key lemma from **[218]** says that this tangency does occur. Yet this lemma, which we state and prove further on, is crucial to the existence proofs in the next two sections. It has the consequence that if $y(t, \alpha, \beta)$ is the solution of (3.1) and (3.2), then the location of the first zero of $y''$ may be a discontinuous function of $(\alpha, \beta)$, and numerical computations indicate that this does happen. Hence a shooting method seems difficult to use in this problem.

[3] Also called the linearization of (3.1) around a given solution $y$.

**3.2.2. An alternative approach.** We now state the lemma from [**218**] referred to in the last section. In its statement we let $r(t) = -\sqrt{t}$. We incorporate a second result from [**218**] which is also needed.

**Lemma 3.1.** *For each* $t_0 > 0$, *let* $y_{t_0}$ *be the unique solution of* (3.1) *such that* $y(t_0) = r(t_0)$, $y'(t_0) = r'(t_0)$. *Then* $y'_{t_0} < 0$ *on* $[0, t_0]$ *and* $y_{t_0} < r$ *on* $(0, t_0)$. *Also, if* $0 < t < t_0$ *then*

$$y_{t_0}(t) > -t^{1/2} - \frac{\sqrt{3}}{2} t^{-3/4}. \tag{3.6}$$

*In particular, if* $t_0 > 1$ *then*

$$y_{t_0}(0) > y_{t_0}(1) > M = -1 - \frac{\sqrt{3}}{2}. \tag{3.7}$$

**Proof.** We defer the proof until Section 3.2.3. □

Instead of shooting, we consider further the solutions discussed in Lemma 3.1. But on the principle that it is easier to think about solutions in the forward direction, we follow [**218**] by letting $x = -t$ and $y(x) = u(t)$, giving the equation

$$u'' = u^2 + x. \tag{3.8}$$

For each $x_0 < 0$ let $u_{x_0}$ denote the solution of (3.8) such that

$$u(x_0) = -\sqrt{-x_0}, \qquad u'(x_0) = \frac{1}{2\sqrt{-x_0}}. \tag{3.9}$$

Then Lemma 3.1 implies that $u_{x_0}(x) < -\sqrt{-x}$ on $(x_0, 0)$ and also gives us a lower bound on $u_{x_0}$. We will show that as $x_0 \to -\infty$, some subsequence of the solutions $u_{x_0}$ tends to the desired solution, which in the variables $u$ and $x$ satisfies $u'' > 0$ on $(-\infty, 0)$.

**Theorem 3.2.** *The Painlevé I equation* (3.8) *has a solution* $u$ *which is strictly increasing and concave up on* $(-\infty, 0]$ *and satisfies*

$$\lim_{x \to -\infty} \frac{u(x)}{\sqrt{-x}} = -1. \tag{3.10}$$

**Proof.** For each $x_0 < -1$, consider as before the unique solution $u = u_{x_0}$ of (3.8) satisfying (3.9). By Lemma 3.1, $u(x) < -\sqrt{-x}$ on $(x_0, 0]$, and in particular, $u\left(-\frac{1}{2}\right) < -\sqrt{\frac{1}{2}}$. Choose $M$ as in (3.7), so that for every $x_0 < -1$, $u(-1) \geq -M$. Then the mean value theorem implies that there is an $x_1 \in \left(-1, -\frac{1}{2}\right)$ such that

$$u'(x_1) = \frac{u\left(-\frac{1}{2}\right) - u(-1)}{\frac{1}{2}} \leq 2\left(M - \sqrt{\frac{1}{2}}\right).$$

Since $u''$ is bounded in $(-1,0)$, $u'(0)$ is bounded as a function of $x_0$. Since $u < -\sqrt{-x}$ on $(-1,0)$, it follows that there is a $\delta > 0$ such that for every $x_0 < -1$,

$$-M \le u(0) \le -\delta. \tag{3.11}$$

Take a sequence $\{x_0^k\}$ of $x_0$'s tending to $-\infty$, getting corresponding solutions $u_k$. From (3.11) and the uniform bound on $u'(0)$ we conclude that some subsequence of the points $\{(u_k(0), u_k'(0))\}$ converges, say to $(\alpha^*, \beta^*)$. We claim that the solution $u_*$ of (3.8) with $u(0) = \alpha^*, u'(0) = \beta^*$ is increasing on $(-\infty, 0]$, with $u_*'' \ge 0$. If not, then this solution crosses $-\sqrt{-x}$, and there is an $x^* < 0$ such that $u_*(x^*) > -\sqrt{-x^*}$. However $(\alpha^*, \beta^*)$ was chosen as the limit of initial conditions $(u_k(0), u_k'(0))$ for solutions $u_k$ such that for sufficiently large $k$, $u_k(x^*) < -\sqrt{-x^*}$. This is a contradiction, and so $u_*'' \ge 0$. We leave the proof of (3.10) as Exercise 3.1, which will complete the proof of the theorem. □

**Remark 3.3.** This proof is a good illustration of why we advocate finite dimensional methods for ode's. It only requires us to show the convergence of a subsequence in $R^2$. The proof in [**218**] needs a fairly sophisticated result, Helly's theorem about the convergence of subsequences in a function space. Use of such results is often unavoidable in pde's but frequently unnecessary in ode problems. As in this example, a finite dimensional approach may be noticeably shorter.

### 3.2.3. Proof using asymptotic analysis.

For this method it is slightly more convenient to consider the problem on $[0, \infty)$, using equation (3.1).

**Lemma 3.4.** *There exists a solution $y$ such that $\lim_{t\to\infty} \frac{y(t)}{\sqrt{t}} = -1$ and $y < -\sqrt{t}$ for sufficiently large $t$.*

**Proof.** Let $\phi = y + \sqrt{t}$. Then

$$\phi'' = -\frac{1}{4}t^{-3/2} + \phi^2 - 2t^{1/2}\phi$$

and

$$\phi'' + 2t^{1/2}\phi = -\frac{1}{4}t^{-3/2} + \phi^2. \tag{3.12}$$

Set $r = \frac{4}{5}t^{5/4}$. Then

$$\frac{d\phi}{dt} = t^{1/4}\frac{d\phi}{dr},$$

$$\frac{d^2\phi}{dt^2} = t^{1/2}\frac{d^2\phi}{dr^2} + \frac{1}{4}\left(\frac{5r}{4}\right)^{-3/5}\frac{d\phi}{dt},$$

so that

$$\frac{d^2\phi}{dr^2} + \frac{1}{5r}\frac{d\phi}{dr} + 2\phi = -\frac{1}{4}t^{-2} + t^{-1/2}\phi^2.$$

Set $\phi = r^{-1/10}\psi$. Then

$$\frac{d^2\psi}{dr^2} + 2\psi = -\frac{1}{4}\left(\frac{4}{5}\right)^{8/5} r^{-3/2} - \frac{9}{100}r^{-2}\psi + \left(\frac{4}{5}\right)^{2/5} r^{-1/2}\psi^2.$$

Using variation of parameters,

$$\psi = -\frac{1}{\sqrt{2}}\int_r^\infty \sin\left\{\sqrt{2}\,(r-\rho)\right\}\left(-A\rho^{-3/2} - \frac{9}{100}\rho^{-2}\psi + \left(\frac{4}{5}\right)^{2/5}\rho^{-1/2}\psi^2\right)d\rho$$

if these integrals converge, where $A = \frac{1}{4}\left(\frac{4}{5}\right)^{8/5}$.

We find a solution by successive approximations, starting with $\psi_0(r) = -\frac{1}{2}Ar^{-3/2}$, which guarantees that the integrals converge. (In Section 3.3.1 details of this are carried out for a similar but more difficult analysis of Painlevé II.) The solution is found to satisfy

$$\psi = -\frac{1}{2}Ar^{-3/2} + O\left(r^{-5/2}\right)$$

as $r \to \infty$. Thus the required solution exists, and it is clearly unique, since any other terms would involve $\sin\sqrt{2}r$ or $\cos\sqrt{2}r$ and so be oscillatory. We observe for use below that $\lim_{r\to\infty}\psi' = 0$, which implies that $\lim_{t\to\infty}\phi' = 0$. □

This result and Lemma 3.1 imply Theorem 3.2 if in Lemma 3.1 we allow the case $t_0 = \infty$, by which we mean exactly the conclusion of Lemma 3.4. Now we complete the proofs of existence in this subsection and the previous one by proving Lemma 3.1 with this expanded interpretation.

**Proof of Lemma 3.1.** As above, let $\phi = y + \sqrt{t}$. Multiply (3.12) by $\phi'$ and integrate over $(t, t_0)$ to give

$$\begin{aligned}\frac{1}{2}\phi'^2 + t^{\frac{1}{2}}\phi^2 - \frac{1}{3}\phi^3 &= -\frac{1}{2}\int_t^{t_0} t^{-1/2}\phi(t)^2\,dt + \frac{1}{4}\int_t^{t_0} t^{-3/2}\phi'(t)\,dt \\ &= -\frac{1}{2}\int_t^{t_0} t^{-1/2}\phi(t)^2\,dt - \frac{1}{4}t^{-3/2}\phi + \frac{3}{8}\int_t^{t_0} t^{-5/2}\phi(t)\,dt, \qquad (3.13)\end{aligned}$$

where in the case $t_0 = \infty$ the integrals are seen to converge from Lemma 3.4, and where we have used either that $\phi(t_0) = \phi'(t_0) = 0$ or the remark at the end of the proof of Lemma 3.4. In addition, if $t_0$ is finite then the hypotheses imply that $\phi'(t_0) = 0$ and $\phi''(t_0) < 0$, so that for either $t_0 < \infty$ or $t_0 = \infty$, $\phi < 0$ in some interval $(t_1, t_0)$. Suppose for contradiction that $t_1$ can be chosen to be positive and with $\phi(t_1) = 0$. Since $\phi < 0$ on $(t_1, t_0)$, we obtain $\frac{1}{2}\phi'(t_1)^2 < 0$, giving the needed contradiction which proves that $y < r$ on $(0, t_0)$. Since $y'' > 0$ if $y < r$, $y' < 0$ on $[0, t_0]$ follows as well. For the other assertions note from (3.13) that if $t \le t_0$ then $-\frac{1}{3}\phi(t)^3 \le -\frac{1}{4}t^{-3/2}\phi$, which gives (3.6) and (3.7). □

## 3.3. Painlevé II—shooting from infinity

**3.3.1. Introduction and existence proof.** In Chapter 2 we discussed the idea of "shooting from infinity". In this chapter we turn again to this technique, for a more difficult problem. One of the points to be made is that some problems may seem at first glance to require two-parameter shooting, which was also introduced in Chapter 2, but are better done using a one-parameter "shooting from infinity" approach. Our example is the second Painlevé transcendent.

Painlevé II can be written as

$$y'' - xy = 2y^3, \tag{3.14}$$

and we consider it on the interval $-\infty < x < \infty$. The $y^3$ term causes most solutions to blow up in finite time. Our problem is to find a nontrivial solution which exists for $-\infty < x < \infty$. The following material is based on **[89]**.

To find a solution which exists on $[0, \infty)$, we look for a solution such that $y(x) \to 0$ as $x$ tends to infinity. We may expect such a solution to exist because Airy's equation

$$y'' - xy = 0 \tag{3.15}$$

is known to have solutions with this behavior, as we proved in Section 2.3.3.

It is harder to find a solution of (3.14) which exists on $[-\infty, 0)$. In (3.15) let $E = y'^2 - xy^2$. Clearly $E > 0$ if $x < 0$. Also,

$$E' = -y^2,$$

so that as $x$ decreases, $E$ increases. While this might seem to rule out a solution to (3.15) which tends to zero at minus infinity, it turns out that such solutions do exist but the derivatives of these solutions are unbounded. Adding the cubic term to get (3.14) seems to make finite $x$ blowup likely for any nontrivial solution as $x$ decreases. However for negative $x$ the terms $-xy$ and $2y^3$ in (3.14) have the same sign and so there is the possibility that they will balance each other to some extent. These observations lead to the following conjectured boundary conditions for a nontrivial solution which exists on $(-\infty, \infty)$:

$$\lim_{x\to\infty} y(x) = 0, \qquad \lim_{x\to-\infty} \frac{y(x)^2}{\frac{1}{2}x} = -1. \tag{3.16}$$

To see further motivation for the second boundary condition, observe that the second condition in (3.16) is equivalent to $y(x) \sim \sqrt{-\frac{1}{2}x}$ as $x \to -\infty$.[4] If $u(x) = \sqrt{-\frac{1}{2}x}$ for $x < 0$ then $u''(x) \to 0$ as $x \to -\infty$, and

$$xu(x) + 2u(x)^3 = 0.$$

Therefore, it is plausible that $y$ has $\sqrt{-\frac{1}{2}x}$ behavior at $-\infty$.

As stated above, there are two shooting techniques one could try on this problem:

(i) two-parameter shooting from $x = 0$,

(ii) one-parameter backward shooting from $\infty$.

In the first approach we consider the initial value problem

$$\begin{gathered} y'' = xy + 2y^3, \\ y(0) = \alpha, \quad y'(0) = \beta. \end{gathered} \tag{3.17}$$

We use three steps:

(a) Show that for each $\alpha > 0$ there is a $\beta < 0$ such that $y(\infty) = 0$. This is not hard to prove, for example by the method of Section 2.3.3.

(b) Show that for each $\alpha > 0$, $\beta$ is unique and $\beta$ is a continuous function of $\alpha$. This step is harder but can be done. Since we will not be using this approach, we leave this as a part of Exercise 3.2 below.

(c) Show that by varying $\alpha$, and always choosing $\beta = \beta(\alpha)$ as in step (a), we can find an $\alpha$ such that, as well, $y \sim \sqrt{-\frac{1}{2}x}$ as $x \to -\infty$. Again, we leave this to Exercise 3.2.

We observe that while we referred to this method as "two-parameter shooting", it does not require complicated topology in the plane. Rather, it is a pair of one-parameter shooting steps carried out in sequence. This approach is simpler than the method of Section 2.6.

To explain our second approach to proving existence of a solution to (3.14) which satisfies (3.16), consider the simple linear equation

$$y'' - y = 0.$$

[4] We say that $f(s)$ is asymptotic to $g(s)$ as $s \to \infty$, and write

$$f(s) \sim g(s),$$

if

$$\lim_{s \to \infty} \frac{f(s) - g(s)}{g(s)} = 0.$$

The general solution is $y = c_1 e^x + c_2 e^{-x}$. To obtain $y(\infty) = 0$, set $c_1 = 0$, while $c_2$ is arbitrary. Hence there is a one-parameter family of solutions which tend to zero at $+\infty$. (The parameter is $c_2$.)

Suppose that we can show that the Painlevé equation (3.14) also has a one-parameter family of solutions tending to 0 at $\infty$. Then we could vary this parameter to try to obtain the desired behavior at $-\infty$. There is often an advantage to shooting from an equilibrium point (or singular point), in this case $y = 0$, as compared with regular or nonequilibrium points. The behavior of possible solutions is usually more restricted at such a point, giving us fewer shooting parameters to consider. As another example we can consider the reduced FitzHugh-Nagumo equation studied in Chapter 2. There we saw that a single parameter, the wave speed $c$, determined the unstable manifold at the origin, which was the only possible candidate for the desired heteroclinic orbit. The same principle occurs in a more complicated setting in Chapter 6.

To use either of these methods we first need to consider the possible behavior of a solution at $+\infty$. Are there solutions which tend to zero there? We could easily show this by the method described in (a) above. But we will adopt a different approach. If $y(x) \to 0$ at $\infty$ then $y^3 << y$ for large $x$, and so we again consider Airy's equation,

$$y'' - xy = 0, \tag{3.18}$$

which was studied well before Painlevé's work [**2**]

**Lemma 3.5.** *There is a one-parameter family of solutions of* (3.18) *which are positive and tend to* 0 *as* $x \to \infty$.

**Proof.** The existence of a solution $y$ such that $y(0) = 1$, $\lim_{x\to\infty} y(x) = 0$ was proved in Section 2.3.3. The multiples $cy(x)$ constitute the desired one-parameter family. □

These solutions have been much studied. For reasons not of interest to us here, we pick out a particular one to be called the Airy function, $Ai(x)$. It is characterized by the following behavior at infinity:

$$Ai(x) \sim \frac{1}{2}\pi^{-1/2}x^{-1/4}e^{-\frac{2}{3}x^{3/2}} \tag{3.19}$$

as $x \to \infty$.

We see the plausibility of this formula by formally solving $y'' = xy$, treating $x$ as though it were constant. We get $y = ce^{\pm\int\sqrt{x}dx}$. For the decaying solution this gives the exponent $-\frac{2}{3}x^{3/2}$. The algebraic term $x^{-1/4}$ in (3.19) requires more work [**2**].

We will also make use of a second solution of the linear equation, denoted by $Bi(x)$ and called an "Airy function of the second kind". Its asymptotic behavior is

$$Bi\,(x) \sim \pi^{1/2} x^{-1/4} e^{\frac{2}{3}x^{3/2}}$$

as $x \to \infty$. In particular, it is unbounded on $[0, \infty)$. It is also useful to give the asymptotic behavior of the derivatives:

$$Ai'\,(x) \sim -\frac{1}{2}\pi^{-1/2} x^{1/4} e^{-\frac{2}{3}x^{3/2}},$$
$$Bi'\,(x) \sim \pi^{1/2} x^{1/4} e^{\frac{2}{3}x^{3/2}}.$$

These formulas then give an asymptotic formula for the Wronskian

$$W\,(x) = Ai'\,(x)\,Bi\,(x) - Ai\,(x)\,Bi'\,(x) \sim -1.$$

However, since there is no $y'$ in the Airy equation, $W$ is constant and so is always equal to $-1$.

Now consider the nonlinear equation (3.14). Again look for solutions which tend to zero at infinity. These could be found using the technique in Lemma 3.5. However we will use a more analytical method which involves substituting the asymptotic formulas for the Airy functions $Ai$ and $Bi$ into a variation of parameters formula which is valid for solutions of (3.14) which converge to zero at $\infty$. As we will show, this technique results in an integral equation for $y$, solutions of which must tend to zero at a certain rate.

To find this integral equation, suppose that $\tilde{y}$ is a continuous function defined on $[0, \infty)$ and tending to zero at $\infty$, with the further property that the improper integral

$$\int_x^\infty \{Ai\,(x)\,Bi\,(t) - Bi\,(x)\,Ai\,(t)\}\,\tilde{y}\,(t)^3\,dt$$

converges for each $x \geq 0$. Assuming that the differentiations involved are valid (with the resulting integrals converging), it is easily shown that for every $\kappa > 0$, the function

$$y\,(x) = \kappa Ai\,(x) + \int_x^\infty 2\,\{Ai\,(x)\,Bi\,(t) - Bi\,(x)\,Ai\,(t)\}\,\tilde{y}\,(t)^3\,dt$$

satisfies

$$y'' - xy = 2\tilde{y}^3$$

and

$$\lim_{x\to\infty} \frac{y\,(x)}{Ai\,(x)} = \kappa.$$

(Here we have used the equation $W\,(x) \equiv -1$.) Hence we look for solutions $y_\kappa$ of the integral equation

$$y_\kappa\,(x) = \kappa Ai\,(x) + \int_x^\infty 2\,\{Ai\,(x)\,Bi\,(t) - Bi\,(x)\,Ai\,(t)\}\,y_\kappa\,(t)^3\,dt. \tag{3.20}$$

If the integrals converge then differentiation shows that a solution to this integral equation also satisfies the differential equation (3.14). The following result is a different way of looking at the existence of solutions which tend to zero and can completely replace the technique used in Lemma 3.5 above.

**Lemma 3.6.** *For each* $\kappa > 0$ *the equation* (3.20) *has a unique solution* $y_\kappa$ *which tends to zero at infinity. This solution depends continuously on* $\kappa$.

**Remark.** This result is something like a stable manifold theorem. The usual stable manifold theorem does not apply to a nonautonomous equation.

**Proof.** We prove the result by successive approximations. Fix $\kappa > 0$ and let $y^0(x) = \kappa Ai(x)$. Also, choose $X$ so large that for $x \geq X$,

$$Ai(x) \leq cx^{-1/4}e^{-\frac{2}{3}x^{3/2}},$$

$$Bi(x) \leq dx^{-1/4}e^{\frac{2}{3}x^{3/2}}$$

for some $c > 0, d > 0$. Then for $t \geq X$,

$$Bi(t)\, y^0(t)^3 \leq dc^3\kappa^3 t^{-1}e^{-\frac{4}{3}t^{3/2}}.$$

Therefore $\int_x^\infty Bi(t)\, y^0(t)^3\, dt$ converges, and for $x \geq X$,

$$Ai(x)\int_x^\infty Bi(t)\, y^0(t)^3\, dt \leq cx^{-1/4}e^{-\frac{2}{3}x^{3/2}}\int_x^\infty dc^3\kappa^3 t^{-1}e^{-\frac{4}{3}t^{3/2}}\,dt.$$

Since the latter integral is convergent, this implies that for every $\rho > 0$ there is an $X_0$ such that if $x \geq X_0$ then

$$Ai(x)\int_x^\infty Bi(t)\, y^0(t)^3\, dt \leq \rho x^{-1/4}e^{-\frac{2}{3}x^{3/2}}.$$

Also, $X_0$ can be chosen so that if $x \geq X_0$ then

$$\begin{aligned} Bi(x)\int_x^\infty Ai(t)\, y^0(t)^3\, dt &\leq dx^{-1/4}e^{\frac{2}{3}x^{3/2}}\int_x^\infty \kappa^3c^4t^{-1}e^{-\frac{8}{3}t^{3/2}}dt \\ &\leq dx^{-1/4}e^{\frac{2}{3}x^{3/2}}\int_x^\infty \kappa^3c^4t^{1/2}e^{-\frac{8}{3}t^{3/2}}dt \\ &= dx^{-1/4}e^{\frac{2}{3}x^{3/2}}\kappa^3c^4\left(\frac{1}{4}\right)e^{-\frac{8}{3}x^{3/2}} \leq \rho x^{-1/4}e^{-\frac{2}{3}x^{3/2}}. \end{aligned}$$

Therefore, if

$$y^1(x) = \kappa Ai(x) + \int_x^\infty 2\{Ai(x)\, Bi(t) - Bi(x)\, Ai(t)\}\, y^0(t)^3\, dt,$$

then $y^1(x)$ is defined for $x \geq X_0$ and

$$\left|y^1(x)\right| \leq (\kappa + 4\rho)\, cx^{-1/4}e^{-\frac{2}{3}x^{3/2}} \leq 2\kappa cx^{-1/4}e^{-\frac{2}{3}x^{3/2}},$$

if $\rho < \frac{\kappa}{4}$.

Now suppose that $y^j(x)$ has been defined on $[X_0, \infty)$ and on this interval,

$$\left|y^j(x)\right| \leq 2\kappa cx^{-1/4}e^{-\frac{2}{3}x^{3/2}}.$$

Then as above, $y^{j+1}$ is well-defined on the same interval by

$$y^{j+1}(x) = \kappa Ai(x) + \int_x^\infty 2\{Ai(x)\,Bi(t) - Bi(x)\,Ai(t)\}\,y^j(t)^3\,dt,$$

and assuming that $\rho$ was chosen small enough (depending only on $\kappa$),

$$y^{j+1}(x) \le 2\kappa c x^{-1/4} e^{-\frac{2}{3}x^{3/2}}$$

for $x \ge X_0$. Further, if $w(x,t) = Ai(x)\,Bi(t) - Bi(x)\,Ai(t)$, $v(x,t) = e^{-\frac{2}{3}x^{3/2}}e^{\frac{2}{3}t^{3/2}} + e^{\frac{2}{3}x^{3/2}}e^{-\frac{2}{3}t^{3/2}}$, and $||\phi|| = \max_{x\ge X_0}|\phi(x)|$, then

$$\begin{aligned}
&\left|y^{j+1}(x) - y^j(x)\right| \\
&\le \left|\int_x^\infty 2w(x,t)\left\{y^j(t)^2 + y^j(t)\,y^{j-1}(t) + y^{j-1}(t)^2\right\}\left\{y^j(t) - y^{j-1}(t)\right\}dt\right| \\
&\le \int_x^\infty 2cdx^{-1/4}t^{-1/4}v(x,t)\left(12\kappa^2c^2t^{-1/2}e^{-\frac{4}{3}t^{3/2}}\right)dt\,\left|\left|y^j - y^{j-1}\right|\right|.
\end{aligned}$$

In the term containing $12\kappa^2$, replace $t^{-1/2}$ by $t^{1/2}$. Then drop the $t^{-1/4}$ term, and do the two integrals. We find that for $X_0$ large enough,

$$\left|\left|y^{j+1} - y^j\right|\right| \le \frac{1}{2}\left|\left|y^j - y^{j-1}\right|\right|.$$

It follows that $\Sigma_{j=1}^\infty\left(y^{j+1} - y^j\right)$ converges uniformly on $[X_0,\infty)$, and hence that $\lim_{j\to\infty} y^j(x)$ converges uniformly on this interval, to a continuous function $y_\kappa(x)$ which solves (3.20). Uniqueness and continuity with respect to $\kappa$ follow in the usual way for proofs using successive approximations. See, for example, Chapter 13 of [**41**] for similar results using this method. □

Now consider the behavior at $-\infty$. As usual we define two sets:

$$A = \left\{\kappa > 0 \,|\, y_\kappa'(x_1) > 0 \text{ for some } x_1 < 0\right\},$$

$$B = \left\{\kappa > 0 \,|\, \text{for some } X > -\infty,\ \lim_{x\searrow X} y_\kappa(x) = \infty \text{ and } y_\kappa' < 0 \text{ on } (X,\infty)\right\}.$$

It is easy to see that $A$ is an open set, since $y_\kappa'(x)$ is a continuous function of $\kappa$. To prove that $A$ is nonempty, consider a small $\kappa > 0$. We first prove that the solution $y_\kappa(x)$, which has only been shown to exist for large values of $x$, in fact exists at $x = 0$. Note that for $\kappa = 0$ the unique solution is $y^0(x) = 0$. This solution exists on $(-\infty,\infty)$, and so by continuity in $\kappa$ it follows that for sufficiently small $\kappa > 0$, $y_\kappa$ exists on $[0,\infty)$. Since

$$y_\kappa'' = xy_\kappa + 2y_\kappa^3,$$

and $y_\kappa \sim \kappa Ai(x)$ as $x \to \infty$, we see that for $\kappa > 0$, $y_\kappa(x) > 0$ and $y_\kappa'(x) < 0$ for sufficiently large $x$. But then, $y_\kappa'' > 0$ while $x > 0$. Hence as $x$ decreases but remains positive, $y_\kappa$ increases and $y_\kappa'$ remains negative. It is convenient to let $\beta(\kappa) = y'_\kappa(0)$, which exists and is negative for sufficiently small $\kappa$. It is easy to show that $\beta(\kappa) \to 0$ as $\kappa \searrow 0$, for if $y'_\kappa(0)$ remained below some negative number, while $y_\kappa(0)$ tended to zero, then $y_\kappa$ would become negative in $x > 0$, a contradiction to what we proved before.

We claim that sufficiently small values of $\kappa > 0$ are in the set $A$. Assuming the opposite, and therefore that $y'_\kappa < 0$ on the maximal negative interval of existence, it is seen that as $\kappa \searrow 0$, $y_\kappa \to 0$ uniformly on the interval $[-100, 0]$. (Any large enough interval will do.) Write the equation as

$$y'' + \left(-x + 2y^2\right) y = 0.$$

For $x < -1$, $\left(-x + 2y^2\right) > 1$. The Sturm comparison theorem then implies that $y$ oscillates on every interval of length greater than $\pi$, which is a contradiction. Hence the set $A$ is nonempty.

To see that $B$ is open and nonempty requires more work. In the following two lemmas we make no effort to find the best possible constants.

**Lemma 3.7.** *Consider a function $Y$ which is defined, along with $Y'$ and $Y''$, on some interval $(x_1 - a, x_1]$ and satisfies the inequality*

$$Y'' \geq Y^3 \tag{3.21}$$

*on this interval. Further, assume that $Y(x_1) > 2$ and $Y'(x_1) < 0$. Then $a < \sqrt{2}$.*

In other words, $Y(x)$ blows up somewhere on the interval $\left(x_1 - \sqrt{2}, x_1\right)$.

**Proof.** Since $Y'(x_1) < 0$ and $Y(x_1) > 2$, $Y$ is decreasing, with $Y' < 0$ and $Y'' > 0$, on the interval $(x_1 - a, x_1)$. Multiply the inequality (3.21) by $Y'$ and integrate from $x$ to $x_1$ ($x \in (x_1 - a, x_1)$) to get

$$Y'(x)^2 \geq \frac{1}{2}\left(Y(x)^4 - Y_1^4\right),$$

where $Y_1 = Y(x_1)$. Since $Y^4 - Y_1^4 \geq (Y - Y_1) Y^3$, we find that

$$\frac{Y'(x)}{Y(x)^{3/2}\sqrt{Y(x) - Y_1}} \leq -\frac{1}{\sqrt{2}}.$$

Integrate this to obtain

$$\int_Y^{Y_1} \frac{dy}{y^{3/2}\sqrt{y - Y_1}} \leq \frac{x - x_1}{\sqrt{2}}.$$

The limit of the integral as $Y \to \infty$ can be computed to be $-\frac{2}{Y_1}$, which proves the lemma. □

**Lemma 3.8.** *A number $\kappa > 0$ is in $B$ if and only if there is an $x_1$ such that $y'_\kappa < 0$ on $[x_1, \infty)$ and*

$$y_\kappa(x_1)^2 > |x_1| + 4. \tag{3.22}$$

**Proof.** It is clear that if $\kappa \in B$ then such an $x_1$ exists. On the other hand, suppose that such an $x_1$ exists and that $y = y_\kappa$ exists on an interval $I = (x_1 - b, x_1]$ with $b \le \sqrt{2}$. Then $|x| \le |x_1| + \sqrt{2}$ on $I$, and as long as $y' < 0$, (3.22) implies that $|x| < y^2$ and so $x + 2y^2 \ge y^2$. Hence $y'' \ge y^3$ and so $y' < 0$ on $I$. Since $y_\kappa(x_1) > 2$, Lemma 3.7 implies that $y_\kappa$ blows up before $x = x_1 - \sqrt{2}$. □

To show that $B$ is nonempty, we will show that for some $\kappa > 0$, $y_\kappa(0)$ exists and is larger than 2, implying that (3.22) is satisfied with $x_1 = 0$. We already know that $y'_\kappa < 0$ on $[0, \infty)$, and so Lemma 3.8 will prove that $\kappa \in B$.

To show that $y_\kappa(0) > 2$ for some $\kappa$, note that $y_\kappa(0)$ exists for small $\kappa$. If, as $\kappa$ increases, $y_\kappa(0)$ remains below 2, then $y_\kappa(0)$ exists for all $\kappa$. Now use the following result.

**Lemma 3.9.** *$\frac{y_\kappa}{Ai}$ is decreasing on $[0, \infty)$.*

**Proof.** Assume here that $y = y_\kappa$. Then $\left(\frac{y}{Ai}\right)' = \frac{y'Ai - Ai'y}{Ai^2}$. Also,

$$(y'Ai - Ai'y)' = 2y^3 Ai > 0$$

on $[0, \infty)$. Further, $\lim_{x\to\infty} y'Ai - Ai'y = 0$. Hence $y'Ai - Ai'y < 0$, and so $\left(\frac{y}{Ai}\right)' < 0$. □

Since $\left(\frac{y_\kappa}{Ai}\right) \to \kappa$ as $x \to \infty$, it follows that if $y_\kappa(0)$ exists then $y_\kappa(0) > \kappa Ai(0)$. It is easy to show that $Ai(0) > 0$, and so $y_\kappa(0) > 2$ for some $\kappa$. Therefore $B$ is nonempty.

Lemma 3.8 also shows that $B$ is open. It is seen from the definitions of $A$ and $B$ that they are disjoint, and so there is a $\kappa^* > 0$ which lies in neither $A$ nor $B$. The corresponding solution exists and is decreasing on $(-\infty, \infty)$. We want to show that $\frac{y_{\kappa^*}(x)}{\sqrt{-x}} \to \frac{1}{\sqrt{2}}$ as $x \to -\infty$.

For this final step, let $z(x) = \frac{y(x)}{\sqrt{-\frac{1}{2}x}}$ for $x < 0$. We then find that

$$z'' + \frac{1}{x}z' = \left(-x\left(z^2 - 1\right) + \frac{1}{4x^2}\right) z.$$

Also, $z(x) > 0$ for $x < 0$. Hence, $z$ has no maximum at which $z^2 - 1 > \frac{1}{4x^3}$ and no minimum at which $z^2 - 1 < \frac{1}{4x^3}$. This argument implies that either $y_{\varkappa^*}$ satisfies (3.16) or $z$ is monotonic for large negative $x$ and so tends to infinity or to a finite limit other than 1. In Exercise 3.3 you are asked to show that $z(-\infty) = 1$, which completes the proof of the following result.

**Theorem 3.10.** *For some* $\kappa^* > 0$ *the solution* $y_{\kappa^*}$ *of* (3.20) *can be extended to* $(-\infty, \infty)$ *as a solution of* (3.14) *which satisfies the conditions* (3.16).

We leave to Exercise 3.4 the proof that the solution just found is unique.

## 3.4. Some interesting consequences

While not strictly on the topic of this text, we wish to point out some further points concerning the solution just found. The origin of these additional results lies in some numerical computations of Rosales [**204**]. They suggested that in Theorem 3.10, $\kappa^* = 1$, and if so, then it is rather surprising, since the equation is certainly not integrable in elementary terms. The challenge, therefore, is two fold: first to prove that $\kappa^* = 1$ and second to understand why we are able to evaluate $\kappa^*$ explicitly. The result $\kappa^* = 1$ is an example of a nonlinear connection problem, since we are connecting the asymptotic behavior of the solution as $x \to \infty$ to the asymptotic behavior as $x \to -\infty$. Connection problems for linear equations were one of the main areas of research in ordinary differential equations throughout the nineteenth and early twentieth centuries, but nonlinear connection problems are rarely solvable.

One reason for this is that the method which is perhaps the most useful for linear problems is not, in general, applicable. This method is to consider the independent variable $x$ as complex and pass from $x = +\infty$ to $x = -\infty$ along a large semicircle in the $x$-plane. Provided that the coefficients in the equation are reasonably simple, it may be possible to construct an asymptotic expansion for the solution at all points on the large semicircle, and so relate a specific behavior as $x \to +\infty$ to a specific behavior as $x \to -\infty$ [**19**].

In nonlinear problems this method fails in general because, even if the coefficients in the equation are reasonable, the solutions may not continue to exist as $|x| \to \infty$. For a linear equation, singularities of the solutions occur only at singularities of the coefficients of the equation. This may not hold for a nonlinear equation. We see this from the trivial example

$$y' = 1 + y^2.$$

The solutions are $y = \tan(x + c)$, for an arbitrary constant $c$, and this has singularities when $(x + c) = (2n + 1)\frac{\pi}{2}$. In this particular example the singularities are all simple poles, but in general they may be branch points or essential singularities, whose position may vary from solution to solution. Under such circumstances one has little hope of being able to extend the solution to large $|x|$ and follow its behavior there.

But, as we described briefly at the beginning of this chapter, there is one important class of equations where something can be said, and the second Painlevé transcendent belongs to this class. Painlevé II has the additional property that its solutions have no singularities at all except for poles. For such an equation it is possible to track a solution for large $|x|$, and it is known that every solution behaves asymptotically like an elliptic function. In [**115**], Joshi and Kruskal managed to use these ideas to prove that $\kappa^* = 1$. This result was also proved in [**89**], but by a method which was very much ad hoc.

However, the Painlevé equations have even richer properties. Let us return once again to the case of linear differential equations. We have said that it is reasonable to expect to be able to work out the asymptotics of a solution $y$ as $|x| \to \infty$. The asymptotics remain the same, at least as far as leading order asymptotics go, throughout a sector, say $\alpha < \arg x < \beta$, but may change from one sector to another [**186**]. This is a classical result which was observed first by Stokes in the context of Airy's equation.

In order to know the full asymptotics of a solution, we need therefore to know how the asymptotics in one sector relate to the asymptotics in the next sector, and the assemblage of all this information is called the monodromy data for the linear equation.

Now to each Painlevé equation there corresponds a linear equation, and its monodromy data. The coefficients of this linear equation (and so its monodromy data) have as parameters the solution $y$ and the independent variable $x$ from the Painlevé equation. The relation between the two is that if $y$ changes as a function of $x$ according to the Painlevé equation, then the monodromy data does not change.

This observation gives us a procedure for solving connection problems for (say) Painlevé II. We first let $x$ be large and positive, with $y \sim \kappa Ai\,(x)$, insert this into the coefficients of the associated linear equation, and compute the monodromy data. We then repeat the procedure with $x$ large and negative, and $y \sim \left(-\frac{1}{2}x\right)^{1/2}$, and again compute the monodromy data. Equating the two sets of monodromy data gives us $\kappa$.

This procedure is systematic and in theory can be applied to solve every connection problem for every Painlevé transcendent. While this has been done for some of these equations [**119**], [**164**], [**14**], [**15**], [**1**], it has not been done for all, and there remain many challenging problems.

## 3.5. Exercises

**Exercise 3.1.** Prove (3.10). Hint: Suppose that there is a $\delta > 0$ and a sequence $x_n \to -\infty$ such that $u^*\left(x_n\right) < -\sqrt{-x_n}\left(1+\delta\right)$.

**Exercise 3.2.** Carry out the details of the two-parameter shooting approach using (3.17) and steps (a), (b), and (c) in Section 3.3.1.

**Exercise 3.3.** Complete the proof of Theorem 3.10 by showing that $z(-\infty) > 1$ (or $z(-\infty) = \infty$) and $z(-\infty) < 1$ are impossible.

**Exercise 3.4.** Prove that the $\kappa^*$ found for the second Painlevé transcendent is unique. Hint: Consider the function $V(x) = \frac{1}{2}y'^2 - \frac{1}{2}xy^2 - \frac{1}{2}y^4$.

*Chapter 4*

# Periodic solutions of a higher order system

## 4.1. Introduction, Hopf bifurcation approach

As we said in Chapter 1, the Brouwer fixed point theorem is often difficult to apply in solving boundary value problems involving systems of ode's. But in this short chapter we will describe a situation in which it proved to be useful. The specific model discussed here is from biochemistry and involves a set of reactions in enzyme chemistry which includes a negative feedback loop. An earlier application is described in [**94**].

The model enzyme equations were derived by Goodwin in [**79**] and have been of continued interest ever since, e.g. [**212**], [**198**], [**206**]. Before describing the approach using the Brouwer fixed point theorem, we will briefly discuss an application of the Hopf bifurcation theorem to the problem. This important method has been widely used to study oscillations in higher order systems.

One of the simplest versions of the Goodwin equations is

$$\begin{aligned} x' &= f(z) - x, \\ y' &= x - y, \\ z' &= y - z. \end{aligned} \tag{4.1}$$

The variables $x, y$, and $z$ represent enzyme concentrations. It is assumed that $f(z)$ is continuously differentiable and bounded in $z \geq 0$. Negative feedback is obtained by requiring that $f'(z) < 0$, so that an increase in $z$ results in a decrease in $x$. To obtain physically reasonable solutions, in which $x, y$, and $z$ are positive, we also assume that $f(z) > 0$. The standard function used for $f$ is of the form $f(z) = \frac{k}{1+z^p}$.

The mathematical issue we will address is to find conditions on $f$ which ensure that (4.1) has periodic solutions [**244**], [**96**], [**8**], [**170**]. One way to approach this is via the theory of Hopf bifurcation. Although we did not list this as one of the prerequisites for this book, it is widely known and accessible, for example in [**97**], [**78**], or [**34**].

In a bifurcation approach, we assume that $f$ depends on a parameter (for example, $p$ in the formula above), and as the parameter changes, a qualitative change occurs in the behavior of the system. The conditions given for $f$ ensure that the system (4.1) has a unique equilibrium point $\mathbf{x}_0 = (x_0, y_0, z_0)$, which is in the positive octant. The linearization of (4.1) around $(x_0, y_0, z_0)$ is

$$\begin{aligned} u' &= f'(z_0)\, w - u, \\ v' &= u - v, \\ w' &= v - w. \end{aligned} \tag{4.2}$$

We write this as

$$\mathbf{u}' = A\mathbf{u},$$

where

$$A = \begin{pmatrix} -1 & 0 & -\delta \\ 1 & -1 & 0 \\ 0 & 1 & -1 \end{pmatrix}$$

with $\delta = -f'(z_0)$, which we are assuming is positive. We will take $\delta$ as the bifurcation parameter, realizing that it may, as in the example above, depend on other quantities ($k$ and $p$) in the definition of $f$.

The eigenvalues of $A$ satisfy

$$(\lambda + 1)^3 + \delta = 0, \tag{4.3}$$

from which it is easily established that for small positive $\delta$, one root is real and negative and the other two are complex with negative real parts. For large $\delta$, on the other hand, the complex roots have positive real parts. There is a critical value $\delta_0$ ($\delta_0 = 8$) such that as $\delta$ is increased from below $\delta_0$ to above $\delta_0$, the two complex roots move from the left half of the complex plane to the right half. This is a key requirement for Hopf bifurcation. The other requirements are more difficult to check, but verifiable in this case. It is found that $\delta_0$ is a "supercritical" Hopf bifurcation point, meaning that as $\delta$ crosses $\delta_0$, a stable periodic solution bifurcates from $\mathbf{x}_0$.[1] Further details of

[1]For $\delta < \delta_0$ there are no periodic solutions, while for $\delta > \delta_0$ and $\delta - \delta_0$ small, there is a periodic solution whose orbit is close to the equilibrium point. The terms sub- and supercritical are discussed in the references on Hopf bifurcation given earlier.

this bifurcation are found in [**107**], where an $n^{\text{th}}$ order system of the same type is studied.

## 4.2. A global approach via the Brouwer fixed point theorem

We stated the one-dimensional version of Brouwer's theorem in Chapter 1. Here is an $n$-dimensional version.

**Theorem 4.1.** *Suppose that $B \subset R^n$ is homeomorphic to the closed unit ball and $f : B \to B$ is continuous. Then $f$ has a fixed point.*

There are many proofs. Readers of this book might particularly like one using Stokes' theorem, as in [**197**]. Here we need only the case $n = 2$.

The specific geometric technique for applying the Brouwer theorem to be discussed here was first used, as far as we know, in [**94**], though the general approach certainly was known considerably earlier. The Goodwin model is as simple a setting as we know of to illustrate this method. See the last section of this chapter for references to work of Hirsch, Smith, and Mallet-Paret on a theory which includes the result below in a much broader setting. We present here what we feel is a particularly simple approach to the Goodwin model, recognizing that a stronger result can be obtained with more machinery.

We analyze the trajectories of the system (4.1) in the positive octant $E^+$ of points $(x, y, z)$ such that $x > 0, y > 0$, and $z > 0$. The first step is to find an attracting box, meaning a region $B$ of the form $0 \leq x \leq X$, $0 \leq y \leq Y$, $0 \leq z \leq Z$ such that every solution starting in $E$ eventually enters, and remains in, $B$.

Such a region is easily constructed. If $f(z) < M$ for all $z \geq 0$, we let $X = Y = Z = M$. In Exercise 4.1 you are asked to prove the following:

**Proposition 4.2.** *The region*

$$B = \{(x, y, z) \mid 0 \leq x \leq M, \ 0 \leq y \leq M, \ 0 \leq z \leq M\}$$

*is an attracting region in $E^+$ as described above.*

Let $\mathring{B}$ denote the interior of $B$. We will prove the following theorem:

**Theorem 4.3.** *If the equilibrium $\mathbf{x}_0$ of (4.1) is unstable and the matrix $A$ has no eigenvalues with zero real part, then (4.1) has a nonconstant periodic solution $\mathbf{x}$ such that $\mathbf{x}(t) \in \mathring{B}$ for all $t$.*

**Proof.** We divide $B$ into eight smaller boxes as follows:

$$B_1 = \{(x, y, z) \mid 0 \le x \le x_0, 0 \le y \le y_0, 0 \le z \le z_0\},$$
$$B_2 = \{(x, y, z) \mid x_0 \le x \le M, 0 \le y \le y_0, 0 \le z \le z_0\},$$
$$B_3 = \{(x, y, z) \mid x_0 \le x \le M, y_0 \le y \le M, 0 \le z \le z_0\},$$
$$B_4 = \{(x, y, z) \mid x_0 \le x \le M, y_0 \le y \le M, z_0 \le z \le M\},$$
$$B_5 = \{(x, y, z) \mid 0 \le x \le x_0, y_0 \le y \le M, z_0 \le z \le M\},$$
$$B_6 = \{(x, y, z) \mid 0 \le x \le x_0, 0 \le y \le y_0, z_0 \le z \le M\},$$
$$B_7 = \{(x, y, z) \mid 0 \le x \le x_0, y_0 \le y \le M, 0 \le z \le z_0\},$$
$$B_8 = \{(x, y, z) \mid x_0 \le x \le M, 0 \le y \le y_0, z_0 \le z \le M\}.$$

Also, let $F$ denote the face between $B_1$ and $B_2$. Thus,

$$F = \{(x, y, z) \,|x = x_0, 0 \le y \le y_0, 0 \le z \le z_0\}.$$

**Lemma 4.4.** *Suppose that a nonconstant solution* $\mathbf{x}$ *starts off with* $\mathbf{x}(0) \in F$. *Then* $\mathbf{x}(t)$ *cycles repeatedly through the boxes* $B_2, B_3, B_4, B_5, B_6, B_1, B_2, \ldots$ *in that cyclic order and never enters* $B_7 \cup B_8$. *For* $t > 0$, $\mathbf{x}(t)$ *does not intersect any of the edges of boxes* $B_1, \ldots, B_6$.

**Remark 4.5.** This lemma is probably at least as interesting physically as the theorem.

**Proof.** Clearly, $x(0) = x_0$. If $\mathbf{x}(0)$ is in the relative interior of $F$, then

$$0 < y(0) < y_0, \qquad 0 < z(0) < z_0.$$

Hence, at $t = 0$, $x' = f(z) - x > f(z_0) - x_0 = 0$. Thus, $\mathbf{x}$ immediately leaves box $B_1$ and enters $\mathring{B}_2$. The remaining possibilities are

(i) $y(0) = 0$ or $z(0) = 0$, and $y(0) < y_0$, $z(0) < z_0$,

(ii) $y(0) = y_0, z(0) < z_0$,

(iii) $z(0) = z_0, y(0) < y_0$.

(Note that $\mathbf{x}(0) \neq \mathbf{x}_0$ because we are considering a nonconstant solution.)

In case (i), if $y(0) = 0$ then $y'(0) = x_0 > 0$. Since $z(0) < z_0$, $x'(0) = f(z(0)) - x_0$, which is also positive. If $0 < z(0) < z_0$ then $\mathbf{x}$ enters $\mathring{B}_2$. If $y(0) = z(0) = 0$ then $z'(0) = 0, z''(0) = y'(0) > 0$ so again $\mathbf{x}$ enters $\mathring{B}_2$. Finally in case (i), if $z(0) = 0$ and $0 < y(0) < y_0$, then $x' > 0$ and $z' > 0$, so the solution again enters $\mathring{B}_2$.

In case (ii), $y'(0) = 0$ and $y''(0) = x'(0) > 0$. In this case, we note that the solution starts in $B_1 \cap B_2$ but immediately enters $\mathring{B}_3$.

In case (iii), $x'(0) = 0$, $z'(0) < 0$, and $x''(0) = f'(z_0)\, z' - x' > 0$. Hence, the solution enters $\mathring{B}_2$.

We thus see that every solution starting on $F$ immediately enters either $\mathring{B}_2$ or $\mathring{B}_3$.

**Lemma 4.6.** *No solution remains in any one of the boxes $B_1, \ldots, B_6$ on an interval $[T, \infty)$.*

**Proof.** Suppose that $\mathbf{x}(t) \in B_2$ for all $t$ in $[T, \infty)$. In this box, $y \leq y_0$ and $x \geq x_0$. Hence, $y' \geq 0$, and $\lim_{t\to\infty} y(t)$ exists. If follows easily that $x(t)$ and $z(t)$ also approach limits, and since there is only one equilibrium point in $B$, $\lim_{t\to\infty} \mathbf{x}(t) = \mathbf{x}_0$.

Our assumptions on $A$ imply that there is one real eigenvalue, $\lambda_1$, which is negative. The vector $\mathbf{v}_1 = \begin{pmatrix} (\lambda_1 + 1)^2 \\ \lambda_1 + 1 \\ 1 \end{pmatrix}$ is an eigenvector of $A$ corresponding to $\lambda_1$. From (4.3), $\lambda_1 + 1 < 0$, so the second component of $\mathbf{v}_1$ is negative while the first and third components are positive. It follows that from the equilibrium point $\mathbf{x}_0$, $\mathbf{v}$ points into $\mathring{B}_8$. Therefore, solutions on the stable manifold of (4.1) at $\mathbf{x}_0$ tend to $\mathbf{x}_0$ from either $B_8$, or $B_7$, which is its reflection across $\mathbf{x}_0$. Hence, a solution entering $B_2$ must exit from $B_2$. In a similar way, no solution can remain in any of the other boxes $B_3, \ldots, B_6$. □

Continuing with the proof of Lemma 4.4, we now consider where a solution $\mathbf{x}$ leaving $\mathring{B}_2$ can go. Since solutions do not leave $B$, this solution exits via either an edge or the interior of a face which is in $\mathring{B}$. Our previous argument shows that $\mathbf{x}$ does not leave $B_2$ across $F$. We first deal with the case where $\mathbf{x}$ leaves $B_2$ across the interior of some other face. One of the following two possibilities must hold at the exit point.

(i) $z = z_0, 0 < y < y_0$, $x_0 < x < M$. In this case $z' < 0$, so that the solution cannot have arrived at this point from $\mathring{B}_2$.

(ii) $y = y_0, 0 < z < z_0, x_0 < x < M$. In this case, $y' > 0$ and the solution enters $\mathring{B}_3$.

Next, we ask if a solution can leave along one of the edges of $B_2$. Having excluded edges of $F$, and those on the boundary of $B$, the only remaining possibility is that at the departure point,

$$y = y_0, \qquad z = z_0, \qquad x_0 < x < M.$$

In this case, $y' > 0$, $z' = 0$, and $z'' = y' > 0$. So this point is a relative minimum of $z$, and the solution could not have arrived at this point directly from $\mathring{B}_2$. Therefore, the only possible exit from $\mathring{B}_2$ is through the interior of the face $y = y_0$ into $\mathring{B}_3$.

The remaining proofs, that from $\mathring{B}_3$ solutions go to $\mathring{B}_4$ and so forth, are similar. At no point after $t = 0$ does the solution intersect an edge of any of the sub-boxes. This completes the proof of Lemma 4.4. $\square$

To prove Theorem 4.3, we consider a map from the face $F$ to itself. Note that $\mathbf{x}_0 \in F$. We have just shown that if a solution starts on $F$ at any point other than $\mathbf{x}_0$, then it circulates through boxes $B_2$ to $B_6$ and back to $B_1$, after which it intersects $F$ again. This defines what is called a "Poincaré map" $P : F \backslash \{\mathbf{x}_0\} \to F \backslash \{\mathbf{x}_0\}$.[2] If we set $P(\mathbf{x}_0) = \mathbf{x}_0$, then $P$ maps the closed rectangle $F$ into itself.

We must ask if $P$ is continuous. This is obvious except at $\mathbf{x}_0$, because Lemma 4.4 implies that if $\mathbf{x}(0) \neq \mathbf{x}_0$, then $P(\mathbf{x}(0))$ is in the interior of $F$, where $x' > 0$. (Use the implicit function theorem to show continuity.) But even if we can show that $P$ is continuous at $\mathbf{x}_0$ we have a problem, because the resulting fixed point might be $\mathbf{x}_0$. Certainly $\mathbf{x}_0$ is a fixed point, but we want a nonconstant periodic solution and so must find another one.

We do this by cutting out an open piece of $F$ near the corner $\mathbf{x}_0$ so that the remaining closed set $\tilde{F}$ is homeomorphic to the unit square and then showing that $P$ maps $\tilde{F}$ into itself. To construct this set, use (4.3) to see that in addition to the real negative eigenvalue, $\lambda_1$, there are two complex eigenvalues. Since $\mathbf{x}_0$ is assumed to be unstable, and no eigenvalue has zero real part, the complex eigenvalues of $A$ have positive real parts. As before, let $\mathbf{v}_1$ be an eigenvector of $A$ corresponding to $\lambda_1$. Let $L$ be the line through $\mathbf{x}_0$ which is parallel to $\mathbf{v}_1$.

**Lemma 4.7.** *There exists a cylinder $C$ with axis $L$ and elliptical cross section such that no solution enters the inside of $C$ from any of the boxes $B_1, B_2, B_3, B_4, B_5$, or $B_6$.*

**Proof.** There is a nonsingular matrix $K$ such that

$$K^{-1}AK = Q,$$

where

$$Q = \begin{pmatrix} \alpha & \beta & 0 \\ -\beta & \alpha & 0 \\ 0 & 0 & \lambda_1 \end{pmatrix},$$

$\alpha$ and $\beta$ are real, $\beta \neq 0$, and $\alpha > 0$. (This is the so-called real normal form of a matrix with complex eigenvalues.) If $\mathbf{x} - \mathbf{x}_0 = K\mathbf{y}$ then $\mathbf{y}$ satisfies a system

$$\mathbf{y}' = G(\mathbf{y}) \tag{4.4}$$

[2]See Chapter 15 for a more general discussion of Poincaré maps.

where $G(\mathbf{0}) = \mathbf{0}$, and the linearized equation for $\mathbf{y}$ at $\mathbf{0}$ is

$$\mathbf{z}' = Q\mathbf{z}. \tag{4.5}$$

If $\tilde{C}$ is a cylinder

$$z_1^2 + z_2^2 = \varepsilon,$$

for some $\varepsilon > 0$, then solutions of (4.5) do not cross $\tilde{C}$ from outside to inside, since

$$\left(z_1^2 + z_2^2\right)' = 2\alpha\left(z_1^2 + z_2^2\right).$$

It follows that if $\varepsilon$ is sufficiently small then there is a neighborhood $\tilde{O}$ of $\mathbf{y} = \mathbf{0}$ such that solutions of (4.4) do not cross $\tilde{C}$ from outside to inside within $\tilde{O}$. Transforming this result back to the original coordinates $\mathbf{x}$ completes the proof of the lemma, because if $\varepsilon$ is small enough then the intersection of $C$, the transformed cylinder, with the boxes $B_1, \ldots, B_6$ lies in a small neighborhood of $\mathbf{x}_0$. This completes the proof of Lemma 4.7. □

We now define the subset $\tilde{F}$ of $F$ to be all of $F$ except the part which is inside the cylinder $C$. Then the Poincaré map $P$ takes $\tilde{F}$ into itself and is continuous on $\tilde{F}$ and so has a fixed point, which corresponds to a periodic solution. This completes the proof of Theorem 4.3. □

## 4.3. Subsequent developments

Theorem 4.3 was generalized to $n$-dimensions in [**96**]. But since then this result has been subsumed into the much more general theory of monotone feedback systems, developed by Hirsch and Smith in a number of papers, with important contributions by Mallet-Paret and others. Monotone feedback systems are of the form

$$x_i' = f_i\left(x_{i-1}, x_i\right), \quad i = 1, \ldots, n,$$

where $x_0 = x_n$, with the requirement that for each $i$, $\frac{\partial f_i(x_1,\ldots,x_n)}{\partial x_{i-1}} \neq 0$ for all positive $x_1, \ldots, x_n$. (This domain can be generalized.) See [**101**] or [**222**] for surveys of this work. From the point of view of this chapter, one of the most important papers was [**161**], where the result proved above was greatly strengthened as part of a general theory. (This had been done for the three-dimensional case earlier, in [**221**].) It was shown that every trajectory entering $B$ except those on the stable manifold tends to some periodic solution. Unfortunately nothing has been proved about uniqueness. If there is only one nonconstant periodic solution then it must be asymptotically orbitally stable.

## 4.4. Exercises

**Exercise 4.1.** Prove that if $\mathbf{x} = (x, y, z)$ is a solution of (4.1) such that $\mathbf{x}(0) \in E^+$, then there is a $T \geq 0$, which depends on $\mathbf{x}(0)$, such that $\mathbf{x}(t) \in B$ for all $t > T$. Prove that the interior $\mathring{B}$ of $B$ has these properties.

Hint: Techniques used in this chapter may be helpful.

**Exercise 4.2.** Theorem 4.3 can be extended to the system $x' = f(z) - \alpha x$, $y' = x - \beta y$, $z' = y - \gamma z$ where $\alpha$, $\beta$, and $\gamma$ are positive constants. Give an algebraic condition on $\alpha, \beta, \gamma$, and $f'(z_0)$ which ensures that the equilibrium in the positive octant is unstable.

# Chapter 5

# A linear example

## 5.1. Statement of the problem and a basic lemma

Most of the problems discussed in this book are nonlinear. However to illustrate further the distinction between finite and infinite dimensional approaches to proving the existence of solutions to a boundary value problem, we will discuss a linear system of equations and boundary conditions where several methods work, each of which is important in the right context.

The model comes from mathematical biology and results from the study of molecular motors, which are nanoscale devices for moving material around within biological cells. See [**55**] or [**36**] for further details. Much of the material in this section is taken from [**35**] or [**87**]. We thank David Kinderlehrer for introducing us to this problem.

A relatively simple form of the equations is the system

$$\begin{aligned} (\rho_1' + f_1\rho_1)' &= \alpha_1\rho_1 - \alpha_2\rho_2, \\ (\rho_2' + f_2\rho_2)' &= -\alpha_1\rho_1 + \alpha_2\rho_2, \end{aligned} \tag{5.1}$$

where $\alpha_1$ and $\alpha_2$ are continuously differentiable functions on $[0,1]$ which are positive and have period one. Also, $f_1$ and $f_2$ are continuously differentiable, with period one and mean value zero. Thus

$$\int_0^1 f_i(x)\ dx = 0, \quad i = 1,2. \tag{5.2}$$

The $\rho_i$ may be viewed as partial probability density functions on $[0,1]$, with the requirement that $\rho_i \geq 0$ for $i = 1,2$ and

$$\int_0^1 (\rho_1(x) + \rho_2(x))\ dx = 1. \tag{5.3}$$

There are several possible boundary value problems for these equations, as noted in [**87**], but here we prove only the existence and uniqueness of positive solutions of period one. These were the solutions considered in [**55**]. By "positive solution", we mean a solution such that $\rho_1 > 0$ and $\rho_2 > 0$ on $R^1$. Because the problem is linear, every positive solution can be multiplied by a positive constant to achieve the normalization condition (5.3).

We present three existence proofs, one using Schauder's fixed point theorem, one using a "continuation" method in an infinite dimensional space, and one using a combination of elementary linear algebra and a finite dimensional continuation (or "homotopy") argument. Understanding the first two methods requires some background in functional analysis. In our discussion we will state the required abstract results, such as Schauder's theorem, but will not give their proofs or all of the precise definitions required for a full understanding. The reader without this background is encouraged to try to read through the proofs anyway as a beginning towards mastering these ideas. Although they turn out not to be necessary for the problem at hand, they are indispensable in other contexts, especially when dealing with inherently infinite dimensional problems. We will cite sources for the details of this material. The third proof is fully within our category of classical methods. Each of the sections containing these proofs can be read independently of the others. At the end we refer briefly to still another (classical) proof which, while rather special to this problem, is perhaps the shortest of the four proofs we discuss.

All four proofs require a maximum principle type lemma which is itself an important tool, of much wider applicability. We begin the discussion with one version of this result.

**Lemma 5.1.** *Let $a$ and $b$ be continuously differentiable functions in a closed interval $\bar{J} = [x_0, x_1]$, with $a(x) > 0$ and $b(x) \geq 0$. Let $u''$ be continuous on $\bar{J}$ and suppose that in the open interval $J = (x_0, x_1)$,*

$$\frac{d}{dx}\left(a(x)\frac{du}{dx}\right) - b(x)u \leq 0. \tag{5.4}$$

*Then*

(i) *the function $u$ does not attain a negative minimum in the open interval $J$ unless $b$ is identically zero and $u$ is constant;*

(ii) *if $u \geq 0$ on $J$ and $u$ is not identically zero on $J$, then $u > 0$ on $J$;*

(iii) *if $u$ is not identically zero and $\inf_J u = 0$, then $u = 0$ at either $x_0$ or $x_1$, and if $u(x_0) = 0$ then $u'(x_0) > 0$ while if $u(x_1) = 0$ then $u'(x_1) < 0$.*

**Proof.** First suppose that the inequality in (5.4) is strict. ($<$ instead of $\leq$. The result in this case is sometimes called a "weak maximum principle".)

Then $u$ does not have a minimum in $(x_0, x_1)$ at which $u \leq 0$, proving (i) and (ii) in this case, and we observe also that $u$ is not constant. For (iii), if for example $u(x_0) = 0$ and $u'(x_0) = 0$, then $u''(x_0) < 0$, so that $u < 0$ just to the right of $x_0$, a contradiction. Similarly, $u'(x_0) < 0$ is ruled out, as is the case $u(x_1) = 0, u'(x_1) \geq 0$, implying (iii).

To prove the full result, with $\leq$ in (5.4), suppose first that $u$ has a negative minimum, say at $x = \zeta \in J$. In every interval $[\zeta, \zeta + \delta)$ where $u$ is negative, the quantity $a(x) u'(x)$ is nonincreasing, so that $u' \leq 0$ to the right of $\zeta$, and likewise, $u' \geq 0$ to the left of $\zeta$. Since $u$ has a minimum at $\zeta$, $u$ must be constant.

For (ii), suppose that $u \geq 0$ in $J$ and for some $\zeta \in J$, $u(\zeta) = \min u = 0$. If $u$ is not constant then choose $\delta \neq 0$ so that $\zeta + \delta \in J$ and $u(\zeta + \delta) > 0$. Consider the case $\delta > 0$, the case $\delta < 0$ being similar. Let

$$v(x) = 1 - e^{\kappa(x-\zeta)}$$

where $\kappa$ is chosen large enough that

$$\frac{d}{dx}\left(a(x)\frac{dv}{dx}\right) - b(x)v < 0 \tag{5.5}$$

on $J$. Let $w(x) = u(x) + \varepsilon v(x)$, where $\varepsilon > 0$ is chosen so small that $w(\zeta + \delta) > 0$. From (5.4) and (5.5) we see that $\frac{d}{dx}\left(a(x)\frac{dw}{dx}\right) - b(x)w < 0$ on $J$. Note that $w(\zeta) = u(\zeta) = 0$ and $w'(\zeta) < 0$. Hence $w$ has a negative minimum in $(\zeta, \zeta + \delta)$. But the weak maximum principle applies to $w$, giving a contradiction which proves (ii). The proof of (iii) is left as Exercise 5.1. □

## 5.2. Uniqueness

In this brief section we will illustrate the application of Lemma 5.1 to prove a uniqueness result for our problem. This result is important in itself, and also useful in the second and third existence proofs below, bringing out the important point that often uniqueness implies existence. This principle is not limited to differential equations. For example, the Fredholm alternative, in both finite and infinite dimensional settings, has the consequence that if a linear homogeneous problem of the form $Lu = 0$ has only the trivial (zero) solution, then a nonhomogeneous problem $Lu = f$ has a solution for every allowable $f$ [**201**]. As we will see in Chapter 12, uniqueness for nonlinear problems is often more difficult to prove than existence. But in this case, the uniqueness will be used in proving existence.

**Theorem 5.2.** *If $f_1$ and $f_2$ are continuously differentiable functions on $R^1$ and there is a positive period one solution $(\rho_1, \rho_2)$ of* (5.1) *which also* *satisfies* (5.3), *then there is no other solution of* (5.1)–(5.3) *(positive or not) with period one.*

**Proof.** Suppose that $(\sigma_1, \sigma_2)$ is a second period one solution of (5.1) which satisfies (5.3). For sufficiently small $\mu$, $(\rho_1 - \mu\sigma_1, \rho_2 - \mu\sigma_2)$ is a positive periodic solution. However, for $\mu > 1$, $\int_0^1 \rho_1 + \rho_2 - \mu(\sigma_1 + \sigma_2)\ dx < 0$ and so $(\rho_1 - \mu\sigma_1, \rho_2 - \mu\sigma_2)$ is not positive. Let

$$\mu^* = \sup\{\mu \mid (\rho_1 - \mu\sigma_1, \rho_2 - \mu\sigma_2) \text{ is a positive solution}\}.$$

Then $(\rho_1 - \mu^*\sigma_1, \rho_2 - \mu^*\sigma_2)$ is a nonnegative solution such that some component $\rho_i - \mu^*\sigma_i$ vanishes at some $\zeta \in [0,1]$. Suppose for convenience that $i = 1$. Let $F_1(x) = \int_0^x f_1(s)\ ds$ and notice that $u = e^{F_1}(\rho_1 - \mu^*\sigma_1)$ satisfies the equation

$$\left(e^{-F_1}u'\right)' - \alpha_1 e^{-F_1}u = -\alpha_2(\rho_2 - \mu^*\sigma_2) \le 0$$

on $[0,1]$. By Lemma 5.1 either $u$ is identically zero or $\zeta$ is one of the end points 0 or 1. In the first case, $\int_0^1 (\rho_1 + \rho_2)\ dx = \mu^* \int_0^1 (\sigma_1 + \sigma_2)\ dx = 1$, implying that $\mu^* = 1$ and $(\rho_1, \rho_2) = (\sigma_1, \sigma_2)$. In the second case, periodicity implies that $u(0) = u(1) = 0$ and that $u'(0) = u'(1)$. By (iii) of Lemma 5.1, $u'(0) \neq 0$. However $u > 0$ on $(0,1)$ implies that $u'(0) \ge 0$ and $u'(1) \le 0$ and a contradiction results. □

## 5.3. Existence using Schauder's fixed point theorem

**Theorem 5.3.** *Suppose that* $f_1, f_2, \alpha_1$, *and* $\alpha_2$ *are continuously differentiable on* $R^1$ *and have period one. Suppose also that* $\alpha_1$ *and* $\alpha_2$ *are positive and that* $f_1$ *and* $f_2$ *have mean value zero on* $[0,1]$. *Then the system* (5.1) *has a unique positive solution with period one satisfying* (5.3).

Uniqueness was proved above. We will outline the proof of existence using Schauder's theorem, which is stated below. First we show how a fixed point theorem is relevant.

Rewrite the system (5.1) as follows:

$$\left(\rho_1' + f_1\rho_1\right)' - \alpha_1\rho_1 = -\alpha_2\rho_2,$$
$$\left(\rho_2' + f_2\rho_2\right)' - \alpha_2\rho_2 = -\alpha_1\rho_1.$$

Looking at the individual equations of this system motivates consideration of a single equation of the form

$$(5.6) \qquad \left(\rho' + f\rho\right)' - \alpha\rho = -\beta\sigma,$$

where $\sigma$ is assumed to be a known periodic function, $\beta$ is a positive function, and we look for a periodic solution $\rho$. (Every periodic function will be assumed to have period one, though this may not be the least period of the function.) Let $F(x) = \int_0^x f(s)\,ds$. If $f$ is periodic and has zero mean over a period, then $F$ is also periodic. Let $\eta = e^F \rho$. Then

$$(e^{-F}\eta')' - \alpha e^{-F}\eta = -\beta\sigma. \tag{5.7}$$

This equation is in selfadjoint form. The corresponding homogeneous equation has no periodic solution, because $\eta$ has no positive maximum or negative minimum. Hence by the Fredholm alternative theory (theory of Green's functions in this case), (5.7) has a unique periodic solution for every continuous periodic $\sigma$ (e.g. [**41**, Chapter 7, Theorem 2.2]). Setting

$$\rho = e^{-F}\eta \tag{5.8}$$

gives a mapping $T_0$ from $\sigma$ to a unique periodic solution $\rho$ of (5.6).

Next, consider a pair of periodic functions $(\sigma_1, \sigma_2)$ and the system

$$\begin{aligned} (\rho_1' + f_1\rho_1)' - \alpha_1\rho_1 &= -\alpha_2\sigma_2, \\ (\rho_2' + f_2\rho_2)' - \alpha_2\rho_2 &= -\alpha_1\sigma_1. \end{aligned} \tag{5.9}$$

Applying the mapping $T_0$ to each equation, with $f$ replaced successively by $f_1$ and $f_2$, $\alpha$ by $\alpha_1$ and $\alpha_2$, and $\beta$ by $\alpha_2$ and $\alpha_1$, we obtain a map $T$ taking $(\sigma_1, \sigma_2)$ to the unique pair $(\rho_1, \rho_2)$ of periodic solutions to the equations in (5.9). By Lemma 5.1, if $\sigma_i \geq 0$ for $i = 1, 2$ and neither $\sigma_i$ is identically zero, then $\rho_i > 0$ for $i = 1, 2$. We look for fixed points of $T$, that is, pairs $(\sigma_1, \sigma_2)$ of continuous periodic functions such that

$$T(\sigma_1, \sigma_2) = (\sigma_1, \sigma_2).$$

Clearly a fixed point for $T$ is a periodic solution to (5.1). We now state the Schauder fixed point theorem.

**Theorem 5.4.** *Let $B$ be a Banach space and $K$ a closed convex subset of $B$. Let $T : K \to B$ be a continuous transformation from $K$ into $B$ such that the image $T(K) \subset K$ and the closure of $T(K)$ is compact. Then $T$ has a fixed point in $K$.*

For the needed definitions, and a proof, see [**62**]. To apply this theorem to the system (5.1) we must specify the Banach space $B$ and the closed convex set $K$. To motivate these definitions, first look in more detail at the periodic solution of (5.6). This solution is constructed by using a Green's function $G(x, y)$, which is bounded and has bounded first partial derivatives at every $(x, y)$ with $y \neq x$. From the equation for $\eta$ in terms of $G$ and $\sigma$, namely

$$\eta(x) = \int_0^1 G(x,y)\,\sigma(y)\,dy, \tag{5.10}$$

one obtains bounds on $\eta$ and $\eta'$. These bounds can be stated in several norms, but it will be seen below that the $\mathcal{L}^1$ norm on $[0,1]$, denoted in this section by $\|\ \|$, is useful here. (See [**62**], Appendix D for background.)

We need to know that $T(K)$ is compact for an appropriate Banach space $B$ and closed convex set $K$. This will follow if the mapping $T_0$ is compact for every $f$ and $\alpha$, meaning that if $K_0$ is a closed bounded subset of $\mathcal{L}^1([0,1])$ then the image $T_0(K_0)$ is a compact set. Equivalently, we need to show that every sequence of functions in $T_0(K_0)$ has a convergent subsequence.[1]

The Ascoli-Arzela lemma tells us that it is sufficient for $T_0(K_0)$ to be an equicontinuous uniformly bounded set of functions. Neither of these properties, equicontinuity and uniform boundedness, is natural in the space $\mathcal{L}^1([0,1])$. They are best defined for subsets of the space $C([0,1])$ of continuous functions on $[0,1]$. Fortunately, the formulas (5.8) and (5.10) give us the bounds which are needed. From these equations and the properties of $G$ it follows that there is a constant $M$ such that if $\rho = T_0(\sigma)$ then for every $x \in [0,1]$,

$$\begin{aligned} |\rho(x)| &\le M\,\|\beta\sigma\|, \\ |\rho'(x)| &\le M\,\|\beta\sigma\|. \end{aligned} \tag{5.11}$$

Since $\beta$ is bounded and both $\beta$ and $\sigma$ are nonnegative, there are uniform estimates for $\rho$ and $\rho'$ in terms of the $\mathcal{L}^1$ norm of $\sigma$, and therefore the mapping $T_0 : \mathcal{L}^1([0,1]) \to \mathcal{L}^1([0,1])$ is compact. Each function $T_0(\sigma)$ is in $\mathcal{L}^1([0,1])$ and has a representative $\rho$ which is continuous and differentiable in $[0,1]$, with $|\rho|$ and $|\rho'|$ bounded in terms of $\|\sigma\|$. Also, from (5.11),

$$\|\rho\| = \int_0^1 |\rho(x)|\,dx \le M\,\|\beta\sigma\|, \tag{5.12}$$

so that the linear mapping $T_0$ is continuous.

To apply the Schauder fixed point theorem, $B$ is chosen to be the space $\mathcal{L}^1([0,1]) \times \mathcal{L}^1([0,1])$, with norm $\|(\sigma_1,\sigma_2)\|_B = \|\sigma_1\| + \|\sigma_2\|$.

[1] The reader should think carefully about what sort of convergence is needed and whether this type of convergence is proved.

Let $K$ be the set of pairs $(\sigma_1, \sigma_2) \in B$ such that (a representative of) each $\sigma_i$ is periodic and the following additional properties hold for $i = 1, 2$:

$$\begin{aligned} &\sigma_i \geq 0 \text{ on } [0,1] \text{ for } i = 1,2, \\ &\int_0^1 (\alpha_1(x)\sigma_1(x) + \alpha_2(x)\sigma_2(x))\, dx = 1, \\ &\int_0^1 (\alpha_1(x)\sigma_1(x) - \alpha_2(x)\sigma_2(x))\, dx = 0. \end{aligned} \tag{5.13}$$

It is easily seen that $K$ is a closed convex set in $B$. The compactness of $T(K)$ follows from the compactness of $T_0$, and continuity of $T$ is also inherited in this way. To complete the proof of Theorem 5.3 it must also be shown that $T$ maps $K$ into $K$. We leave this as Exercise 5.2.

## 5.4. Existence using a continuation method

Continuation methods have been used in a variety of types of problems, including ode's, pde's, and integral equations [**61**], [**160**]. They require more functional analysis than was needed for the Schauder theorem, and we will be content here to outline a proof of Theorem 5.3 assuming this background, with appropriate citations where it can be found. This method also relies on Lemma 5.1 and Theorem 5.2. Continuation is closely related to Leray-Schauder degree theory, which allows extension of the method to partial differential equations. See [**163**] for an application to swirling flow in fluid mechanics.

A continuation method begins by finding a system similar to the given problem but for which it is known that a unique solution of the desired type exits. In the case of (5.1) we start with the system

$$\begin{aligned} (\rho_1' + f_1\rho_1)' &= \alpha_1^0\rho_1 - \alpha_2^0\rho_2, \\ (\rho_2' + f_2\rho_2)' &= -\alpha_1^0\rho_1 + \alpha_2^0\rho_2, \end{aligned}$$

where $\alpha_i^0(x) = e^{F_i}$ for $i = 1, 2$. A positive periodic solution to this system is given by $\rho_i = e^{-F_i}$.

The method then continues by considering systems of the form

$$\begin{aligned} (\rho_1' + f_1\rho_1)' &= \alpha_1^\lambda\rho_1 - \alpha_2^\lambda\rho_2, \\ (\rho_2' + f_2\rho_2)' &= -\alpha_1^\lambda\rho_1 + \alpha_2^\lambda\rho_2, \end{aligned} \tag{5.14}$$

where for $i = 1, 2$, $\alpha_i^\lambda(x) = \lambda\alpha_i(x) + (1-\lambda)\alpha_i^0(x)$.

We will show that the solution $\rho^0 = (\rho_1^0, \rho_2^0)$ is part of a continuous family of solutions of (5.14) obtained as $\lambda$ increases from 0 to 1. The method is often challenging for a nonlinear equation, but the current problem is linear, which simplifies the analysis.

We now suppose that a positive periodic solution $\left(\rho_1^{\lambda_0}, \rho_2^{\lambda_0}\right)$, necessarily unique, has been found for some $\lambda_0 \geq 0$ and show how to continue this solution to a nearby $\lambda_1$. Denote the solution which we seek at $\lambda = \lambda_1$ by $\left(\rho_1^{\lambda_1}, \rho_2^{\lambda_1}\right)$. If this solution exists then $\phi = \rho^{\lambda_1} - \rho^{\lambda_0}$ satisfies

$$\phi_i'' + (f_i \phi_i)' - \alpha_i^{\lambda_0} \phi_i = -\alpha_j^{\lambda_0} \phi_j + S_i \tag{5.15}$$

for $i = 1, 2$ and $j \neq i$, where $S_i = \left(\alpha_i^{\lambda_1} - \alpha_i^{\lambda_0}\right) \rho_i^{\lambda_1} - \left(\alpha_j^{\lambda_1} - \alpha_j^{\lambda_0}\right) \rho_j^{\lambda_1}$. The unknown function $\rho^{\lambda_1}$ appears on both sides of (5.15). Therefore we consider

$$\phi_i'' + (f_i \phi_i)' - \alpha_i^{\lambda_0} \phi_i = -\alpha_j^{\lambda_0} \phi_j + S_i^* \tag{5.16}$$

where $S_i^* = \left(\alpha_i^{\lambda_1} - \alpha_i^{\lambda_0}\right) \rho_i^* - \left(\alpha_j^{\lambda_1} - \alpha_j^{\lambda_0}\right) \rho_j^*$ for some known pair $\rho^* = (\rho_1^*, \rho_2^*)$ of positive continuous periodic functions on $[0,1]$. The plan is to show that (5.16) has a unique periodic solution $\phi$ for each $\rho^*$, giving a map from $\rho^*$ to $\rho^{\lambda_1} = \phi + \rho^{\lambda_0}$, and showing that this map has a fixed point. This fixed point will solve (5.14) for $\lambda = \lambda_1$. A fixed point exists because, for $|\lambda_0 - \lambda_1|$ sufficiently small, the map $\rho^* \to \rho^{\lambda_1}$ is a contraction in an appropriate space.

Write (5.16) symbolically as

$$-T_i \phi_i = -\alpha_j^{\lambda_0} \phi_j + S_i^*. \tag{5.17}$$

The operator $T_i$ can be put in the form of a selfadjoint operator on the Hilbert space $\mathcal{P}$ of periodic functions in $\mathcal{L}^2[0,1]$, by setting

$$\eta_i = e^{F_i} \phi_i.$$

Then for $\eta$ in $C^2([0,1])$,

$$T_i \phi_i = -\frac{d}{dx}\left(e^{-F_i(x)} \eta_i'\right) + \alpha_i^{\lambda_0} e^{-F_i} \eta_i = M_i \eta_i, \text{ say,} \tag{5.18}$$

where $M_i$ has the standard form for a selfadjoint second order differential operator in $L^2(0,1)$; see Chapter 7 of [**41**]. Further, $M_i$ is what is called a positive operator, meaning that for all $\eta_i \neq 0$, $\int_0^1 \eta_i (M_i \eta_i)\, ds > 0$. To show this we integrate by parts:

$$\int_0^1 \eta_i (M_i \eta_i)\, ds = -e^{-F_i(x)} \eta_i' \eta_i |_0^1 + \int_0^1 e^{-F_i} \eta_i'^2 \, ds + \int_0^1 \alpha_i^{\lambda_0} e^{-F_i} \eta_i^2 ds,$$

which is positive because the boundary terms cancel due to periodicity.

The general theory of positive operators is invoked at this stage. See [**127**]. This theory tells us that the inverse $M_i^{-1}$ exists in $\mathcal{P}$, and (5.17) becomes

$$\eta_i = M_i^{-1}\left(\alpha_j^{\lambda_0} e^{-F_j}\eta_j\right) - M_i^{-1}S_i^*. \tag{5.19}$$

$M_i^{-1}$ is an integral operator with a symmetric kernel (a Green's function). We then see that $M_i^{-1}$ is a compact operator, meaning (as in the Schauder theorem) that if $\Omega$ is a closed bounded set in $\mathcal{P}$ then the image $M_i^{-1}(\Omega)$ is a compact set, and sequences in this set always have convergent subsequences. The compactness of $M_i^{-1}$ follows ultimately from the Ascoli-Arzela theorem.

The operator $M_i^{-1}$ is positive (from the general theory), but it has the additional property that it takes positive functions to positive functions. (Notice that this is not the same as being a positive operator, which is a property involving inner products.) To see this, observe that if $M_i\eta_i$ is positive then $\frac{d}{dx}\left(e^{-F_i}\eta_i'\right) - \alpha_i^{\lambda_0} e^{-F_i}\eta_i < 0$. Also, $\eta_i(0) > 0$ and $\eta_i(1) > 0$, for otherwise $\eta_i$ could not be periodic. Then $\eta_i > 0$ in $[0,1]$ by Lemma 5.1.

Now consider the pair of equations (5.19) with $i = 1, 2$ and $j \neq i$, which we can write as

$$\eta = P\eta + QS^*, \tag{5.20}$$

where $\eta$ is in the space $H$ of periodic functions in $L^2(0,1) \times L^2(0,1)$. Here

$$(QS^*)_i = -M_i^{-1}\left(\left(\alpha_i^{\lambda_1} - \alpha_i^{\lambda_0}\right)\rho_i^* - \left(\alpha_j^{\lambda_1} - \alpha_j^{\lambda_0}\right)\rho_j^*\right). \tag{5.21}$$

The linear operator $P$ is necessarily compact and takes pairs of positive functions to pairs of positive functions. Furthermore, $P$ has 1 as an eigenvalue, since a corresponding eigenfunction has the positive components $\eta_i = e^{F_i}\rho_i^{\lambda_0}$. It is then a further standard result from the theory of positive operators that the eigenvalue 1 is simple, meaning that there are no other eigenfunctions [**127**]. This fact is deducible also from the uniqueness result that we proved earlier. Further, this eigenvalue is the eigenvalue of largest modulus. Thus if, in (5.20), $QS^*$ is orthogonal to the unique positive eigenfunction corresponding to the eigenvalue 1 for the adjoint operator $P^*$, then by the Fredholm alternative [**201**] we can solve (5.20) for the $\eta_i$, and the solution is unique modulo an additive multiple of $\left(e^{F_1}\rho_1^{\lambda_0}, e^{F_2}\rho_2^{\lambda_0}\right)$.

To complete the proof, we need to investigate $P^*$. By definition,

$$(P\psi)_i = M_i^{-1}\left(\alpha_j^{\lambda_0} e^{-F_j}\psi_j\right).$$

To find the adjoint $P^*$, we recall that on $H$ the inner product is $\langle\psi,\omega\rangle = \int_0^1 (\psi_1\omega_1 + \psi_2\omega_2)\,dx$ and the adjoint $P$ is defined by $\langle P\psi,\omega\rangle = \langle\psi, P^*\omega\rangle$.

We write

$$\int_0^1 (\omega_1 (P\psi)_1 + \omega_2 (P\psi)_2)\, dx$$
$$= \int_0^1 \left(\omega_1 M_1^{-1}\left(\alpha_2^{\lambda_0} e^{-F_2}\psi_2\right) + \omega_2 M_2^{-1}\left(\alpha_1^{\lambda_0} e^{-F_1}\psi_1\right)\right) dx$$
$$= \int_0^1 \left(\alpha_2^{\lambda_0} e^{-F_2}\psi_2 M_1^{-1}(\omega_1) + \alpha_1^{\lambda_0} e^{-F_1}\psi_1 M_2^{-1}(\omega_2)\right) dx,$$

since $M_i$ and $M_i^{-1}$ are selfadjoint. Hence,

$$(P^*\omega)_i = \alpha_i^{\lambda_0} e^{-F_i} M_j^{-1}(\omega_j).$$

It follows that the solution to $\omega = P^*\omega$ is given by

$$\omega_i = M_i(1), \tag{5.22}$$

as is seen by substitution. From (5.18), $\omega_i = \alpha_i^{\lambda_0} e^{-F_i}$.

We now have to check that $QS^*$ in (5.20) is orthogonal to $\omega$. From equation (5.21), we must show that

$$\sum_{\substack{i=1\\ j\neq i}}^{2} \int_0^1 M_i^{-1}\left(\left(\alpha_i^{\lambda_1} - \alpha_i^{\lambda_0}\right)\rho_i^* - \left(\alpha_j^{\lambda_1} - \alpha_j^{\lambda_0}\right)\rho_j^*\right)\alpha_i^{\lambda_0} e^{-F_i} dx = 0.$$

Since $M_i$ is selfadjoint, this is

$$\int_0^1 \sum_{\substack{i=1\\ j\neq i}}^{2} \left(\left(\alpha_i^{\lambda_1} - \alpha_i^{\lambda_0}\right)\rho_i^* - \left(\alpha_j^{\lambda_1} - \alpha_j^{\lambda_0}\right)\rho_j^*\right) M_i^{-1}\left(\alpha_i^{\lambda_0} e^{-F_i}\right) dx = 0,$$

but as we saw in (5.22),

$$M_i^{-1}\left(\alpha_i^{\lambda_0} e^{-F_i}\right) = 1$$

for $i = 1, 2$, which gives the result because the integrand is now zero.

We now have a solution $(\eta_1, \eta_2)$ to

$$\eta = P\eta + QS^*$$

for each $\rho^*$, and it is unique modulo an additive multiple of $\left(e^{F_1}\rho_1^{\lambda_0}, e^{F_2}\rho_2^{\lambda_0}\right)$. We specify the multiple by insisting that if $\phi_i = e^{-F_i}\eta_i$ and $\rho^1 = \phi + \rho^0$, then

$$\int_0^1 (\rho_1^1 + \rho_2^1) = \int_0^1 (\rho_1^0 + \rho_2^0) = 1.$$

We thus have a mapping $L : \rho^* \to \rho^1$. From (5.21) it can be shown that for sufficiently small $|\lambda_1 - \lambda_0|$ this map is a contraction on the orthogonal complement of the eigenspace corresponding to the eigenvalue 1, giving the desired solution $\rho^{\lambda_1}$.

The final step is to show that the set $\Lambda$ of $\lambda$ for which a solution exists includes the entire interval $[0,1]$. This step follows by showing that $\Lambda$ is both open and closed in $[0,1]$ and nonempty. We will give a similar argument in the next section (in a finite dimensional setting) and so omit it here. This completes our outline of a proof by continuation for Theorem 5.3.

As we are seeing in this chapter, there are several existence proofs, and the continuation method as outlined above requires the most sophisticated functional analysis of the three we will give. However, in more global problems, such as integral equations and differential-difference equations, this degree of sophistication is required, and it is perhaps in these areas that the method of continuation really comes into its own, as in [**61**] and [**160**].

## 5.5. Existence using linear algebra and finite dimensional continuation

This proof of Theorem 5.3 is divided into two parts. It turns out that existence can be proved independently of the uniqueness and positivity of the solution. To do so, we use a standard ode approach and write our system (5.1) of two second order equations as a first order system of four equations. Define new independent variables $\phi_1$ and $\phi_2$, by setting

$$\phi_i = \rho_i' + f_i\rho_i, \quad i = 1, 2.$$

The system (5.1) is then equivalent to

$$\begin{aligned} \rho_1' &= \phi_1 - f_1\rho_1, \\ \rho_2' &= \phi_2 - f_2\rho_2, \\ \phi_1' &= \alpha_1\rho_1 - \alpha_2\rho_2, \\ \phi_2' &= \alpha_2\rho_2 - \alpha_1\rho_1. \end{aligned} \tag{5.23}$$

This system is nonautonomous and linear with continuous coefficients on $[0,1]$ and so has a fundamental solution $Z$, a $4\times4$ matrix function whose columns form a fundamental set of solutions for the system. Further, we can assume that $Z$ satisfies $Z(0) = I$, the $4\times4$ identity matrix.

Every solution of (5.23) can be written as

$$\begin{pmatrix} \rho_1 \\ \rho_2 \\ \phi_1 \\ \phi_2 \end{pmatrix} = Z\mathbf{c} \tag{5.24}$$

where $\mathbf{c}$ is a constant vector. A solution is periodic if and only if

$$Z(1)\,\mathbf{c} = \mathbf{c}. \tag{5.25}$$

This system of four algebraic equations in four unknowns $c_1, \ldots, c_4$ has a nontrivial solution if and only if the matrix

$$Z(1) - I$$

is singular.

From (5.23) it is seen that $\phi_1' + \phi_2' = 0$. Hence, the sum of the last two rows of $Z'(x)$ is zero. Since $Z(0) = I$, the last two rows of $Z(0) - I$ are zero. Therefore the sum of the last two rows of $Z(x) - I$ is zero for all $x$, which implies that $Z(x) - I$ is singular for each $x$. Therefore $Z(1) - I$ is singular, and so (5.1) has a nontrivial periodic solution.

**Remark 5.5.** This proof shows that the existence of a periodic solution to (5.23) is a problem in linear algebra. The signs of the terms in (5.1) play no role. In contrast, it was found in Section 5.2 that the uniqueness of a positive solution depends on these signs, and we will see now that existence of such a solution is also dependent on these signs.

To prove existence of a positive solution we observe as in the previous section that this is easy in the special case

$$\alpha_1^0 = e^{F_1}, \qquad \alpha_2^0 = e^{F_2} \tag{5.26}$$

where

$$F_i(x) = \int_0^x f_i(s)\, ds.$$

A positive periodic solution of (5.23) with (5.26) is given by

$$\rho_i^0 = e^{-F_i}, \qquad \phi_i^0 = \rho_i' + f_i \rho_i \tag{5.27}$$

for $i = 1, 2$. The corresponding solution to (5.23) is $\left(e^{-F_1}, e^{-F_2}, 0, 0\right)^T$, and by (5.24) a solution $\mathbf{c}^0$ to (5.25) is $(1, 1, 0, 0)^T$. Theorem 5.2 implies that if $\alpha_1$ and $\alpha_2$ in (5.23) are positive, then a positive periodic solution is the only periodic solution up to a constant multiple, and so the solution space of (5.25) is one dimensional.

It is convenient now to use a different normalization from (5.3), by considering solutions in the form (5.24) with

$$\|\mathbf{c}\| := \max_{i=1,\ldots,4} |c_i| = 1. \tag{5.28}$$

If (5.23) has a positive periodic solution, then there is a unique solution to (5.25) with $c_1 > 0$ and satisfying (5.28).

We prove existence of a positive solution for given positive, continuous, and periodic functions $\alpha_1$ and $\alpha_2$ by showing that (5.25) is satisfied by a nonzero vector $\mathbf{c}$ such that the solution with initial condition $Z(0)\,\mathbf{c}$ is positive on $[0, 1]$. It is useful to phrase the problem in terms of a homotopy between the pairs $\left(\alpha_1^0, \alpha_2^0\right)$ and $(\alpha_1, \alpha_2)$. Introduce a parameter $\lambda$ taking values in the interval $[0,1]$. Let $\alpha_i^\lambda = \lambda\alpha_i + (1-\lambda)\,\alpha_i^0$, $i=1,2$. For each $\lambda$ we consider the system

$$\begin{aligned} \rho_1' &= \phi_1 - f_1\rho_1, \\ \rho_2' &= \phi_2 - f_2\rho_2, \\ \phi_1' &= \alpha_1^\lambda\rho_1 - \alpha_2^\lambda\rho_2, \\ \phi_2' &= \alpha_2^\lambda\rho_2 - \alpha_1^\lambda\rho_1. \end{aligned} \tag{5.29}$$

There is a corresponding system of linear equations

$$Z_\lambda\,(1)\,\mathbf{c}^\lambda = \mathbf{c}^\lambda, \tag{5.30}$$

where $Z_\lambda$ is the fundamental matrix solution of (5.29). We have shown that for every $\lambda \in [0,1]$ there is a nontrivial solution to (5.30). We can assume that for this solution $c_1 \geq 0$ and (5.28) holds. If the corresponding solution of (5.29) is positive, then this solution is unique and $c_1 > 0$.

We now show that if a periodic solution of (5.29) and (5.28) is positive, and therefore unique, when $\lambda = \lambda_0$, then there is a positive periodic solution to these two equations for $\lambda$ in a neighborhood of $\lambda_0$, and this solution (necessarily unique) is continuous in $\lambda$ at $\lambda_0$. In fact, all this follows if we prove that for every sequence $\{\lambda_i\} \subset [0,1]$ converging to $\lambda_0$ and corresponding nonzero solutions $\mathbf{c}^{\lambda_i}$ of (5.30) with $\|\mathbf{c}^{\lambda_i}\| = 1$ and $c_1 \geq 0$, it is the case that $\lim_{i\to\infty} \mathbf{c}^{\lambda_i} = \mathbf{c}^{\lambda_0}$.

Otherwise there is a sequence $\{\lambda_i\} \subset [0,1]$ which converges to $\lambda_0$ such that some sequence $\{\mathbf{c}^{\lambda_i}\}$ of solutions to (5.30) satisfying (5.28) and $c_1 \geq 0$ does not converge to $\mathbf{c}^{\lambda_0}$. It follows that some subsequence of $\{\mathbf{c}^{\lambda_i}\}$ converges to some $\mathbf{d} \neq \mathbf{c}^{\lambda_0}$ which also satisfies (5.30) with $\lambda = \lambda_0$, (5.28), and $d_1 \geq 0$. However $Z_{\lambda_0}\,(t)\,\mathbf{d}$ is a periodic solution of (5.29) for $\lambda = \lambda_0$, and $\mathbf{d} \neq -\mathbf{c}^{\lambda_0}$, which contradicts the uniqueness of $Z_{\lambda_0}\,(t)\,\mathbf{c}^{\lambda_0}$. The idea that uniqueness implies continuity is rather common in analysis.

Now let $\Lambda$ be the set of $\lambda \in [0,1]$ such that (5.29) has a positive periodic solution. Clearly $0 \in \Lambda$. We will show that $1 \in \Lambda$ by showing that $\Lambda$ is both open and closed within $[0,1]$.

To show that $\Lambda$ is open, recall that for each $\lambda$, the matrix $Z_\lambda\,(1) - I$ is singular. If $\lambda_0 \in \Lambda$ then the null space of $Z_{\lambda_0}\,(1) - I$ has dimension 1, since the nontrivial solution of (5.30) is unique to within a multiplicative constant. Hence the rank of $Z_{\lambda_0}\,(1) - I$ is three. In other words, there is a nonzero three-by-three subdeterminant of this matrix. Since $Z_\lambda\,(1)$ is a continuous function of $\lambda$, the same subdeterminant is nonzero for $\lambda$ in some neighborhood of $\lambda_0$, while $Z_\lambda\,(1) - I$ remains singular. Hence the set of solutions of (5.30) is still one dimensional. There is a unique solution to (5.30) satisfying (5.28) and with $c_1^\lambda > 0$ for $\lambda$ in this neighborhood. Hence, in this neighborhood $\mathbf{c}^\lambda$ is a continuous function of $\lambda$, implying that $(\rho_1^\lambda, \rho_2^\lambda)$ remains positive in some neighborhood of $\lambda_0$.

To show that $\Lambda$ is closed, suppose that $\lambda_1$ is a limit point of $\Lambda$ which is not in $\Lambda$. The set of all functions which are periodic solutions of (5.29) for some $\lambda$ and with initial conditions satisfying (5.28) is equicontinuous and uniformly bounded, since both the solutions and their derivatives must be bounded by a constant independent of $\lambda$. Hence corresponding to $\lambda_1$ there is a nontrivial periodic solution of (5.29) which is a limit of positive solutions. For this solution $\rho_i \geq 0$ for $i = 1, 2$, and at least one of the $\rho_i$ is zero somewhere in $[0, 1]$. Applying Lemma 5.1 to $\eta = e^{F_i}\rho_i$ shows that $\rho_i$ is identically zero and it follows easily from (5.29) that both components are identically zero. But then, $\phi_1 = \phi_2 = 0$, so that $\mathbf{c}^{\lambda_1} = 0$, a contradiction.

Since $\Lambda$ is open and closed in $[0, 1]$ and nonempty, $\Lambda = [0, 1]$, and in particular, $1 \in \Lambda$. This completes the proof of Theorem 5.3.

## 5.6. A fourth proof

In [**35**] there is still another proof, entirely classical and perhaps shorter than the Schauder proof above.[2] It is based on the observation that in (5.23), $\phi_1' + \phi_2' = 0$. Hence, $\phi_2 = -\phi_1 + c$, and the system reduces to a set of three equations with an extra parameter. As with the other proofs, important use is made of Lemma 5.1 and Theorem 5.2. However we did not give this proof here, because it is rather special to the particular system (5.1). In a subsequent paper, [**87**], the theory was extended to a system of $n$ equations, in variables $\rho_1, \ldots, \rho_n$. Each of the three methods for proving existence which we gave above applies to the larger system, but the special fourth proof used in [**35**] does not.

Finally, we mention that the main emphasis in both [**35**] and [**87**] is not on the existence and uniqueness of solutions, but rather on physically important properties of the solutions. We refer the reader to these papers, particularly [**87**], for details.

## 5.7. Exercises

**Exercise 5.1.** Complete the proof of Lemma 5.1 by proving (iii) more generally.

**Exercise 5.2.** Complete the proof of existence by Schauder's fixed point theorem by showing that if $(\sigma_1, \sigma_2)$ is in $K$ and $(\rho_1, \rho_2)$ is the unique periodic solution to (5.9), then $(\rho_1, \rho_2)$ is in $K$. Hint: Show that in $K$, neither $\sigma_1$ nor $\sigma_2$ is identically zero. Also, note from (5.15) that $\int_0^1 \alpha_i(x)\, \rho_i(x)\, dx \leq 1$ for $i = 1, 2$, and use (5.11).

[2] Different boundary conditions are treated in [**35**], but the proof can be adapted to the case of periodic solutions.

Chapter 6

# Homoclinic orbits of the FitzHugh-Nagumo equations

## 6.1. Introduction

The physical background of this chapter lies in neurobiology. One of the most important mathematical models in biology is the set of equations derived in 1953 by A. Hodgkin and A. Huxley [**103**], which give good quantitative predictions of electrical activity in the giant nerve axon of a squid. A simpler version was developed by FitzHugh in 1961 [**69**], and an equivalent electrical circuit was studied by Nagumo et al. [**176**]. See [**59**] for background in mathematical neuroscience. We introduced the equations in Section 2.4 and analyzed the reduced case, where the small parameter $\varepsilon$ is zero. Now we consider the full system, with $\varepsilon > 0$.

Recall that the FitzHugh-Nagumo model equations are

$$\begin{aligned} \frac{\partial u}{\partial t} &= \frac{\partial^2 u}{\partial x^2} + f(u) - w, \\ \frac{\partial w}{\partial t} &= \varepsilon (u - \gamma w), \end{aligned} \tag{6.1}$$

where

$$f(u) = u(1-u)(u-a).$$

(See Figure 6.1.) Here $a \in \left(0, \frac{1}{2}\right)$, $\varepsilon > 0$, and $\gamma > 0$. The variable $u$ represents the trans-membrane potential along the axon, while $w$ lumps together several variables in the Hodgkin-Huxley equations which determine the conductances of the axonal membrane to different ionic species [**58**]. A traveling wave substitution $s = x + ct$ leads to the system

$$\begin{aligned} u' &= v, \\ v' &= cv - f(u) + w, \\ w' &= \tfrac{\varepsilon}{c}(u - \gamma w). \end{aligned} \tag{6.2}$$

Let $M = \max_{a \le u \le 1} \frac{f(u)}{u}$. It will be assumed that $\frac{1}{\gamma} > M$, so that

$$u - \gamma f(u) > 0 \quad \text{if } u > 0, \tag{6.3}$$

and hence, $(0,0,0)$ is the only equilibrium point of (6.2). We look for values of $c > 0$, the "wave speed", such that there is a solution with

$$\lim_{s \to -\infty} (u(s), v(s), w(s)) = \lim_{s \to \infty} (u(s), v(s), w(s)) = (0,0,0). \tag{6.4}$$

The trajectory of such a solution is called a "homoclinic orbit". We wish to prove that (6.2) has nontrivial solutions of this type for some values of $c$.

An important physical consideration is the stability of the traveling waves as solutions of (6.1). It has been shown that the fast wave is stable. See [**108**] or [**257**], the latter being more "classical" in flavor.

In Chapter 2, we showed that if $\varepsilon = w = 0$ then there is a unique speed $c = c_0^*(a)$ such that there is a monotonic traveling wave connecting $(0,0)$ to $(1,0)$ in the $(u,v)$-plane. In this chapter we will prove the following theorem.

**Theorem 6.1.** *Suppose that $0 < a < \frac{1}{2}$. For sufficiently small positive values of $\gamma$ there is an $\varepsilon_1 > 0$ such that if $0 < \varepsilon < \varepsilon_1$ then (6.2) has homoclinic orbits for at least two positive values of $c$, say $c^* > c_*$, where $c^*$ and $c_*$ depend on $\varepsilon$ and $a$. Further, $c^*$ and $c_*$ can be chosen so that $\lim_{\varepsilon \to 0^+} c^* = c_0^*(a)$ and $\lim_{\varepsilon \to 0^+} c_* = 0$.*[1]

For the fast wave speed $c = c^*$, $\gamma < \frac{1}{M}$ is sufficient. We will assume a little more for the slow wave.

We will describe two approaches to proving the existence of a "fast" wave, with speed $c^*$, one using the modern theory of geometric perturbation and one by shooting. The geometric proof uses techniques applicable to a wide variety of singularly perturbed boundary value problems.[2] It is relatively simple to outline but requires more background to understand completely. Our presentation will sketch some of the background but leave out most details. These may be found in the paper [**110**] by Jones, Kopell, and Langer, on which this section is based. A drawback to this method for the problem in this chapter is that except for $a$ close to $\frac{1}{2}$ it only gives the

---

[1] What is actually proved is a little stronger. It is shown that for each sufficently small $\varepsilon > 0$, there is an interval $I_\varepsilon = (\rho_\varepsilon, \sigma_\varepsilon)$ such that there is no homoclinic orbit if $c \in I_\varepsilon$, and $\lim_{\varepsilon \to 0^+} \rho_\varepsilon = 0$, $\lim_{\varepsilon \to 0} \sigma_\varepsilon = c_0^*(a)$. Further, $0 < c_* < \rho_\varepsilon$ and $\sigma_\varepsilon < c^* < c_0^*(a)$.

[2] Typically, a singularly perturbed boundary value problem involves a small parameter, say $\varepsilon$, and no solution exists when $\varepsilon = 0$.

faster wave, but this solution is much more important physically than the slower wave.

A third method which has been used on this problem is what is known as the "Conley index". This powerful method, developed by Conley in the 1970s, uses a type of topological degree. We refer the reader to works of Conley, such as [**44**], or to [**211**] for expositions of this theory. See our index for further citations. In [**27**] Carpenter used this method to prove the existence of the faster of the two FitzHugh-Nagumo traveling waves. In this paper she credits Conley with a proof of the same result in unpublished notes. The shooting method, by contrast, uses more details about the behavior of solutions of (6.2). As usual in this method, there is a "nontangency" lemma, Lemma 6.16 below.

**6.1.1. Preliminary results.** First we repeat what was proved in Section 2.4 about the case $\varepsilon = 0$, $w = 0$. Let $F(u) = \int_0^u f(s)\,ds$.

**Proposition 6.2.** (A) *For every $c > 0$, the system*

$$u' = v,\\ v' = cv - f(u) \tag{6.5}$$

*has three equilibrium points, at $(0,0)$, $(a,0)$, and $(1,0)$. The points $(0,0)$ and $(1,0)$ are saddle points, while $(a,0)$ is either an unstable node or an unstable spiral point. The unstable manifolds at $(0,0)$ and $(1,0)$ have positive slope at the equilibrium point, while the stable manifolds have negative slope.*

(B) *When $c = 0$ and $0 < a < \frac{1}{2}$, there is a nonconstant solution $p_0 = (u_0, v_0)$ of (6.5) such that $p_0(\pm\infty) = (0,0)$. The trajectory of this solution lies on the curve $E(u,v) = 0$, where $E(u,v) = \frac{1}{2}v^2 + F(u)$.*

(C) *For every $c > 0$, let $\mathcal{U}_c^+$ denote the branch of the unstable manifold at $(0,0)$ which points into the region $u > 0$, $v > 0$. A solution on $\mathcal{U}_c^+$ exists at least until $u = a$, with $v > 0$ up to that point. Let $p_c = (u_c, v_c)$ be the unique solution of (6.5) such that $p_c(0) \in \mathcal{U}_c^+$, $v_c(s) > 0$ if $s \le 0$, and $u_c(0) = a$. Then there is a unique $c_0^*(a) > 0$ such that $p_{c_0^*(a)}(s)$ exists on $(-\infty, \infty)$ and $p_{c_0^*(a)}(s) \to (1,0)$ as $s \to \infty$.*

(D) *If $c \in (0, c_0^*(a))$ then $v_c(s) = 0$ exactly once, at a point $s = s_1$ with $a < u_c(s_1) < 1$. In this case the solution exists on a semi-infinite interval $(-\infty, \omega_c)$, with $\omega_c < \infty$, and both $u_c(s)$ and $v_c(s)$ approach $-\infty$ as $s \to \omega_c$. Further, $v_c'(s_1) < 0$, and $u_c(s_1)$ is a continuous function of $c$ which tends monotonically to 1 as $c \to c_0^*(a)^-$.*

(E) *If $c > c_0^*(a)$ then the solution $p_c$ exists on a semi-infinite interval $(-\infty, \omega_c)$, and $v_c > 0$. Also, $u_c$ and $v_c$ tend to $\infty$ as $s \to \omega_c$.*

Next we give some basic facts about the full system, (6.2), which are obtained from the linearization around the origin and from routine analysis of the vector field in $R^3$. The next proposition involves a common term which, however, we have not used so far in this book:

**Definition 6.3.** Let $\Omega$ be a subset of $R^3$ such that *if* $q$ is a solution of (6.2) and $q(0) \in \Omega$, *then* $q(t) \in \Omega$ *for all* $t > 0$ in the maximal interval of existence of $q$. *Then* $\Omega$ *is called "positively invariant" for* (6.2).

**Proposition 6.4.** (A) *For each* $\varepsilon > 0$, *the system* (6.2) *has a one-dimensional unstable manifold* $\mathcal{U}_{c,\varepsilon}$ *and a two-dimensional stable manifold* $\mathcal{S}_{c,\varepsilon}$ *at* $(0,0,0)$. $\mathcal{U}_{c,\varepsilon}$ *has a component* $\mathcal{U}^+_{c,\varepsilon}$ *which is initially in the positive octant. If* $(u,v,w)$ *is a solution on* $\mathcal{U}^+_{c,\varepsilon}$ *then* $u, v,$ *and* $w$ *are positive on every interval* $(-\infty, T]$ *in which* $u \leq a$. *Further,* $w > 0$ *in every interval* $(-\infty, \tau]$ *in which* $u \geq 0$.

(B) *The regions*

$$E^+ = \{(u,v,w) \mid u > 1, v > 0, w > f(u)\},$$
$$E^- = \{(u,v,w) \mid u < 0, v < 0, w < f(u)\} \tag{6.6}$$

*are each positively invariant for* (6.2). *Solutions* $(u(s), v(s), w(s))$ *entering* $E^-$ *tend to* $-\infty$ *in* $u$ *and* $v$ *as* $s$ *increases, while solutions entering* $E^+$ *tend to* $+\infty$ *in both variables. Conversely, a solution such that* $u(s) \to -\infty$ *eventually enters* $E^-$, *while a solution such that* $u(s) \to \infty$ *enters* $E^+$.

**Proof.** A routine linearization analysis proves the first two sentences of (A). (See Exercise 6.3.) For each positive $c$ and $\varepsilon$, $v$ is bounded on $\mathcal{U}^+_{c,\varepsilon}$ as long as $0 < u \leq 1$, since $f$ is bounded there. For the third sentence of (A), integrate the equation for $w'$ and observe that for a solution on $\mathcal{U}^+_{c,\varepsilon}$, $w(t)e^{\frac{\varepsilon\gamma}{c}t} = \int_{-\infty}^{t} \frac{\varepsilon}{c} u(s) e^{\frac{\varepsilon\gamma}{c}s} ds$. Hence, $w > 0$ as long as $u > 0$. Note also that as long as $v > 0$, $v$ and $w$ can be considered as functions of $u$,[3] with

$$\frac{dv}{du} = c + \frac{w - f(u)}{v},$$
$$\frac{dw}{du} = \frac{\varepsilon}{c}\frac{u - \gamma w}{v}, \tag{6.7}$$
$$\lim_{u \to 0^+} (v, w) = (0,0).$$

Since $f(u) < 0$ for $0 < u < a$, $u' = v(u)$ is positive as long as $0 < u \leq a$, completing the proof of (A). For (B), observe that at a point on the boundary of $E^-$ where $u < 0$, $v < 0$, and $w = f(u)$, (6.2) implies that $w > 0$, $w' < 0$, and $f'(u)u' > 0$. Hence, $(u,v,w)$ enters $E^-$ at this point. Similarly, at $(1,0,0)$, which is on the boundary of $E^+$, $u' = 0$, $u'' = v' = 0$, and by (6.3), $u''' = \frac{\varepsilon}{c}(u - \gamma f(u)) > 0$. Therefore $(u,v,w)$ enters $E^+$. The rest of the proof uses similar arguments. $\square$

Let $q_{c,\varepsilon} = (u_{c,\varepsilon}, v_{c,\varepsilon}, w_{c,\varepsilon})$ denote the unique solution of (6.2) on $\mathcal{U}^+_{c,\varepsilon}$ such that $v > 0$ on $(-\infty, 0]$ and $u(0) = a$. Then $\lim_{\varepsilon \to 0^+} q_{c,\varepsilon} = q_{c,0} = (u_c, v_c, 0)$,

[3] See Exercise 2.7.

where $(u_c, v_c)$ is the solution $p_c$ found in Proposition 6.2(C), and the limit is uniform on intervals of the form $(-\infty, T]$ where $p_c$ exists.

In the next section we warm up by proving something far easier than Theorem 6.1, namely that, for sufficiently small $\varepsilon > 0$, there are at least two values of $c$ where $\mathcal{U}_{c,\varepsilon}^+$ is bounded. However, those interested primarily in the methods of geometric perturbation may wish to skip to Section 6.3.

## 6.2. Existence of two bounded solutions

Our first step is based on (C), (D), and (E) of Proposition 6.2. We wish to show that similar behavior occurs when $\varepsilon > 0$, with solutions on $\mathcal{U}_{c,\varepsilon}^+$ blowing up positively for some positive values of $c$ and blowing up negatively for others. This result will give us values of $c$ where neither happens, and these are candidates for a bounded, indeed homoclinic, orbit.

In (6.6) we defined positively invariant sets $E^+$ and $E^-$. These are important in several places, starting with the following result, in which $c_0^*(a)$ is the speed of the traveling front when $\varepsilon = 0$. In the following lemma $u_{\max}$ denotes the point in $(a, 1)$ where $f$ has a maximum.

**Lemma 6.5.** *For each $c_1 \in (0, c_0^*(a))$ there is an $\varepsilon_1 > 0$ such that if $0 < \varepsilon < \varepsilon_1$ then $v_{c_1,\varepsilon}$ has exactly one zero, which is at a point where $a < u_{c_1,\varepsilon} < 1$, and later, $q_{c_1,\varepsilon}$ enters $E^-$. If $c_0^*(a) - c_1$ is positive but sufficiently small, then $\varepsilon_1$ can be chosen so that in addition, $u_{c_1,\varepsilon} > u_{\max}$ at the point where $v_{c_1,\varepsilon} = 0$. On the other hand, for any $\varepsilon > 0$ there is a $c_2(\varepsilon) > 0$ such that if $0 < c < c_2(\varepsilon)$ then $v_{c,\varepsilon} > 0$ and $q_{c,\varepsilon}$ enters $E^+$. Finally, if $c \geq c_0^*(a)$ then for every $\varepsilon > 0$, $q_{c,\varepsilon}$ enters $E^+$, with $v_{c,\varepsilon} > 0$ on its interval of existence.*

**Proof.** Suppose that $c_1 \in (0, c_0^*(a))$. By Proposition 6.2, part (D), $q_{c_1,0}$ enters the region $u < 0, v < 0$, after a single zero of $v_{c_1,0}$, this zero being at a point where $a < u_{c_1,0} < 1$. Further, $u_{c_1,0}$ and $v_{c_1,0}$ are unbounded, with both tending to $-\infty$. Choose $\tau > 0$ such that $u_{c_1,0}(\tau) < -1$, $v_{c_1,0}(\tau) < -1$, and $w_{c_1,0}(\tau) = 0$. Since $f$ has cubic growth as $u$ decreases from 0, $\tau$ can also be chosen so that $f(u_{c_1,0}(\tau)) > 1$. Then for sufficiently small $\varepsilon > 0$, $u_{c_1,\varepsilon}(\tau) < -1$, $v_{c_1,\varepsilon}(\tau) < -1$, and $w_{c_1,\varepsilon}(\tau) < 1 < f(u_{c_1,\varepsilon}(\tau))$. This proves the existence of an $\varepsilon_1$ as stated in the first sentence of the lemma.

Again using part (D), if $c_1$ is close to $c_0^*(a)$ then the zero of $v_{c_1,0}$ occurs with $u_{\max} < u_{c_1,0} < 1$. Therefore $\varepsilon_1$ can further be chosen so that if $0 \leq \varepsilon < \varepsilon_1$ then $u_{c_1,\varepsilon} \in (u_{\max}, 1)$ at the unique zero of $v_{c_1,\varepsilon}$.

Next consider small $c > 0$, for a fixed $\varepsilon > 0$, still assuming that $q = q_{c,\varepsilon}$. It follows from (6.7) that as long as $v > 0$ and $0 < u \leq 1$, $v$ is bounded independently of $c \in (0, c_0^*(a))$. We see from the equation for $\frac{dw}{du}$ in (6.7) and the definition of $q_{c,\varepsilon}$ that $u - \gamma w \to 0$ as $c \to 0^+$, uniformly in the maximal interval $[0, T(c, \varepsilon)]$ where $v \geq 0$ and $a \leq u \leq 1$. Hence, (6.3) implies that for each $\varepsilon > 0$, if $c$ is positive and sufficiently small then $w > f(u)$ as long as $v > 0$. In that case, (6.2) implies that $v > 0$ as long as the solution exists. Therefore, $q_{c,\varepsilon}$ enters $E^+$.

For $c \geq c_0^*(a)$, $v_{c,0} > 0$. Write the first equation of (6.7) as $\frac{1}{2}\frac{d}{du}v^2 = cv + w - f(u)$. Since $w > 0$ as long as $u > 0$, it follows that $v_{c,\varepsilon}(u) > v_{c,0}(u)$ as long as $v_{c,0} > 0$, implying that if $\varepsilon > 0$ and $c > c_0^*(a)$, then $q_{c,\varepsilon}$ enters $E^+$. If $c = c_0^*(a)$, then $v_{c,\varepsilon} > 0$ as long as $u_{c,\varepsilon} < 1$. There is no equilibrium of (6.2) in the region $u > 0$, and since $w > 0$, $q_{c_0^*(a),\varepsilon}$ cannot cross $v = 0$ at $u = 1$. It follows that the $q_{c_0^*(a),\varepsilon}$ cannot remain in $u \leq 1$ and so enters $E^+$. This completes the proof of Lemma 6.5.[4] □

**Theorem 6.6.** *For $\varepsilon_1$ as defined in Lemma 6.5 and each $\varepsilon \in (0, \varepsilon_1)$, there are at least two positive values of $c$ such that $\mathcal{U}_{c,\varepsilon}^+$ is bounded.*

**Proof.** We use a standard shooting argument. Fix $\varepsilon$ in $(0, \varepsilon_1)$ and let $q_c = (u_c, v_c, w_c)$ be the unique solution of (6.2) on $U_{c,\varepsilon}^+$ such that $u(0) = a$ and $v > 0$ on $(-\infty, a]$. Then, define two subsets of the semi-axis $c > 0$:

$$A = \{c > 0 \mid \text{there is a } t_1 \text{ such that } q_c(t_1) \text{ is in the interior of } E^+\},$$
$$B = \{c > 0 \mid \text{there is a } t_2 \text{ such that } q_c(t_2) \text{ is in the interior of } E^-\}.$$

Since $q_c$ is continuous with respect to $c$, these two sets are open. Lemma 6.5 shows that $c$ is in $A$ if $c$ is sufficiently large or sufficiently small, while $c_1 \in B$. The two sets are disjoint, because $E^+$ and $E^-$ are positive invariant and disjoint. Hence, there are at least two positive values of $c$ which are not in $A \cup B$. Say these are $c_* < c_1$ and $c^* > c_1$. We will show later that $c_*$ and $c^*$ can be chosen so that both $\mathcal{U}_{c_*,\varepsilon}$ and $\mathcal{U}_{c^*,\varepsilon}$ are homoclinic orbits. At this stage, we will prove that both of these trajectories are bounded.

Suppose, for example, that $q_{c^*} = (u_{c^*}, v_{c^*}, w_{c^*})$ is unbounded. It is easy to show that if the variables are eventually monotonic, then $q_{c^*}$ enters $E^+$ or $E^-$, contradicting how $c^*$ was chosen. Hence this solution must oscillate, with bigger and bigger oscillations. To eliminate these large oscillations, consider again the $(u, w)$-plane and the curve $w = f(u)$. We construct a certain rectangle $H$, shown in Figure 6.1.

To define this rectangle, consider some point $(u_R, f(u_R))$ with $u_R > 1$. Also consider the points $P_1 = \left(u_R, \frac{u_R}{\gamma}\right)$ on the $w$-nullcline, $P_2 = \left(u_L, \frac{u_R}{\gamma}\right)$ on the curve $w = f(u)$ to the left of $P_1$, and $P_3 = \left(u_L, \frac{u_L}{\gamma}\right)$ on the $w$-nullcline below $P_2$. These are, respectively, the upper right, upper left, and

[4] To be more careful about this, first analyze the linearizations for $\varepsilon = 0$ and $\varepsilon > 0$ to show that $v_{c,\varepsilon}(u) > v_{c,0}(u)$ for small positive $u$. See Exercise 6.5. Also, only part of Lemma 6.5 is needed in this section, with the rest being applied later in the chapter.

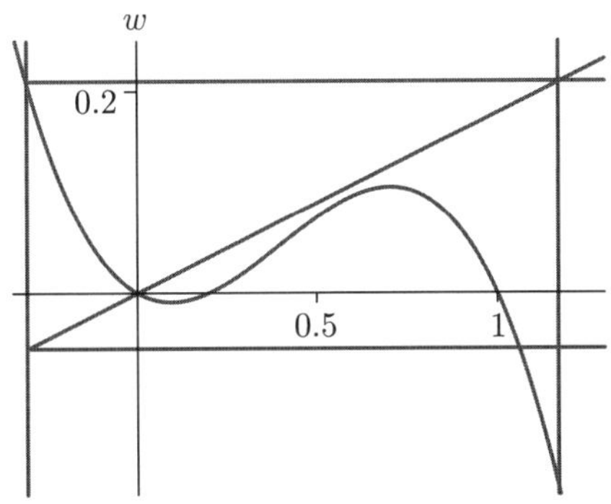


**Figure 6.1.** Graphs of $w = f(u)$ and $w = \gamma u$, and also the rectangle $H$.

lower left corners of a unique rectangle $H$, and we let $P_4$ be the fourth corner of this rectangle, $\left(u_R, \frac{u_L}{\gamma}\right)$. The cubic growth of $f(u)$ for $u < 0$ and for $u > 1$ ensures that if $u_R$ is large enough then $P_4$ lies above the original point chosen, $(u_R, f(u_R))$.

If $q = (u, v, w)$ is a solution of (6.2) such that $(u(0), w(0)) \in H$, then $(u, w)$ does not leave $H$ along its top or bottom edges, excluding the corners, because along the top, $w' < 0$ and along the bottom, $w' > 0$. If a solution leaves along the right side, including the corners, then it does so with $u > 1, v \geq 0$, and $w > f(u)$, so that $v' > 0$. Hence the solution enters $E^+$. Similarly, if a solution leaves along the left side of $H$ then it enters $E^-$.[5] Hence the solution $q_{c^*}$ is bounded, and similarly, $q_{c_*}$ is also bounded. ☐

## 6.3. Existence of homoclinic orbits using geometric perturbation theory

We will discuss two proofs of the existence of homoclinic orbits. In this section we outline the proof by geometric perturbation theory, including citations where the details can be found. This method gives good intuition about why the result is true. We include an elementary proof of one key lemma, thereby avoiding the differential forms which were used originally.

In Section 6.4 we use a shooting approach, which has the advantage that it shows the existence of two homoclinic orbits for each $a$ in $\left(0, \frac{1}{2}\right)$ and

[5] At the upper left corner $P_2$, if $v = 0$ then $v' = 0$ and $v'' = w' < 0$, so the solution still enters $E^-$.

applies to larger values of $\varepsilon$, as described briefly in Section 6.5. This allows some phenomena which don't occur if $\varepsilon$ is too small relative to $a$. We give a complete and elementary existence proof.

Both methods depend on analyzing $\mathcal{U}^+_{c,\varepsilon}$ as $c$ changes. This analysis can only be done for small $\varepsilon$. In the rest of this subsection we outline the main steps in the geometric perturbation proof. Following this we give further details.

Consider a solution $q_{c,\varepsilon}(t) = (u_{c,\varepsilon}, v_{c,\varepsilon}, w_{c,\varepsilon})$ on $\mathcal{U}^+_{c,\varepsilon}$. For large negative $t$, $q_{c,\varepsilon}(t)$ is in the positive octant, near the origin. In the geometric perturbation method, $\varepsilon$ is chosen so small that for an appropriate range of $c$, $u_{c,\varepsilon}$ increases until $q_{c,\varepsilon}(t)$ is close to $(1,0,0)$. It is shown that eventually, for an even more restricted range of $c$, $q_{c,\varepsilon}$ returns to a region close to the origin. The final step in the process is to show that for some $c^*$ close to $c_0^*(a)$, $q_{c^*}$ intersects the stable manifold $\mathcal{S}_{c^*,\varepsilon}$.

To carry out this plan we choose a small interval $I$ of values of $c$ containing $c_0^*(a)$ and consider the union $\mathcal{U}_\varepsilon$ of the set of curves $\mathcal{U}^+_{c,\varepsilon}$ with $c \in I$. Then $\mathcal{U}_\varepsilon$ is a two-dimensional manifold in $R^3$. We want $\mathcal{U}_\varepsilon$ to intersect $\mathcal{S}_{c,\varepsilon}$ for some $c$. In order to discuss both $\mathcal{U}_\varepsilon$ and $\mathcal{S}_{c,\varepsilon}$ simultaneously, it is convenient to add $c$ as a new dependent variable in the system, with equation $c' = 0$. Thus, we consider

$$\begin{aligned} u' &= v, \\ v' &= cv - f(u) + w, \\ w' &= \tfrac{\varepsilon}{c}(u - \gamma w), \\ c' &= 0, \end{aligned} \tag{6.8}$$

with solutions $(q, c)$ where, as before, $q = (u, v, w)$. We now use $\mathcal{U}^+_{c,\varepsilon}$ to denote the orbit of $(q_{c,\varepsilon}, c)$. For each $(c, \varepsilon)$, $\mathcal{U}^+_{c,\varepsilon}$ is still a one-dimensional manifold, now in $R^4$. We also revise the definition of $\mathcal{U}_\varepsilon$ to be the union of the manifolds $\mathcal{U}^+_{c,\varepsilon} \subset R^4$ with $c \in I$. Then $\mathcal{U}_\varepsilon$ is a two-dimensional invariant manifold for (6.8).

Further, consider the union $\mathcal{S}_\varepsilon$ of the two-dimensional manifolds $\mathcal{S}_{c,\varepsilon}$ with $c \in I$, now also in $R^4$. Thus, $\mathcal{S}_\varepsilon$ is an invariant three-dimensional manifold for (6.8). In the next section we will outline how to use what is known about $\mathcal{U}^+_{c,\varepsilon}$ when $\varepsilon = 0$ to show that for sufficiently small $\varepsilon$, $\mathcal{U}_\varepsilon$ intersects $\mathcal{S}_\varepsilon$ along $\mathcal{U}_{c^*,\varepsilon}$ for some $c^*$ close to $c_0^*(a)$, and this intersection is transverse.

To understand better what this means, we consider a possible nonzero point of intersection of $\mathcal{U}_\varepsilon$ and $\mathcal{S}_\varepsilon$. At this point, since $\mathcal{U}_\varepsilon$ and $\mathcal{S}_\varepsilon$ are smooth manifolds, they have tangent spaces, which are of dimensions two and three, respectively. To say that the intersection of $\mathcal{U}_\varepsilon$ and $\mathcal{S}_\varepsilon$ is transverse is to say that these two tangent spaces intersect transversely. But the tangent spaces are linear, and so transversality and their respective dimensions of two and three imply that their intersection is one dimensional.[6] Thus, $\mathcal{U}_\varepsilon \cap \mathcal{S}_\varepsilon$ is a one-dimensional invariant submanifold of $\mathcal{U}_\varepsilon$ in $R^4$, which must be $\mathcal{U}^+_{c^*,\varepsilon}$ for some $c^*$.

One can then show the existence of a locally unique homoclinic orbit and traveling wave. In other words, there is a $\delta > 0$ such that $\mathcal{U}_\varepsilon \cap \mathcal{S}_\varepsilon$ contains no points with $c^* - \delta < c < c^* + \delta$ except those on the one-dimensional submanifold $\mathcal{U}^+_{c^*,\varepsilon}$.

The proof starts by considering the equations when $\varepsilon = 0$. In this case, there is no smooth solution to the problem, but there is an approximation to a solution which can be analyzed. This is the so-called "singular" solution, which we describe in the next subsection. For this reduced problem the unstable manifold $\mathcal{U}^+_{c,0}$ still exists, as a continuous, though not smooth, curve. An appropriate transverse intersection of invariant surfaces is found for the $\varepsilon = 0$ problem, and because it is transverse, this intersection can be shown to exist also for small $\varepsilon > 0$.

**6.3.1. The singular solution.** Geometric perturbation theory was developed principally by Fenichel in the 1970s [**64**]. Subsequent applications and important additions were made by Jones, Kopell, and others [**109**]. Fundamental to this technique is the idea of a "singular solution". Briefly, a singular solution is a finite ordered set of functions, each of which satisfies an appropriate scaled version of the equations when $\varepsilon = 0$, with the further requirement that the trajectories of these solutions form, together, a continuous piecewise smooth curve, either in the three-dimensional phase space or, when $c$ is appended, in $R^4$. If the goal is to find a homoclinic orbit, then this curve should be simple and closed. We will denote it by $\Pi$.

The singular solution of the problem (6.2) and (6.4) has four pieces. The first and third are solutions of the system obtained by setting $\varepsilon = 0$ in (6.2), which means that $w$ is constant. This gives

$$\begin{aligned} u' &= v, \\ v' &= cv - f(u) + w. \end{aligned} \tag{6.9}$$

The first piece of the graph of the singular solution is the heteroclinic orbit given in Proposition 6.2 when $w = 0$. Recall that this exists for a particular positive value of $c$, which we are denoting by $c_0^*(a)$. With a shift $t \to t + T$ of the independent variable, the solution $\left(u_{c_0^*(a)}, v_{c_0^*(a)}\right)$ is characterized by the properties

$$u(0) = a, \qquad \lim_{t\to-\infty} (u(t), v(t)) = (0,0), \qquad \lim_{t\to\infty} (u(t), v(t)) = (1,0).$$

[6] These tangent spaces are the null spaces of $4 \times 4$ matrices of ranks 2 and 1 respectively. Transversality means that their intersection is the set of solutions of three linearly independent equations in four unknowns. As a a lower dimensional analogue, think of the intersection of the tangent spaces to a surface and a transverse curve in $R^3$.

This piece of the singular solution is called the "upjump".

The third piece of the singular solution is similar, except that it is a "downjump". To define it, we first note that the equilibria of (6.9) for a given constant $w$ are points $(u, 0)$ where

$$w = f(u). \tag{6.10}$$

This equation defines $u$ as a function of $w$, so long as $u$ remains in one of the three maximal intervals in which $f$ is monotonic. We will be interested in the two such intervals where $f$ is decreasing. They are of the form $(-\infty, u_{\min})$ and $(u_{\max}, \infty)$, where $0 < u_{\min} < a < u_{\max} < 1$.

Let $f_L^{-1}$ and $f_R^{-1}$ be the two functions of $w$ found by solving (6.10) for $u$ in those intervals. The third piece of the singular solution is described in the following result.

**Proposition 6.7.** *There exists a unique* $w^* \in (0, f(u_{\max}))$ *such that if in* (6.9) $c = c_0^*(a)$ *and* $w = w^*$, *then there is a heteroclinic orbit* $(u, v)$ *such that*

$$\lim_{t \to -\infty} (u(t), v(t)) = \left(f_R^{-1}(w^*), 0\right),$$
$$\lim_{t \to +\infty} (u(t), v(t)) = \left(f_L^{-1}(w^*), 0\right).$$

We omit the proof, which is similar to that of the upjump given in Chapter 2. (But see Exercise 6.3.)

To obtain the second and fourth pieces of the singular solution, use a change of variables in (6.2), by letting $s = \varepsilon t$, $(U(s), V(s), W(s)) = (u(t), v(t), w(t))$. Then $(U, V, W)$ satisfy

$$\begin{aligned} \varepsilon U' &= V, \\ \varepsilon V' &= c_0^*(a) V - f(U) + W, \\ W' &= \tfrac{1}{c_0^*(a)} (U - \gamma W). \end{aligned} \tag{6.11}$$

Setting $\varepsilon = 0$, without regard to whether this represents some sort of limit, results in two algebraic equations, $V = 0$ and $W = f(U)$. The second and fourth pieces of the singular solution are found by substituting $U = f_R^{-1}(W)$ and $U = f_L^{-1}(W)$ into (6.11), to give

$$W' = \frac{1}{c_0^*(a)} \left(f^{-1}(W) - \gamma W\right), \tag{6.12}$$

where $f^{-1} = f_R^{-1}$ for the second piece and $f_L^{-1}$ for the fourth. Each of these equations, together with the algebraic equations defining $f^{-1}$, is called a "slow system" associated with (6.2). For the second piece, $W(0) = 0$, $W' > 0$, and the solution is continued until $W = w^*$. The interval of $W$ for the fourth piece is the same, but now $W' < 0$ in (6.12).

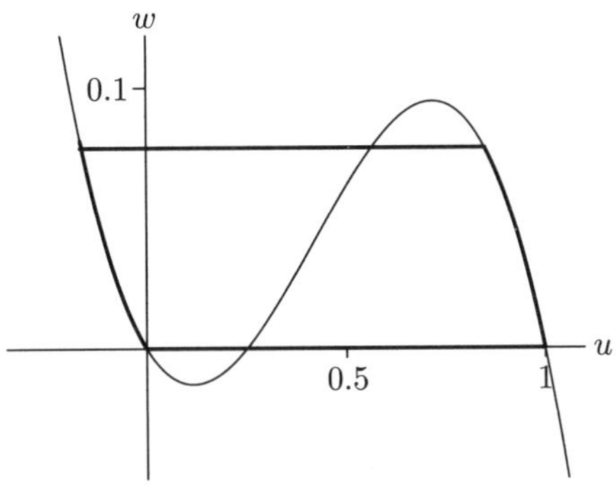


**Figure 6.2.** The orbit of the singular solution.

We see that the singular solution is defined as a sequence of four curves, say $\Pi_1, \Pi_2, \Pi_3, \Pi_4$, where $\Pi_1$ and $\Pi_3$ are solutions of (6.9) with $w$ constant and $\Pi_2$ and $\Pi_4$ are solutions of (6.12) with $V = 0$, $U = f^{-1}(W)$. It is a continuous piecewise smooth curve in $R^3$. The projection of the singular solution onto the $(u, w)$-plane is shown in Figure 6.2. The pieces $\Pi_2$ and $\Pi_4$ are called "slow manifolds" for (6.2).

**6.3.2. The first transverse intersection.** Our goal is to show that the singular solution just constructed is near to a real solution, with $\varepsilon > 0$. A second key idea in the geometric approach is that of transverse intersections of invariant manifolds. This concept is applied several times in the argument and is easiest to understand in its first appearance, which we now describe. At this stage we are considering (6.5). Again we add $c$ as a dependent variable, to get

$$\begin{aligned} u' &= v, \\ v' &= cv - f(u), \\ c' &= 0. \end{aligned} \tag{6.13}$$

Use is made of the theory of "center manifolds", which we have not encountered before in this book. Center manifolds were developed to analyze nonhyperbolic equilibrium points—those where at least one eigenvalue has zero real part. The theory is now well developed. See, for example, [**28**] or [**81**]. For this outline, no prior knowledge of center manifolds is necessary.

For every $c$, $(0, 0, c)$ is an equilibrium point of (6.13), and since it is non-isolated, it is nonhyperbolic. A center manifold for (6.13) at the equilibrium point $(0, 0, c_0^*(a))$ is an invariant manifold which is tangent to, and of the same dimension as, the eigenspace of the linearized matrix corresponding to the zero eigenvalue. The $c$-axis trivially has these properties and so is a center manifold in this case. While center manifolds in general are not unique, the trivial nature of the equation $c' = 0$ ensures uniqueness here.

As before, for each $c$ we consider the positive branch $\mathcal{U}_{c,0}^{+}$ of the unstable manifold of (6.5) at $(0,0)$ to be embedded in a plane of constant $c$ in the coordinates $(u,v,c)$ in $R^3$. We will still refer to this as the "unstable manifold" of (6.13) at $(0,0,c)$. There is also a unique branch of the stable manifold at $(1,0,c)$, which points into the region $u < 1$, which we denote by $\mathcal{S}_{c,0}^{-}$.[7] When $c = c_0^*$, and only then, $\mathcal{U}_{c,0}^{+} = \mathcal{S}_{c,0}^{-}$.

The union of the one-dimensional trajectories $\mathcal{U}_{c,0}^{+}$ in $R^3$ over an interval of $c$ values around $c_0^*(a)$ is an invariant manifold for (6.13), called the "center-unstable manifold" at $(0,0,c_0^*(a))$ and denoted by $\mathcal{W}^{cu}(0,0,c_0^*(a))$.[8]

Similarly, at $(1,0,c_0^*(a))$ define a center-stable manifold $\mathcal{W}^{cs}(1,0,c_0^*(a))$, which consists of the union of all of the stable manifolds $\mathcal{S}_{c,0}^{-}$ of (6.9) at $(1,0,c)$, for $c$ in a neighborhood of $c_0^*(a)$. Then $\mathcal{W}^{cu}(0,0,c_0^*(a))$ and $\mathcal{W}^{cs}(1,0,c_0^*(a))$ intersect exactly along the heteroclinic orbit connecting $(0,0,c_0^*(a))$ to $(1,0,c_0^*(a))$. The basic transversality result in this stage of the argument is

**Lemma 6.8.** *The manifolds $\mathcal{W}^{cu}(0,0,c_0^*(a))$ and $\mathcal{W}^{cs}(1,0,c_0^*(a))$ intersect transversely along the heteroclinic orbit from $(0,0,c_0^*(a))$ to $(1,0,c_0^*(a))$. In other words, the tangent planes to these two-dimensional smooth manifolds at any point along this orbit are not identical.*

**Proof.** The proof in [**110**] uses differential forms. For a short elementary proof, recall from Chapter 2 that for (6.9), with $w = 0$, the slope of $\mathcal{U}_{0,c}^{+}$ at $(0,0)$ is an increasing function of $c$. It follows that for sufficiently small $\delta > 0$, if $v_0(c)$ is the value of $v$ on $\mathcal{U}_{0,c}^{+}$ when $u = \delta$ for the first time, then $\frac{dv_0(c)}{dc} \geq 0$. Now consider a fixed $u_0 \in (0,1)$, and assume that $\delta < u_0$. The intersection in $R^3$ of $\mathcal{W}^{cu}(0,0,c_0^*(a))$ with the plane $u = u_0$ includes a smooth curve $v = h(c)$, defined for $c$ near $c_0^*(a)$, and we wish to determine the sign of $h'(c_0^*(a))$.

Consider the initial value problem

$$\begin{aligned} \frac{dv}{du} &= c - \frac{f(u)}{v}, \\ v(\delta) &= v_0(c). \end{aligned}$$

[7]To maintain consistency with previous notation, in $\mathcal{S}_{c,0}^{-}$ the subscript 0 refers to $\varepsilon = 0$.

[8]By adopting $c$ for the wave speed, we cause some ambiguity of notation here. We hope the reader will understand that "$cu$" stands for "center unstable", while "$c$" is the wave speed parameter in the equations.

If $p = \frac{\partial v}{\partial c}$ then as long as $v > 0$,

$$\frac{dp}{du} = 1 + \frac{f(u)}{v^2} p,$$
$$p(\delta) = \frac{dv_0}{dc} \geq 0.$$

But this shows that $p(u_0) > 0$, and if $c = c_0^*(a)$ then $p(u_0) = h'(c_0^*(a))$. Hence $h'(c_0^*(a)) > 0$.

Now consider the intersection of $\mathcal{W}^{cs}(1, 0, c_0^*(a))$ with the same plane $u = u_0$. Say that near $c = c_0^*(a)$ this is a curve $k(c)$. From the linearization of (6.5) at $(1, 0)$ it is seen that the slope of $\mathcal{S}_{c,1}^-$ at this equilibrium point is a negative increasing function of $c$, which means that near to $u = 1$, $v$ is a decreasing function of $c$ on this manifold. Repeating the argument above shows that $k'(c_0^*(a)) < 0$, and this proves the lemma. □

A similar result is needed with $w = w^*$, again taking $\varepsilon = 0$. Let $u^- = f_R^{-1}(w^*)$ and $u^+ = f_L^{-1}(w^*)$. Then consider the union $\mathcal{W}^{cu}(u^-, 0, c_0^*(a))$ of the unstable manifolds $\mathcal{U}_{c,\varepsilon}^-$ at $(u^-, 0, c)$, in the $(u, v, c)$ coordinates, as $c$ varies in a neighborhood of $c_0^*(a)$. (Here $\mathcal{U}_{c,\varepsilon}^-$ is the left pointing branch of the unstable manifold.) Similarly, let $\mathcal{W}^{cs}(u^+, 0, c_0^*(a))$ be the union of the stable manifolds $\mathcal{S}_{c,\varepsilon}^+$ at $(u^+, 0, c)$, again as $c$ varies in a neighborhood around $c_0^*(a)$. Then a second lemma, with the same proof, asserts that these two smooth surfaces also intersect transversely, along the orbit found in Proposition 6.7.

One important additional technique is used in the existence proof in [**110**]. This result, called the "exchange lemma", enables one to understand better the details of the flow of the augmented system (6.8) near the "slow manifolds". These are the two surfaces $u = f_L^{-1}(w)$ and $u = f_R^{-1}(w)$ described earlier, considered in $R^4$, where they are three dimensional. In [**110**], use is made of differential forms to sketch the proof of the exchange lemma. An elementary proof, in the style of this book, was given by Brunovsky [**26**].

We will not give a precise statement of the exchange lemma here. Instead, we give a linear example. This result would be the one needed in our application to the FitzHugh-Nagumo equations if the functions $f_L^{-1}$ and $f_R^{-1}$ were linear. Unfortunately, they are not.

**6.3.3. Example of the exchange lemma.** Consider the system

$$\begin{aligned} x' &= (1+c)\,x, \\ y' &= -(1+c)\,y, \\ z' &= \varepsilon, \\ c' &= 0. \end{aligned} \tag{6.14}$$

For small $\delta > 0$, let

$$\Delta = \{(x, y, z, c) \mid x = y = 0,\ 0 \le z \le 1, -\delta \le c \le \delta\}$$

and

$$B = \left\{(x, y, z, c) \mid |x| < 1, |y| < 1, -\delta < z < 1 + \delta,\ -\frac{1}{2} < c < \frac{1}{2}\right\}.$$

Choosing $\delta < \frac{1}{2}$, $B$ is a neighborhood of $\Delta$.

For $\varepsilon = 0$, each point $(0, 0, z, c)$ is an equilibrium point of (6.14), and there are one-dimensional center-stable and center-unstable manifolds

$$\mathcal{U}^{cs}(0, 0, z, c) = \{(x, y, z, c) \ |x = 0\},$$
$$\mathcal{U}^{cu}(0, 0, z, c) = \{(x, y, z, c) \ |y = 0\}.$$

Let

$$\begin{aligned}\mathcal{V}^{cs} &= \bigcup \{\mathcal{U}^{cs}(0, 0, z, c) \mid -\delta < z < 1 + \delta, -\delta < c < \delta\} \\ &= \{(0, y, z, c) \mid -\delta < z < 1 + \delta, -\delta < c < \delta\}.\end{aligned}$$

Then $\mathcal{V}^{cs}$ does not depend on $\varepsilon$. For $\varepsilon = 0$, $\mathcal{V}^{cs}$ can be considered as the global center-stable manifold of the set of equilibrium points in $B$. It is three dimensional. Its projection onto $(x, y, z)$-space is a strip in the plane $x = 0$.

For small $\varepsilon > 0$, consider a curve $\Gamma$ of initial conditions $(x, y, z, c) = (\alpha, 1, 0, \alpha)$, where $-\alpha_0 < \alpha < \alpha_0$ for some small $\alpha_0$ independent of $\varepsilon$. This curve lies on the boundary $\partial B$ and intersects $\mathcal{V}^{cs}$ transversely at $(0, 1, 0, 0)$.[9] Its $c$ component ranges over an interval containing 0. If $0 < \alpha_0 < \frac{1}{2}$ then all solutions starting on $\Gamma$ immediately enter $B$. But eventually they exit from $B$ again, because $|x|$ grows beyond 1 or because $z$ increases beyond $1 + \delta$.

We also define a global center-unstable manifold when $\varepsilon = 0, c = 0$:

$$\mathcal{V}^{cu} = \bigcup \{\mathcal{U}^{cu}(0, 0, z, 0) \mid -\delta < z < 1 + \delta\}.$$

The manifold $\mathcal{V}^{cu}$ is two dimensional. Its projection onto $(x, y, z)$-space is a strip in the plane $y = 0$.

Fix $z_0 \in (0, 1)$ and choose $z_1 \in (z_0, 1]$. The point $p_1 = (1, 0, z_1, 0)$ is on $\mathcal{V}^{cu}$. A preliminary result is that for small $\varepsilon$ there is a point $y_\varepsilon$ on $\Gamma$ such that the corresponding exit point from $B$ is close to $p_1$. Here, we can verify this by using the exact solution. Let $t_1 = \frac{z_1}{\varepsilon}$. For a solution of (6.14) starting at $(\alpha, 1, 0, \alpha)$ on $\Gamma$, $z(t_1) = z_1$ and $x(t_1) = \alpha e^{(1+\alpha)\frac{z_1}{\varepsilon}}$. For each sufficiently small $\varepsilon > 0$ there is a unique $\alpha = \alpha_1 \in (0, \alpha_0)$ such that $x(t_1) = 1$. Further,

[9] See Exercise 6.4.

$|\alpha_1| = O\left(e^{-\frac{z_0}{\varepsilon}}\right)$ as $\varepsilon \to 0$.[10] Also, $y(t_1) = e^{-(1+\alpha_1)\frac{z_1}{\varepsilon}}$ is small, so that $\pi(t_1) := (x(t_1), y(t_1), z(t_1), \alpha_1)$ is close to $p_1$, and hence close to $\mathcal{V}^{cu}$.

The exchange lemma is stronger. Choose a small $\eta_1 > 0$ (which will depend on $\varepsilon$). Then consider the two-dimensional manifold $G_{\eta_1}$ of points

$$\left(\alpha e^{(1+\alpha)t}, e^{-(1+\alpha)t}, \varepsilon t, \alpha\right) \tag{6.15}$$

with $\alpha \in (\alpha_1 - \eta_1, \alpha_1 + \eta_1)$, $t \in (t_1 - \eta_1, t_1 + \eta_1)$. The exchange lemma says that for sufficiently small $\varepsilon$, each point $p$ of the smooth manifold $G_{\eta_1}$ is close to a point $q$ of the two-dimensional manifold $\mathcal{V}^{cu}$, and a pair of linearly independent unit tangent vectors to $G_{\eta_1}$ at $p$ is close to a pair of unit tangent vectors to $\mathcal{V}^{cu}$ at $q$. In other words, $G_{\eta_1}$ is "$C^1$ close" to $\mathcal{V}^{cu}$.

In our case, unit tangents to $\mathcal{V}^{cu}$ at $(1, 0, z_1, 0)$ are $n_1 = (1, 0, 0, 0)^T$ and $n_2 = (0, 0, 1, 0)^T$ while unit tangents to $G_{\eta_1}$ are in the directions of

$$\tau_1 = \left(\alpha (1+\alpha) e^{(1+\alpha)t}, -(1+\alpha) e^{-(1+\alpha)t}, \varepsilon, 0\right)^T,$$
$$\tau_2 = \left((1+\alpha t) e^{(1+\alpha)t}, -te^{-(1+\alpha)t}, 0, 1\right)^T.$$

(These are obtained by differentiating (6.15) with respect to $t$ and $\alpha$, respectively.) We will evaluate these tangents when $\alpha = \alpha_1$, $t = t_1$, in which case the solution exits $B$ at $\pi(t_1) = (1, \alpha_1, z_1, \alpha_1)$, a point exponentially close to $p_1$. Then, $\tau_1 = (1+\alpha_1, -(1+\alpha_1)\alpha_1, \varepsilon, 0)^T$ and $\tau_2 = \left(\frac{1+\alpha_1 t_1}{\alpha_1}, -\alpha_1 t_1, 0, 1\right)^T$. Since $|\alpha_1| = O\left(e^{-\frac{z_0}{\varepsilon}}\right)$ as $\varepsilon \to 0$, a unit vector in the direction of $\tau_1$ is exponentially close to $\nu_1 = \frac{1}{\sqrt{1+\varepsilon^2}}(1, 0, \varepsilon, 0)$ for small $\varepsilon$.[11] Also, a unit vector in the direction of $\tau_2$ is exponentially close to $\nu_2 = (1, 0, 0, 0)$. Note that as $\varepsilon \to 0$, $\nu_1 - \nu_2 = \left(O(\varepsilon^2), 0, \frac{\varepsilon}{\sqrt{1+\varepsilon^2}}, 0\right)$, and so a unit vector in the direction of $\nu_1 - \nu_2$ is of the form $\left(O(\varepsilon), 0, 1 + O(\varepsilon^2), 0\right)$.

The exchange lemma tells us that solutions starting on the curve $\Gamma$ close to $x = 0$ leave $B$ at about the prescribed level $z = z_1$ and going approximately parallel to the $(x, z)$-plane. As the solution continues on outside $B$, the range of $c$, now very small, will reassert itself and give a widening two-dimensional surface of trajectories. Transferring this idea to the FitzHugh-Nagumo system, we expect to be able to restrict $c$ even further and know a lot about the rest of the trajectory.

[10] This is our first use of "big $O$" notation. In this book, $f(\alpha) = O(g(\alpha))$ as $\alpha \to 0$ if there is a $K$ and an $\alpha_0 > 0$ such that $|f(\alpha)| \le K|g(\alpha)|$ for all $\alpha$ with $0 < |\alpha| < \alpha_0$. Also, $p(\alpha) = O(q(\alpha))$ as $\alpha \to \infty$ if $p\left(\frac{1}{\beta^2}\right) = O\left(q\left(\frac{1}{\beta^2}\right)\right)$ as $\beta \to 0$.

[11] That is, if $n_1$ is the unit vector then $||n_1 - \nu_1|| = O\left(e^{-\sigma/\varepsilon}\right)$ as $\varepsilon \to 0$ for some $\sigma > 0$.

**6.3.4. Completion of the proof (outline).** We will not repeat the complete proof in [**110**], which in any case may be a little over the line of classical techniques which we have set for ourselves in most of this book. The main complication is that the geometrical arguments are in four dimensions and so are not easy to picture. Our hope is that the reader will get the flavor of the approach and, if interested, find the complete proof more accessible as a result of going over our outline. The final steps are like what went before. Because the manifold $\mathcal{W}^{cu}(0,0,c_0^*(a))$ is transverse to $\mathcal{W}^{cs}(1,0,c_0^*(a))$, the exchange lemma implies that its extension up the right branch of $w=f(u)$ includes a region almost tangent to $\mathcal{W}^{cu}(u^-,0,c_0^*(a))$. Following the solutions from that region back to the left branch, $W^u$ is shown to intersect $\mathcal{W}^{cs}(u^+,0,c_0^*(a))$, also transversely, and another use of the exchange lemma brings the surface down the left branch to a transverse intersection (as $c$ varies) with $W^s$, and this completes the proof. See [**110**] for further details.

## 6.4. Existence of homoclinic orbits by shooting

The material in this section is adapted from [**86**] and can be read independently of Section 6.3. However it assumes familiarity with Sections 6.1 and 6.2. Turning again to the FitzHugh-Nagumo traveling wave system,

$$\begin{aligned} u' &= v, \\ v' &= cv - f(u) + w, \\ w' &= \tfrac{\varepsilon}{c}(u-\gamma w), \end{aligned} \tag{6.16}$$

we consider whether the bounded solutions found in Section 6.2 are actually homoclinic. To see that this is a challenging problem, consider the possibility that there is a periodic solution to the system. It might be a small periodic solution, near $(0,0,0)$. The analysis in Section 6.2 leaves open the possibility that a bounded solution $q_{c,\varepsilon}$ on $\mathcal{U}_{c,\varepsilon}^+$ tends to such a periodic solution and so is not homoclinic. However for the FitzHugh-Nagumo system, this possibility is eliminated by using a Lyapunov function which was discovered by Conley in the 1970s (unpublished notes).

Recall that $F(u)=\int_0^u f(s)\,ds$. In Section 2.4.2 we used the function $E(u,v)=\frac{1}{2}v^2+F(u)$ and saw that when $\varepsilon=0$, $w=0$, and $c>0$ this function is increasing, since then $\dot{E}=cv^2$. By experimenting with the equations, Conley found a modification of $E$ which is useful when $\varepsilon>0$. Let

$$G=\frac{1}{2}v^2+F(u)+\frac{1}{2}c^2u^2+\frac{c}{2\varepsilon}\left(c-\frac{\varepsilon\gamma}{c}\right)w^2-uw-cuv.$$

If $(u,v,w)=(u(s),v(s),w(s))$ is a solution of (6.2), then

$$\dot{G}=cuf(u)-\frac{\varepsilon}{c}u^2+\gamma^2c\left(\frac{\varepsilon}{c^2}-\frac{1}{\gamma}\right)w^2. \tag{6.17}$$

If $\frac{\varepsilon}{c^2} < \frac{1}{\gamma}$ and $u \le a$ then $\dot{G} < 0$ unless $u = w = 0$. Therefore, if $\frac{\varepsilon}{c^2} < \frac{1}{\gamma}$ then there is no periodic solution to (6.16) which remains entirely in the region $u \le a$.

Further, if $q = (u, v, w)$ is bounded and the solution satisfies $u(s) \le a$ for all large $s$, then $\lim_{s\to\infty} G(s)$ exists. Also, $\ddot{G}$ is bounded, implying that $\lim_{s\to\infty} \dot{G}(s) = 0$. (See Exercise 2.5(d).) Since we are considering a parameter range and solution such that all terms in $\dot{G}$ are negative for large $s$, $\lim_{s\to\infty} (u(s), w(s)) = (0,0)$, and from this it follows that $\lim_{s\to\infty} q(s) = (0,0,0)$. Hence, to prove Theorem 6.1 we have to show that for sufficiently small $\varepsilon > 0$ there are two values of $c$ such that $\frac{\varepsilon}{c^2} < \frac{1}{\gamma}$, $q_{c,\varepsilon} = (u_{c,\varepsilon}, v_{c,\varepsilon}, w_{c,\varepsilon})$ is bounded, and $u_{c,\varepsilon}(s) \le a$ for large $s$.

Next, we identify a specific region in parameter space where the solutions on $\mathcal{U}^+_{c,\varepsilon}$ tend to $+\infty$ in $u$ and $v$. As before, let $M = \max_{u>0} \frac{f(u)}{u}$.

**Lemma 6.9.** *If $M < \frac{\varepsilon}{c^2} < \frac{1}{\gamma}$ then* (6.2) *has no solution which is bounded other than $q = 0$.*

**Proof.** This proof also uses the function $G$. Note from (6.17) that in the parameter range of the lemma, $\dot{G} \le 0$ for all $(u, v, w)$. If the solution is bounded then both $\lim_{s\to\infty} G(s)$ and $\lim_{s\to-\infty} G(s)$ exist. Because $G$ is monotonic, the two limits are different unless $q = 0$. But in each case $\dot{G} \to 0$, and therefore $(u, w) \to (0,0)$. It easily follows that $v \to 0$ as well. Thus the limits at both $\pm\infty$ of $G$ are zero. This is a contradiction, showing that there are no nontrivial bounded solutions. □

**Lemma 6.10.** *There is an $\varepsilon_2 > 0$ such that if $c > 0$ and $\varepsilon \ge \varepsilon_2$, then $u$ and $v$ tend to $+\infty$ on $\mathcal{U}^+_{c,\varepsilon}$.*

**Proof.** This result follows from (6.3), (6.7), and Lemma 6.5. When $c > c_0^*(a)$, the conclusion follows from Lemma 6.5. When $0 < c \le c_0^*(a)$, choose $\varepsilon_2$ so that $\frac{\varepsilon_2}{c_0^*(a)}$ is large and follow the argument used for the last conclusion of the same lemma. □

Now let

$$\Omega = \left\{(c, \varepsilon) \mid c > 0 \text{ and } M < \frac{\varepsilon}{c^2} < \frac{1}{\gamma}\right\}.$$

**Corollary 6.11.** *If $(c, \varepsilon) \in \Omega$ then $u_{c,\varepsilon}$ and $v_{c,\varepsilon}$ tend to $+\infty$.*

**Proof.** This is implied by Lemma 6.10 for $(c, \varepsilon)$ in $\Omega$ with $\varepsilon \ge \varepsilon_2$. Everywhere in $\Omega$ the solutions on $\mathcal{U}^+_{c,\varepsilon}$ are unbounded, and so enter either $E^+$ or $E^-$. If there is a point $(\bar{c}, \bar{\varepsilon})$ in $\Omega$ such that $q_{\bar{c},\bar{\varepsilon}}$ enters $E^-$, then consider the curve $\mathcal{K} = \left\{(c, \varepsilon) \mid c \ge \bar{c}, \frac{\varepsilon}{c^2} = \frac{\bar{\varepsilon}}{\bar{c}^2}\right\}$, which is contained in $\Omega$. For $(c, \varepsilon) \in \mathcal{K}$ and $\varepsilon \ge \varepsilon_2$, $q_{c,\varepsilon}$ enters $E^+$. But as in Section 6.2, this implies that there is a point $(c, \varepsilon) \in \mathcal{K}$ such that $q_{c,\varepsilon}$ is bounded, contradicting Lemma 6.9. □

Now, following the procedure in Section 6.2, use Lemma 6.5 to choose $c_1 \in (0, c_0^*(a))$ and $\varepsilon_1 > 0$ so that if $0 < \varepsilon \leq \varepsilon_1$ then $v_{c_1,\varepsilon}$ has a unique zero, and at this zero $u_{\max} < u_{c_1,\varepsilon} < 1$. Further, $\varepsilon_1$ can be chosen so that after this zero $q_{c_1,\varepsilon}$ eventually enters $E^-$. Note from Lemma 6.5 that we can choose $c_1$ as close as we wish to $c_0^*(a)$, after which we choose $\varepsilon_1$.

Then consider an $\hat{\varepsilon}$ with $0 < \hat{\varepsilon} \leq \varepsilon_1$. If $c > c_0^*(a)$ or $(c, \hat{\varepsilon}) \in \Omega$, then $q_{c,\hat{\varepsilon}}$ enters $E^+$, while $q_{c_1,\hat{\varepsilon}}$ enters $E^-$. We will improve on the result in Section 6.2 by showing that if $\hat{\varepsilon}$ is sufficiently small then there are two values of $c$, one in $(c_1, c_0^*(a))$ and one in $(0, c_1)$, such that $\mathcal{U}_{c,\hat{\varepsilon}}$ is homoclinic.

**6.4.1. Existence of the fast wave.** Let $u_{\min}$ be the point in $(0, a)$ where $f(u)$ has its minimum, and let

(6.18)
$$\begin{aligned}\Lambda = \{c \in [c_1, \infty) \mid \text{ if } q_{c,\hat{\varepsilon}} = (u, v, w) \text{ then there are } t_2 > t_1 > 0 \\ \text{with } v > 0 \text{ on } (-\infty, t_1),\, v'(t_1) < 0,\ \ v < 0 \text{ on } (t_1, t_2], \\ u(t_2) = u_{\min}, \text{ and } u < u_{\min} \text{ for } t > t_2\}.\end{aligned}$$

Then $c_1 \in \Lambda$ and no $c \geq c_0^*(a)$ is in $\Lambda$. Let $c^* = \sup \Lambda$.

**Theorem 6.12.** *For sufficiently small $\hat{\varepsilon} > 0$, $\mathcal{U}^+_{c^*,\hat{\varepsilon}}$ is a homoclinic orbit.*

**Proof.** The proof proceeds by a number of lemmas. These gradually restrict the possible behavior of $q_{c^*,\varepsilon}$.

**Lemma 6.13.** *The following statements hold for each solution $q = q_{c,\hat{\varepsilon}}$.*

(i) *If $q$ enters $E^+$ then $c$ is not in the closure $\bar{\Lambda}$ of $\Lambda$.*

(ii) *If $c \in \bar{\Lambda}$ then $t_1$ and $t_2$ are defined ($t_2$ defined as the second point where $u = u_{\min}$), and $v(t_2) < 0$.*

(iii) *If $c \in \bar{\Lambda}$ then $v < 0$ on $(t_1, t_2]$.*

**Proof.** (i) This follows since $E^+$ is an open set.

(ii) If $t_1$ is not defined, then $v_{c,\varepsilon} \geq 0$ as long as the solution exists, and $q_{c,\varepsilon}$ enters $E^+$, contradicting (i). If $t_1$ is defined but $t_2$ is not, then $q_{c,\varepsilon}$ crosses $v = 0$ and does so again before reaching $u = u_{\min}$.[12] But solutions with $c \in \Lambda$ do not enter $v > 0$ in $(t_1, t_2)$, and so $c \notin \bar{\Lambda}$, a contradiction. Also, if $v(t_2) = 0$ then $v'(t_2) = w(t_2) - f(u(t_2)) > 0$, so that $q$ reenters $v > 0, u > u_{\min}$. Hence, nearby solutions either cross $u_{\min}$ twice in a neighborhood of $t_2$, and so enter $u < u_{\min}$ and then go back to $u > u_{\min}$, or else do not enter $u < u_{\min}$ at all near $t = t_2$ but do reenter $v > 0$, $u > u_{\min}$. Either case implies that $c$ is not in $\bar{\Lambda}$.

(iii) We showed that $v(t_2) < 0$. Suppose that there is a $\bar{t} \in (t_1, t_2)$ with $v(\bar{t}) = 0, v'(\bar{t}) = 0$. Then $w(\bar{t}) = f(u(\bar{t}))$ and $v''(\bar{t}) = w'(\bar{t}) =$

[12]Otherwise, $v \leq 0$ after $t_1$, and $t_2$ is defined.

$\frac{\varepsilon}{c}(u - \gamma f(u))$. By (6.3), $u - \gamma f(u) > 0$ for all $u > 0$, so that $q$ reenters $v > 0$. □

The following result is more delicate.

**Lemma 6.14.** *If $c \in \bar{\Lambda}$ and $q_{c,\hat{\varepsilon}}$ first crosses $v = 0$ at $t = t_1$, then $v > 0$ on $(-\infty, t_1)$.*

**Proof.** Since $E^+$ is open, conclusion (i) of Lemma 6.13 implies that $c_1 < c < c^*(a)$. Consider again the function $E(u,v) = \frac{1}{2}v^2 + F(u)$. For solutions of (6.2), $\dot{E} = cv^2 + vw$. Recall that $w > 0$ on $(-\infty, t_1)$ and, indeed, on every interval $(-\infty, T]$ where $u \geq 0$. It follows that on $(-\infty, t_1)$, with $(u,v,w) = q_{c,\hat{\varepsilon}}$, $(u,v)$ lies outside the curve $\mathcal{U}^+_{c,0}$ in the $(u,v)$-plane, and so the first zero of $v$ occurs with $u_{\max} < u < 1$.[13] Suppose that the lemma is false and this zero occurs at $\tau \in (0, t_1)$. Since $v \geq 0$ on $(0, t_1)$, we see that $v(\tau) = v'(\tau) = 0$ and $v''(\tau) = w'(\tau) > 0$. (Compare the proof of (iii) above.) For $t > \tau$, $v'' = cv' - f'(u)v + w' > 0$ as long as $v' > 0$, since $f'(u) < 0$ for $u > u_{\max}$ and $w' > 0$ as long as $u$ is increasing. Hence $v$ has no further zeros, and the solution enters $E^+$. By (i) above, $c \notin \bar{\Lambda}$. □

**Lemma 6.15.** *If $c \in \Lambda$ and $q$ is unbounded, then $c$ is not in $\partial\Lambda$.*

**Proof.** To see this, suppose that the solution $q_{c,\hat{\varepsilon}}$ enters $E^-$ at some $t_3 > t_2$. Then there is a $\delta > 0$ such that in $[t_2, t_2 + \delta]$, $v < 0$, while in $[t_2 + \delta, t_3 + \delta]$, $u < u_{\min}$. Lemma 6.14 and the definition of $\Lambda$ then imply that $c$ is in the interior of $\Lambda$. □

The next result is the most difficult, and its proof will be deferred to the end of this subsection. It is the nontangency condition for this problem. Since it is also used to obtain the slow wave, we do not assume that $c > c_1$.

**Lemma 6.16.** *If $\hat{\varepsilon}$ is sufficiently small and $\frac{\hat{\varepsilon}}{c^2} < M$, then there is no solution $(u,v,w)$ of (6.16) such that $w(0) > 0$ and for some $s_0 > 0$,*

$$
\begin{aligned}
u(s_0) &= u_{\min}, \\
v(s_0) &= 0, \\
v'(s_0) &\leq 0.
\end{aligned}
\tag{6.19}
$$

To prove Theorem 6.12 assuming Lemma 6.16, recall that $c^* = \sup \Lambda \geq c_1$, where $\Lambda$ is defined by (6.18). Let $q^* = (u^*, v^*, w^*) = q_{c^*,\hat{\varepsilon}}$. By (i) of Lemma 6.13, $u^*$ is bounded above. By (ii), $t_2$ exists and $v^*(t_2) < 0$. Also, $q^*$ does not reenter $u > u_{\min}$ after $t_2$. If $q^*$ is bounded then Conley's function $G$ shows that the orbit is homoclinic, so suppose that $q^*$ is unbounded and eventually enters $E^-$. By Lemma 6.15, $c^* \notin \Lambda$. By Lemma 6.13(iii) and

[13] The initial choice of $c_1$ guarantees that if $c_1 \leq c < c^*(a)$, then $\mathcal{U}_{c,0}$ intersects $v = 0$ in $(u_{\max}, 1)$. Now see Exercise 6.5.

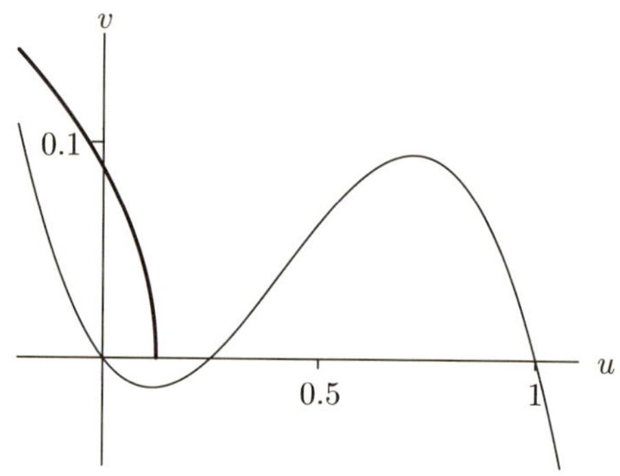


**Figure 6.3.** For the solution shown, $u(s_0) = u_{\min}$, $v(s_0) = 0$, and $v'(s_0) \le 0$.

Lemma 6.14, $v^* \neq 0$ on $(-\infty, t_1)$ and on $(t_1, t_2)$. Hence, there is an $s_0 > t_2$ such that $u^*(s_0) = u_{\min}$ and $v^*(s_0) = 0$. Also, $v^{*\prime}(s_0) = -f(u_{\min}) + w^*(s_0) \le 0$. Note from the definition of $q^*$ that $w^*(0) > 0$. If $\hat{\varepsilon}$ is sufficiently small then this contradicts Lemma 6.16, and so the orbit of $(u^*, v^*, w^*)$ is homoclinic. □

**Proof of Lemma 6.16.** See Figure 6.3 for a graph of a solution satisfying (6.19). The lemma implies that for this solution, $w \le 0$ for all $s < s_0$.

**Lemma 6.17.** *Suppose that $\bar{u}$ is some point in $(0, u_{\min})$. If a solution as described in Lemma 6.16 exists then there are $\sigma_0 \le \sigma_1 < \sigma_2 < s_0$ such that*

(i) $w(\sigma_0) = 0$ *and* $w < 0$ *on* $(\sigma_0, s_0]$,

(ii) $u(\sigma_1) = 0, u > 0$ *on* $(\sigma_1, s_0]$,

(iii) $u(\sigma_2) = \bar{u}$ *and* $u > \bar{u}$ *on* $(\sigma_2, s_0]$,

(iv) $w(s_0) \le f_{\min}$, $w < f_{\min}$ *on* $[\sigma_1, s_0)$,

(v) $v > 0$ *on* $[\sigma_1, s_0)$,

(vi) $v' < 0$ *on* $[\sigma_1, s_0)$.

**Proof.** Since $w(0) > 0$ and $w(s_0) < 0$, there is some largest $s < s_0$, say $s = \sigma_0$, where $w = 0$. Also, $w'(\sigma_0) \le 0$ so that $u(\sigma_0) \le 0$. Since $u(s_0) = u_{\min} > 0$, there is a largest $s < s_0$, say $s = \sigma_1 \ge \sigma_0$, where $u = 0$. Further, there is a largest $s \in (\sigma_1, s_0)$, say $s = \sigma_2$, where $u = \bar{u}$. These choices lead easily to (i)–(iv). For (v), observe that if $v \le 0$ at $\tau \in [\sigma_1, s_0)$ then from (iv), $v' < 0$ on $(\tau, s_0)$, making (6.19) impossible. Similarly, for (vi) we note that if $v' \ge 0$ somewhere in $[\sigma_1, s_0)$ then $v'' > 0$, because $w' > 0$. This again rules out (6.19). □

Since $f_{\min} - f(\bar{u}) < 0$, there is an $\alpha > 0$ such that

$$\left(c + \frac{f_{\min} - f(\bar{u})}{\alpha}\right)\bar{u} < -\alpha \tag{6.20}$$

for every $c \in (0, c_0^*(a)]$. We distinguish two cases: (a) $0 < v(\sigma_1) < \alpha$ and (b) $v(\sigma_1) \geq \alpha$.

If (a) holds then property (vi) above implies that $0 < v < \alpha$ on $[\sigma_1, \sigma_2]$. We can express $v$ as a function of $u$ as long as $v > 0$. Also,

$$\frac{dv}{du} = c + \frac{w - f(u)}{v} \leq c + \frac{f_{\min} - f(\bar{u})}{\alpha}$$

on $(\sigma_1, \sigma_2)$, with the last inequality holding because in this interval $w < f_{\min}$ and $f(u) > f(\bar{u})$. Since $u(\sigma_1) = 0$ and $u(\sigma_2) = \bar{u}$, (6.20) implies that $v(\bar{u}) \leq \alpha + \left(c + \frac{f_{\min} - f(\bar{u})}{\alpha}\right)\bar{u} < 0$, contradicting (v) above.

Now assume (b). From (vi), $v'(\sigma_1) < 0$, while $v'(s) > 0$ for large negative $s$. Let $\sigma_{-1}$ be the largest value of $s < \sigma_1$ where $v'(s) = 0$. Then $v''(\sigma_{-1}) \leq 0$. A contradiction will be obtained by estimating $v''(\sigma_{-1})$ for small $\hat{\varepsilon}$.

From (b), $v(\sigma_1) \geq \alpha$, and $v' < 0$ on $(\sigma_{-1}, \sigma_1)$, so that $v(\sigma_{-1}) \geq \alpha$. Since $v(\sigma_1) \geq 0$ and since $v' < 0$ on $(\sigma_{-1}, \sigma_1)$, $v > 0$ on $(\sigma_{-1}, \sigma_1)$. Since $u(\sigma_1) = 0$, $u(\sigma_{-1}) < 0$. Note that $f'(u) \leq f'(0) = -a$ for $u \leq 0$. (All that is needed is that $f'(u) < 0$ if $u \leq 0$.) Therefore, at $\sigma_{-1}$,

$$v'' = cv' + \frac{\hat{\varepsilon}}{c}(u - \gamma w) - f'(u)v \geq \frac{\hat{\varepsilon}}{c}(u - \gamma w) + \alpha a.$$

We obtained bounds on $u$ and $\gamma w$ in the proof that there were bounded solutions. In particular, from the rectangle $H$ constructed there (proof of Theorem 6.6), we see that $u(\sigma_{-1}) \geq u_L$ and $w(\sigma_{-1}) \leq \frac{u_R}{\gamma}$. Therefore, $v''(\sigma_{-1}) > 0$ if $\frac{\hat{\varepsilon}}{c} < \frac{a\alpha}{u_R - u_L}$. The hypotheses of Lemma 6.16 imply that $\frac{\hat{\varepsilon}}{c} \leq \sqrt{\hat{\varepsilon}M}$, so that $v''(\sigma_{-1}) > 0$ if $\hat{\varepsilon}$ is sufficiently small.

This contradiction completes the proofs of Lemma 6.16 and Theorem 6.12. □

We have therefore given an existence proof for the "fast wave", with speed $c^*$ close to $c_0^*(a)$. We now consider the existence of a wave with speed $c_* \in (0, c_1)$.

**6.4.2. Existence of the slow wave.** The technique here is a little different, because we cannot prove the conclusion of Lemma 6.14 for $c < c_1$. In fact, this result is false in this region. We make the following additional assumption on $\gamma$:

$$\max_{0\leq u\leq 1} f(u) < \frac{u_{\min}}{\gamma}. \tag{6.21}$$

With more work, this can be relaxed.

Recall the properties of the solution $q_{c1,\varepsilon}$ introduced in Section 6.2. If $(u,v,w) = q_{c1,\varepsilon}$ then $v$ has exactly one zero, after which $v$ is negative, $u$ decreases to below zero, and the solution enters $E^-$. Also recall that for every $(c,\varepsilon)$, $u_{c,\varepsilon}(0) = a > u_{\min}$. In addition to $\Lambda$ we consider

$$\Sigma = \{c > 0 \mid \text{if } (u,v,w) = q_{c,\hat{\varepsilon}} \text{ then there is a } t_1 \text{ such that } u(t_1) = u_{\min}, \\ v(t_1) < 0,\ u > u_{\min} \text{ on } (0,t_1), \text{ and after } t_1,\ v < 0 \text{ at least until } u = 0\}$$

and

$$\Sigma_1 = \{c \in \Sigma \mid \text{ after } q_{c,\hat{\varepsilon}} \text{ crosses } u = u_{\min} \text{ with } v < 0, \\ u \text{ remains in } u < u_{\min}\}.$$

Note that $\Sigma_1 = \Sigma \cap \Lambda$. In particular, $(c_1,\hat{\varepsilon}) \in \Sigma_1$.

We saw in Lemma 6.5 that there is a $c_2(\hat{\varepsilon}) > 0$ such that if $0 < c < c_2(\hat{\varepsilon})$ then on $\mathcal{U}_{c,\hat{\varepsilon}}$, $v > 0$ and $q$ enters $E^+$. Hence the interval $(0, c_2(\hat{\varepsilon}))$ does not intersect $\Sigma$. Let $c_3 = \sup\{c < c_1 \mid (c,\hat{\varepsilon}) \notin \Sigma\}$.

**Lemma 6.18.** *There exist numbers $t_1$ and $t_2$ such that $t_1 < t_2$ and*

(a) (i) $u_{c3,\hat{\varepsilon}}(t_1) = u_{\min}$,
   (ii) $u_{c3,\hat{\varepsilon}} > u_{\min}$ on $(0,t_1)$,
   (iii) $v_{c3,\hat{\varepsilon}}(t_1) < 0$,

(b) $u_{c3,\hat{\varepsilon}}(t_2) = 0$, $v_{c3,\hat{\varepsilon}} < 0$ on $(t_1,t_2)$,

(c) $v_{c3,\hat{\varepsilon}}(t_2) = 0$ and $w_{c3,\hat{\varepsilon}}(t_2) > 0$.

**Proof.** The solution $q_{c3,\hat{\varepsilon}}$ does not enter $E^+$ before $u_{c3,\hat{\varepsilon}} = 0$, for otherwise, nearby solutions could not reach $u = 0$ and so nearby $c$ could not be in $\Sigma$. Hence $(u_{c3,\hat{\varepsilon}}, w_{c3,\hat{\varepsilon}})$ remains in the rectangle $H$ constructed in Section 6.2 at least until a first $t_2$ such that $u_{c3,\hat{\varepsilon}}(t_2) = 0$.

We now show that $q_{c3,\hat{\varepsilon}}$ does not remain in $u > u_{\min}$ for all $t > 0$. If, on the contrary, $u_{c3,\hat{\varepsilon}}(t) > u_{\min}$ for all large $t$, then from (6.16) it follows that $\liminf_{t\to\infty} \gamma w(t) \geq u_{\min}$. From (6.21) and the equation for $v'$ we see that for large $t$, $v' > 0$ when $v > 0$, and $q_{c3,\hat{\varepsilon}}$ enters $E^+$ without reaching $u = 0$, a contradiction. Hence $t_1$ exists such that (a)(i) and (a)(ii) are satisfied.

Now suppose that for some $t \geq t_1$, $v_{c3,\hat{\varepsilon}}(t) = 0$ and $u > 0$ on $[t_1,t]$. Then $w_{c3,\hat{\varepsilon}}(t) > 0$ and $f(u_{c3,\hat{\varepsilon}}(t)) < 0$, and so $v'_{c3,\hat{\varepsilon}} > 0$. Therefore $v_{c3,\hat{\varepsilon}}$ becomes positive before $u_{c3,\hat{\varepsilon}} = 0$, and $c_3$ could not be a boundary point for $\Sigma$. Hence (a)(iii) is satisfied and a $t_2$ satisfying (b) must exist. If $v_{c3,\hat{\varepsilon}}(t_2) < 0$ then once again $c_3$ could not be on the boundary of $\Sigma$. The positivity of

$w_{c_3,\hat{\varepsilon}}(t_2)$ follows because $u_{c_3,\hat{\varepsilon}} > 0$ on $(-\infty, t_2)$, completing the proof of Lemma 6.18. □

This result implies that $c_3$ is not in $\Sigma$, since at the first intersection of $q_{c_3,\hat{\varepsilon}}$ with $u = 0$ we also have $v = 0$. But the entire interval $(c_3, c_1)$ does lie in $\Sigma$. We will now show that points near $c_3$ in this interval do not lie in $\Sigma_1$.

To see this, note first that because $w_{c_3,\hat{\varepsilon}}(t_2) > 0$, $v'_{c_3,\hat{\varepsilon}}(t_2)$ is positive and $q_{c_3,\hat{\varepsilon}}$ reenters $u > 0, v > 0, w > 0$, forcing $u$ to increase beyond $u_{\min}$. Therefore, if $c > c_3$ and $c - c_3$ is sufficiently small, then $q_{c,\hat{\varepsilon}}$ must also reenter the region $u > u_{\min}$. Such values of $c$ lie in $\Sigma$ but not in $\Sigma_1$.

On the other hand, $c_1 \in \Sigma_1$. Let $c_* = \inf\{c > c_3 | \ (c, \hat{\varepsilon}) \in \Sigma_1\}$. We investigate the behavior of $q_{c_*,\hat{\varepsilon}}$. We see the following:

(a) $q_{c_*,\varepsilon}$ does not enter $E^+$, for otherwise, nearby solutions would also, and so nearby $c$ could not be in $\Sigma_1$.

(b) Since $c_* \in \Sigma$, choose $t_1$ as in the definition of $\Sigma$, as the first point where $u_{c^*,\hat{\varepsilon}} = u_{\min}$ with $v < 0$. For $t > t_1$, $q_{c_*,\hat{\varepsilon}}$ cannot enter $E^-$ before first reentering $u \geq u_{\min}$, for otherwise, nearby points would all be in $\Sigma_1$. Also, $q_{c_*,\hat{\varepsilon}}$ cannot enter $u > u_{\min}$ after $t_1$, for otherwise nearby points could not be in $\Sigma_1$.

It follows that if $\mathcal{U}^+_{c_*,\hat{\varepsilon}}$ is not homoclinic, then after $t_1$ there must be a first point where $u_{c^*,\hat{\varepsilon}} = u_{\min}$, $v_{c^*,\varepsilon} = 0$, and $v'_{c^*,\hat{\varepsilon}} \leq 0$. This possibility is eliminated by Lemma 6.16. Hence, $\mathcal{U}^+_{c_*,\hat{\varepsilon}}$ is homoclinic, and together with Theorem 6.12, this completes the proof of Theorem 6.1.

## 6.5. Advantages of the two methods

Geometric perturbation theory is a general method which extends to many other boundary value problems of singular perturbation type, at least formally. Constructing singular solutions is usually not difficult, and this alone gives important structure to the problem. From then on, as far as we know, each case is different and requires its own special techniques to prove the needed transverse intersections, as in Lemma 6.8.

In the case of the FitzHugh-Nagumo equations, however, there are relatively few technical difficulties particular to the problem, so that once Fenichel's theories and the exchange lemma have been proved, this method is fairly quick. A singular solution is even more accessible.

As pointed out earlier, the shooting method requires special attention to the detailed behavior of solutions at several spots, most notably in the proof of Lemma 6.16. However, there is a return for the effort. First, the method gives both the fast and slow waves at once, for each $a$ in $\left(0, \frac{1}{2}\right)$.[14] Also, the proof by shooting that there are at least two waves which are bounded (Theorem 6.6) is relatively simple.

But perhaps of more interest is a result illustrating how sometimes shooting methods give results which in some sense are "more global" than other perturbation techniques. For any problem which we have seen there seems to be no possibility of giving an estimate on the size of $\varepsilon$ using geometric perturbation, but with shooting, this sometimes can be done.[15] Our presentation above did not give such an estimate, but in [**86**] it was proved, roughly, that for the fast wave, $\varepsilon$ can be at least as large as $\frac{1}{2}a^2$, as both $\gamma$ and the "threshold" $a$ approach zero in a certain way. It turns out that this allows the existence proof by shooting to extend into the region where the stable eigenvalues of the linearization of (6.2) around $(0,0,0)$ are complex.

This fact, in turn, brings into play what is called "Shilnikov theory" from dynamical systems [**214**]. Shilnikov theory is one approach to "chaos", based on the existence of homoclinic orbits with oscillations on the stable manifold $\mathcal{S}_{c,\varepsilon}$. Using this, it follows that for the FitzHugh-Nagumo equations with $\varepsilon \simeq \frac{1}{2}a^2$ and $\gamma$ and $a$ small, there are infinitely many homoclinic orbits and many other bounded orbits with complicated structure [**86**]. We will see such behavior elsewhere in this book, obtained by other shooting methods, but will not go further into Shilnikov's results. See [**81**] for a description.

In summary, if one is mainly interested in the existence of the fast FitzHugh-Nagumo pulse and is willing either to skip the technical details of the proof or learn geometric perturbation theory, then that approach is to be recommended. On the other hand, for a proof which covers a wider parameter range, thereby revealing some interesting new properties of solutions, shooting has proved useful.

In the next chapter we consider another singular perturbation problem, where the ode's involved are nonautonomous. The main goal is to give a rigorous asymptotic expansion of the solution to this problem. Nonautonomous equations sometimes seem an unnatural setting for dynamical systems methods, and so while we appreciate the geometrical insight obtained from such methods, we feel that the ratio of results obtained to effort expended in this case is on the side of the classical approach.

---

[14]Krupa, Sandstede, and Szmolyan [**128**] also showed that there are two waves, using dynamical systems techniques, but only locally, that is, for $a$ sufficiently close to $\frac{1}{2}$.

[15]An extreme example is given in the existence and uniqueness theorem in the next chapter. See also remarks at the end of Chapter 15.

## 6.6. Exercises

**Exercise 6.1.** (Routine; needed later in chapter) Complete the proof of Proposition 6.4. For (A), start by analyzing the characteristic polynomial of the linearized system. For (B), be sure to check points where the vector field is tangent to $E^+$ or $E^-$. Recall (6.3).

**Exercise 6.2.** Referring to (6.5), is the set of values of $c$ which are positive, and have $v_c(t) > 0$ as long as the solution exists, an open set? What about the same set for (6.2)? In either case, what about the set of $c$ such that $u_c > 0$ as long as the solution exists?

**Exercise 6.3.** Prove that $w^*$ as found in Proposition 6.7 exists and is unique. Hint: This proof is related to the proof in Chapter 2 that $c_0^*(a)$ is unique. Therefore, you may want to start with Exercise 2.8.

**Exercise 6.4.** (Needed in chapter) Show that the tangent vector to the curve $\Gamma$ at $(0,1,0,0)$ is transverse to $\mathcal{V}^{cs}$.

**Exercise 6.5.** (Needed in chapter) The proofs of Lemmas 6.5 and 6.14 both require that immediately to the right of $(0,0)$ the projection of $\mathcal{U}_{c,\varepsilon}^+$ onto the $(u,v)$-plane is above the projection of $\mathcal{U}_{c,0}^+$. To analyze this, let $\lambda_{c,\varepsilon}$ denote the positive eigenvalue of the linearization of (6.2) at $(0,0,0)$. We need to show that $\lambda_{c,\varepsilon} > \lambda_{c,0}$. Do this by considering the characteristic polynomial $p_{c,\varepsilon}(\lambda)$ of the linearized matrix and showing that $\frac{\partial p_{c,\varepsilon}}{\partial \varepsilon}$ evaluated at the positive root of $p_{c,\varepsilon}$ is negative. Hint: Show that $\frac{\partial p_{c,\varepsilon}}{\partial \varepsilon} < 0$ at $\lambda = 0$ and that if $\lambda > 0$ and $\frac{\partial p_{c,\varepsilon}}{\partial \varepsilon} = 0$ then $p > 0$.

Chapter 7

# Singular perturbation problems—rigorous matching

## 7.1. Introduction to the method of matched asymptotic expansions

We begin this chapter with a brief introduction to the method of matched asymptotic expansions, illustrating this technique on a simple linear problem.[1] Then we consider a model problem proposed by Kaplun and Lagerstrom [**118**] to show some of the difficulties which arise in nonlinear problems and how to deal with them. This model problem has been considered by a number of authors, more recent examples including [**194**] and [**195**], where it was studied using methods of geometric perturbation theory. We will outline this approach and then follow [**93**] to show how classical methods can be used in this area. The method involves converting the problem to an integral equation which is then solved by an iterative process. The chapter ends with two shorter sections. The first describes another model problem to which the integral equation method has been applied. The second short section discusses a method of "blowing up" a degenerate singularity in an autonomous system which arose in a dynamical systems approach to the Kaplun-Lagerstrom problem and is of independent interest.

The method of matched asymptotic expansions is a technique developed by applied mathematicians to handle singular perturbation problems. See

[1]If you are unfamiliar with singularly perturbed boundary value problems, you may want to start by doing Exercise 7.1. Also see footnote 2 in Chapter 6.

[**245**], [**183**], or [**99**] for general presentations. Consider, for example, the following simple boundary value problem [**99**]:

$$\varepsilon y'' + y' = 2, \tag{7.1}$$

$$y(0) = 0, \quad y(1) = 1. \tag{7.2}$$

The differential equation is easily solved, and so we can find an exact solution of the boundary value problem if a solution exists. But before doing so, we point out the features which make this a problem in singular perturbations.

First, there is a parameter $\varepsilon$, and our interest is the behavior of the solution for small positive values of this parameter. Secondly, if $\varepsilon = 0$ then the ode is first order, and since there are two boundary conditions, we do not expect to be able to solve the resulting boundary value problem. A simple integration verifies that this "reduced" problem,

$$\begin{gathered} y' = 2, \\ y(0) = 0, \quad y(1) = 1, \end{gathered} \tag{7.3}$$

has no solution. It is this property which usually characterizes singular perturbation problems.[2] But notice that if the second boundary condition in (7.3) is changed to $y(1) = 2$ then there is a unique solution, so it may not be obvious whether a problem is really singular in this sense. Later in this chapter we will study a problem where the reduced problem does have a solution which satisfies both of the boundary conditions and which approximates the exact solution for small nonzero $\varepsilon$, and yet this is still a singularly perturbed problem, because when one attempts to find a more accurate approximation, surprising features are revealed.

To illustrate further the main properties of most singular perturbation problems, consider the exact solution of (7.1)–(7.2), which is easily determined to be

$$y(x) = 2x + \frac{e^{-\frac{x}{\varepsilon}} - 1}{1 - e^{-\frac{1}{\varepsilon}}}. \tag{7.4}$$

The graph of this solution for $\varepsilon = .01$ is shown in Figure 7.1.

The region of steep slope seen near $x = 0$ is called a "boundary layer" and immediately indicates a problem of some interest, however simple the actual solution is.

[2]It is not necessary for the small parameter to multiply the highest derivative in the equation, resulting in a lower order ode when this term is dropped. Singular perturbation problems can arise even when setting the small parameter equal to zero does not reduce the order of the equation. These are most commonly boundary value problems on an infinite or semi-infinite interval.

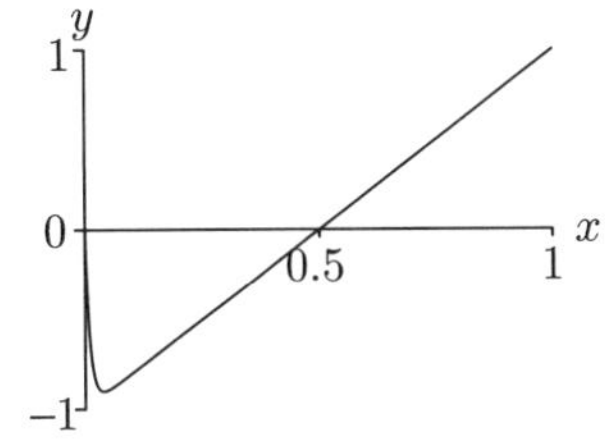


**Figure 7.1**

We continue our discussion of this problem by ignoring the fact that we have solved it exactly. Singular boundary value problems frequently arise for equations where no exact solution has been found. Numerical calculations may be difficult or impossible.[3] A standard perturbation technique in this situation is to try to expand the solution in powers of $\varepsilon$, setting

$$y(x) = y_0(x) + \varepsilon y_1(x) + \varepsilon^2 y_2(x) + \cdots.$$

In some cases it is possible to find a solution in terms of a convergent series. In more interesting cases, however, one finds that the resulting series do not converge. When this happens, one may invoke the theory of asymptotic series [**60**] to find information about the solution.

We do not have space here to give an exposition of this theory, which in any case is widely available. Instead, we will simply try this method on our model problem (7.1) and see what happens. We first address the question of how to find the functions $y_i(x)$.

We will assume that the $y_i$ are all smooth functions and that

$$y'(x) = \sum_{i=0}^{\infty} \varepsilon^i y_i'(x), \qquad y''(x) = \sum_{i=0}^{\infty} \varepsilon^i y_i''(x).$$

Significant questions about the convergence of these series have been ignored. Plunging ahead, by substituting into (7.1), leads to

$$\sum_{i=0}^{\infty} \varepsilon^{i+1} y_i''(x) + \sum_{i=0}^{\infty} \varepsilon^i y_i'(x) = 2.$$

This equation holds for all small values of $\varepsilon$ only if, for each $j$, the coefficient of $\varepsilon^j$ on the left is equal to its coefficient on the right. Here, the coefficient of $\varepsilon^j$ on the right is zero except for $j = 0$.

[3]This is not the case with our model equation, however. A computer solution is easy to find, using numerical shooting.

This procedure yields an infinite system of ordinary differential equations:

$$y_0' = 2,$$
$$y_i' = -y_{i-1}'', \quad i \geq 1.$$

Since the boundary conditions do not depend on $\varepsilon$, they are most easily satisfied for all $\varepsilon$ if they are satisfied by $y_0$ and for $i \geq 1$, $y_i$ satisfies homogeneous boundary conditions. Thus,

$$y_0(0) = 0, \qquad y_0(1) = 1,$$
$$y_i(0) = y_i(1) = 0 \quad \text{for } i \geq 1.$$

We are at the same impasse as before, with $y_0(x) = 2x + c$ for some $c$, so that only one boundary condition can hold. Furthermore, all of the other $y_i$ are forced to be identically zero. Apparently the problem cannot be solved by this method.

This difficulty is characteristic of singular perturbation problems. To make further progress in this specific instance, we cheat and examine the graph of our exact solution more carefully. It looks like $y = 2x - 1$, except in a small region close to 0. Hence we are led to choose $c = -1$ in the solution $y_0$. The resulting approximation, which appears valid in most of the interval $(0, 1]$ and does not show any dramatic change (a steep rise or fall) in this interval, is called the "outer" solution of (7.1)–(7.2).

Were it not for our knowledge of the exact solution, we would have a choice as to which boundary condition $y_0$ should satisfy. It turns out that for this particular problem a consistent argument results if $c$ is chosen so that $y_0(1) = 1$ but not if it is chosen so that $y_0(0) = 0$. (See Exercise 7.2.) Usually one cannot tell in advance which boundary condition the outer solution should satisfy.

Now we need a family of approximate solutions in a small interval $[0, h(\varepsilon)]$ which satisfy the boundary condition at 0, one of which can somehow be matched to $y_0$ to give a uniform approximation to the actual solution $y$ on $[0, 1]$. This family is found by choosing a new scale on the horizontal axis, in which the region close to zero is expanded. Let $\zeta = \frac{x}{\varepsilon}$ and $y(x) = z(\zeta) = z\left(\frac{x}{\varepsilon}\right)$. Then $z$ satisfies

$$\frac{d^2 z}{d\zeta^2} + \frac{dz}{d\zeta} = 2\varepsilon. \tag{7.5}$$

Since $z$ is to be used near $x = 0$, we require that $z(0) = 0$. Once again ignore the fact that we are able to solve this problem exactly, and expand in powers of $\varepsilon$, to get

$$z = z_0 + \varepsilon z_1 + \varepsilon^2 z_2 + \cdots.$$

Substitute this expression into (7.5), giving

$$z_0'' + z_0' = 0, \quad z_1'' + z_1' = 2, \quad z_i'' + z_i' = 0, \quad i \geq 2. \tag{7.6}$$

Also, $z_i(0) = 0$ for all $i$.

These conditions yield

$$z_0(\zeta) = c\left(e^{-\zeta} - 1\right), \tag{7.7}$$

$$\frac{dz_0}{d\zeta} = -ce^{-\zeta}, \tag{7.8}$$

and

$$z_1(\zeta) = 2\zeta + d\left(e^{-\zeta} - 1\right)$$

for some new constants $c$ and $d$. Note the nonuniqueness of these solutions, since the initial conditions are satisfied for every $c$ and $d$. Further, set $z_i = 0$ for $i > 1$. We then have what is called the "inner" solution of (7.1)–(7.2),

$$\bar{z}\left(\frac{x}{\varepsilon}\right) = z_0 + \varepsilon z_1 = c\left(e^{-\frac{x}{\varepsilon}} - 1\right) + 2x + d\varepsilon\left(e^{-\frac{x}{\varepsilon}} - 1\right).$$

In this example $\bar{z}$ happens to be an exact solution to (7.5) satisfying $\bar{z}(0) = 0$, for each $c$ and $d$, but we continue to ignore this. The oversupply of constants will be taken care of by matching with $y_0$.

Choose $Z > 0$. Since $\bar{z}$ is a continuous function of $\varepsilon$, we expect for each $c$ and $d$ that whether or not it is an exact solution, $\bar{z}$ tends as $\varepsilon \to 0$ to a solution $z$ of (7.5) on the interval $0 \leq \zeta \leq Z$. That is, in terms of $x$ we expect that

$$\lim_{\varepsilon \to 0} \max_{0 \leq x \leq \varepsilon Z} \left|\bar{z}\left(\frac{x}{\varepsilon}\right) - z\left(\frac{x}{\varepsilon}\right)\right| = 0.$$

The "outer" solution, $y_0 = 2x - 1$, may be expected to be a good approximation to $y$ on intervals of the form $\alpha \leq x \leq 1$, because $y_0'' = 0$. Unfortunately, the intervals $[0, \varepsilon Z]$ and $[\alpha, 1]$ do not overlap if $\varepsilon$ is small. We must hope, and assume, that the situation is better and that the intervals of validity of the two solutions are larger.

In discussing how far to extend these intervals, we will assume that $Z = 1$. In matching, one generally wants the two solutions to agree on overlapping intervals, as we will discuss further on. At this stage we somewhat arbitrarily match only at one point, $x = \sqrt{\varepsilon}$. In this case it is clearly necessary to show that there is matching in the derivatives at that point as well.

Thus, we will assume that $\bar{z}$ is close to the actual solution on the interval $0 \leq \zeta \leq \frac{1}{\sqrt{\varepsilon}}$, or $0 \leq x \leq \sqrt{\varepsilon}$, while $y_0$ is a valid approximation on $\sqrt{\varepsilon} \leq$

$x \leq 1$. We then choose the constants $c$ and $d$ so that $\left|\bar{z}\left(\frac{1}{\sqrt{\varepsilon}}\right) - y_0\left(\sqrt{\varepsilon}\right)\right|$ and $\left|\frac{1}{\varepsilon}\bar{z}'\left(\frac{1}{\sqrt{\varepsilon}}\right) - y_0'\left(\sqrt{\varepsilon}\right)\right|$ are as small as possible. Our goal is to have these expressions bounded by a constant times $\varepsilon^n$ as $\varepsilon \to 0$, for every integer $n$. If this holds then we will use "big $O$" notation and write that $z\left(\frac{1}{\sqrt{\varepsilon}}\right) - y_0\left(\sqrt{\varepsilon}\right) = O\left(\varepsilon^n\right)$ as $\varepsilon \to 0$, and a similar expression for the derivatives.

Substitute $x = \sqrt{\varepsilon}$, $\zeta = \frac{1}{\sqrt{\varepsilon}}$, to get

$$\begin{aligned} y_0\left(\sqrt{\varepsilon}\right) &= 2\sqrt{\varepsilon} - 1, \\ \bar{z}\left(\frac{1}{\sqrt{\varepsilon}}\right) &= c\left(e^{-\frac{\sqrt{\varepsilon}}{\varepsilon}} - 1\right) + 2\sqrt{\varepsilon} + d\varepsilon\left(e^{-\frac{\sqrt{\varepsilon}}{\varepsilon}} - 1\right). \end{aligned}$$

We see that if $c = 1$ and $d = 0$, and only then, $\bar{z}\left(\frac{1}{\sqrt{\varepsilon}}\right) - y_0\left(\sqrt{\varepsilon}\right) = O\left(e^{-\frac{1}{\sqrt{\varepsilon}}}\right)$ as $\varepsilon \to 0$. The derivatives $\frac{1}{\varepsilon}\bar{z}'\left(\frac{x}{\varepsilon}\right)$ and $y_0'(x)$ satisfy $\frac{1}{\varepsilon}\bar{z}'\left(\frac{1}{\sqrt{\varepsilon}}\right) - y_0'\left(\sqrt{\varepsilon}\right) = O\left(\frac{e^{-\frac{1}{\sqrt{\varepsilon}}}}{\varepsilon}\right)$. Thus, the matching condition determines the unknown coefficients in the approximate solution in such a way that the difference between the inner and outer solutions at the matching point $x = \sqrt{\varepsilon}$ and the difference between their derivatives are both smaller than every power of $\varepsilon$ as $\varepsilon \to 0$.

For this linear problem we get a solution which is valid in the entire interval $[0, 1]$ by adding the two expansions and subtracting the lowest order term, here $2x - 1$, since it has been added twice. Let

$$\begin{aligned} y_u(x) &= y_0(x) + \bar{z}\left(\frac{x}{\varepsilon}\right) - (2x - 1) \\ &= 2x - 1 + e^{-\frac{x}{\varepsilon}}. \end{aligned}$$

Note that $y_u$ satisfies the left boundary condition exactly but is off by $e^{-\frac{1}{\varepsilon}}$ at $x = 1$. Considering again the exact solution,

$$y(x) = 2x + \frac{e^{-\frac{x}{\varepsilon}} - 1}{1 - e^{-\frac{1}{\varepsilon}}} = y_u(x) + O\left(e^{-\frac{1}{\varepsilon}}\right)$$

as $\varepsilon \to 0$, uniformly in $0 \leq x \leq 1$. Hence, our approximate solution from matching is correct to order $e^{-\frac{1}{\varepsilon}}$, a better result than our original goal of $O\left(\varepsilon^n\right)$ for every $n$.

Finally, it is of interest to determine more precisely the regions of validity of the inner and outer expansions. Usually these are difficult to find, which is why most applications of this method are not rigorously justified. In this simple case, both expansions are known completely, and we also know the exact solution, so that it is easy to find intervals where these are valid to all polynomial orders. We see from (7.4) that for every $\alpha \in (0,1)$ and every $n$,

$$\max_{\varepsilon^\alpha \le x \le 1} |y(x) - (2x-1)| = O(\varepsilon^n) \quad \text{as } \varepsilon \to 0.$$

Also, for every $n$, as $\varepsilon \to 0$,

$$\max_{0 \le x \le \varepsilon^\beta} \left| \bar{z}\left(\frac{x}{\varepsilon}\right) - y(x) \right| = O(\varepsilon^n)$$

for every $\beta \in (0,1)$. Choosing $0 < \beta < \alpha < 1$ ensures that the intervals $[0, \varepsilon^\beta]$ and $[\varepsilon^\alpha, 1]$ overlap, allowing matching to work. It is not hard to see that the constants $c$ and $d$ are the same if matching is done at $x = \varepsilon^\delta$, for each $\delta \in (0,1)$, and as we said earlier, the usual technique is to look for matching in overlapping intervals, instead of at just one point. Then, the derivative takes care of itself.

It should be apparent that this method is not rigorous. We had no right to think that we would be so close to the correct solution by matching the inner and outer expansions, since neither has been proved a priori (i.e. without using the known solution) to be valid at the point of matching, or in overlapping intervals. One of our goals below is to give a method whereby matching can be made rigorous in some problems.

## 7.2. A problem of Kaplun and Lagerstrom

This boundary value problem will be our interest for the next five sections. First proposed by Kaplun and Lagerstrom in 1957 [**118**], it has been studied by a number of authors, with varying degrees of rigor. The equations are

$$y'' + \frac{n-1}{r} y' + \varepsilon y y' = 0, \tag{7.9}$$

$$y(1) = 0, \quad y(\infty) = 1. \tag{7.10}$$

The physical relevance of this problem is, to quote Lagerstrom and Casten, "vague" [**137**]. The first two terms are the radial Laplacian operator in dimension $n = 2$ or 3. The variable $y$ might represent temperature, in which case $\varepsilon y y'$ is some fictitious heat loss.[4] The boundary conditions represent a temperature of 0 at the surface of a sphere of radius 1 and a temperature of 1 at infinity. Thus, we may think roughly of a ball of ice contained in a large vessel of water which is in a room with (scaled) temperature 1. However there is no physical justification for the "heat loss" term. Despite this tenuous physical background, the problem has been studied often as an interesting example in the theory of matched asymptotic expansions. Several references were given at the beginning of this chapter.

[4]The original setting of Kaplun and Lagerstrom is from fluid mechanics. The more elementary heat conduction interpretation is in the paper by Lagerstrom and Casten.

We will first show how the problem is handled using the (nonrigorous) matching technique. This presentation follows [**99**]. Then, in Section 7.3, we will give an outline of the approach in [**194**], [**195**] using the modern technique of geometric perturbation theory, which was previously discussed in Chapter 6. Then we will give our classical approach, adapted from [**93**].

In this section, we skip over the questions of existence and uniqueness for (7.9)–(7.10). The method of matched asymptotic expansions rarely yields rigorous results, and our goal is simply to get a reasonable approximation of a possible solution for small $\varepsilon$. We will deal with the problem rigorously in later sections of this chapter.

The matching technique starts, as before, with an expansion

$$y = y_0 + \varepsilon y_1 + \varepsilon^2 y_2 + \cdots . \tag{7.11}$$

Substituting this into (7.9) and equating powers of $\varepsilon$ gives the equations

$$y_0'' + \frac{n-1}{r} y_0' = 0, \tag{7.12}$$

$$y_1'' + \frac{n-1}{r} y_1' + y_0 y_0' = 0, \tag{7.13}$$

$$y_2'' + \frac{n-1}{r} y_2' + y_1 y_0' + y_0 y_1' = 0,$$

$$y_j'' + \frac{n-1}{r} y_j' = p_j\left(y_1, \ldots, y_{j-1}, y_1', \ldots, y_{j-1}'\right), \quad j \geq 3,$$

for certain polynomial functions $p_i$. The boundary conditions are

$$\begin{aligned} &y_0(1) = 0, \quad y_0(\infty) = 1, \\ &y_i(1) = y_i(\infty) = 0 \quad \text{for } i \geq 1. \end{aligned} \tag{7.14}$$

The general solution to (7.12) is

$$y_0 = \begin{cases} a \log r + b & \text{if } n = 2, \\ a - \frac{b}{r^{n-2}} & \text{if } n > 2. \end{cases}$$

If $n > 2$, then both boundary conditions in (7.14) are satisfied by setting $a = b = 1$. In the case $n = 2$, this is not possible. We can satisfy either $y(1) = 0$, by setting $b = 0$, or $y(\infty) = 1$, by setting $a = 0$, $b = 1$.

We deal here only with the cases $n = 2$ and $n = 3$. See [**93**] for greater generality. We start with the easier of these.

**7.2.1. The case $n = 3$.** Observe first that for $n = 3$, the function $y_0(r) = 1 - \frac{1}{r}$ is an exact solution to (7.9)–(7.10) if $\varepsilon = 0$. This fact suggests that the solution for small $\varepsilon$, if one exists, is close to $y_0$, with no boundary layer. We will see that this is true.

Having found $y_0$, we solve (7.13), which is now

$$y_1'' + \frac{2}{r}y_1' = \frac{1}{r^3} - \frac{1}{r^2},$$

giving

$$y_1(r) = C_1 - \log r - \frac{1}{r} + \frac{C_2}{r} - \frac{1}{r}\log r + 1 \tag{7.15}$$

for some constants $C_1, C_2$. If $y_1(1) = 0$ then

$$y_1(r) = -\left(1 + \frac{1}{r}\right)\log r + C\left(1 - \frac{1}{r}\right), \tag{7.16}$$

for some constant $C$. Now, however, it appears that this problem is not so easy after all, because there is no way to satisfy the boundary condition at infinity. As pointed out in [**99**], one may doubt that the problem even has a solution. Fortunately we will be able to prove (in Section 7.4) that it does. But (7.15) suggests that there is something fundamentally wrong with the expansion (7.11).

The way around this is to notice that if $y_1$ is given by (7.15) then there is a term $\varepsilon \log r$ in the solution. For fixed $\varepsilon$ this is unbounded as $r \to \infty$, but if we choose $r = \frac{1}{\varepsilon}$ and let $\varepsilon \to 0$, the resulting term $\varepsilon \log \frac{1}{\varepsilon}$ is bounded. For this reason, buttressed by much experience in this sort of thing, Kaplun and Lagerstrom decided to add a term involving $\varepsilon \log \frac{1}{\varepsilon}$ to the expansion. This term has come to be called a "switchback", being of lower order than the originally assumed second term.

Hence, we rewrite our expansion as

$$y = y_0 + \varepsilon \log \frac{1}{\varepsilon} y_1 + \varepsilon y_2 + \cdots. \tag{7.17}$$

Substituting this into (7.9)–(7.10) gives, first, the same $y_0$ as before, namely $y_0(r) = 1 - \frac{1}{r}$. The equation for $y_1$ turns out to be the same as that for $y_0$, namely

$$y_1'' + \frac{2}{r}y_1' = 0$$

with

$$y_1(1) = y_1(\infty) = 0.$$

We cannot satisfy both conditions with a nontrivial solution, so we have a family of solutions

$$y_1 = A_1\left(1 - \frac{1}{r}\right)$$

which satisfy the condition at 1. We will determine $A_1$ by matching with a solution satisfying the boundary condition at $\infty$. The terms "inner" and "outer" get a little confused here, so following [**99**], we will not use them in this section. The expansion (7.17) will be assumed at this stage to converge to the actual solution in intervals of the form $[1, R]$ as $\varepsilon \to 0$.

The term $y_2$ turns out to satisfy the earlier equation for $y_1$, where there was no $\varepsilon \log \frac{1}{\varepsilon}$ term in the expansion, and so, as in (7.16) but renaming the constant,

$$y_2 = -\left(1 + \frac{1}{r}\right) \log r + A_2 \left(1 - \frac{1}{r}\right).$$

Then, the function $y_0 + \varepsilon \log \frac{1}{\varepsilon} y_1 + \varepsilon y_2$ satisfies $y(1) = 0$. Now change variables by letting $\rho = \varepsilon r$ and $y(r) = z(\rho)$. Then

$$z'' + \frac{2}{\rho} z' + z z' = 0,$$

and we look for $z$ such that (i) $z(\infty) = 1$ and (ii) $z$ and $z'$ match with $y$ and $y'$ at some intermediate point, which again we will choose to be $\rho = \sqrt{\varepsilon}$, or $r = \frac{1}{\sqrt{\varepsilon}}$.

In order to match, we need the same sort of expansion for $z$ as for $y$, namely

$$z = 1 + \varepsilon \log \frac{1}{\varepsilon} z_1(\rho) + \varepsilon z_2(\rho) + \cdots.$$

Substituting, it turns out that $z_1$ and $z_2$ both satisfy the same equation,

$$z'' + \left(\frac{2}{\rho} + 1\right) z' = 0.$$

The solutions such that $z_i(\infty) = 0$ for $i = 1, 2$ are found to be of the form

$$z_i = B_i \int_\rho^\infty \frac{e^{-\tau}}{\tau^2} d\tau, \tag{7.18}$$

where the $B_i$ are constants to be determined.

Integrate by parts twice to get

$$\begin{aligned} \int_\rho^\infty \frac{e^{-\tau}}{\tau^2} d\tau &= \frac{e^{-\rho}}{\rho} + e^{-\rho} \log \rho - \int_\rho^\infty e^{-\tau} \log \tau d\tau \\ &= \frac{e^{-\rho}}{\rho} + e^{-\rho} \log \rho - \int_0^\infty e^{-\tau} \log \tau d\tau + \int_0^\rho e^{-\tau} \log \tau d\tau, \end{aligned}$$

where both of these improper integrals are convergent. The first is one formula for Euler's constant $\gamma = .57722\cdots$. Using the Taylor series for $e^{-\tau}$ one finds that

$$\int_\rho^\infty \frac{e^{-\tau}}{\tau^2} d\tau = \frac{1}{\rho} + \log \rho + \gamma - 1 - \frac{1}{2}\rho + o(\rho) \quad \text{as } \rho \to 0. \tag{7.19}$$

This formula is needed in later sections.

We now try to match $y$ and $z$, by setting $r = \frac{1}{\sqrt{\varepsilon}}$, $\rho = \sqrt{\varepsilon}$. We obtain

$$-\sqrt{\varepsilon} + A_1 \log\frac{1}{\varepsilon}\left(\varepsilon - \varepsilon\sqrt{\varepsilon}\right) + \varepsilon\left(A_2\left(1-\sqrt{\varepsilon}\right) - \frac{1}{2}\left(1+\sqrt{\varepsilon}\right)\log\frac{1}{\varepsilon}\right)$$
$$= B_1 \varepsilon \log\frac{1}{\varepsilon}\left(\frac{1}{\sqrt{\varepsilon}} - \frac{1}{2}\log\frac{1}{\varepsilon} + \gamma - 1 - \frac{1}{2}\sqrt{\varepsilon}\right)$$
$$+ B_2\varepsilon\left(\frac{1}{\sqrt{\varepsilon}} - \frac{1}{2}\log\frac{1}{\varepsilon} + \gamma - 1 - \frac{\sqrt{\varepsilon}}{2}\right).$$

As $\varepsilon \to 0$, the five largest terms are of order $\sqrt{\varepsilon}\log\frac{1}{\varepsilon}$, $\sqrt{\varepsilon}$, $\varepsilon\left(\log\frac{1}{\varepsilon}\right)^2$, $\varepsilon\log\frac{1}{\varepsilon}$, and $\varepsilon$, and these balance if $B_1 = 0$, $B_2 = -1$, $A_1 - \frac{1}{2} = -\frac{1}{2}B_2$, and $A_2 = (\gamma - 1)\, B_2$. Note that $B_1 = 0$ eliminates both the $\sqrt{\varepsilon}\log\frac{1}{\varepsilon}$ and $\varepsilon\left(\log\frac{1}{\varepsilon}\right)^2$ terms. We then get $A_1 = 1$, $A_2 = 1-\gamma$, $B_1 = 0$, and $B_2 = -1$. Hence the expansion for fixed $r$ is given by

$$y(r) = 1 - \frac{1}{r} + \varepsilon\log\frac{1}{\varepsilon}\left(1 - \frac{1}{r}\right) \tag{7.20}$$
$$+\varepsilon\left((1-\gamma)\left(1-\frac{1}{r}\right) - \left(1+\frac{1}{r}\right)\log r\right) + o(\varepsilon) \quad \text{as } \varepsilon \to 0.$$

This equation lends credence to our earlier claim that $y$ is close to $1 - \frac{1}{r}$. The expansion for fixed $\rho = \varepsilon r$ is

$$z(\rho) = 1 - \varepsilon\int_\rho^\infty \frac{e^{-\tau}}{\tau^2}\,d\tau + o(\varepsilon) \quad \text{as } \varepsilon \to 0,$$

but we hasten to emphasize that nothing has been proved yet. Proofs will come later.

**7.2.2. The case $n = 2$.** Now the equations are

$$y'' + \frac{1}{r}y' + \varepsilon y y' = 0, \qquad y(1) = 0, \quad y(\infty) = 1. \tag{7.21}$$

Again, start with the expansion $y = y_0 + \varepsilon y_1 + \cdots$, which yields the equations

$$\begin{aligned} y_0'' + \tfrac{1}{r}y_0' &= 0,\\ y_1'' + \tfrac{1}{r}y_1' + y_0 y_0' &= 0,\\ y_2'' + \tfrac{1}{r}y_2' + y_1 y_0' + y_0 y_1' &= 0,\\ y_j'' + \tfrac{1}{r}y_j' = p_j\left(y_1, \ldots, y_{j-1}, y_1', \ldots, y_{j-1}'\right) &\quad \text{for } j \geq 3 \end{aligned} \tag{7.22}$$

and boundary conditions

$$\begin{aligned} y_0(1) = 0, &\quad y_0(\infty) = 1,\\ y_i(0) = y_i(\infty) = 0 &\quad \text{for } i \geq 1. \end{aligned} \tag{7.23}$$

Setting

$$y_0(r) = A_0 \log r \tag{7.24}$$

satisfies the boundary condition at 0, but no $A_0$ will give $y_0(\infty) = 1$.

Again Kaplun and Lagerstrom found a method that works [**118**]. If an expansion in functions satisfying (7.22)–(7.23) exists, then the solution behaves like $\log r$ near $r = 1$. We are not expecting that this approximation will be valid for all $r \geq 1$, but it should be valid far enough out to match with another approximation which satisfies the second boundary condition in (7.23). Therefore a leading term is sought which satisfies $y\left(\frac{1}{\varepsilon}\right) \to 1$ as $\varepsilon \to 0$. (The tricky part is why $\frac{1}{\varepsilon}$ is chosen, rather than some other function of $\varepsilon$.) With this assumption, it is argued (and will be proved later) that the coefficient of the $\log r$ term needs to be on the order of $\frac{1}{\log \frac{1}{\varepsilon}}$. Therefore we look for an expansion of the form

$$y(r) = \frac{1}{\log \frac{1}{\varepsilon}} y_1(r) + \frac{1}{\left(\log \frac{1}{\varepsilon}\right)^2} y_2(r) + \cdots.$$

Observe that there is no leading term $y_0(r)$ with coefficient independent of $\varepsilon$.

Substituting this expansion into (7.21), one obtains the same equation for $y_1$ and $y_2$, namely

$$y_i'' + \frac{1}{r} y_i' = 0,$$

and with the requirement that $y_i(1) = 0$ we get

$$y_i = A_i \log r, \quad i = 1, 2.$$

Therefore,

$$y(r) = \frac{1}{\log \frac{1}{\varepsilon}} A_1 \log r + \frac{1}{\left(\log \frac{1}{\varepsilon}\right)^2} A_2 \log r + O\left(\frac{1}{\left(\log \frac{1}{\varepsilon}\right)^3}\right). \tag{7.25}$$

Initially, this expansion is assumed to be valid in intervals of the form $[1, R]$ as $\varepsilon \to 0$.

We now rescale, by setting $\rho = \varepsilon r$ and $z(\rho) = y(r)$. The equation for $z$ is

$$z'' + \frac{1}{\rho} z' + z z' = 0. \tag{7.26}$$

For fixed $\rho$, $r \to \infty$ as $\varepsilon \to 0$. The leading term in $z$ is assumed to match the boundary condition at $\infty$, giving the unique solution $z_0 = 1$. The subsequent terms are chosen to have a chance of matching $y$ in some intermediate region, and so it is assumed that

$$z(\rho) = 1 + \frac{1}{\log \frac{1}{\varepsilon}} z_1 + \frac{1}{\left(\log \frac{1}{\varepsilon}\right)^2} z_2 + \cdots.$$

Letting $L = \log \frac{1}{\varepsilon}$, we substitute this into (7.26), and the left-hand side becomes, to order $\frac{1}{L^2}$, the expression

$$\frac{1}{L}z_1'' + \frac{1}{L^2}z_2'' + \frac{1}{\rho}\left(\frac{1}{L}z_1' + \frac{1}{L^2}z_2'\right) + \left(1 + \frac{1}{L}z_1 + \frac{1}{L^2}z_2\right)\left(\frac{1}{L}z_1' + \frac{1}{L^2}z_2'\right).$$

Setting the coefficient of $\frac{1}{L}$ to zero then gives

$$z_1'' + \left(\frac{1}{\rho} + 1\right) z_1' = 0.$$

Solutions with $z_1(\infty) = 0$ are of the form

$$z_1(\rho) = B_1 \int_\rho^\infty \frac{e^{-\tau}}{\tau} d\tau \tag{7.27}$$

for some constant $B_1$.

From the terms in $\frac{1}{L^2}$ we obtain

$$z_2'' + \left(\frac{1}{\rho} + 1\right) z_2' + z_1 z_1' = 0.$$

With an integrating factor $\rho e^\rho$ and (7.27), this becomes

$$\left(\rho e^\rho z_2'\right)' = B_1^2 \int_\rho^\infty \frac{e^{-\tau}}{\tau} d\tau. \tag{7.28}$$

The function

$$E_1(\rho) = \int_\rho^\infty \frac{e^{-\tau}}{\tau} d\tau$$

will play a prominent role in the rest of this chapter. For example, the solution to (7.28) which tends to 0 as $\rho \to \infty$ is found to be

$$B_2 E_1(\rho) + B_1^2 \left(2E_1(2\rho) - e^{-\rho} E_1(\rho)\right). \tag{7.29}$$

We then have the approximation

(7.30)

$$\bar{z}(\rho) = 1 + \frac{B_1}{\log \frac{1}{\varepsilon}} E_1(\rho) + \frac{1}{\left(\log \frac{1}{\varepsilon}\right)^2} \left\{B_2 E_1(\rho) + B_1^2 \left(2E_1(2\rho) - e^{-\rho} E_1(\rho)\right)\right\},$$

and it would be reasonable to assume initially that this expansion is valid on intervals of the form $[R_1, \infty)$ as $\varepsilon \to 0$.

In preparation for matching, we need to know the asymptotic behavior of $E_1(\rho)$ for small $\rho$. Using integration by parts, it can be shown that as

$\rho \to 0$,

$$\begin{aligned} E_1(\rho) &= -\log\rho - \gamma + \rho + O\left(\rho^2 \log\rho\right), \\ 2E_1(2\rho) - e^{-\rho}E_1(\rho) &= -\log\rho - \gamma - 2\log 2 - \rho\log\rho \\ &\quad + (3-\gamma)\rho + O\left(\rho^2\log\rho\right). \end{aligned} \tag{7.31}$$

(See Exercise 7.3.)

We now try matching. A more subtle technique than before is needed if we are to use only the terms found so far. Once again, we extend (without rigorous justification) the expansions for $y(r)$ and $z(\rho)$ to the point $r = \frac{1}{\varepsilon^\alpha}$, where $\alpha \in (0,1)$. (Previously we used only $\alpha = \frac{1}{2}$.) Then $\rho = \varepsilon^{1-\alpha}$. Substituting these values into the relations (7.25) and (7.30), and equating powers of $\log\frac{1}{\varepsilon}$, gives

$$\alpha A_1 = 1 + (1-\alpha)B_1.$$

The terms in $\frac{1}{\log\frac{1}{\varepsilon}}$ are found in a similar way to satisfy

$$\alpha A_2 = -B_1\gamma + (1-\alpha)B_2 + (1-\alpha)B_1^2.$$

If these are to hold for all $\alpha \in (0,1)$, then from the first equation,

$$B_1 = -1, \qquad A_1 = 1,$$

and, from the second equation,

$$A_2 = \gamma, \qquad B_2 = -1-\gamma.$$

As mentioned earlier, our presentation follows [**99**], where further details may be found.

This completes our study of the formal method of matched asymptotic expansions. See the references cited above for other aspects of this subject.

## 7.3. A geometric approach

Geometric perturbation theory was discussed earlier, in Chapter 6. It is now well developed, and we will not even attempt to be complete in this section. But since it has been applied to the problem of Kaplun and Lagerstrom, using features not encountered in Chapter 6, we will outline this approach. See the papers [**194**] and [**195**] by Popovic and Szmolyan for details.

Once again, consider the problem (7.9)–(7.10) :

$$y'' + \frac{n-1}{r}y' + \varepsilon yy' =, 0 \tag{7.32}$$

$$y(1) = 0, \quad y(\infty) = 1. \tag{7.33}$$

We will outline part of the proof of existence and local uniqueness for this problem using geometric perturbation theory, as given in [**194**].

The ode (7.32) is nonautonomous, while the dynamical systems techniques of geometric perturbation theory are most easily applied to autonomous systems. The standard method for circumventing this difficulty is to turn the equation into an autonomous system by introducing a new dependent variable, say $\zeta$, with $\zeta' = 1$. Then, the independent variable, $r$ in (7.32), is replaced by $\zeta$. However, in this problem, one of the boundary conditions is at $r = \infty$, and it is better to bring this boundary condition to a finite point in the phase space. This change is easily accomplished, as in [**194**] and [**195**], by letting $\eta(r) = \frac{1}{r}$.[5] We will follow the notation in [**194**] and use $u$ to denote the dependent variable in (7.32), instead of $y$. We then write this equation as a first order system, with $u' = v$, giving

$$\begin{aligned} u' &= v, \\ v' &= -(n-1)\,\eta v - \varepsilon u v, \\ \eta' &= -\eta^2, \end{aligned} \tag{7.34}$$

with boundary conditions

$$u(1) = 0, \qquad \eta(1) = 1, \qquad u(\infty) = 1. \tag{7.35}$$

It is easily seen that (7.34) has a line of equilibrium points in $R^3$, namely all points of the form $(u, v, \eta) = (u, 0, 0)$. We are looking for a solution of (7.34) which starts, at $r = 1$, on the line of points of the form $(0, v, 1)$ and tends to the point $Q = (1, 0, 0)$ as $r \to \infty$. The initial point will be chosen on a line segment $\mathcal{V}$ defined by $u = 0$, $\eta = 1$, $\underline{v} \leq v \leq \bar{v}$ for some suitably chosen $\underline{v}$ and $\bar{v}$. Thus, as the authors of [**194**] point out, it is a type of shooting method, though hardly classical in its execution.

Linearize (7.34) around the equilibrium point $(1, 0, 0)$, giving the matrix

$$A = \begin{pmatrix} 0 & 1 & 0 \\ 0 & -\varepsilon & 0 \\ 0 & 0 & 0 \end{pmatrix}.$$

For each $\varepsilon > 0$ this matrix has $\lambda = 0$ as an eigenvalue of multiplicity 2, and associated eigenvectors $\begin{pmatrix} 1 \\ 0 \\ 0 \end{pmatrix}$ and $\begin{pmatrix} 0 \\ 0 \\ 1 \end{pmatrix}$. The other eigenvalue is $-\varepsilon$, with eigenvector $\begin{pmatrix} 1 \\ -\varepsilon \\ 0 \end{pmatrix}$.

[5]This is not a change of independent variable. The goal is to make the equation autonomous by introducing a new dependent variable $\eta$.

Since 0 is an eigenvalue, the usual stable manifold theorem for ode's does not apply. Instead, the extension of this theorem to include what are called "center manifolds" must be considered.[6] We do not have space here to delve into this topic but refer readers to, for example, [**81**] or [**28**]. In this case a center manifold is easily found, namely the plane $v = 0$. This plane has the properties required of a center manifold: it is invariant, because all solutions starting on $v = 0$ remain there as long as they exist; it contains the equilibrium point $Q$; and at $Q$ it is tangent (indeed, equal) to the eigenspace of $A$ corresponding to $\lambda = 0$.

We consider two specific trajectories which tend to the point $Q$ as $r \to \infty$. One is in the plane $v = 0$, namely

$$u = 1, \qquad v = 0, \qquad \eta = \frac{1}{r}.$$

This solution approaches $Q$ at an algebraic rate. The second known solution tending to $Q$ is in the plane $\eta = 0$. In this plane the equations become

$$\begin{aligned} u' &= v, \\ v' &= -\varepsilon u v. \end{aligned} \tag{7.36}$$

Writing this as

$$\frac{dv}{du} = -\varepsilon u$$

enables us to give the trajectory of this solution as the curve

$$\begin{aligned} \eta &= 0, \\ v &= \tfrac{\varepsilon}{2}\left(1 - u^2\right). \end{aligned} \tag{7.37}$$

From (7.36) it is seen that for every $\varepsilon > 0$, the solution with the trajectory (7.37) tends to $Q$ at an exponential rate in $r$.

It follows from general center manifold theory that there is a smooth two-dimensional manifold, of the form $u = f(v, \eta)$ such that solutions on this manifold tend to $Q$ as $r \to \infty$. (We can construct this manifold by shooting. See Exercise 7.4.) Denote this manifold by $\mathcal{W}^s_\varepsilon$. It is defined in some neighborhood of $Q$.

A second two-dimensional manifold, $\mathcal{M}_\varepsilon$, is formed by those solutions starting on the line segment $\mathcal{V}$. To prove existence of a solution of (7.32)–(7.33), it must be shown that $\mathcal{M}_\varepsilon \cap \mathcal{W}^s_\varepsilon$ is nonempty.

To obtain such a result using geometric perturbation theory, one considers the case $\varepsilon = 0$ and tries to construct a so-called "singular solution".

---

[6] Center manifolds also appear briefly in Chapter 6.

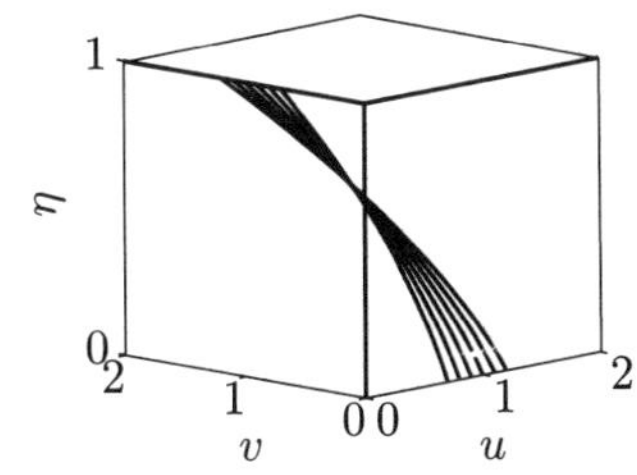


**Figure 7.2.** The surface $\mathcal{M}_0$ for $n > 2$.

For (7.34) this is easier if $n > 2$.[7] In that case, there is an explicit solution to the complete boundary value problem (7.32)–(7.33) when $\varepsilon = 0$, namely $u(r) = 1 - \frac{1}{n-2}\frac{1}{r^{n-2}}$, $v = u', \eta = \frac{1}{r}$. It must be shown that this solution is part of a continuous family which includes solutions for small $\varepsilon > 0$.

The usual way of doing this is to show that for $\varepsilon = 0$, the manifolds $\mathcal{M}_\varepsilon$ and $\mathcal{W}_\varepsilon^s$ intersect transversely. The manifold $\mathcal{M}_0$ is easy to construct. It consists of the orbits of solutions to

$$\begin{aligned} u' &= v, \\ v' &= -(n-1)\eta v, \\ \eta' &= -\eta^2 \end{aligned}$$

with $u(0) = 0$, $\underline{v} \leq v(0) \leq \bar{v}$, $\eta(0) = 1$. For $n > 2$, all of these solutions tend to the center manifold $v = 0$ and to the plane $\eta = 0$, and moreover, $v/\eta$ tends to 0. Hence, the intersection of $\mathcal{M}_0$ and the $(\eta, v)$-plane is parallel to the $v$-axis, while the intersection of $\mathcal{M}_0$ and a plane $\eta = \delta$, where $\delta$ is small and positive, is nearly parallel to the $u$-axis. We observed that for small $\varepsilon$, $\mathcal{W}_\varepsilon^s$ is a surface of the form $u = f(\eta, v)$, with $u$ close to 1 and $(\eta, v)$ close to $(0,0)$, and this suggests that we should expect a transverse intersection of $\mathcal{M}_\varepsilon$ and $\mathcal{W}_\varepsilon^s$. (See Figure 7.2.)

But you will notice that we constructed no manifold $\mathcal{W}_0^s$. When $\varepsilon = 0$, there is only one nonconstant trajectory tending to $Q$ such that $(u, v) \to (1, 0)$ at an exponential rate as $r \to \infty$, namely the orbit of $u = 1, v = 0, \eta = \frac{1}{r}$. This degeneracy makes the problem nonstandard within the class of problems usually considered in geometric perturbation theory.

The case $n = 2$ is even more complicated, because we cannot find an exact solution to (7.32)–(7.33). Indeed, the singular solution is not the union of a finite number of solutions with $\varepsilon = 0$. It is constructed in two parts, as follows.

[7]In **[194]** the details are given only for the physically significant cases $n = 2, 3$.

First set $\varepsilon = 0$ in (7.32) and solve the initial value problem

$$\begin{aligned} u' &= v, \\ v' &= -(n-1)\,\eta v, \\ \eta' &= -\eta^2, \\ u(1) = 0, &\quad v(0) = 0, \quad \eta(0) = 1. \end{aligned}$$

The solution is $u = 0, v = 0, \eta = \frac{1}{r}$, which tends to $(0,0,0)$ as $r \to \infty$. Then in (7.32) make the change of variables $\rho = \varepsilon r$, which gives the new equations

$$\begin{aligned} \frac{du}{d\rho} &= \varepsilon v, \\ \frac{dv}{d\rho} &= -\varepsilon(n-1)\,\eta v - \varepsilon^2 uv, \\ \frac{d\eta}{d\rho} &= -\frac{\varepsilon}{\eta^2}. \end{aligned}$$

For $\varepsilon = 0$ we get the trivial "slow system" where all variables are constants. Hence, all points on the interval $\eta = v = 0$, $0 \le u \le 1$ are equilibrium points. The trajectory of the singular solution is the curve in $R^3$ consisting of the interval $0 \le \eta \le 1$ on the $\eta$-axis followed by the interval $0 \le u \le 1$ on the $u$-axis. This situation is highly degenerate, since in one region the slow system has no nonconstant solutions. Contrast this with the singular solutions found in Chapter 6.

Fortunately, the dynamical systems community has encountered this situation before. The first step is to introduce $\varepsilon$ as an independent variable, with equation $\varepsilon' = 0$. This technique is standard in bifurcation theory, in conjunction with the center manifold ([**81**, Section 3.2]; [**34**, Section 1.8]). But now, adding this equation to (7.34), the setting is in $R^4$, and not surprisingly, the geometry becomes complicated. The particularly singular nature of the problem requires use of another technique, called "blow-up". This method is interesting, and we will include a brief discussion of some two-dimensional examples, as well as some exercises. However we will give no further details of the geometric approach to Lagerstrom's problem, referring the reader to [**194**] and [**195**], and we put off the examples of the blow-up method to the end of the chapter.

## 7.4. A classical approach

When attention shifts from the existence of solutions to their asymptotic expansions, one may not expect that geometric techniques will be useful. But surprisingly, they have been applied to this problem, as in [**195**]. However, for the problem of Kaplun and Lagerstrom we prefer our classical methods, to which we now turn. We begin this approach with a proof that there is a unique solution of Lagerstrom's model problem. This section, which is based on [**93**], does not depend on either of the previous sections.

In [**93**] a wider class of equations is considered. All the results below carry over, with much the same proof, to problems of the form

$$y'' + \frac{n-1}{r}y' + \varepsilon yy' + f(y)\,y'^2 = 0,$$
$$y(1) = 0, \quad y(\infty) = 1,$$

for any continuous function $f$ which is nonnegative on $0 \le u \le 1$. The case $f(u) = 1$ was introduced in [**118**], where it was related to compressible flow. It was also studied formally in [**99**]. In those references it appears considerably harder than the case $f(u) = 0$, but in [**93**] it was seen to be a fairly routine extension of the method we will present below.

**7.4.1. Existence and uniqueness.** The existence proof outlined in the previous section is valid only for small $\varepsilon$ and gives just local uniqueness. A much stronger result is easily proved. A general existence result seems to have been first given in [**236**], while uniqueness was proved in [**43**]. See [**93**] for further references. Our approach is similar to these in some respects but carried out more efficiently.

**Theorem 7.1.** *For each $\varepsilon > 0$ and $n \ge 1$ the problem* (7.9)–(7.10) *has a unique solution.*

**Proof.** We use a shooting method, by considering the initial value problem

$$y'' + \frac{n-1}{r}y' + \varepsilon yy' = 0, \tag{7.38}$$

$$y(1) = 0, \quad y'(1) = c \tag{7.39}$$

for each $c > 0$. We write the equation in the form

$$\left(r^{n-1}y'\right)' + \varepsilon y\left(r^{n-1}y'\right) = 0, \tag{7.40}$$

so that

$$y'(r) = \frac{c}{r^{n-1}}e^{-\int_1^r \varepsilon y(s)ds}. \tag{7.41}$$

Hence, $y$ is increasing and exists for all $r \ge 1$. Integrating again,

$$y(r) = \int_1^r \frac{c}{s^{n-1}}e^{-\int_1^s \varepsilon y dt}ds. \tag{7.42}$$

For $n \ge 1$, (7.42) implies that either $y(2) > 1$ or $y(2) \ge c\kappa e^{-\varepsilon}$ where $\kappa = \int_1^2 \frac{1}{s^{n-1}}ds$. Therefore, if

$$c > \frac{e^{\varepsilon}}{\kappa} \tag{7.43}$$

then $y(2) > 1$. Also, from (7.38), $y'' + \varepsilon yy' \le 0$, and using (7.39) we get $y' + \frac{1}{2}\varepsilon y^2 \le c$. Since $y' \ge 0$, $y \le \sqrt{\frac{2c}{\varepsilon}}$ and so for sufficiently small $c$, $y(\infty) < 1$. We have thus shown that there are $c_1$ and $c_2$ with $0 < c_1 < c_2$ such that if $c = c_1$ then $y(\infty) < 1$, while if $c = c_2$ then $y(\infty) > 1$. Further, for every $r > R > 2$,

$$y(r) = y(R) + c\int_R^r \frac{1}{s^{n-1}} e^{-\int_1^s \varepsilon y(t)dt}.$$

If $n > 2$ the second term is bounded by $\frac{c}{(n-2)R^{n-2}}$, while if $1 \le n \le 2$, it is bounded by

$$c\int_R^\infty e^{-\varepsilon\mu(s-2)}ds$$

where $\mu = \min\{1, c\kappa e^{-\varepsilon}\}$. Hence, for every $n \ge 1$ this term tends to zero as $R \to \infty$ uniformly for $c_1 \le c \le c_2$. Since $y(R)$ is a continuous function of $c$ for every $R$ where the solution exists, it follows that $y(\infty)$ also is continuous in $c$. This implies that there is a $c$ with $y(\infty) = 1$, giving a solution to (7.9)–(7.10).

For uniqueness we again use the form (7.40), which we integrate to obtain

$$r^{n-1}y'(r) = c - \frac{1}{2}\varepsilon r^{n-1}y^2 + \frac{\varepsilon(n-1)}{2}\int_1^r s^{n-2}y^2 ds. \tag{7.44}$$

From this it follows that if $c_1 > c_2$ then the corresponding solutions satisfy $y_1 > y_2$ on $(0, \infty)$. However this does not quite show the desired uniqueness. Suppose that there are two solutions of (7.9)–(7.10), say $y_1$ and $y_2$, with $y_1'(1) = c_1 > y_2'(1) = c_2$. Then from (7.44),

$$\begin{aligned} &r^{n-1}(y_1' - y_2') \\ &\quad = (c_1 - c_2) - \frac{1}{2}\varepsilon r^{n-1}(y_1^2 - y_2^2) + \frac{\varepsilon(n-1)}{2}\int_1^r s^{n-2}(y_1^2 - y_2^2)\,ds. \end{aligned} \tag{7.45}$$

Equation (7.41) implies that as $r \to \infty$, $y_1' \to 0$ and $y_2' \to 0$, and we are assuming that $y_1 \to 1$, $y_2 \to 1$, all limits being at an exponential rate. Further, $c_1 > c_2$, and $y_1 > y_2 > 0$ on $(1, \infty)$. It follows that as $r \to \infty$, the left side of (7.45) tends to zero and the right side to a positive limit. This contradiction proves uniqueness. □

**7.4.2. Asymptotic expansion.** As in Section 7.2, the problem is to investigate the asymptotic behavior as $\varepsilon \to 0$ of the unique function $y = y_\varepsilon$ such that

$$y'' + \frac{n-1}{r}y' + \varepsilon yy' = 0 \tag{7.46}$$

and

$$y(1) = 0, \quad y(\infty) = 1. \tag{7.47}$$

In this section, $n \geq 2$.

From (7.41) we obtain

$$r^{n-1} y' = c e^{-\varepsilon \int_1^r y(t) dt} \tag{7.48}$$

for some constant $c > 0$. Since $y(\infty) = 1$, (7.48) implies that $y'(r)$ is exponentially small as $r \to \infty$. It follows that $\int_r^\infty (1 - y(t))\, dt$ converges, and so we may rewrite (7.48) in the form

$$r^{n-1} y' = C e^{-\varepsilon r - \varepsilon \int_\infty^r (y-1) dt}. \tag{7.49}$$

This step is key to the whole method, for it allows us to obtain an integral equation for $y$ with solutions which automatically satisfy the boundary condition at infinity.

Integration of (7.49) gives

$$y - 1 = C \int_\infty^r \frac{1}{t^{n-1}} e^{-\varepsilon t - \varepsilon \int_\infty^t (y-1) ds} dt.$$

Setting $\varepsilon r = \rho$, $y(r) = z(\rho)$, $\varepsilon t = \tau$, and $\varepsilon s = \sigma$, we obtain

$$z(\rho) - 1 = C \varepsilon^{n-2} \int_\infty^\rho \frac{1}{\tau^{n-1}} e^{-\tau} e^{-\int_\infty^\tau (z(\sigma) - 1) d\sigma} d\tau. \tag{7.50}$$

We will use this equation to obtain an asymptotic expansion for the solution of (7.46)–(7.47) for small $\varepsilon$.

It is also possible to use this equation to prove existence of a solution to (7.46)–(7.47), by setting up a mapping from a pair $(C, z)$ to a pair $\left(\tilde{C}, \tilde{z}\right)$ and looking for a fixed point. However this involves some functional analysis, and so we will not present it here.

Note also that

$$\begin{array}{c} z'' + \frac{n-1}{\rho} z' + z z' = 0, \\ z(\varepsilon) = 0, \quad z(\infty) = 1. \end{array} \tag{7.51}$$

Hence in (7.50), $C$ is a constant satisfying

$$-1 = C \varepsilon^{n-2} \int_\infty^\varepsilon \frac{1}{\tau^{n-1}} e^{-\tau} e^{-\int_\infty^\tau (z-1) d\sigma} d\tau. \tag{7.52}$$

Since there is a unique solution $z$ for each $\varepsilon > 0$, this determines a unique $C$, dependent on $\varepsilon$.

We now consider the integral

$$\int_\infty^\tau (z(\sigma) - 1)\, d\sigma = \int_\tau^\infty (1 - z(\sigma))\, d\sigma, \tag{7.53}$$

which appears in the exponent in (7.52). This integral has been seen to converge for each $\varepsilon$, but we need a bit more, namely, that it is bounded uniformly in $\varepsilon \le \tau < \infty$ as $\varepsilon \to 0$. To see this, we note that as a function of $\sigma$, $z$ satisfies (7.51). Denoting the unique solution of this boundary value problem by $z_\varepsilon(\sigma)$, we claim that if $0 < \varepsilon_1 < \varepsilon_2$ then $z_{\varepsilon_1} > z_{\varepsilon_2}$ on $[\varepsilon_2, \infty)$. If this is false then there exist $\varepsilon_1$ and $\varepsilon_2$ such that $z_{\varepsilon_1}(\sigma_0) = z_{\varepsilon_2}(\sigma_0)$ for some $\sigma_0 \ge \varepsilon_2$. But then the problem

$$z'' + \frac{n-1}{\sigma} z' + zz' = 0,$$
$$z(\sigma_0) = z_{\varepsilon_1}(\sigma_0), \quad z(\infty) = 1$$

has two solutions, which contradicts our earlier uniqueness proof.

Now fix $\varepsilon_2 > 0$. A consequence of the argument above is that

$$\int_{\varepsilon_2}^{\infty} (1 - z_{\varepsilon_1}(\sigma))\, d\sigma < \int_{\varepsilon_2}^{\infty} (1 - z_{\varepsilon_2}(\sigma))\, d\sigma,$$

which implies that the integral in the exponent in (7.52) is bounded independently of $\varepsilon \in (0, \varepsilon_2)$ and $\tau \ge \varepsilon$. Because $n \ge 2$, the $\tau$-integral in (7.52) approaches $-\infty$ as $\varepsilon \to 0$. Hence,

$$\lim_{\varepsilon \to 0} C\varepsilon^{n-2} = 0.$$

Since $\int_\infty^\tau (z-1)\, d\sigma > 0$, it follows from (7.50) that if

$$E_{n-1}(\rho) = \int_\rho^\infty \frac{1}{\tau^{n-1}} e^{-\tau}\, d\tau \tag{7.54}$$

then

$$|z(\rho) - 1| < C\varepsilon^{n-2} E_{n-1}(\rho). \tag{7.55}$$

For future estimates, we observe that

$$E_{n-1}(\rho) = \begin{cases} O\left(\rho^{2-n}\right) & \text{as } \rho \to 0 \quad \text{if } n > 2, \\ O(\log \rho) & \text{as } \rho \to 0 \quad \text{if } n = 2, \\ O\left(\rho^{1-n} e^{-\rho}\right) & \text{as } \rho \to \infty. \end{cases} \tag{7.56}$$

The method now is to work from (7.50). As remarked before, since $y'(r)$ is exponentially small as $r \to \infty$, the integral term $\int_\rho^\infty (z-1)\, d\sigma$ converges. Hence for each $\varepsilon > 0$ and $\rho_0 > 0$, and every $\rho \ge \rho_0$,

(7.57) $z(\rho) - 1 =$

$$C\varepsilon^{n-2} \int_\infty^\rho \frac{1}{\tau^{n-1}} e^{-\tau} \left\{ 1 - \int_\infty^\tau (z-1)\, d\sigma + \frac{1}{2} \left( \int_\infty^\tau (z-1)\, d\sigma \right)^2 - \cdots \right\} d\tau,$$

where the series in the integrand converges uniformly for $\rho_0 \le \tau < \infty$.

In fact, we will need to use this series for all $\tau \geq \varepsilon$. Thus we need to check its convergence in this interval. Let

$$\Phi = \Phi(\varepsilon) = C\varepsilon^{n-2} \int_{\varepsilon}^{\infty} E_{n-1}(s)\, ds. \tag{7.58}$$

Then (7.55) and (7.56) imply that for every $\tau \geq \varepsilon$, if $n \geq 2$ then

$$\left| \int_{\tau}^{\infty} (z(s) - 1)\, ds \right| < \Phi. \tag{7.59}$$

Also,

$$\varepsilon^{n-2} \int_{\varepsilon}^{\infty} E_{n-1}(s)\, ds = \begin{cases} o(1) \text{ as } \varepsilon \to 0 & \text{if } n > 2, \\ O(1) \text{ as } \varepsilon \to 0 & \text{if } n = 2. \end{cases}$$

Hence for each $n \geq 2$ and $C$ in a compact interval, the series in the integrand of (7.57) converges uniformly on $[\varepsilon, \infty)$.

Note that if $n > 2$ then for each $C$, $\Phi \to 0$ as $\varepsilon \to 0$, while if $n = 2$ then for fixed $\varepsilon$, $\Phi \to 0$ as $C \to 0$. We proceed to solve (7.57) by iteration. From (7.57) and (7.59) the first approximation is

$$\begin{aligned} z(\rho) - 1 &= C\varepsilon^{n-2} \int_{\infty}^{\rho} \frac{e^{-\tau}}{\tau^{n-1}} d\tau \,(1 + O(\Phi)) \\ &= -C\varepsilon^{n-2} E_{n-1}(\rho)\,(1 + O(\Phi)) \end{aligned} \tag{7.60}$$

as $\Phi \to 0$.

In order to substitute into (7.57) we integrate (7.60), which gives

$$\int_{\infty}^{\tau} (z(\sigma) - 1)\, d\sigma = -C\varepsilon^{n-2} \int_{\infty}^{\tau} E_{n-1}(\sigma)\, d\sigma + O\left(\Phi^2\right)$$

uniformly in $\tau \geq \varepsilon$ as $\Phi \to 0$. We will use this for all $\tau \geq \varepsilon$, although it is very crude for large $\rho$, since the two integrals tend to zero as $\rho \to \infty$, while $\Phi$ is independent of $\rho$. Note that the absolute value of the first term on the right is bounded by $\Phi$.

We obtain the second approximation by substituting this back into (7.57). Again we integrate the resulting expression, to get

$$\begin{aligned} \int_{\infty}^{\tau} (z(\sigma) - 1)\, d\sigma &= -C\varepsilon^{n-2} \int_{\infty}^{\tau} E_{n-1}(\sigma)\, d\sigma \\ &+ \left(C\varepsilon^{n-2}\right)^2 \int_{\infty}^{\tau} \int_{\infty}^{\sigma} \frac{1}{s^{n-1}} e^{-s} \int_{\infty}^{s} E_{n-1}(t)\, dt ds d\sigma + O\left(\Phi^3\right). \end{aligned}$$

By substituting once more into (7.57) we obtain

(7.61)

$$z(\rho) - 1 = -C\varepsilon^{n-2}E_{n-1} + \left(C\varepsilon^{n-2}\right)^2 \int_\infty^\rho \tfrac{1}{\tau^{n-1}} e^{-\tau}\left(\int_\infty^\tau E_{n-1} d\sigma\right) d\tau$$
$$+\tfrac{1}{2}C\varepsilon^{n-2} \int_\infty^\rho \tfrac{1}{\tau^{n-1}} e^{-\tau}\left\{\left(C\varepsilon^{n-2}\right)^2 \left(\int_\infty^\tau E_{n-1} d\sigma\right)^2\right\} d\tau$$
$$-C\varepsilon^{n-2} \int_\infty^\rho \tfrac{1}{\tau^{n-1}} e^{-\tau}\left\{\left(C\varepsilon^{n-2}\right)^2 \int_\infty^\tau \int_\infty^\sigma \tfrac{1}{s^{n-1}} e^{-s}\left(\int_\infty^s E_{n-1}(t)\, dt\right) ds d\sigma\right\} d\tau$$
$$+C\varepsilon^{n-2} \int_\infty^\rho \tfrac{1}{\tau^{n-1}} e^{-\tau} d\tau \left(O\left(\Phi^3\right)\right),$$

as $\Phi \to 0$. It should be observed that the two expressions in curly brackets $\{\ \}$ in (7.61) are bounded by $\Phi^2$, ensuring that the last term is of higher order than the others in this expansion.

To obtain $C$, we need to evaluate each of these terms for small $\rho$ (in particular, for $\rho = \varepsilon$), and this is a matter of integration by parts. In the next two sections we carry the process through in the most interesting cases, $n = 2, 3$. For this purpose we need the following Taylor series expansions for $E_1$ and $E_2$, which we repeat from Section 7.2:

(7.62) $$E_1(\rho) = \int_\rho^\infty \frac{1}{\tau} e^{-\tau} d\tau = -\log\rho - \gamma + \rho + O\left(\rho^2 \log\rho\right) \quad \text{as } \rho \to 0,$$

(7.63)
$$E_2(\rho) = \frac{e^{-\rho}}{\rho} - E_1(\rho) = \frac{1}{\rho} + \log\rho + (\gamma - 1) - \frac{1}{2}\rho + O\left(\rho^2\log\rho\right) \quad \text{as } \rho \to 0.$$

## 7.5. The case $n = 3$

The calculations and results are simpler when $n > 2$. For $n = 3$ we are interested in $E_2$, and thus in (7.63). From this we see that the first term on the right of (7.61) evaluated at $\rho = \varepsilon$ is

$$-C\left(1 + \varepsilon\log\varepsilon + \varepsilon(\gamma - 1) + O\left(\varepsilon^2\right)\right) \quad \text{as } \varepsilon \to 0.$$

The second term is

$$(C\varepsilon)^2 \int_\infty^\rho \frac{1}{\tau^2} e^{-\tau}\left(\int_\infty^\tau E_2(\sigma)\, d\sigma\right) d\tau$$
$$= (C\varepsilon)^2 \left\{\left[-E_2(\tau) \int_\infty^\tau E_2(\sigma)\, d\sigma\right]_\infty^\rho + \int_\infty^\rho E_2^2(\tau)\, d\tau\right\}$$
$$= (C\varepsilon)^2 \left\{-E_2(\rho) \int_\infty^\rho E_2(\tau)\, d\tau + \int_\infty^\rho E_2^2(\tau)\, d\tau\right\}.$$

Again using (7.63) we obtain that as $\rho \to 0$,

$$\int_\infty^\rho E_2(\tau)^2\, d\tau = -\frac{1}{\rho} + \log^2 \rho + O(\log \rho),$$
$$E_2(\tau) \int_\infty^\rho E_2(\tau)\, d\tau = \frac{1}{\rho} \log \rho + \frac{\gamma+1}{\rho} + O\left(\log^2 \rho\right).$$

In all, the second term of (7.61) is

$$(C\varepsilon)^2 \left\{ -\frac{1}{\rho} \log \rho - \frac{\gamma+2}{\rho} + O\left(\log^2 \rho\right) \right\}.$$

It is readily verified that the third and fourth terms in (7.61) give $O\left\{C^3\varepsilon^3\left(\frac{1}{\rho}\log^2 \rho\right)\right\}$, which is of higher order than the first two terms when $\rho = \varepsilon$ and $\varepsilon$ is small. Evaluating (7.61) at $\rho = \varepsilon$,

$$\begin{aligned} -1 = -C\varepsilon \left( \frac{1}{\varepsilon} + \log \varepsilon + \gamma - 1 \right) + (C\varepsilon)^2 \left( -\frac{1}{\varepsilon} \log \varepsilon - \frac{\gamma+2}{\varepsilon} \right) \\ + O\left(C^3 \varepsilon^2 \log^2 \varepsilon\right), \end{aligned}$$

so that

$$C = 1 - 2\varepsilon \log \varepsilon - \varepsilon(2\gamma+1) + O\left(\varepsilon^2 \log^2 \varepsilon\right).$$

Then, for fixed $r$, $\rho$ of order $\varepsilon$,

$$\begin{aligned} y(r) - 1 = -\varepsilon(1 - 2\varepsilon\log\varepsilon - \varepsilon(2\gamma+1)) \left( \frac{1}{\varepsilon r} + \log \varepsilon + \log r + \gamma - 1 \right) \\ + \varepsilon^2 \left( -\frac{1}{\varepsilon r} (\log \varepsilon + \log r) - \frac{\gamma+2}{\varepsilon r} \right) + O\left(\varepsilon^2 \log^2 \varepsilon\right), \end{aligned}$$

or

$$\begin{aligned} y(r) = 1 - \frac{1}{r} - \varepsilon \log \varepsilon \left(1 - \frac{1}{r}\right) - \varepsilon \left( \log r + \frac{\log r}{r} \right) + \varepsilon(1-\gamma)\left(1 - \frac{1}{r}\right) \\ + O\left(\varepsilon^2 \log^2 \varepsilon\right). \end{aligned}$$

As we would hope, this agrees with the result obtained formally in (7.20).

For fixed $\rho$, $r$ of order $\varepsilon^{-1}$,

$$\begin{aligned} z(\rho) - 1 = -\varepsilon(1 - 2\varepsilon\log\varepsilon - \varepsilon(2\gamma+1)) E_2(\rho) \\ + \varepsilon^2 \left( -E_2(\rho) \int_\infty^\rho E_2(\tau)\, d\tau + \int_\infty^\rho E_2^2(\tau)\, d\tau \right) + O\left(\varepsilon^3\right). \end{aligned} \tag{7.64}$$

These results are in agreement with those of Hinch [**99**] and others, although (7.64) gives one term further.[8]

## 7.6. The case $n = 2$

The analysis here is more complicated. When $n = 2$, we are interested in $E_1(\rho)$, for which we use the expansion (7.62). Looking also at (7.61) with $\rho = \varepsilon$, we see that as $\varepsilon \to 0$,

$$C \log \varepsilon \to -1$$

and

$$C = \frac{1}{\log \frac{1}{\varepsilon}} + O\left(\frac{1}{\left(\log \frac{1}{\varepsilon}\right)^2}\right).$$

Hence the series in (7.61) is in powers of $\frac{1}{\log \frac{1}{\varepsilon}}$.

Also, we will find our approximations (in order to compare the results with those of Hinch) to order $\frac{1}{\log^2\left(\frac{1}{\varepsilon}\right)}$, so that (for example)

$$y(r) = \frac{a(r)}{\log\left(\frac{1}{\varepsilon}\right)} + \frac{b(r)}{\log^2\left(\frac{1}{\varepsilon}\right)} + O\left(\log^{-3} \frac{1}{\varepsilon}\right) \tag{7.65}$$

for each fixed value of $r$ ($\rho$ of order $\varepsilon$). As we shall see, this necessitates finding

$$C = \frac{1}{\log\left(\frac{1}{\varepsilon}\right)} \left\{1 + \frac{A}{\log\left(\frac{1}{\varepsilon}\right)} + \frac{B}{\log^2\left(\frac{1}{\varepsilon}\right)} + O\left(\log^{-3}\left(\frac{1}{\varepsilon}\right)\right)\right\}$$

and requires use of all the terms in (7.61).

With this in mind, we look at the second term of (7.61). From (7.62),

$$\int_\rho^\infty E_1 d\tau = -\rho E_1 + e^{-\rho}, \tag{7.66}$$

so that the second term is

$$\begin{aligned} C^2 \int_\infty^\rho \frac{1}{\tau} e^{-\tau}\left(\tau E_1 - e^{-\tau}\right) d\tau &= C^2 \left\{\int_\infty^\rho e^{-\tau} E_1 d\tau - \int_\infty^\rho \frac{e^{-2\tau}}{\tau} d\tau\right\} \\ &= C^2 \left\{\left[-e^{-\tau} E_1\right] |_\infty^\rho - 2\int_\infty^\rho \frac{e^{-2\tau}}{\tau} d\tau\right\} \\ &= C^2 \left(-e^{-\rho} E_1(\rho) + 2E_1(2\rho)\right). \end{aligned} \tag{7.67}$$

[8] An integration by parts shows that

$$-E_2(\rho) \int_\infty^\rho E_2(\tau)\, d\tau = E_2 E_1 - \rho E_2^2,$$

which is the form used by Hinch.

From (7.62) the second term is therefore

$$\begin{aligned} & C^2 (\log \rho + \gamma - 2 \log 2\rho - 2\gamma + O(\rho)) \\ & \quad = C^2 (-\log \rho - \gamma - 2 \log 2 + O(\rho)) \end{aligned} \tag{7.68}$$

as $\rho \to 0$.

In the third and fourth terms of (7.61) we need only the leading terms, and so we ignore the equivalent of $\gamma + 2\log 2$ in (7.68). Using (7.66) the third term becomes

$$\frac{1}{2} C^3 \int_{\infty}^{\rho} \frac{e^{-\tau}}{\tau} \left(e^{-\tau} - \tau E_1\right)^2 d\tau = \frac{1}{2} C^3 (\log \rho + O(1)) \text{ as } \rho \to 0. \tag{7.69}$$

Finally, in the fourth term, the integrand in the $\tau$-integral is just the second term (as a function of $\sigma$), so that from (7.67), the fourth term is

$$M = -C^3 \int_{\infty}^{\rho} \frac{e^{-\tau}}{\tau} \left[ \int_{\infty}^{\tau} \left\{ -e^{-\sigma} E_1(\sigma) + 2E_1(2\sigma) \right\} d\sigma \right] d\tau. \tag{7.70}$$

It is seen from (7.66) that for every positive $\tau \leq \infty$, $\int_0^{\tau} E_1(\sigma)\, d\sigma$ converges. Hence we can write the inner integral above in the form $\int_{\infty}^{0} + \int_0^{\tau}$, and it follows that

$$M = -C^3 \int_{\infty}^{\rho} \frac{e^{-\tau}}{\tau} \{ K + k(\tau) \} \, d\tau$$

where $K$ is a constant, $k$ is bounded, and $k(\tau) = O(\tau \log \tau)$ as $\tau \to 0$. It further follows that

$$M = C^3 (K E_1(\rho) + O(1)) \quad \text{as } \rho \to 0.$$

From (7.70) it is seen that

$$K = \int_{\infty}^{0} \left\{ -e^{-\sigma} E_1(\sigma) + 2E_1(2\sigma) \right\} d\sigma,$$

and we can evaluate this using (7.66) and (7.62):

$$\begin{aligned} & \int_0^{\infty} E_1(2\sigma)\, d\sigma = \frac{1}{2} \int_0^{\infty} E_1(y)\, dy = \frac{1}{2}, \\ & \int_0^{\infty} e^{-\sigma} E_1(\sigma)\, d\sigma = \left[ -\left(e^{-\sigma} - 1\right) E_1 \right]_0^{\infty} - \int_0^{\infty} \left(e^{-\sigma} - 1\right) \frac{e^{-\sigma}}{\sigma} d\sigma \\ & \quad = \lim_{\sigma \to 0} \{ -E_1(2\sigma) + E_1(\sigma) \} = \lim_{\sigma \to 0} (\log 2\sigma - \log \sigma) = \log 2. \end{aligned} \tag{7.71}$$

Hence, $K = \log 2 - 1$, and so from (7.70) the fourth term of (7.61) is

$$C^3 \{ E_1(\rho) (\log 2 - 1) + O(1) \} = -C^3 \{ (\log 2 - 1) \log \rho + O(1) \} \text{ as } \rho \to 0. \tag{7.72}$$

Now setting $\rho = \varepsilon$ and using (7.61), (7.62), (7.68), (7.69), and (7.72), we obtain that

$$-1 = -C\left(-\log\varepsilon - \gamma + O(\varepsilon)\right) + C^2\left(-\log\varepsilon - \gamma - 2\log 2 + O(\varepsilon)\right)$$
$$+\frac{1}{2}C^3\left(\log\varepsilon + O(1)\right) - C^3\left\{(\log 2 - 1)\log\varepsilon + O(1)\right\}$$

as $\varepsilon \to 0$. Hence,

$$\frac{1}{\log\left(\frac{1}{\varepsilon}\right)} = C\left(1 - \frac{\gamma}{\log\left(\frac{1}{\varepsilon}\right)}\right) - C^2\left(1 - \frac{\gamma + 2\log 2}{\log\left(\frac{1}{\varepsilon}\right)}\right) + C^3\left(\frac{3}{2} - \log 2\right) + O\left(\log^{-4}\left(\frac{1}{\varepsilon}\right)\right), \tag{7.73}$$

and

$$C = \frac{1}{\log\left(\frac{1}{\varepsilon}\right)} + \frac{A}{\log^2\left(\frac{1}{\varepsilon}\right)} + \frac{B}{\log^3\left(\frac{1}{\varepsilon}\right)} + O\left(\frac{1}{\log^4\left(\frac{1}{\varepsilon}\right)}\right),$$

where

$$-\gamma + A - 1 = 0,$$
$$B - \gamma A - 2A + (\gamma + 2\log 2) + \frac{3}{2} - \log 2 = 0.$$

Hence,

$$A = \gamma + 1,$$
$$B = \gamma^2 + 2\gamma + \frac{1}{2} - \log 2.$$

Thus, for fixed $r$, $\rho$ of order $\varepsilon$, and with $\lambda = \log\left(\frac{1}{\varepsilon}\right)$,

$$y - 1 = \left(\frac{1}{\lambda} + \frac{\gamma+1}{\lambda^2} + \frac{(\gamma+1)^2 - \frac{1}{2} - \log 2}{\lambda^3}\right)(\log r + \log\varepsilon + \gamma)$$
$$+\frac{1}{\lambda^2}\left(1 + \frac{2(\gamma+1)}{\lambda}\right)(-\log r - \log\varepsilon - \gamma - 2\log 2)$$
$$+\frac{1}{\lambda^3}\left(\frac{3}{2} - \log 2\right)(\log r + \log\varepsilon) + O\left(\lambda^{-4}\right),$$

so that after cancellation,

$$y = \frac{\log r}{\lambda} + \frac{\gamma \log r}{\lambda^2} + O\left(\lambda^{-3}\right).$$

This is the "inner expansion". For the "outer expansion", i.e. fixed $\rho$, $r$ of order $\frac{1}{\varepsilon}$, we obtain

$$z - 1 = -E_1(\rho)\left(\frac{1}{\lambda} + \frac{\gamma+1}{\lambda^2}\right) + \frac{1}{\lambda^2}\left(2E_1(2\rho) - e^{-\rho}E_1(\rho)\right) + O\left(\lambda^{-3}\right).$$

These results also agree with those of Hinch and others on this problem.

**7.6.1. Remark on switchback.** Starting with Lagerstrom, the terms involving $\log \varepsilon$ in the inner expansions have been considered strange, and difficult to explain. They are often called "switchback" terms, because an expansion in powers of $\varepsilon$ gives inconsistent results which are only resolved by adding terms of lower order, that is, powers of $\varepsilon \log \frac{1}{\varepsilon}$. The recent approach to the problem by geometric perturbation theory explains this by reference to a "resonance phenomenon" which is too complicated for us to describe here [**194**], [**195**].

In our approach, the necessity for such terms is seen already from the equation (7.50) and the resulting expansion (7.57) :

$$z(\rho) - 1 =$$
$$C\varepsilon^{n-2} \int_{\infty}^{\rho} \frac{1}{\tau^{n-1}} e^{-\tau} \left\{ 1 - \int_{\infty}^{\tau} (z-1)\, d\sigma + \frac{1}{2} \left( \int_{\infty}^{\tau} (z-1)\, d\sigma \right)^2 - \cdots \right\} d\tau.$$

The first term is simply $C\varepsilon^{n-2} E_{n-1}(\rho)$, and the simple expansions given for $E_1$ and $E_2$ show immediately the need for the logarithmic terms. No "switchback" is seen, because the procedure does not start with any assumption about the nature of the expansion.

## 7.7. A second application of the method

**7.7.1. Introduction.** A second boundary value problem to which the integral equation method of the last three sections has been applied is

$$\varepsilon y'' + 2y' + f(y) = 0, \tag{7.74}$$

$$y(0) = 0, \quad y(A) = 0, \tag{7.75}$$

for $A > 0$, $\varepsilon > 0$, and a class of positive functions $f$.

An example of this problem, with $f(y) = e^y$ and $A = 1$, appears in the books of Verhulst [**245**] and Bender and Orszag [**18**], where the method of matched asymptotic expansions is used formally to get inner and outer solutions. We thank Professor Verhulst for helpful correspondence and for sending us his recent paper on this problem [**13**]. It was apparently first noticed by Sadhu ([**208**]) that in this example there are exactly two solutions to (7.74)–(7.75). In [**165**] this result was proved more generally, for a class of functions $f$ and for $A$ in an interval $(0, A^*(\varepsilon))$ depending on $f$. Our goal here is to prove existence of solutions satisfying certain bounds and then use these bounds to confirm that the asymptotic expansion of this solution which is obtained by the methods of [**18**] or [**245**] is rigorously correct. As in the previous sections, this will be done using an integral equation.

We begin by finding the formal (nonrigorous) uniform asymptotic expansion for a solution of (7.74)–(7.75). This is routine and so is left to Exercise

7.5. As part of this calculation it is shown that the outer solution is

$$y_0(x) = \Phi^{-1}\left(\frac{1}{2}(A-x)\right)$$

where

$$\Phi(z) = \int_0^z \frac{1}{f(s)}ds. \tag{7.76}$$

Note that $\Phi$ is increasing, and $\lim_{y\to\infty}\Phi(y)$ may be finite or infinite, depending on $f$.

In the rest of this section, if $p$ is continuous on $[0,A]$ then $||p|| = \max_{0\le x\le A}|p(x)|$. The number $A$ will vary from one use to another, but this will not always be indicated in the notation.

**Theorem 7.2.** *Suppose that for $0 \le y < \infty$, $f(y) > 0$, $f'(y) \ge 0$, and $f''(y) \ge 0$. Then for any $A_0 \in (0, \Phi(\infty))$ there is an $\varepsilon(A_0) > 0$ such that if $0 < \varepsilon < \varepsilon(A_0)$ and $0 < A \le A_0$, then (7.74)–(7.75) has a nonnegative solution $y$, and this solution can be chosen to satisfy $\|y\| \le 3\Phi^{-1}\left(\frac{A_0}{2}\right)$ and $\varepsilon y'(0) \le 3\Phi^{-1}\left(\frac{A_0}{2}\right)$.*

**Proof.** Let $K_0 = 3\Phi^{-1}\left(\frac{A_0}{2}\right)$. The proof starts with a uniqueness theorem.[9]

**Theorem 7.3.** *There is an $\varepsilon_1 > 0$ such that if $0 < \varepsilon < \varepsilon_1$ and $A \in (0, A_0]$, then there is at most one solution to (7.74)–(7.75) satisfying $||y|| \le 2K_0$.*

**Proof.** Suppose that there are two solutions, say $y_1$ and $y_2$, and let $z = y_1 - y_2$. Then using the mean value theorem we get

$$\varepsilon z'' + 2z' + f'(\xi)z = 0,$$
$$z(0) = z(A) = 0,$$

where, for each $x$, $\xi$ lies between $y_1(x)$ and $y_2(x)$. If $z = e^{-\frac{x}{\varepsilon}}\zeta$, then $\zeta(0) = \zeta(A) = 0$ and

$$\varepsilon\zeta'' + \left(f'(\xi) - \frac{1}{\varepsilon}\right)\zeta = 0. \tag{7.77}$$

But $f'(y)$ is bounded in the region $|y| \le 2K_0$, and so for sufficiently small $\varepsilon$, independent of $A \in (0, A_0]$, the coefficient of $\zeta$ in (7.77) is negative on $[0,A]$. It is then easy to see that the only solution of (7.77) which vanishes at 0 and $A$ is $\zeta = 0$. This proves Theorem 7.3. □

Next, we derive an integral equation motivated by the formal asymptotic expansion found in Exercise 7.5. In that exercise the reader is asked to

[9] This is not the first time we have proved existence using a uniqueness theorem. See Chapter 5 for a brief discussion.

use the method of Section 7.1, matched asymptotic expansions, to obtain (nonrigorously) the approximate solution to (7.74)–(7.75) given by

$$y_u(x) = -\Phi^{-1}\left(\frac{A}{2}\right)e^{-\frac{2x}{\varepsilon}} + y_0(x). \tag{7.78}$$

After proving existence our goal is to show that $y - y_u = O(\varepsilon)$ uniformly on $[0, A]$ as $\varepsilon \to 0$. We find, however, that it is more convenient, and equivalent, to work with the function

$$g(x) = y(x) - y_0(x) + \frac{1}{2}\varepsilon C e^{-\frac{2x}{\varepsilon}} \tag{7.79}$$

where $C = y'(0)$. We will derive an integral equation for $g$.

**Lemma 7.4.** *With $A_0$ chosen as above, let $K_0 = 3\Phi^{-1}\left(\frac{A_0}{2}\right)$. Suppose that for some $\varepsilon_0 > 0$ and each $A \le A_0$ (7.74)–(7.75) has a nonnegative solution $y$ for each $\varepsilon \in (0, \varepsilon_0)$ and each of these solutions satisfies $||y|| \le K_0$ and $\varepsilon y'(0) \le K_0$. Then there is an $\varepsilon_1 \in (0, \varepsilon_0)$ such that for each $\varepsilon \in (0, \varepsilon_1)$ and each $A \in (0, A_0]$, the function $g$ as defined by (7.79) with $C = y'(0)$ satisfies the bound $\|g\| \le 2K_0$. Further, $g$ satisfies the equation*

$$g(x) = h_2(x, \varepsilon, \varepsilon C) + \frac{1}{\varepsilon}\int_x^A e^{-\frac{2t}{\varepsilon}} \int_0^t e^{\frac{2s}{\varepsilon}} \left(f\left(y_0 + g - \frac{\varepsilon C}{2}e^{-\frac{2s}{\varepsilon}}\right) - f(y_0)\right) ds\ dt \tag{7.80}$$

*where $h_2$ is a known function (given below). This equation can be put in the form*

$$g(x) = h(x, \varepsilon, \varepsilon C, g(x)) + \frac{1}{\varepsilon}\int_x^A e^{-\frac{2t}{\varepsilon}} \int_0^t e^{\frac{2s}{\varepsilon}} \left(f(y_0 + g) - f(y_0)\right) ds\ dt \tag{7.81}$$

*where $h(x, \varepsilon, \varepsilon C, g(x)) = O(\varepsilon)$ as $\varepsilon \to 0$ uniformly in $A$, $x$, $\varepsilon C$, and $g$ so long as $0 < A \le A_0$, $0 \le x \le A$, $\varepsilon C \le K_0$, and $||g|| \le 2K_0$. Conversely, if for some $C$ the equation (7.80) has a solution $g$, then the function $y$ defined by (7.79) satisfies (7.74), $y(A) = 0$, and $y'(0) = C$.*[10]

**Proof.** Under the assumptions in the first two sentences of the lemma the stated bound for $g$ is obvious from (7.79) and the definitions of $K_0$ and $y_0$. Now consider the following equations satisfied by $y$ and $y_0$ (see Exercise 7.5):

$$\varepsilon y'' + 2y' + f(y) = 0, \quad y(0) = y(A) = 0,$$
$$\varepsilon y_0'' + 2y_0' + f(y_0) = \varepsilon y_0'', \quad y_0(A) = 0.$$

[10] The converse is not used in our proof of Theorem 7.2, but plays an important role in the alternative proof in [**165**].

Subtract these equations and use variation of parameters to get

$$y - y_0 = h_1(x,\varepsilon) - \frac{1}{2}\varepsilon C\left(e^{-\frac{2x}{\varepsilon}} - e^{-\frac{2A}{\varepsilon}}\right)$$
$$- \frac{1}{\varepsilon}\int_A^x e^{-\frac{2t}{\varepsilon}} \int_0^t e^{\frac{2s}{\varepsilon}} \left(f(y) - f(y_0)\right) ds dt,$$

where $h_1(x,\varepsilon)$ is a known function, namely

$$h_1(x,\varepsilon) = \int_x^A e^{-\frac{2t}{\varepsilon}} \int_0^t e^{\frac{2s}{\varepsilon}} y_0''(s)\, ds dt + \varepsilon \frac{y_0'(0)}{2}\left(e^{-\frac{2x}{\varepsilon}} - e^{-\frac{2A}{\varepsilon}}\right).$$

Hence, the function $g$ defined by (7.79) satisfies (7.80) with

$$h_2(x,\varepsilon,\varepsilon C) = h_1(x,\varepsilon) + \frac{1}{2}\varepsilon C e^{-\frac{2A}{\varepsilon}}.$$

It is not hard to show that $h_1(x,\varepsilon) = O(\varepsilon)$ as $\varepsilon \to 0$ uniformly for $0 < A \le A_0$ and $0 \le x \le A$. It is then clear that $h_2(x,\varepsilon,\varepsilon C) = O(\varepsilon)$ as $\varepsilon \to 0$ uniformly for $x$ and $A$ in the same region and $\varepsilon C \le K_0$. Now write

$$f\left(y_0 + g - \frac{\varepsilon C}{2} e^{-\frac{2s}{\varepsilon}}\right) - f(y_0) = f(y_0 + g) - f(y_0) + Q\left(y_0, g, \varepsilon C e^{-\frac{2s}{\varepsilon}}\right)$$

for some continuous function $Q$. Under our hypotheses $||g||$ is bounded independent of $y$ and $\varepsilon$, and so there is a $K_1$ such that if $0 < A \le A_0$ and $\varepsilon C \le K_0$ then

$$\left|Q\left(y_0, g, \varepsilon C e^{-\frac{2s}{\varepsilon}}\right)\right| \le K_1 \varepsilon C e^{-\frac{2s}{\varepsilon}}.$$

Therefore the contribution of $Q$ to the double integral in (7.80) is bounded by a constant times $\frac{1}{\varepsilon}\int_x^A e^{-\frac{2t}{\varepsilon}} \int_0^t e^{\frac{2s}{\varepsilon}} \varepsilon C e^{-\frac{2s}{\varepsilon}}\, ds\, dt$. In Exercise 7.6 you are asked to show that because $\varepsilon C \le K_0$, this expression is $O(\varepsilon)$ as $\varepsilon \to 0$. Therefore we obtain the integral equation (7.81), proving the first three assertions of Lemma 7.4. The converse follows easily from (7.79), (7.80), and the definitions of $h_1$ and $h_2$. □

The next step is to obtain a much stronger bound on $||g||$.

**Lemma 7.5.** *There are positive numbers $M$ and $\varepsilon_2$, depending on $A_0$ and $f$, such that if, for some $A \in (0, A_0]$ and a pair $(\varepsilon, C)$ with $\varepsilon < \varepsilon_2$ and $\varepsilon C \le K_0$, there is a solution $g$ to (7.80) such that $||g|| \le 2K_0$, then this solution satisfies $||g|| \le M\varepsilon$.*

**Proof.** Let

$$L = \max_{0 \le w \le 3K_0} \left|f'(w)\right|.$$

If $g$ solves (7.80) and $||g|| \le 2K_0$, then from (7.81),

$$|g(x)| \le O(\varepsilon) + \frac{1}{\varepsilon}\int_x^A e^{-\frac{2t}{\varepsilon}} \int_0^t e^{\frac{2s}{\varepsilon}} L|g(s)|ds\, dt \tag{7.82}$$

$$\le O(\varepsilon) + ||g|| \frac{1}{\varepsilon}\int_x^A e^{-\frac{2t}{\varepsilon}} \int_0^t e^{\frac{2s}{\varepsilon}} L ds\, dt$$

$$= O(\varepsilon) + \frac{L}{2}||g||(A-x)$$

as $\varepsilon \to 0$. Substituting back into (7.82) gives

$$|g(x)| \le O(\varepsilon) + \frac{1}{\varepsilon}||g|| \int_x^A e^{-\frac{2t}{\varepsilon}} \int_0^t e^{\frac{2s}{\varepsilon}} \frac{L^2}{2}(A-s)\, ds\, dt$$

$$\le O(\varepsilon) + \left(\frac{L}{2}\right)^2 \frac{1}{2!}(A-x)^2 ||g||.$$

Choose $n_0$ so that $\frac{1}{n_0!}\left(\frac{L}{2}\right)^{n_0} = r < 1$. Continue back-substituting for $n_0$ steps to get

$$|g(x)| \le O(\varepsilon) + r\,||g||$$

or

$$(1-r)\,||g|| = O(\varepsilon),$$

where the constant in the $O(\varepsilon)$ term depends only on $K_0$, $L$, and $n_0$. Lemma 7.5 follows. □

Let

$$\varepsilon_3 = \min\left\{\varepsilon_1, \varepsilon_2, \frac{K_0}{24M}\right\}. \tag{7.83}$$

Consider a fixed $\varepsilon < \varepsilon_3$. We now look at the problem in a very different way, using basically a shooting method. We consider solutions to the equation (7.74) which satisfy the initial conditions

$$y(0) = 0, \qquad y'(0) = \alpha. \tag{7.84}$$

Denote the solution by $y = y(x, \alpha)$. Extend $f$ to $(-\infty, \infty)$ by setting $f(y) = f(0)$ if $y < 0$. In Exercise 7.7 you are asked to show that for any $\alpha > 0$, $y$ has exactly one local maximum say at $\xi(\alpha)$, and then exactly one zero, say at $\eta(\alpha)$. Hence, $y(x, \alpha)$ satisfies (7.74)–(7.75) with $A = \eta(\alpha)$.

Also, $y'(\eta(\alpha), \alpha) < 0$, and so $\eta(\alpha)$ is continuous, by the implicit function theorem. Note that

$$y(\eta(\alpha), \alpha) = 0. \tag{7.85}$$

Since $f(y) > 0$, we see that $y(x, 0) < 0$ for all $x > 0$. It follows that

$$\lim_{\alpha \to 0+} \eta(\alpha) = 0.$$

Also, $\lim_{\alpha\to 0^+} ||y(x,\alpha)|| = 0$. Hence for small $\alpha$, $y$ satisfies the bounds in the hypotheses of Lemma 7.4 and (7.80) has a solution $g$ for $A = \eta(\alpha)$.

We wish to study in more detail how $y(x,\alpha)$ varies with $\alpha$, and for this we turn to the equation of first variation, that is, the equation satisfied by $\phi = \frac{\partial y}{\partial \alpha}$. This is

$$\varepsilon\phi'' + 2\phi' + f'(y)\,\phi = 0. \tag{7.86}$$

Also,

$$\phi(0) = 0, \qquad \phi'(0) = 1. \tag{7.87}$$

It follows that $\phi > 0$ on some initial interval $[0, x_0]$. For a given $\varepsilon$, $x_0$ can be chosen independent of bounded $\alpha$, because $f'(y)$ is bounded as long as $||y|| \le K_0$. Thus, for sufficiently small $\alpha$, $\phi > 0$ on $[0, \eta(\alpha)]$.

By differentiating (7.85) with respect to $\alpha$ we obtain

$$y'(\eta(\alpha),\alpha)\frac{d\eta}{d\alpha} + \phi(\eta(\alpha)) = 0. \tag{7.88}$$

Thus, $\eta$ is an increasing function of $\alpha$ for small $\alpha$. (We continue to consider a fixed $\varepsilon \in (0,\varepsilon_3)$.) As $\alpha$ increases, we anticipate that $||y||_\alpha := \max_{0\le x\le\eta(\alpha)} |y(x,\alpha)|$ may increase. Recall that $y'(0,\alpha) = \alpha$. We consider three possibilities: (i) $\eta(\alpha)$ has a first critical point, say at $\alpha_1$, and for $0 < \alpha \le \alpha_1$, $||y||_\alpha < K_0$ and $\varepsilon\alpha < K_0$, or (ii) at some first $\alpha_2$, before $\eta$ either exceeds $A_0$ or has a critical point, $||y||_{\alpha_2} = K_0$ or $\varepsilon\alpha_2 = K_0$, or (iii) $\eta(\alpha)$ has no critical point on $0 < \alpha < \infty$ and the solutions all satisfy the required bounds.

Consider possibility (ii). It assumes that $\eta(\alpha_2) \le A_0$. Then by Lemma 7.4, the function $g$ defined by (7.79) with $y = y(x,\alpha_2)$ satisfies $||g|| \le 2K_0$. Then, by Lemma 7.5, $||g|| \le M\varepsilon$. From (7.79), noting that $C = \alpha$ and $\varepsilon < \varepsilon_3$ and using (7.83),

$$||y||_{\alpha_2} \le M\varepsilon + ||y_0|| + \left|\frac{1}{2}\varepsilon C\right| \le \frac{K_0}{24} + \frac{K_0}{3} + \frac{1}{2}K_0 < K_0.$$

Also, from (7.84)

$$\left|\frac{1}{2}\varepsilon C\right| \le |g(0)| + ||y_0|| \le \frac{K_0}{24} + \frac{K_0}{3} < \frac{1}{2}K_0,$$

and so

$$|\varepsilon y'(0)| = |\varepsilon C| < K_0.$$

Hence, possibility (ii) does not occur. Now consider possibility (i). We claim that in this case, $\eta(\alpha_1) > A_0$.

To show this we must look more carefully at $\phi$, which satisfies (7.86)–(7.87). Differentiating (7.74) shows that another solution to (7.86) is given by $y'$, which satisfies $y'(0) > 0$ and so is not a multiple of $\phi$. Since $y'$ has a unique zero in $[0, \eta(\alpha)]$, at $\xi(\alpha)$, the Sturm comparison theorem implies that $\phi$ has no zero in $[0, \xi(\alpha)]$ and at most one zero in $[\xi(\alpha), \eta(\alpha)]$. If $\eta(\alpha_1)$ is a maximum for $\eta$, then (7.88) implies that $\phi(\eta(\alpha_1)) = 0$.

**Lemma 7.6.** *Let* $\psi = \frac{d\phi}{d\alpha} = \frac{d^2y}{\partial\alpha^2}$. *If* $\phi(\eta(\alpha_1)) = 0$ *then* $\psi(\eta(\alpha_1)) < 0$.

**Proof.** We leave this to Exercise 7.8. □

But this result implies that there are $\alpha_3 < \alpha_1$ and $\alpha_4 > \alpha_1$, both as close as we wish to $\alpha_1$, with $\eta(\alpha_3) = \eta(\alpha_4)$. Hence, for $A$ slightly below $\eta(\alpha_1)$ there are two solutions of (7.74)–(7.75). We assumed that $||y||_{\alpha_1} \leq K_0$. Therefore, for $A$ sufficiently close to $\eta(\alpha_1)$ each of the two solutions satisfies $||y|| \leq 2K_0$. This contradicts Theorem 7.3. Hence $\eta(\alpha_1) > A_0$.

We are left with possibility (iii) above. If $\eta$ ever reaches $A_0$ we are done. If $\eta$ continues to increase but never reaches $A_0$, then we obtain a contradiction when $\alpha$ reaches $\frac{K_0}{\varepsilon}$, because that puts us in case (ii).[11] The solutions satisfy the needed bounds as long as $\eta(\alpha) \leq A_0$.

We have thus shown that for any $\varepsilon < \varepsilon_3$ there is a solution of (7.74)–(7.75) satisfying the bounds in Theorem 7.2 for any $A \in (0, A_0]$. We have therefore proved this theorem. □

**Corollary 7.7.** *The solutions found in Theorem 7.2 satisfy* $y - y_u = O(\varepsilon)$ *uniformly on* $[0, A]$ *as* $\varepsilon \to 0$.

**Proof.** Lemma 7.5 implies that $||g|| = O(\varepsilon)$, and Exercise 7.5 completes the proof. □

**Remark 7.8.** The corollary depends primarily on Lemma 7.5, which has the existence of a solution with certain bounds as a hypothesis. If one can prove this existence by some other means, the conclusion would still hold. For example, in [**165**] existence was proved for any positive continuous function $f$ such that $f(y) \leq e^y$ for all $y$. No monotonicity conditions are imposed. The proof was by the method of upper and lower solutions, which is discussed in Chapter 18.

## 7.8. A brief discussion of blow-up in two dimensions

The blow-up technique (due to Takens [**235**]) is a way of handling nonlinear perturbations of linear systems when an equilibrium point is nonhyperbolic. Thus, some eigenvalues of the linear system have zero real part. Blow-up can be used in conjunction with center manifold analysis, as in [**194**] and [**195**]. We will discuss two examples. The first is from [**34**].

[11] In [**165**] conditions on $f$ are given which ensure that $\eta$ does have a local maximum value, say $A^*(\varepsilon)$, and that there are exactly two solutions for $0 < A < A^*(\varepsilon)$.

The system is

$$
\begin{aligned}
x' &= x^2 - 2xy, \\
y' &= y^2 - 2xy.
\end{aligned}
\tag{7.89}
$$

The only equilibrium is $(0,0)$, and the linearized system at this point is $u' = v' = 0$. The blow-up procedure is to put (7.89) in polar coordinates $(r,\theta)$ and then view the resulting system as one in the $(r,\theta)$-plane, with the entire $\theta$-axis as a set of equilibrium points.[12]

Setting $x = r\cos\theta$, $y = r\sin\theta$, the differential equations in $r$ and $\theta$ are found to be

$$
\begin{aligned}
r' &= r^2\left(\cos^3\theta - 2\cos^2\theta\sin\theta - 2\cos\theta\sin^2\theta + \sin^3\theta\right), \\
\theta' &= 3r\left(\cos\theta\sin^2\theta - \cos^2\theta\sin\theta\right).
\end{aligned}
\tag{7.90}
$$

As stated above, this is considered on the entire $(r,\theta)$-plane, with no regard to the singular nature of its relation to the $(x,y)$ system at $r=0$. The final goal is to describe the $(x,y)$ phase space in a neighborhood of $r=0$. It is important to note that the trajectories of (7.90) are unchanged away from $r=0$ if each right-hand side is divided by $r$, because $\frac{dr}{d\theta}$ does not change. The resulting system has isolated equilibria on the $\theta$-axis, at $\theta = 0, \frac{\pi}{4}, \frac{\pi}{2}, \pi, \frac{5\pi}{4}$, and $\frac{3\pi}{2}$, and then at subsequent points by periodicity.

It turns out that each of these equilibrium points of (7.90) on $r=0$ is hyperbolic, and they are alternating stable and unstable nodes. It is then easy to translate this behavior back to $x$, $y$ to get the existence of three orbits which tend to $(0,0)$ as $t\to\infty$ and three which tend to $(0,0)$ as $t\to-\infty$, with other orbits leaving a small neighborhood of the origin for both increasing and decreasing $t$. Some pictures of these are in [**34**]. Also, see Exercise 7.9.

The second system to be considered here was shown to us by our colleague Jon Rubin. The equations are

$$
\begin{aligned}
x' &= x^2 - y^2, \\
y' &= x\left(y + x^2\right).
\end{aligned}
\tag{7.91}
$$

Once again, $(0,0)$ is a nonhyperbolic equilibrium point. Setting $x = r\cos\theta$, $y = r\sin\theta$ and dividing by a common factor $r$, we obtain the system

$$
\begin{aligned}
r' &= r\cos^3\theta + r^2\sin\theta\cos^3\theta = F(r,\theta), \\
\theta' &= r\cos^4\theta + \sin^3\theta = G(r,\theta).
\end{aligned}
\tag{7.92}
$$

The equilibrium points along the $\theta$-axis are at $\theta = k\pi$, $k = 0, \pm1, \pm2, \ldots$.

[12] A more modern technique is to consider the system on a cylinder, thus restricting $\theta$ to $[0,2\pi]$ and identifying the lines $\theta=0$ and $\theta=2\pi$. While this geometry is standard in dynamical systems theory, we do not wish to get into a detailed discussion here, and when stating the results of the theory (as opposed, perhaps, to the proofs), we do not have to consider it.

But now we encounter a further difficulty. Linearizing (7.92) around $k\pi$ shows that these points are still nonhyperbolic. Indeed, the linearized matrix is

$$\begin{pmatrix} (-1)^k & 0 \\ 1 & 0 \end{pmatrix}.$$

Hence the behavior around these equilibrium points is still not clear, though it is easy to determine the flow in the $\theta$-axis, which is invariant. We find alternating stable and unstable nodes for the resulting one-dimensional system.

This hiccup is common when applying the blow-up method. The usual remedy is first to determine the Taylor series for the functions $F$ and $G$ in (7.92) around the equilibrium points and truncate at an appropriate level. The theory of "normal forms" [**81**] tells us what level of truncation is appropriate but is well beyond what we can cover here, so we will simply give a plausible looking truncated system for $k = 0$, in which terms of order 4 and higher have been dropped. This system is

$$\begin{aligned} r' &= r + r^2\theta - \frac{3}{2}r\theta^2, \\ \theta' &= r - 2r\theta^2 + \theta^3. \end{aligned}$$

Then one does a second blow-up, letting $r = \rho\cos\phi$, $\theta = \rho\sin\phi$. One might find that again the equilibrium points are nonhyperbolic and still another blow-up is required. See [**235**] or [**81**] for an example where three blow-ups are used to give the desired answer.

One may ask if this is the best way to analyze this system. Professor Rubin has pointed out that an intermediate strategy may be better, in which one blow-up is done and then standard phase plane analysis is applied to the resulting system (7.92). In Exercise 7.10 the reader is asked to give a complete phase plane description for the system (7.91).

## 7.9. Exercises

**Exercise 7.1.** (a) Find the solution of each of the following equations which satisfies $y(0) = 0$, $y(1) = 1$: (i) $\varepsilon^2 y'' = 1$, (ii) $\varepsilon^2 y'' = y$, (iii) $\varepsilon^2 y'' = y - 1$. (b) Plot each of the solutions with $\varepsilon = 0.1$ and decide which has what might be called a "boundary layer" and where this layer is located.

**Exercise 7.2.** Show that if, in problem (7.1)–(7.2), we let $y_0(x) = 2x$ and set $\eta = \frac{1-x}{\varepsilon}$, then matching is not possible.

**Exercise 7.3.** Use integration by parts to verify the asymptotic relations in (7.31).

**Exercise 7.4.** Construct the manifold $\mathcal{W}_\varepsilon^s$ by shooting. Consider initial conditions for (7.34) of the form

$$u(0) = \alpha, \qquad v(0) = \beta, \qquad \eta(0) = \delta.$$

Hold $\delta$ fixed as a small positive number. Show that for each positive $\beta$ sufficiently close to 0 there is a unique $\alpha < 1$ such that

$$\lim_{r \to \infty} (u(r), v(r), \eta(r)) = (1, 0, 0).$$

Hint: Use techniques from Section 7.4.1.

**Exercise 7.5.** (Needed in chapter; not difficult) Derive the uniform expansion $y_u$ in (7.78) by following the method in Section 7.1. Show that if $g = O(\varepsilon)$ as $\varepsilon \to 0$ ($g$ defined by (7.79)), then $y - y_u = O(\varepsilon)$ as $\varepsilon \to 0$.

**Exercise 7.6.** (Needed in chapter; not difficult) Prove that $\int_x^1 te^{-\frac{2t}{\varepsilon}} dt = O(\varepsilon^2)$ as $\varepsilon \to 0$ uniformly for $0 \le x \le 1$.

**Exercise 7.7.** (Needed in chapter; not difficult) Show that for any $\alpha > 0$ the solution $y(x, \alpha)$ of (7.74) and (7.84) has a single maximum and then decreases to its first and only zero, at $\eta(\alpha)$. Hint: Putting the linear part in selfadjoint form is helpful.

**Exercise 7.8.** (Needed in chapter; more challenging; see [**165**]) Complete the proof of Lemma 7.5 by showing that if $\phi(\eta(\alpha)) = 0$ then $\psi(\eta(\alpha)) < 0$. Hint: Find a differential equation for $\phi\psi' - \psi\phi'$ and solve it. Note that $\phi > 0$ on $(0, \eta(\alpha_1))$. The case $f''(0) > 0$ is easier, and the size of $\varepsilon$ plays no role in this problem.

**Exercise 7.9.** Analyze the system (7.89) directly in the $(x, y)$-plane, using nullclines. Obtain the results described above for this system. Hint: For part of the analysis it may be helpful to consider $x - y$. Give a global description of the flow using the symmetry of the system.

**Exercise 7.10.** Analyze the system (7.91) in the $(x, y)$-plane, and the system (7.92) in the $(r, \theta)$-plane, using nullclines. Show that for (7.91) there are three families of orbits which are homoclinic to the point $(0, 0)$, characterized as follows: (a) orbits which tend to $(0, 0)$ in both directions and are entirely in the half-plane $y < 0$, (b) orbits which do the same except that they are entirely in the half-plane $y > 0$, (c) orbits which also tend to $(0, 0)$ in both directions but are partly in the upper half-plane and partly in the lower. Continue on to obtain a description of the global flow. Compare the partial use of the blow-up method with an analysis done entirely in the $(x, y)$-plane.

Chapter 8

# Asymptotics beyond all orders

## 8.1. Introduction

The term "beyond all orders" generally refers to an expression involving some parameter, say $\varepsilon$, which tends to zero faster than any power of $\varepsilon$ as $\varepsilon$ tends to zero. (An explicit example is given in formula (8.9) below.) If a solution of a boundary value problem involves such an expression, then standard asymptotic expansions may be invalid. In this chapter we study another problem in singular perturbation:

$$\varepsilon u''' + u' = 1 - u^2, \tag{8.1}$$

$$u(\pm\infty) = \pm 1. \tag{8.2}$$

We ask whether there is a solution which is monotonic on $(-\infty, \infty)$. This problem arises when looking for traveling wave solutions of the Kuramoto-Sivashinsky equations, which are an important model for phase turbulence in reaction-diffusion systems [**131**], [**217**]. In connection with turbulence, most interest centers on the existence of periodic solutions, as in [**240**], but it is desirable to know if, as well, there are heteroclinic orbits. By looking at the linearized system around the two equilibrium points, we will see that existence of a solution to (8.1)–(8.2) is likely to be a delicate question. Singular perturbation analysis suggests that there is a solution, but its monotonicity is in doubt. The main result in the chapter is that no monotonic solution exists. The proof involves complex variables and is quite different from anything else in this book. The problem (8.1) was discussed by Rhagavan,

McLeod, and Troy in [**203**], and this chapter is largely modeled on that paper, with some simplifications. We will also cite other papers where similar phenomenon have been studied.

As a first step in analyzing the problem we linearize (8.1) around $u = 1$, obtaining the eigenvalue equation

$$\varepsilon\lambda^3 + \lambda + 2 = 0. \tag{8.3}$$

For $\varepsilon > 0$ there are two complex roots, $m_1$ and $\bar{m}_1$, with positive real parts and one real negative root $m_2$. Clearly, $m_2 \to -2$ as $\varepsilon \to 0$. When (8.1) is linearized around $u = -1$ there is one real positive eigenvalue and two with negative real parts.

Writing (8.1) as a system of first order equations, say

$$\begin{aligned} u_1' &= u_2, \\ u_2' &= u_3, \\ \varepsilon u_3' &= 1 - u_1^2 - u_2, \end{aligned} \tag{8.4}$$

we conclude that for every $\varepsilon > 0$ there is a one-dimensional unstable manifold at $(-1, 0, 0)$ and a one-dimensional stable manifold at $(1, 0, 0)$. We do not expect that they will intersect, at least not for every $\varepsilon > 0$, but there is the possibility that some hidden symmetry will allow this to happen. If so then there is a heteroclinic orbit connecting the two equilibrium points, and correspondingly, a solution to (8.1)–(8.2). If this solution does exist, we ask whether it is monotonic. It is asserted in [**104**] that numerical evidence supports the existence of a monotonic solution for small $\varepsilon > 0$. As mentioned already, we will show that no such solution exists. In [**131**] a nonmonotonic solution is exhibited for a specific positive value of $\varepsilon$.[1]

While the argument just given makes us sceptical that there is a heteroclinic orbit, an approach using perturbation methods suggests that such an orbit might exist for small $\varepsilon > 0$. First, observe that if $\varepsilon = 0$ then the problem (8.1)–(8.2) has the monotonic solution $u_0(x) = \tanh x$. Therefore we consider a series solution of the form

$$u_\varepsilon(x) = \sum_{n=0}^{\infty} \varepsilon^n u_n(x). \tag{8.5}$$

If we make this substitution in (8.1), we obtain

$$\begin{gathered} u_0' = 1 - u_0^2, \\ u_1' + 2u_0u_1 + u_0''' = 0, \\ \dots, \end{gathered} \tag{8.6}$$

[1]Physically, $\varepsilon$ can be positive or negative. In [**242**] there is an extended discussion of the case $\varepsilon < 0$. There is also a proof that there is no monotonic solution if $\varepsilon \geq \frac{2}{9}$.

and based on the boundary conditions (8.2), we require that

$$u_0(\pm\infty) = \pm 1,$$
$$u_i(\pm\infty) = 0 \quad \text{for } i \geq 1.$$

As above, $u_0(x) = \tanh x$.

For $u_1$, the solution is

$$u_1(x) = \frac{12\tanh x - 8x}{\cosh 2x + 1}. \tag{8.7}$$

Since both $u_0$ and $u_1$ are odd functions which satisfy the boundary conditions, they appear to buttress the argument for existence of a solution to (8.1)–(8.2). Plotting $u_1(x)$ shows that it is not monotonic. However, the nonmonotonicity consists of just a few oscillations, over a limited range of $x$, so that a monotonic solution for small $\varepsilon$ is not ruled out.

As far as we know, rigorous analysis of the remaining terms has not been done, or at least not been published, but in [**130**] this analysis was done for a similar problem, one which has been studied by a number of authors. The equations are

$$\begin{cases} \varepsilon u''' + u' = \cos u, \\ u(0) = 0, \ u(\pm\infty) = \frac{\pi}{2}. \end{cases} \tag{8.8}$$

This alternative system was suggested as a model of pattern formation in solidification problems. The monotonicity requirement is necessary for the solution to represent a physical pattern, but it was recognized quickly that existence of a monotonic solution is doubtful [**141**].[2]

In [**130**], Kruskal and Segur analyzed the result of substituting (8.5) into (8.8).[3] It turns out that each $u_i$ exists and is an odd function. However they showed, as well, that the problem (8.8) has no monotonic solution. Other proofs of this were given in [**9**] and [**84**]. Subsequently, the problem (8.1)–(8.2) was studied, and again, nonexistence of a monotonic solution was proved [**112**], [**203**]. This chapter is based largely on the last of these references. The basic technique, however, was developed in [**9**]. It was also applied in [**10**].

Kruskal and Segur proved more than just nonexistence of a solution. In their analysis of the differential equation in (8.8), they showed that the series (8.5) is an asymptotic series for a monotonic solution $u_\varepsilon$ of the following

[2]The physical consequence of nonexistence is that some asymmetry must be added to the model.

[3]Although this paper appeared in print later than the others we will cite, it seems to have been the first rigorous analysis of the problem.

problem on a half-line:[4]

$$\varepsilon u''' + u' = \cos u,$$
$$u(\infty) = \frac{\pi}{2}, \quad u(0) = 0.$$

While we will not discuss the theory of asymptotic series [**41**], we note that such a series need not converge. The existence of a monotonic heteroclinic orbit depends on whether $u_\varepsilon$ can be extended to $(-\infty, \infty)$ as an odd solution, which requires that $u''_\varepsilon(0) = 0$. Since all the terms in the asymptotic series are odd functions, this suggests that if $u_\varepsilon$ is not odd then as $\varepsilon \to 0$, $u''_\varepsilon(0)$ must be smaller than every term in the series expansion and so smaller than every power of $\varepsilon$.

All of the authors who have written on this topic have gone further than the observations above and have found a precise asymptotic formula for $u''_\varepsilon(0)$. It is in this formula that we see the "beyond all orders" phenomenon, and so we will present it here, from [**203**]: For some positive constant $A$,

$$u''_\varepsilon(0) \sim -A\varepsilon^{-2} e^{-\frac{\pi}{2\sqrt{\varepsilon}}} \tag{8.9}$$

as $\varepsilon \to 0$.[5] Here we will be content with proving that $u''_\varepsilon(0) \neq 0$, as an illustration of an interesting technique introduced in [**9**]. This technique is to extend the problem into the complex plane. While complex variables are frequently encountered in the classical study of linear ode's, their use in nonlinear equations is much less common.

## 8.2. Proof of nonexistence

The nonexistence result is Corollary 8.3 below. Its proof does not depend on $\varepsilon$ being small, and so for simplicity, we will assume that $\varepsilon = 1$. The changes necessary when $\varepsilon \neq 1$ are trivial. Consider again the one-dimensional stable manifold for the system (8.4) at the equilibrium point $(1, 0, 0)$. On one branch of this manifold, solutions approach equilibrium from the region $u_1 < 1$, $u_2 > 0$, $u_3 < 0$, and so a corresponding solution of (8.1) satisfies $u(x) < 1$, $u'(x) > 0$, $u''(x) < 0$ for large $x$. We then have

**Lemma 8.1.** *If $u$ is a solution of (8.1) which, for large positive $x$, satisfies $0 < u(x) < 1$ and $u'(x) > 0$, then for some $X$,*

$$u(X) = 0, \tag{8.10}$$

$$u' > 0 \text{ on } [X, \infty), \quad u(\infty) = 1. \tag{8.11}$$

[4] Kruskal and Segur consider the left half of the solution, on $(-\infty, 0]$, but a simple reflection $y(x) \to -y(-x)$ transforms to $[0, \infty)$.

[5] The ratio of the two sides tends to one.

**Proof.** Clearly, $u(x) \to 1$, at an exponential rate, as $x \to \infty$, and so the variation of parameters formula shows that

$$u'(x) = \int_x^\infty \sin(x-t)\left(u(t)^2 - 1\right) dt.$$

Integrating by parts gives

$$u'(x) = 2\int_x^\infty (1-\cos(x-t))\ u(t)\, u'(t)\, dt. \tag{8.12}$$

If $u'(x_1) = 0$ at some largest $x_1$, with $u(x_1) \geq 0$, then (8.12) gives a contradiction when $x = x_1$, because on $(x_1, \infty)$, $0 < u < 1$ and $u' > 0$. The existence of $X$ follows, because there are no equilibrium points in $(-1, 1)$. □

Using a translation in $x$ if necessary, we assume, for the solution in this lemma, that $X = 0$. It is then unique, by the stable manifold theorem.

**Theorem 8.2.** *Let $u$ be the unique solution of* (8.1) *satisfying* (8.10) *and* (8.11) *with $X = 0$. Then $u''(0) \neq 0$.*

**Corollary 8.3.** *There is no monotonic solution of* (8.1)–(8.2).

We leave the proof of the corollary to the reader, in Exercise 8.3.

**Proof of Theorem 8.2.** The proof of Lemma 8.1 suggests an obvious first step: Differentiate (8.12) to get an integral equation for $u''$. After another integration by parts this leads to

$$u''(x) = 2\int_x^\infty (1-\cos(x-t))\left(u'^2 + uu''\right) dt. \tag{8.13}$$

It was seen above that for large $x$, $u'' < 0$ and $u > 0$, so that (8.13) does not immediately lead to a conclusion about $u''(0)$. However, this equation will prove useful below.

If $m_2$ is the negative root of (8.3) then for some $c \neq 0$,

$$u(x) = 1 - ce^{m_2 x} + o\left(e^{m_2 x}\right) \quad \text{as } x \to \infty. \tag{8.14}$$

Changing $c$ is equivalent to translation along the $x$-axis, and so $c$ is uniquely determined by setting $X = 0$ in (8.10). Since $u$ approaches 1 from below, $c > 0$. Henceforth in this proof, $c$ is unchanged.

While it follows from Theorem 8.1 in Chapter 1 of [**41**] that $u$ can be extended as a solution in the complex plane, in a neighborhood of each point on the nonnegative real axis, this is insufficient here. From now on, let $z = x + iy$, and let $y_1 = -\frac{\pi}{m_2}$, so that $e^{m_2(x+iy_1)} = -e^{m_2 x}$.

**Lemma 8.4.** *The solution $u$ can be extended as an analytic solution $u(z)$ of* (8.1) *in the complex plane, existing in a half-strip*

$$S_A = (x + iy \mid x > A,\ 0 \leq y \leq y_1)$$

*for some sufficiently large A. This solution satisfies*

$$u(z) = 1 - ce^{m_2 z} + o\left(e^{m_2 x}\right) \tag{8.15}$$

*as $x$ tends to infinity. This approximation is uniform in $y$, meaning that*

$$\lim_{x\to\infty} \frac{u(z) - 1 + ce^{m_2 z}}{e^{m_2 x}} = 0$$

*uniformly for $0 \le y \le y_1$.*

**Proof.** The existence of $u(x)$ for $x$ real follows from the stable manifold theorem for the system (8.4). However, to extend into the complex plane we need a direct existence proof which is valid in $S_A$. Such proofs were given, either for (8.1) or (8.8), by several authors, [**9**], [**203**], and [**130**], but only Appendix A of [**130**] includes details for the extension into $\operatorname{Im} z > 0$. We will give only an outline here.

The variation of parameters formula applied to (8.1) results in the following integral equation, which is satisfied by the solution $u(x)$ defined just after Lemma 8.1:

$$\begin{aligned} u(x) &= 1 - ce^{m_2 x} + a_1 \int_x^\infty e^{m_2(x-t)} (u-1)^2\,dt \\ &\quad + a_2 \int_x^\infty e^{m_1(x-t)} (u-1)^2\,dt + a_3 \int_x^\infty e^{\bar{m}_1(x-t)} (u-1)^2\,dt. \end{aligned} \tag{8.16}$$

Here $a_1, a_2$, and $a_3$ can be calculated explicitly [**203**], but we will not need their precise forms. To extend this solution to the complex plane, define a sequence of functions by

$$u_0(z) = 1 - ce^{m_2 z},$$

$$\begin{aligned} u_{n+1}(z) &= 1 - ce^{m_2 z} + a_1 \int_z^\infty e^{m_2(z-\zeta)} (u_n - 1)^2\,d\zeta \\ &\quad + a_2 \int_z^\infty e^{m_1(z-\zeta)} (u_n-1)^2\,d\zeta + a_3 \int_z^\infty e^{\bar{m}_1(z-\zeta)} (u_n-1)^2\,d\zeta. \end{aligned}$$

The path of integration is from $z$ along the line $\operatorname{Im}\zeta = \operatorname{Im} z$ to some point $\operatorname{Re}\zeta = \beta$, for a large $\beta$, then vertically along $\operatorname{Re}\zeta = \beta$ to the real axis, then along the real axis to $\infty$. A convergence argument, which includes choosing larger and larger $\beta$, shows the convergence of this sequence in $S_A$ if $A$ is sufficiently large. For further details see [**130**]. □

Now, assume that $u = v + iw$, where $v$ and $w$ are real-valued functions of $x$ and $y$. We will be interested in $v$ and $w$ along lines $y = by_1$ for $0 \le b \le 1$. Along such a line, $v$ and $w$ satisfy

$$v_{xxx} + v_x = 1 - v^2 + w^2, \tag{8.17}$$

$$w_{xxx} + w_x = -2vw. \tag{8.18}$$

Referring to Lemma 8.4, let

$$x_0 = \inf \{A \mid u(z) \text{ is analytic in } S_A\}.$$

It will be clear below that $x_0 > -\infty$. We now analyze $v$ and $w$ in more detail in $S_{x_0}$.

Equation (8.15) implies that on the line $y = y_1$, $u$ is real (see Exercises 8.1 and 8.2), and so $u(z) = v(x, y_1)$. The solution exists on this line at least in the interval $(x_0, \infty)$, and by using the Cauchy Riemann equations, we obtain a set of inequalities which are satisfied by $v$ and $w$.

**Lemma 8.5.** *On the interval $(x_0, \infty)$, $v(x, y_1)$ and $w(x, y_1)$ satisfy the following relations:*

$$\begin{gathered} v_x < 0, \quad v_{xx} > 0, \quad v_y = 0, \quad v_{yy} < 0, \\ w = w_x = w_{xx} = 0, \quad w_y < 0, \quad w_{yy} = 0. \end{gathered}$$

**Proof.** From (8.15) it is seen that on $y = y_1$, $w = w_x = w_{xx} = 0$. The proof of Lemma 8.4 shows that (8.15) can be differentiated, and it follows that $v_x < 0$ and $v_{xx} > 0$ for large $x$. The proof that $v_x < 0$ on $(x_0, \infty)$ is essentially the same as the proof of Lemma 8.1. This result implies that $v > 1$ in this interval. Since $w = 0$ on $y = y_1$, (8.17) is the same as (8.1). Hence, the equation (8.13), derived earlier for $u''(x)$, is also valid for $v_{xx}(x, y_1)$. Now, however, with $v_{xx} > 0$ for large $x$, (8.13) and the argument of Lemma 8.1 show that $v_{xx}(x, y_1) > 0$.

The other inequalities in the lemma follow from the Cauchy-Riemann equations and the fact that $v$ and $w$ are harmonic functions of $(x, y)$. For example, $w_x < 0$ implies that $v_y > 0$ by the Cauchy-Riemann equations, and $v_{xx} > 0$ implies that $v_{yy} < 0$. □

**Lemma 8.6.** *Let $x_1 = \max(0, x_0)$. On the $x$-axis, $v$ and $w$ satisfy the following relations for $x > x_1$:*

$$\begin{gathered} v_x > 0, \quad v_{xx} \le 0, \quad v_y = 0, \quad v_{yy} \ge 0, \\ w = w_x = w_{xx} = 0, \quad w_y > 0, \quad w_{yy} = 0. \end{gathered}$$

**Proof.** It has been shown that $v_x(x, 0) > 0$, and hence, from the Cauchy Riemann equations, $w_y(x, 0) > 0$. Further, $w = w_x = w_{xx} = 0$ on the $x$-axis, and therefore $v_y = w_{yy} = 0$. There remains to show that for $x > x_1$,

$$v_{yy}(x, 0) \ge 0, \qquad v_{xx}(x, 0) \le 0. \tag{8.19}$$

To establish (8.19), we use the following lemma.

**Lemma 8.7.** *In the set $S_{x_1}$,*

$$v_y(x, y) \ge 0. \tag{8.20}$$

**Proof.** In $S_{x_1}$, for a fixed $x$, $v$ and $w$ satisfy the following equations, in which $y$ is the independent variable:

$$-v_{yyy} + v_y = 2vw, \tag{8.21}$$

$$-w_{yyy} + w_y = 1 - v^2 + w^2. \tag{8.22}$$

For each $x > x_1$, let $\Gamma_x$ denote the line segment $\{(x, y) \mid 0 \le y \le y_1\}$. It follows from (8.15) that there is an $x_2 > x_1$ such that if $x \ge x_2$ and $(x, y) \in \Gamma_x$ then $v(x, y) > 0$. Now consider a fixed $\bar{y} \in (0, y_1)$. Again, (8.15) implies that $w(x, \bar{y}) > 0$ for sufficiently large $x$. Then use the following integral equation for $w$, which is easily checked:

$$w(x, \bar{y}) = 2\int_x^\infty (1 - \cos(x - t))\, v(t, \bar{y})\, w(t, \bar{y})\, dt. \tag{8.23}$$

By the argument used in Lemma 8.1, $w(x, \bar{y}) > 0$ for $x \ge x_2$.

Since $v_y = 0$ on $y = 0$ and $y = \bar{y}$, it follows from (8.21) that for $x \ge x_2$, $v_y$ does not have a negative minimum on $\Gamma_x$. Hence (8.20) holds for $x \ge x_2$.

Now suppose that there is a point in the interior of $S_{x_1}$ where $v_y < 0$. Let

$$x_3 = \sup\{x > x_1 \mid v_y < 0 \text{ at some point of } \Gamma_x\}.$$

Then for every $x \ge x_3$, $v_y \ge 0$ on $\Gamma_x$. Since $v(x, 0) > 0$ for $x > 0$, we integrate $v_y$ vertically from a point $(x, 0)$ with $x \ge x_3$ and see that $v > 0$ on $\Gamma_x$. Then (8.23) implies that $w > 0$ on $\mathring{\Gamma}_x = \Gamma_x \cap \{0 < y < y_1\}$.

If $v_{yy}(x_3, 0) < 0$ then $v_y(x_3, y) < 0$ for small $y > 0$, a contradiction to the definition of $x_3$. If $v_{yy}(x_3, 0) > 0$, but $v_y$ has a zero in $\mathring{\Gamma}_{x_3}$, then that point is a minimum of $v_y$ on $\Gamma_{x_3}$. But then (8.21) gives a contradiction. Finally, if $v_{yy}(x_3, 0) = 0$ then differentiate (8.21), giving

$$-v_{yyyy} + v_{yy} = 2v_y w + 2v w_y.$$

At $(x_3, 0)$, $v > 0$, $w = 0$, and $w_y = v_x > 0$, and so the right side is positive. Hence at $(x_3, 0)$ we have $v_y = v_{yy} = v_{yyy} = 0$, $v_{yyyy} < 0$, and $v_y$ is therefore negative somewhere on $\Gamma_{x_3}$, again a contradiction. Further, $v_{yy}(x, 0) > 0$ and $v_{yy}(x, y_1) < 0$ on some interval $[x_3 - \delta, x_3]$. Consequently there is no sequence of points in $S_{x_1}$ tending to $\Gamma_{x_3}$ on which $v_y < 0$, contradicting the definition of $x_3$ and proving Lemma 8.7. □

The inequalities in (8.19) follow, so that Lemma 8.6 is proved as well. □

Another consequence of (8.20) is that if, for some $\bar{y} \in (0, y_1]$ and some $\bar{x} > 0$, the solution $v(x, \bar{y})$ is defined for $x \ge \bar{x}$, then this is also true for $0 \le y < \bar{y}$. We see this because $v(x, y) \le v(x, \bar{y})$, so that as $x$ decreases from $\infty$, $v(x, y)$ does not blow up before $v(x, \bar{y})$. The existence of $w$ in the same region follows from (8.18). It is then seen, from the definition of $x_0$, that

$$\lim_{x \to x_0^+} v(x, y_1) = \infty. \tag{8.24}$$

We will now prove that $S_{x_0}$ includes the line segment $\Gamma_0$, so that $x_1 = 0$.

**Lemma 8.8.** $x_0 < 0$.

**Proof.** Suppose that $x_0 \geq 0$, and consider some $x > x_0$. For simplicity of notation, let $\phi(y) = v(x, y)$ and $\psi(y) = w(x, y)$. Then $\phi$ and $\psi$ are defined in the interval $0 \leq y \leq y_1$. Equations (8.21)–(8.22) become

$$\begin{aligned} \phi''' &= \phi' - 2\phi\psi, \\ \psi''' &= \psi' + \phi^2 - \psi^2 - 1. \end{aligned} \tag{8.25}$$

Since we have chosen $x > x_0 \geq 0$, $\phi(0) = v(x, 0) > 0$. The inequalities in Lemma 8.6 are valid in the region $x > x_1 = \max(0, x_0) = x_0$, and so

$$\begin{aligned} &\phi(0) > 0, \quad \phi'(0) = 0, \quad \phi''(0) \geq 0, \\ &\psi(0) = 0, \quad \psi'(0) > 0, \quad \psi''(0) = 0. \end{aligned} \tag{8.26}$$

Further, we saw in the proof of Lemma 8.7 that $v(x, y) > 0$ and $w(x, y) > 0$ for $x > x_1, 0 < y < y_1$, and therefore

$$\phi > 0, \qquad \psi > 0 \tag{8.27}$$

on $(0, y_1)$.

We now wish to get bounds on $\phi$ and $\psi$, assuming that $x - x_0$ is small. If $x_0 \geq 0$ then $v(x_0, 0) \geq 0$ and $v_x(x_0, 0) > 0$. We choose $x$ in an interval $(x_0, x_0 + \delta]$ in which

$$\begin{aligned} &0 < \phi(0) < v(x_0, 0) + 1, \quad \phi'(0) = 0, \\ &\phi''(0) < v_{yy}(x_0, 0) + 1, \\ &\psi(0) = 0, \quad \psi'(0) < v_x(x_0, 0) + 1, \quad \psi''(0) = 0. \end{aligned}$$

Also, from (8.26) and (8.27),

$$\phi''' \leq \phi', \tag{8.28}$$

$$\psi''' \leq \psi' + \phi^2 \tag{8.29}$$

in $[0, y_1]$. We compare the solution $(\phi, \psi)$ with the solution of

$$\begin{aligned} &p''' = p' + 1, \\ &q''' = q' + p^2, \\ &p(0) = v(x_0, 0) + 1, \quad p'(0) = 1, \quad p''(0) = v_{yy}(x_0, 0) + 1, \\ &q(0) = 1, \quad q'(0) = v_x(x_0, 0) + 1, \quad q''(0) = 1. \end{aligned}$$

Then it is easily shown that $(p,q)$ exists on the interval $[0,y_1]$, and

$$\phi(y) < p(y),$$
$$\psi(y) < q(y).$$

Hence, $\phi < p(y_1)$ and $\psi < q(y_1)$ on $[0,y_1]$, independently of $x \in (x_0, x_0+\delta]$.

However, (8.24) implies that for $x$ sufficiently close to $x_0$, $\phi(y_1) = v(x,y_1) > p(y_1)$. This contradiction proves Lemma 8.8. □

We now complete the proof of Theorem 8.2. We have shown that the solution $u(z)$ is analytic on the line segment $\Gamma_0$, where $x=0$, $0 \le y \le y_1$. We also know that in this range of $y$, $v(0,y) = \operatorname{Re} u(iy)$ satisfies the equation

$$-v_{yyy} + v_y = 2vw,$$

with $v(0,0) = 0$, $v_y(0,0) = 0$, and $v_{yy}(0,0) = -v_{xx}(0,0)$. If $u''(0) = v_{xx}(0,0) = 0$ then $v$ is identically zero on this line segment. However, we know that $v(0,y_1) > 0$, a contradiction which proves the theorem. □

**Remark 8.9.** The key lemma is that $x_0 < 0$. The proofs of this in [**112**] and [**203**] are different, and two further proofs are given in [**9**] (for the problem (8.8).) The proof we gave is a modification of one of those in [**9**] and avoids some of the complications of the earlier proofs for the equation (8.1). This simplification also works for (8.8).

## 8.3. Exercises

**Exercise 8.1.** (Needed in this chapter) Show that if $q$ is a possibly complex-valued solution of $u' = -u + f(u)$, where $f$ is analytic in a neighborhood of $u=0$, and where $f(u) = o(u)$ as $u \to 0$, and if $q(x)$ exists for large positive $x$, with $q(x) = e^{-x} + o(e^{-x})$ as $x \to \infty$, then $q$ is real.

**Exercise 8.2.** (Also needed in this chapter) Prove that if a solution $u(z)$ of (8.1) satisfies (8.15) on the line $y = y_1$, then it is real on its entire interval of existence.

**Exercise 8.3.** Prove that a monotonic solution of (8.1)–(8.2) must be an odd function of $x$, so that $u''(0) = 0$.

Chapter 9

# Some solutions of the Falkner-Skan equation

## 9.1. Introduction

The Falkner-Skan equation is important in the theory of boundary layers in fluid mechanics. See, for example, Section 6.3 of [**34**] for a brief derivation of the equation, discussion of its significance, and further references. The equation is

$$f''' + ff'' + \beta\left(1 - f'^2\right) = 0, \tag{9.1}$$

where $\beta$ is a parameter. Here $f'$ is the velocity of a fluid in a thin region above a surface. The variable $f$ is a "stream function". In the most basic physical problem it is assumed that $f'(0) = 0$, while at $\infty$ the velocity reaches that of the flow outside of the boundary layer, here rescaled to give the equation $f'(\infty) = 1$. The mathematical problem is to show that there is a solution satisfying the boundary conditions

$$f(0) = f'(0) = 0, \qquad f'(\infty) = 1. \tag{9.2}$$

The most important parameter range for applications is $0 \leq \beta \leq 2$, but here we will consider solutions of (9.1) for every $\beta \geq 0$. An extended study of the possible solution behaviors was given by Coppel in [**45**].

Except for the case $\beta = 0$, our first result was stated as an exercise in Chapter 2. For completeness we will include the proof here.

**Theorem 9.1.** *For every $\beta \geq 0$ the problem* (9.1)–(9.2) *has a solution such that $f' \geq 0$ on $[0, \infty)$.*

**Proof.** The first proof of this result was by Weyl [**253**] in the 1940s. He used the degree theory of Leray and Schauder, which was relatively recent at the time. The argument is complicated. A more elementary proof, essentially the shooting argument we will use, is in Chapter XIV of [**83**] but the details are still slightly cumbersome. It is interesting to note that once the idea of shooting is considered, it becomes possible to give a short and elementary proof which could have been understood at least in the nineteenth century, if not even earlier.[1]

Suppose that $\beta > 0$. For each $\alpha \geq 0$ let $f = f_\alpha$ denote the unique solution to the equation (9.1) such that

$$f(0) = f'(0) = 0, \qquad f''(0) = \alpha. \tag{9.3}$$

**Proposition 9.2.** *If $f$ is the solution to (9.1) and (9.3) for some $\alpha \neq 0$, then there is no $z$ such that $f'(z) = 1$ and $f''(z) = 0$.*

**Proof.** The uniqueness theorem for ode's implies that a solution satisfying these two conditions must be of the form $f(x) = x + c$ for some constant $c$, and therefore $f'(0) \neq 0$. □

**Remark 9.3.** This result is the "nontangency condition" for this problem.

Observe that for a solution of (9.1) and (9.3), $f'''(0) < 0$, and so if $\alpha = 0$ then $f'', f'$, and $f$ are negative on some interval $(0, x_0]$. By the continuity of $f''_\alpha(x_0)$ with respect to $\alpha$, $f''_\alpha(x_0) < 0$ for sufficiently small $\alpha > 0$, even though $f_\alpha$, $f'_\alpha$, and $f''_\alpha$ are initially positive. Further, $f'_\alpha < 1$ on $[0, x_0]$ for sufficiently small $\alpha > 0$.

Suppose now that $\alpha$ is large and positive. Write (9.1) in the form

$$\left(f'' e^{\int_0^x f}\right)' = \beta e^{\int_0^x f}\left(f'^2 - 1\right) \geq -\beta e^{\int_0^x f}.$$

This equation implies that for sufficiently large $\alpha$, if $0 \leq f'_\alpha \leq 1$ on $[0,1]$ then $f''_\alpha \geq \frac{1}{2}\alpha$ on this interval, leading to a contradiction. Hence, if $\alpha$ is sufficiently large then $f'_\alpha > 1$ somewhere in $[0,1]$. Also, if $f'_\alpha(x_1) = 1$ for some first $x_1$ then $f''_\alpha(x_1) > 0$, for otherwise $f'_\alpha \equiv 1$, contradicting (9.3). Further, $f'_\alpha > 1$ and $f''_\alpha > 0$ on the maximal interval $[x_1, \omega)$ where the solution exists (Exercise 9.1).

We now define our shooting sets, letting

$$A = \left\{\alpha > 0 \mid f''_\alpha(y) < 0 \text{ for some } y > 0, \text{ and } 0 < f'_\alpha < 1 \text{ on } [0, y]\right\},$$
$$B = \left\{\alpha > 0 \mid f'_\alpha(z) > 1 \text{ for some } z > 0, \text{ and } f''_\alpha > 0 \text{ on } [0, z]\right\}.$$

[1] In Chapters 6 and 7 we discussed the method of geometric perturbation theory, a modern technique in dynamical systems. The Falkner-Skan equation is another example where this theory has been applied. In Exercise 6.19 of [**34**] a proof of existence using this method is outlined. We believe that the proof here is much simpler.

These sets are each open, by continuity of solutions with respect to $\alpha$. The remarks above show that each is nonempty, with some interval $(0, \alpha_0)$ contained in $A$ and some interval $(\alpha_1, \infty)$ contained in $B$. Their definitions imply that they are disjoint. Hence there is an $\alpha^* \in (0, \infty)$ such that $\alpha^* \notin A \cup B$. From this we conclude that $0 \leq f'_{\alpha^*} \leq 1$ and $f''_{\alpha^*} \geq 0$ on $(0, \infty)$. By Proposition 9.2, $f'_{\alpha^*} - 1$ and $f''_{\alpha^*}$ do not vanish simultaneously, and it easily follows that $f_{\alpha^*}$ satisfies (9.1) and (9.2).

There remains to consider $\beta = 0$. This special case is called the Blasius equation, and a simple existence proof was given by Serrin in [**213**]. It is illustrative of a slightly different form of shooting, similar to one used for the problem of Lagerstrom studied in Chapter 7. We will outline this proof.

Write the equation as $\left(f'' e^{\int_0^x f}\right)' = 0$, so that $f'' = \alpha e^{-\int_0^x f}$. It follows as in Chapter 7 that $\lim_{x \to \infty} f'(x)$ exists for each $\alpha > 0$. It is not hard to show that $f'(\infty)$ is a continuous function of $\alpha$, with $f'(\infty) < 1$ for small $\alpha$ and $f'(\alpha) > 1$ for large $\alpha$. Hence there is an $\alpha$ such that (9.2) is satisfied. (Uniqueness is also not hard to prove.) The details are left to the reader and complete the proof of Theorem 9.1. □

## 9.2. Periodic solutions

It was asked in [**45**] whether there are solutions of the Falkner-Skan equation such that $f' - 1$ is oscillatory. The answer is yes, as shown in [**95**].

**Theorem 9.4.** *If $\beta > 1$ then (9.1) has a periodic solution.*

**Proof.** We will consider solutions $f$ of (9.1) which are odd functions and so satisfy initial conditions of the form

$$f(0) = 0, \quad f'(0) = \gamma, \quad f''(0) = 0, \tag{9.4}$$

where we restrict $\gamma$ to lie in the interval $(0,1)$. Thus, $f$ and $f'$ are positive on some initial interval $(0, x_\gamma)$, while $f'''$ and $f''$ are initially negative.

The idea is to show that $f$ has a second point of symmetry, besides $x = 0$. If, for some $x^* > 0$, $f(x^*) = f''(x^*) = 0$, then the symmetry of the equation (9.1) implies that $f(x^* - s) = -f(x^* + s)$ for all $s$, and coupled with the symmetry around 0, this implies that $f$ is a periodic solution of period $2x^*$. A shooting argument is used to establish the existence of such an $x^*$, for some $\gamma$.

**Proposition 9.5.** *If $|f'| < 1$ on $[0, X]$ then $f'' < 0$ on this interval.*

**Proof.** If $f''(x_1) = 0$ for some first $x_1$ in $(0, X]$, then $f'''(x_1) < 0$, a contradiction which completes the proof. □

If $\gamma = 0$ then $f'''(0) = -\beta$, and so there is an $x_0$ such that $f, f'$, and $f''$ are all negative on $(0, x_0]$, with $|f'| < 1$ on this interval. If $\gamma$ is small and positive then still, $f(x_0) < 0$, $|f'| < 1$ on $(0, x_0]$, and $f'' < 0$ on this interval. However, $f > 0$ on some smaller interval $(0, x_1]$, and so the first sign change of $f$ occurs before any sign change of $f''$.

We now define two subsets of the interval $0 < \gamma < 1$. Let

$$A = \{\gamma \in (0,1) \mid f \text{ has a sign change before } f''\},$$

$$B = \{\gamma \in (0,1) \mid f'' \text{ has a sign change before } f\}.$$

To be more precise, recall that if $0 < \gamma < 1$ then initially, $f > 0$ and $f'' < 0$. Hence, $\gamma \in A$ if (i) $0 < \gamma < 1$ and (ii) if $f$ is the solution of (9.1) and (9.4) then there is an $x_1 > 0$ with $f(x_1) < 0$ and $f'' \leq 0$ on $(0, x_1]$. $B$ is defined similarly.

We have shown that $A$ is nonempty. To show that $A$ is open, suppose that $\gamma \in A$. Then there is an $x_\gamma$ such that $f \geq 0$ on $(0, x_\gamma)$ and $f < 0$ on some interval $(x_\gamma, x_\gamma + \eta)$ in which $f'' \leq 0$. (This uses Exercise 9.4.) Further, $f'' \leq 0$ in $(0, x_\gamma]$. If at some point in $(0, x_\gamma]$, $f''$ has a zero without changing sign, then at that point, $f''' = 0$, so that $f'^2 \equiv 1$. But $f'(0) \in (0,1)$, a contradiction. Hence $f'' < 0$ on $(0, x_\gamma]$, and openness of $A$ follows.

Suppose next that $\gamma \in B$ and $f''$ has its first sign change at $y_\gamma$. If $f$ has its first zero at $z \in (0, y_\gamma)$, then $f > 0$ on $(0, z)$ and $f'' < 0$ on $(0, z]$. These imply that $f$ changes sign at $z$, a contradiction. If $f(y_\gamma) = 0$ then we are done, by the earlier symmetry argument. Hence we assume that $f > 0$ on $(0, y_\gamma]$, and so $\gamma$ is an interior point of $B$. Therefore $B$ is open.

We have yet to show that $B$ is nonempty, but before doing this, we explore the consequences of this statement. From the other properties of $A$ and $B$ shown above it follows that if $B$ is nonempty then $A \cup B \not\supseteq (0,1)$.

In Exercise 9.3 you are asked to show that, for any $\gamma \in (0,1)$, at least one of $f$ and $f''$ must have a sign change in $x > 0$. If there is some point $\gamma^* \in (0,1)$ which is in neither $A$ nor $B$ then for the corresponding solution to (9.1) and (9.4), $f$ and $f''$ have their first sign changes at the same point, say at $x^* > 0$, and as noted above, this implies that $f$ is periodic.

Hence, to prove Theorem 9.1 we must show that $B$ is nonempty. We will consider values of $\gamma$ close to 1. When $\gamma = 1$, the solution to (9.1)–(9.2) is $f(x) = x$. Therefore, $\lim_{\gamma \to 1^-} f'(x) = 1$ uniformly on $[0, X]$ for every $X > 0$. We now let $\gamma = 1 - \varepsilon$ for some $\varepsilon \in (0,1)$.

**Proposition 9.6.** *For $0 < \varepsilon < 1$, $f''' < 0$ as long as $f'' < 0$ and $f' \geq 0$.*

**Proof.** Differentiate (9.1) to give $f^{(iv)} + ff''' = (2\beta - 1) f'f''$. It follows that $f'''$ remains negative as long as $f'f'' < 0$. Hence, $f'$ has a first zero at some $x_\varepsilon > 0$ and $\lim_{\varepsilon \to 0^+} x_\varepsilon = \lim_{\varepsilon \to 0^+} f(x_\varepsilon) = \infty$. Also, $f''(x_\varepsilon) < 0$.

Suppose that $f'''(x_\varepsilon) = 0$. Then $f^{(iv)}(x_\varepsilon) = 0$, and a further differentiation shows that $f^{(v)}(x_\varepsilon) > 0$. Hence $f''' > 0$ to the left of $x_\varepsilon$, a contradiction. □

We need a simple lemma:

**Lemma 9.7.** *As long as* $f > 0$, $f' \geq -2$.

**Proof.** Let
$$H = \frac{1}{2}f''^2 + \beta\left(f' - \frac{1}{3}f'^3\right).$$
Then $H' = -ff''^2$, so that $H$ is decreasing as long as $f > 0$. Also, $H(0) = \beta\gamma\left(1 - \frac{1}{3}\gamma^2\right) \leq \frac{2}{3}\beta$, since $0 < \gamma < 1$. If $f' \leq -2$ then $H \geq \frac{2}{3}\beta$, a contradiction. □

To proceed further, rescale the equation, letting $m_\varepsilon = f(x_\varepsilon)$ and
$$g(y) = \frac{f(m_\varepsilon y + x_\varepsilon)}{m_\varepsilon}.$$
We find that if $\alpha_\varepsilon = \frac{1}{m_\varepsilon}$ then
$$\alpha_\varepsilon^2 g''' + gg'' + \beta\left(1 - g'^2\right) = 0 \tag{9.5}$$
and
$$g(0) = 1, \quad g'(0) = 0, \quad g''(0) = \frac{1}{\alpha_\varepsilon}f''(x_\varepsilon) < 0. \tag{9.6}$$

Since $a_\varepsilon \to 0$ as $\varepsilon \to 0$, (9.5)–(9.6) is a singularly perturbed initial value problem. We wish to show that for sufficiently small $\varepsilon$, $g'' = 0$ before $g = 0$.

Proposition 9.6 implies that $g'''(0) < 0$, and so from (9.5) and (9.6), $-\beta < g''(0) < 0$. We also note that $g' < 0$ as long as $g'' < 0$. Since $g$ is decreasing and starts at $g(0) = 1$, it is convenient to let $t = 1 - g$ be the independent variable. Setting
$$S(t) = -g'(y), \qquad P(t) = -g''(y),$$
we obtain
$$S' = \frac{P}{S}, \qquad \alpha^2 P' = \frac{(t-1)P + \beta\left(1 - S^2\right)}{S} \tag{9.7}$$
with
$$\lim_{t\to 0+} S(t) = 0, \qquad \lim_{t\to 0+} P(t) = -g''(0). \tag{9.8}$$
We wish to show that for sufficiently small $\alpha$, either $P = 0$ before $t = 1$ or else $P(1) = 0$. Either of these implies the existence of a periodic solution of (9.1). We will therefore assume that there is a sequence $\alpha_j \to 0$ such that $P > 0$ on $(0, 1]$.

From our study of singularly perturbed problems in Chapter 7 it is natural to suppose that solutions of (9.1) are close to solutions of the so-called "outer" equation, which is

$$gg'' + \beta\left(1 - g'^2\right) = 0.$$

This equation is exactly solvable. But it is easier to consider the solutions of (9.7)–(9.8), with $\alpha$ as the small parameter. The outer equation is then obtained by formally setting $\alpha = 0$, giving $P = \frac{\beta(1-S^2)}{1-t}$, so that

$$S' = \frac{\beta\left(1 - S^2\right)}{(1-t)\,S}.$$

This first order equation is easily solved to show that there is a unique solution on $(0,1]$ with $\lim_{t\to 0^+} S(t) = 0$, namely $S_0(t) = \left(1 - (1-t)^{2\beta}\right)^{1/2}$. Further, let

$$P_0(t) = \frac{\beta\left(1 - S_0(t)^2\right)}{1-t} = \beta\,(1-t)^{2\beta-1}. \tag{9.9}$$

**Lemma 9.8.** *For every interval $[t_0, t_1] \subset (0,1)$, $P$ converges to $P_0$ and $S$ converges to $S_0$ uniformly on $[t_0, t_1]$ as $\alpha \to 0$.*

**Proof.** The proof uses the bound $S \le 2$ obtained from Lemma 9.7. Under our assumptions and the results above,

$$S > 0, \qquad S' > 0, \qquad 0 < P(0) < \beta.$$

From (9.7), $P' > 0$ if and only if $P < \beta\frac{1-S^2}{1-t}$. Note also that

$$S(t)^2 = 2\int_0^t P(u)\,du,$$

$$S_0(t)^2 = 2\int_0^t P_0(u)\,du.$$

The rest of the proof is fairly routine and is left as Exercise 9.2. □

**Lemma 9.9.** *There is an $\varepsilon_0 > 0$ and a $t_1 \in (0,1)$ such that if $0 < \varepsilon < \varepsilon_0$ and $t_1 < t < 1$, then*

$$2(\beta - 1) + \frac{(t-1)\,P + \beta\left(1 - S^2\right)}{S^2} > 0. \tag{9.10}$$

**Proof.** Choose $t_1 \in (0,1)$ and $\mu \in (0,1)$ such that

$$2(\beta - 1) + \beta\left(\frac{t_1 - 1}{2\left(1 - \frac{\mu}{2}\right)^2} + \mu^2 - 2\mu\right) > 0. \tag{9.11}$$

This step is accomplished by choosing $t_1$ close to 1 and $\mu$ small. By choosing $t_1$ even closer to 1, if necessary, without changing $\mu$, we can also assume that

$$\frac{4}{3}(1-t_1) < \frac{1}{1-\mu} - 1, \tag{9.12}$$

that $P_0(t_1) < \min\left\{\frac{1}{2}, \frac{\beta}{4}\right\}$, and that $S_0(t_1) > 1 - \frac{\mu}{4}$. Now choose $\varepsilon_0 > 0$ such that if $0 < \varepsilon < \varepsilon_0$ then $P(t_1) \le \min\left\{\frac{2}{3}, \frac{\beta}{2}\right\}$ and $1 - \frac{\mu}{2} \le S(t_1) \le 1$. It follows that on $(t_1, 1)$, $S' = \frac{P}{S} < \frac{4}{3}$, and so from (9.12), $S < \frac{1}{1-\mu}$ on $[t_1, 1)$.

From (9.9) and Lemma 9.8 it follows that for sufficiently small $\varepsilon$,

$$\lim_{t \to 0^+} P(t) > \frac{\beta}{2}.$$

Hence, $P' < 0$ somewhere in $(0, t_1)$. Differentiating (9.7) shows that if $P' = 0$ then

$$\alpha^2 P'' = \frac{P(1-2\beta)}{S} < 0.$$

This inequality implies that $P' < 0$ on $(t_1, 1)$, and so $P < \frac{\beta}{2}$ on this interval. Then (9.10) follows from (9.11), proving Lemma 9.9. □

We now introduce the functions $R(t) = -\frac{1}{1-t}$ and $r(t) = \frac{P'(t)}{P(t)}$. Then

$$R' = -R^2$$

and

$$r' = -r^2 - M(t, r),$$

where

$$M(t,r) = \left(\frac{1-t}{\alpha^2 S} + \frac{P}{S^2}\right) r + \frac{2\beta - 1}{\alpha^2 S}.$$

We claim that $r(t_1) < R(t_1)$ for sufficiently small $\varepsilon$. If not then there are arbitrarily small values of $\varepsilon$ such that because $\beta > 1$,

$$\frac{P'(t_1)}{P(t_1)} \ge R(t_1) > -\frac{2\beta - 1}{1 - t_1} = \frac{P_0'(t_1)}{P_0(t_1)}.$$

Hence there are numbers $\delta > 0$ and $\tau$ in $(0, t_1)$ such that

$$\frac{P'}{P} \ge \frac{P_0'}{P_0} + \delta$$

in $[\tau, t_1]$. Integrating this from $\tau$ to $t_1$ contradicts Lemma 9.8 for small $\varepsilon$.

Hence $r(t_1) < R(t_1)$ for small $\varepsilon$. If $r(t_2) = R(t_2)$ for some first $t_2 > t_1$, then $R'(t_2) \le r'(t_2)$. However,

$$R'(t_2) - r'(t_2) = M(t_2, R(t_2)),$$

and using (9.7) and Lemma 9.9 shows that $M(t_2, R(t_2)) > 0$, another contradiction.

We have thus shown that $r < R$ as far beyond $t_1$ as they both are defined. Therefore, $r(t) \to -\infty$ as $t \to \omega \leq 1$. But this conclusion is only possible if $P(\omega) = 0$, completing the proof of Theorem 9.4. □

## 9.3. Further periodic and other oscillatory solutions

The Falkner-Skan equation, (9.1), has no equilibrium points, meaning no constant solutions. However, there are two families of solutions which are linear functions of the independent variable $x$, namely $f_1(x) = x + c$ and $f_2(x) = -x + c$, for any constant $c$. In other words, the lines $f'' = 0, f' = 1$ and $f'' = 0, f' = -1$ in the three-dimensional $(f, f', f'')$ phase space are invariant sets for (9.1). However, these sets are awkward from the dynamical systems point of view, because each is unbounded. Efforts have been made to "compactify" the phase space, and thereby gain insight into solutions in a neighborhood of these linear solutions, but this approach does not seem to have given much new information.

However if $\beta > 1$ then it is still useful to consider the linearizations of (9.1) around $f_1$ and $f_2$. The behavior of solutions in a neighborhood of $f_1(x) = x$ is straightforward. Because $f''' > 0$ if $ff'' \leq 0$ and $f' > 1$, $f'$ does not oscillate around 1, and as we have seen, there are solutions where $f'$ tends to 1 monotonically. The linearization of (9.1) around the line $f'' = 0, f' = -1$ is more interesting, and important in the analysis of this equation, as was recognized in [**45**]. Setting $f(x) = -x$, the linearized equation is $v''' - xv'' + 2\beta v' = 0$, and if $w = v'$ then

$$w'' - xw' + 2\beta w = 0. \tag{9.13}$$

This second order linear ode is known as Weber's equation and is used in mathematical physics.

In looking for further periodic solutions, it is a natural extension of the proof of Theorem 9.4 to look for a solution of the initial value problem (9.1) and (9.4) such that the first positive zero of $f$ coincides with the $k^{\text{th}}$ zero of $f''$, for some $k > 1$. Solutions of (9.13) are approximations to $f'$ in a neighborhood of the solution $f' = -1$, and so to consider zeros of $f''$, we examine $w'$, which satisfies

$$u'' - xu' + (2\beta - 1)\, u = 0, \tag{9.14}$$

again Weber's equation, with a different coefficient of the $u$ term. One family of solutions of (9.14) may be expressed in terms of a set of classic special functions of mathematical physics called the parabolic cylinder functions and is denoted by $D_k$, where $k = 2\beta - 1$. The standard reference for parabolic cylinder functions is [**2**], but most books on special functions cover this topic.

The relation between solutions of (9.14) and $D_k$ is given by

$$u(x) = e^{x^2/4} D_k(x),$$

and from this it follows that $D_k$ satisfies another form of Weber's equation, namely

$$D'' + \left(k - \frac{x^2}{4}\right) D = 0. \tag{9.15}$$

If $k > 0$ then we see that the coefficient of $D$ is positive in the interval where $|x| < 2\sqrt{k}$. Sturm's comparison theorem implies that $D$ has some zeros if $k$ is large enough, but for each $k > 0$, oscillations will cease once $x$ is outside this interval.

It turns out that we can be quite specific about when oscillations appear, because at special values of $\beta$ the equation (9.14) has polynomial solutions. The reader may check that if $2\beta - 1$ is an integer then (9.14) has a polynomial solution of degree $k = 2\beta - 1$. These are the well-known Hermite polynomials $H_k(x)$,[2] an important family of orthogonal polynomials [**2**]. Each $H_k(x)$ has exactly $k$ real distinct roots.[3] The same is therefore true for $D_k(x)$, and by using the Sturm oscillation theorem in (9.15) we can show that the number of zeros of $D_{2\beta-1}$, and of $u$, increases by two each time $\beta$ crosses an integer.[4]

To relate this to (9.1) with initial conditions (9.4), one must show that if $\gamma$ is close to 1 then the solution stays close to $f(x) = x$ until $f$ is large, and then $f'$ decreases to close to $-1$, with $f''$ then small, and furthermore, this happens while $f$ is still positive. The idea is that $f'$ is then close to $-1$ over some interval, during which $f''$ has several oscillations, still before $f = 0$. The larger $\beta$ is, the more oscillations are possible for $f''$, since in an interval $f''$ is close to a solution $u$ of (9.14). This property suggests that the solutions become more complicated for $\beta > 2$ than they were for $\beta < 2$, and each time $\beta$ increases past another integer, more periodic solutions appear. Moreover, a wide array of new bounded solutions, typical of "chaos", is generated.

Proving all this is quite complicated. In [**95**] the first branch of new periodic solutions was shown to exist for $2 < \beta < 2.05$. The small range of validity of the proof already suggests that it involves highly specific calculations, as is true. We will not go into details here, both because they are complicated and because the approach has been superseded by that in [**232**].

---

[2]The "probabalist's Hermite polynomial".

[3]This can be shown, for example, by considering (9.15) as an eigenvalue problem, looking for nontrivial solutions which are in $L^2(-\infty, \infty)$. These only exist at integer values of $k$, and combining results about such eigenvalue problems in [**41**] and [**239**] proves that the $k^{\text{th}}$ eigenfunction has $k$ real roots.

[4]See Exercise 9.5. This exercise is also intended to give more insight into the branching that occurs when new symmetric periodic solutions appear.

In [**95**] it was conjectured (a) that a new branch of solutions bifurcates from infinity every time $\beta$ crosses a positive integer $n$ and (b) that these branches all continue for $\beta > n$. The first conjecture was verified in a deep paper by Swinnerton-Dyer and Sparrow [**232**]. But conjecture (b) turned out to be false, as shown in a second extensive paper by the same authors [**233**]. Only the first branch of periodic solutions, found in Theorem 9.4 above, exists for arbitrarily large $\beta$. The other branches are shown to have turning points and only exist over a finite range of $\beta$. In fact, this was seen numerically in a paper which appeared shortly after [**95**] was accepted for publication, in which Botta et al. studied exactly this problem in [**23**] and saw some of the behavior eventually described in [**233**].

Despite the extensive work on the problem in [**232**] and [**233**], the Falkner-Skan equation is far from completely understood. In [**233**] the authors raise a number of questions which they do not answer and which suggest that this fascinating equation has secrets still to be revealed by future researchers. They also conjecture that the properties of this equation may hold for a wider class of models yet to be explored.

## 9.4. Exercises

**Exercise 9.1.** (Routine, but needed) Prove that if $f$ is a solution of (9.1) on some interval $I = [x_1, x_2]$, $f'(x_1) \geq 1$, and $f''(x_1) > 0$, then $f'' > 0$ on $I$.

**Exercise 9.2.** Prove Lemma 9.8. Hint: Show that for each interval $[t_0, t_1] \subset (0,1)$, $\lim_{\alpha\to 0^+} \left\{(t-1)P + \beta\left(1 - S^2\right)\right\} = 0$ uniformly in $[t_0, t_1]$. You may also wish to consider Gronwall's lemma.

**Exercise 9.3.** Show that if $0 < \gamma < 1$ and $f$ satisfies (9.1) and 9.4), then at least one of the functions $f$ and $f''$ has a sign change in $(0, \infty)$.

**Exercise 9.4.** Show that if $f$ is a solution to (9.1) such that $f'(0) \neq 1$, then neither $f$ nor $f''$ can have infinitely many zeros in any compact interval. (See Exercise 1.5.) Explain how this is used in the proof of Theorem 9.4.

**Exercise 9.5.** Explain why new periodic solutions are not expected when $\beta$ crosses $\frac{2n+1}{2}$ where $n$ is a positive integer, while they are when $\beta$ crosses $n$. Hint: The graphs in Figures 9.1 and 9.2 of $f'$ for two periodic solutions for different $\beta$ may be helpful. How many zeros does $f''$ have between zeros of $f'$?

Here $\beta = 2.1$ in Figure 9.1 and 3 in Figure 9.2. The scales on the two graphs illustrate how the bifurcation of the second solution occurs at infinity in the $f$ variable.

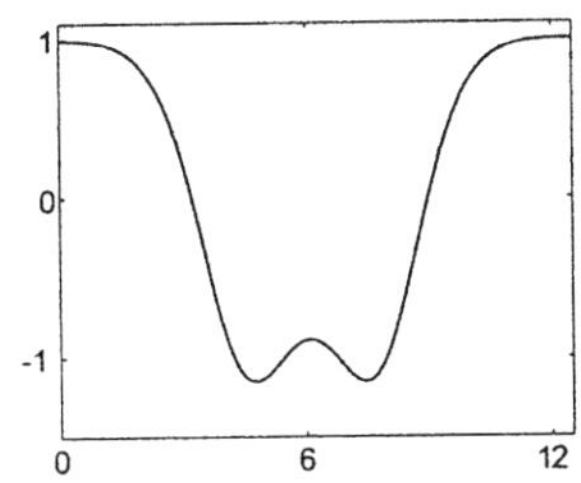


**Figure 9.1**

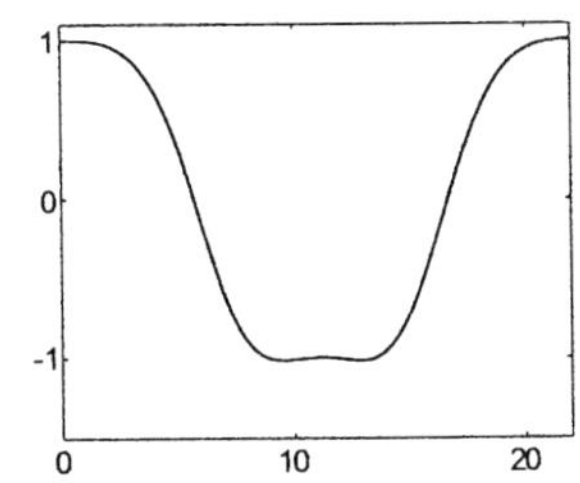


**Figure 9.2**

Chapter 10

# Poiseuille flow: Perturbation and decay

## 10.1. Introduction

In this chapter we consider two related problems arising from the Navier-Stokes equations of fluid mechanics [**138**]. The two-dimensional steady flow of an incompressible fluid with homogeneous boundary conditions and no external force ("zero data") is described by the equations

$$-\nu\Delta\mathbf{u}+(\mathbf{u}\cdot\nabla)\,\mathbf{u}+\nabla\pi=\mathbf{0}, \tag{10.1}$$

$$\nabla\cdot\mathbf{u}=0, \tag{10.2}$$

in a region $\Omega$ of $R^2$, with boundary condition

$$\mathbf{u}=\mathbf{0} \text{ on the boundary } \partial\Omega. \tag{10.3}$$

Here $\mathbf{u}=(u_1,u_2)$ is the velocity of the fluid, $\nu$ is the viscosity, and $\pi$ is the pressure. (Our notation largely follows that in [**74**] and [**199**].)

The term of particular interest is the acceleration due to convection, which we write as $(\mathbf{u}\cdot\nabla)\,\mathbf{u}$, though other notations are also used. It may be helpful to write (10.1) in coordinate form, as

$$\begin{aligned}
-\nu\left(\frac{\partial^2 u_1}{\partial x^2}+\frac{\partial^2 u_1}{\partial y^2}\right)+u_1\frac{\partial u_1}{\partial x}+u_2\frac{\partial u_1}{\partial y}+\frac{\partial\pi}{\partial x}&=0,\\
-\nu\left(\frac{\partial^2 u_2}{\partial x^2}+\frac{\partial^2 u_2}{\partial y^2}\right)+u_1\frac{\partial u_2}{\partial x}+u_2\frac{\partial u_2}{\partial y}+\frac{\partial\pi}{\partial y}&=0.
\end{aligned} \tag{10.4}$$

We certainly cannot undertake a detailed description of these equations or the well-developed theory about them. See [**74**] for an overview. Among the exact solutions known for various regions $\Omega$ is that of Poiseuille, where $\Omega$ is the infinite strip $-\infty < x < \infty$, $-1 < y < 1$. Following the notation in [**199**], Poiseuille's solution is

$$\mathbf{u}_{\Phi}(x,y) = (\nu R)\left(1-y^2,0\right), \qquad \pi_{\Phi}(x,y) = -2R\nu^2 x, \tag{10.5}$$

where $R = \frac{3\Phi}{4\nu}$ is called the "Reynolds number" of the fluid. The factor of $\frac{3}{4}$ ensures that the solution $\mathbf{u}_{\Phi}$ has flux $\Phi$ across vertical cross sections of $\Omega$.[1]

The first result to be proved in this chapter extends work by Rabier in [**199**] and [**200**] on the existence of solutions to the Navier-Stokes equations in the Poiseuille region $\Omega$ with small nonzero data. We give some background,[2] describe Rabier's work, and show how one important aspect of the problem can be handled by ode methods. Our second problem is concerned with the spatial decay as $|x| \to \infty$ of steady solutions which are small perturbations of the Poiseuille flow.

## 10.2. Solutions for small data

In the first of our two problems we consider nonhomogeneous perturbations of the homogeneous system (10.1)–(10.3), of the form

$$\begin{gathered} -\nu\Delta\mathbf{u} + (\mathbf{u}\cdot\nabla)\,\mathbf{u} + \nabla\pi = \mathbf{g}, \\ \nabla\cdot\mathbf{u} = 0, \end{gathered} \tag{10.6}$$

$$\mathbf{u} = \mathbf{v}^* \text{ on the boundary } \partial\Omega. \tag{10.7}$$

We write the solution as

$$\mathbf{u} = \mathbf{u}_{\Phi} + \delta\mathbf{u}, \qquad \pi = \pi_{\Phi} + \nu p.$$

On an appropriate function space one defines a linear operator

$$L_R(\delta\mathbf{u},p) := -\Delta(\delta\mathbf{u}) + \left(R\left(1-y^2,0\right)\cdot\nabla\right)(\delta\mathbf{u}) + R\,(\delta\mathbf{u}\cdot\nabla)\left(1-y^2,0\right) + \nabla(p)\,.$$

This operator is the linearization of the left side of (10.6) (or (10.4)) around the Poiseuille solution. In [**200**], Rabier showed that the invertibility of the operator $L_R$ implies the existence of a locally unique solution to (10.6)–(10.7) for small $\mathbf{g}$ and $\mathbf{v}^*$. For this he used the implicit function theorem in the function spaces involved. It is the invertibility of $L_R$ that will concern us here.

[1] The flux of $\mathbf{u}$ across a vertical cross section is $\int_{-1}^{1} u_1(x,y)\,dy$.

[2] The background material is aimed at those whose graduate-level preparation is mostly in ode's. Those who have had a graduate pde course including Sobolev spaces will want at most to skim this material. Those who want to learn it in depth will have to consult the listed references.

**10.2.1. Orr-Sommerfeld operators.** We now explain why the invertibility of $L_R(\cdot,\cdot)$ is equivalent to a problem in ordinary differential equations, and hence a suitable topic for this book. In this section the explanation will be formal, without regard to a careful formulation of the function space setting. In Section 10.3 we will, for those interested, outline a few of the technical details, including definitions of some of the required spaces, and discuss the approach to invertibility via degree theory. In Section 10.4 we will give a short ode proof of an important step.

First, we introduce a stream function. Since a solution to (10.6) is divergence free and $\Omega$ is simply connected (has no holes), there is a scalar function $\phi(x,y)$ such that $\delta\mathbf{u} = (\phi_y, -\phi_x)$. Also, $\phi_x(x,\pm 1) = \phi_y(x,\pm 1) = 0$, and we can assume that $\phi(x,-1) = 0$. Further, $\phi(x,1)$ is independent of $x$, and

$$\phi(x,1) = \int_{-1}^{1} \phi_y(x,y)\,dy = \int_{-1}^{1} \delta\mathbf{u}_1(x,y)\,dy. \tag{10.8}$$

We will see later that this leads to the conclusion that $\phi$ satisfies

$$\phi(x,\pm 1) = 0. \tag{10.9}$$

For $L_R$ to be invertible there must be a unique solution $(\delta\mathbf{u},p)$ to

$$L_R(\delta\mathbf{u},p) = \mathbf{f} \tag{10.10}$$

for each $\mathbf{f}$ in an appropriate space, and the linear operator $\mathbf{f} \to (\delta\mathbf{u},p)$ must be bounded in appropriate norms.

Substitute $\delta\mathbf{u} = (\phi_y, -\phi_x)$ into the two equations of (10.10), differentiate, and subtract the first equation from the second. In this way $p$ is eliminated and we obtain a single fourth order equation for $\phi$, namely

$$\Lambda_R\phi := \Delta^2\phi - R\left(1-y^2\right)\left(\partial^3_{xyy}\phi + \partial^3_{xxx}\phi\right) - 2R\partial_x\phi = \partial_x f_2 - \partial_y f_1, \tag{10.11}$$

where we have adopted a more concise partial derivative notation and the bilaplacian operator $\Delta^2 = \partial^4_{xxxx} + 2\partial^4_{xxyy} + \partial^4_{yyyy}$. The goal now is to show that for every $R > 0$ and all $\mathbf{f}$ in an appropriate space (which incorporates the boundary conditions), equation (10.11) has a unique solution $\phi$, and to obtain bounds on this solution.

This goal is accomplished by showing that the linear operator $\Lambda_R$ is invertible in the relevant function spaces. To do this we take the partial Fourier transform of (10.11). (See [**62**, Section 4.3] for an introduction to Fourier transform methods in partial differential equations.) Thus, if

$$v(y) = \frac{1}{2\pi i}\int_{-\infty}^{\infty} e^{i\xi x}\phi(x,y)\,dx,$$

then we obtain an ode of the form

(10.12)

$$D_{\xi,R\xi}v := v^{(iv)} - 2\xi^2 v'' + \xi^4 v + iR\xi\left[\left(1-y^2\right)v'' + 2v - \xi^2\left(1-y^2\right)v\right] = \hat{g}$$

for a suitable $\hat{g}$ as made more precise below.[3] From the boundary conditions on $\phi$,

$$v(\pm 1) = v'(\pm 1) = 0. \tag{10.13}$$

Here $\xi$ is a real parameter. It will be seen that in the function spaces involved the problem can be solved with the usual variation of parameters formula, which also yields bounds on the solution, provided that the corresponding homogeneous problem

$$\begin{gathered} D_{\xi,R\xi} v = 0, \\ v(\pm 1) = v'(\pm 1) = 0 \end{gathered} \tag{10.14}$$

has no nontrivial solution.[4] Proving that (10.14) has no nontrivial solution for all $R \geq 0$ and all real $\xi$ is one of the few linear problems considered in this book. (Another is in Chapter 5.) It is shown in [**200**] that this is equivalent to the invertibility of $\Lambda_R$, and so of $L_R$, and therefore implies the solvability of (10.6)–(10.7) for small data.

## 10.3. Some details

**10.3.1. Relevant Sobolev spaces.** We now define some of the spaces needed for a more thorough study of the operator $D_{\xi,R\xi}$. These definitions will be familiar to some readers. One reason for including this material here is that we will need these concepts in some later chapters as well.

Let $I$ denote the interval $(-1,1)$. With $\mathcal{C}$ as the complex plane, let $L^2(I,\mathcal{C})$ denote the Hilbert space of complex-valued functions $v$ which are Lebesgue measurable on $I$ and such that $\int_{-1}^{1} |v(t)|^2 \, dt < \infty$. Let $C_0^\infty(I,\mathcal{C})$ denote the set of all infinitely differentiable functions $\phi : I \to \mathcal{C}$ with compact support in $I$ ($\phi = 0$ outside some compact subset of $I$). Then $v$ has a "weak derivative" $v' = w \in L^2(I,\mathcal{C})$ if for every $\phi \in C_0^\infty(I,\mathcal{C})$,

$$\int_{-1}^{1} v(t)\, \bar{\phi}'(t)\, dt = -\int_{-1}^{1} w(t)\, \bar{\phi}(t)\, dt,$$

where $\bar{\phi}$ is the complex conjugate of $\phi$.

Further, for integers $k \geq 1$ we define the Sobolev space $H^k(I,\mathcal{C})$ to be the subset of $L^2(I,\mathcal{C})$ consisting of all functions $v$ with weak $j^{\text{th}}$ derivatives

[3] We obtained (10.12) by taking a Fourier transform, and from that point of view, $\hat{g}$ is a function of $\xi$ and $y$. However, we will need to consider the invertibility of $D_{\xi,R\xi}$ as an operator between certain spaces of functions of "$y$", with $\xi$ a parameter. Hence we will consider $\hat{g}$ as a function of $y$ alone.

[4] The operator $D_{\xi,R\xi}$ is "Fredholm of index zero" (see [**202**], for example, and [**200**]). This implies that uniqueness of the zero solution of (10.14) is enough to show that $D_{\xi,R\xi}$ is an isomorphism. However to remain as elementary as possible we give a separate proof of surjectivity, and boundedness of the inverse, below.

$v^{(j)} \in L^2(I,\mathcal{C})$ for each $j \leq k$. We obtain a Hilbert space by using the complex inner product

$$(u,v)_k = \sum_{j=0}^{k} \int_{-1}^{1} u^{(j)}(t)\,\bar{v}^{(j)}(t)\,dt.$$

For our application to the operator $D_{\xi,R\xi}$ we will be interested in $H^4(I,\mathcal{C})$. But this space does not incorporate information about boundary conditions, such as (10.13). To include these, we let $H_0^2(I,\mathcal{C})$ denote the closure in $H^2(I,\mathcal{C})$ of the subset $C_0^\infty(I,\mathcal{C})$. It turns out that $H_0^2(I,\mathcal{C})$ consists exactly of those functions $v$ in $H^2(I,\mathcal{C})$ which satisfy the boundary conditions (10.13).[5] (More precisely, since our underlying space is $L^2(I,\mathcal{C})$, $v$ is equal almost everywhere to a function which satisfies these conditions.)

**10.3.2. Application to the operator $D_{\xi,R\xi}$.** We then consider the operator $D_{\xi,R\xi}$ on the set $H^4(I,\mathcal{C}) \cap H_0^2(I,\mathcal{C})$. With the inner product $(u,v)_2$, this is a dense subset of $H_0^2(I,\mathcal{C})$. Techniques involving Sobolev spaces of functions of several variables are of crucial importance in studying partial differential equations. Many books give the required exposition for this purpose, such as [**62**].[6] At this stage of the problem in this chapter, we are considering an ode, with only one independent variable. The further application of the result to the operators $\Lambda_R$ and $L_R$ is inherently infinite dimensional, and Sobolev spaces are indispensable. The reader is again referred to [**199**] and [**200**] for this theory.

In the approach in [**199**] and [**200**] it is necessary also to consider the so-called negative Sobolev space $H^{-2}(I,\mathcal{C})$. Elements of this space are linear functionals on $H_0^2(I,\mathcal{C})$, and this complication leads to more technical difficulty than the eigenvalue approach discussed in the next section. We will not discuss the negative Sobolev spaces here.

As will be outlined in the next subsection, "weak" solutions (defined below) will be found for

$$D_{\xi,R\xi}v = \hat{g}$$

which lie in $H^4(I,\mathcal{C}) \cap H_0^2(I,\mathcal{C})$. It is well known that the Fourier transform is an isometry from $L^2(-\infty,\infty)$ onto itself, which means that the corresponding inverse Fourier transform $\phi(x,y)$ of $v$ is in $L^2(-\infty,\infty)$ in the $x$ variable. As a result it is not hard to show that for each $y$, $\lim_{x\to\infty}\phi(x,y) = 0$. But since $\phi(x,1)$ is independent of $x$, the boundary condition (10.9) follows. Once again, see [**199**] for further details.

[5] As examples, you can check that $v(t) = t^2(1-t)^2$ is in $H_0^2(I,\mathcal{C})$, while $w(t) = 1-t^2$ is not. Note that $v$ satisfies (10.13) but does not satisfy $v''(\pm 1) = 0$, while $w$ does not satisfy the derivative conditions in (10.13).

[6] [**62**] considers only second order operators, but the theory is easily extended to higher orders.

To motivate the definition of a weak solution of $D_{\xi,R\xi}v = \hat{g}$, where $\hat{g} \in L^2(I,\mathcal{C})$, we first consider functions $\phi$ and $\psi$ in $C_0^2(I,\mathcal{C})$ and the quadratic form

$$B(\phi,\psi) = \int_{-1}^{1} (D_{\xi,R\xi}\phi)(y)\,\bar{\psi}(y)\,ds =$$

$$\int_{-1}^{1} \left(\phi^{(iv)} - 2\xi^2\phi'' + \xi^4\phi + iR\xi\left[\left(1-y^2\right)\phi'' + 2\phi - \xi^2\left(1-y^2\right)\phi\right]\bar{\psi}(y)\right) dy.$$

Using integration by parts several times, we obtain

$$\begin{aligned} B(\phi,\psi) = \int_{-1}^{1} & \{\bar{\psi}''\phi'' + 2\xi^2\bar{\psi}'\phi' + \xi^4\bar{\psi}\phi \\ & + iR\xi\left[-\left(1-y^2\right)\bar{\psi}'\phi' + 2y\bar{\psi}'\phi + 2\bar{\psi}\phi - \xi^2\left(1-y^2\right)\bar{\psi}\phi\right] dy\}. \end{aligned}$$

This leads to the following definition:

**Definition 10.1.** For $\hat{g} \in L^2(I,\mathcal{C})$ we say that $v$ is a weak solution of

$$D_{\xi,R\xi}v = \hat{g} \quad \text{in } I, \tag{10.15}$$

$$v = v' = 0 \quad \text{on } \partial I \tag{10.16}$$

if $v \in H_0^2(I,\mathcal{C})$ and

$$\begin{aligned} \int_{-1}^{1} & \{\bar{u}''v'' + 2\xi^2\bar{u}'v' + \xi^4\bar{u}v \\ & + iR\xi\left[-\left(1-y^2\right)\bar{u}'v' + 2y\bar{u}'v + 2\bar{u}v - \xi^2\left(1-y^2\right)\bar{u}v\right]\}dy \\ & = \int_{-1}^{1} \hat{g}(s)\,\bar{u}(s)\,ds \end{aligned}$$

for all $u \in H_0^2(I,\mathcal{C})$.

However, to use the inverse Fourier transform and obtain the invertibility of the operator $\Lambda$ discussed earlier, it is seen in [**200**] that one looks for solutions in $H^4(I,\mathcal{C}) \cap H_0^2(I,\mathcal{C})$.

Our outline to this point shows that, based on results in [**200**], the existence of weak solutions to (10.6)–(10.7) for small data follows if it is shown that for every $\xi$ and every $R > 0$ the operator $D_{\xi,R\xi}$ is an isomorphism from $H^4(I,\mathcal{C}) \cap H_0^2(I,\mathcal{C})$ onto $L^2(I,\mathcal{C})$. In [**199**] this was proved for a special case in which the data is symmetric in a certain sense. In particular, $\mathbf{g}$ exhibits the same symmetry as the Poiseuille solution, in that $g_1$ is even in $y$ and $g_2$ is odd in $y$. In [**200**], the isometry was proved for a range of $R$ and every $\xi$ without symmetry assumptions.

## 10.4. A classical eigenvalue approach

We now give a short ode proof of the following result.

**Theorem 10.2.** *If $\xi$ is real and $R \geq 0$ then problem* (10.14) *has no nontrivial solution.*

**Proof.** Let $\xi R = k$. If $v$ is a solution for some $\xi$ then its complex conjugate $\bar{v}$ is a solution for $-\xi$, which implies that we can assume $k \geq 0$ with no loss of generality. Integrating (10.14), with an integration by parts, gives

(10.17)

$$v''' - 2\xi^2 v' + \xi^4 \int_{-1}^{y} v(t)\,dt + ik\left[\left(1-y^2\right)v' + 2yv - \xi^2 \int_{-1}^{y}\left(1-t^2\right)v(t)\,dt\right] = c,$$

for some constant $c$. Set $v = \left(1-y^2\right)w$. From the boundary condition $v'(\pm 1) = 0$ it follows that[7]

$$w(\pm 1) = 0. \tag{10.18}$$

In terms of $w$, (10.17) becomes

(10.19)

$$\begin{aligned}&\left(1-y^2\right)w''' - 6yw'' - 6w' - 2\xi^2\left(1-y^2\right)w' + 4\xi^2 wy \\ &\quad + \xi^4 \int_{-1}^{y}\left(1-t^2\right)w\,dt + ik\left[\left(1-y^2\right)^2 w' - \xi^2 \int_{-1}^{y}\left(1-t^2\right)^2 w\,dt\right] = c.\end{aligned}$$

Now multiply by $\bar{w}'$ and integrate over $(-1,1)$, using the boundary conditions (10.18). Note in particular that the constant $c$ disappears, because $\int_{-1}^{1}\bar{w}'dt = \bar{w}|_{-1}^{1} = 0$. Therefore, after integrating by parts we obtain

(10.20)

$$\begin{aligned}&-\int_{-1}^{1}\left(1-t^2\right)\left|w''\right|^2 dt - 4\int_{-1}^{1} t\bar{w}'w''dt - 6\int_{-1}^{1}\left|w'\right|^2 dt \\ &\quad - 2\xi^2 \int_{-1}^{1}\left(1-t^2\right)\left|w'\right|^2 dt + 4\xi^2 \int_{-1}^{1} tw\bar{w}'dt - \xi^4 \int_{-1}^{1}\left(1-t^2\right)|w|^2 dt \\ &\quad + ik\left[\int_{-1}^{1}\left(1-t^2\right)\left|w'\right|^2 dt + \xi^2 \int_{-1}^{1}\left(1-t^2\right)^2 |w|^2 dt\right] = 0.\end{aligned}$$

If $z_1$ and $z_2$ are the real and imaginary parts of $w'$, then

$$\operatorname{Re}\bar{w}'w'' = z_1 z_1' + z_2 z_2'.$$

[7] See Exercise 10.1.

Thus,

$$\operatorname{Re}\left(\int_{-1}^{1} t\bar{w}'w''dt\right) = \left[\frac{1}{2}t\left|w'\right|^2\right]_{-1}^{1} - \frac{1}{2}\int_{-1}^{1}\left|w'\right|^2 dt,$$

and taking the real parts of (10.20) gives

$$\begin{aligned} &-\int_{-1}^{1}\left(1-t^2\right)\left|w''\right|^2 dt - 2\left[t\left|w'\right|^2\right]_{-1}^{1} - 4\int_{-1}^{1}\left|w'\right|^2 dt \\ &\quad - 2\xi^2\int_{-1}^{1}\left(1-t^2\right)^2\left|w'\right|^2 dt + 2\xi^2\left[t\left|w\right|^2\right]_{-1}^{1} - 2\xi^2\int_{-1}^{1}\left|w\right|^2 dt \\ &\quad - \xi^4\int_{-1}^{1}\left(1-t^2\right)\left|w\right|^2 dt = 0. \end{aligned}$$

Since each term on the left is negative or zero, we conclude that $w = 0$, so $v = 0$, proving the theorem. □

It is now easy to show that for each real $\xi$, and every $R \geq 0$, the mapping $D_{\xi,R\xi}$ is an isomorphism from $H^4(I,\mathcal{C}) \cap H_0^2(I,\mathcal{C})$ to $L^2(I,\mathcal{C})$. We reconsider the equations (10.15)–(10.16), namely,

$$\begin{aligned} D_{\xi,R\xi}v &= \hat{g}, \\ v(\pm 1) &= v'(\pm 1) = 0, \end{aligned}$$

to be solved for $v$ in terms of $\hat{g}$, where $\hat{g} \in L^2(I,\mathcal{C})$. The corresponding homogeneous equation

$$D_{\xi,R\xi}v = 0$$

is fourth order and so has four linearly independent solutions, say $v_1, v_2, v_3$ and $v_4$, each satisfying three of the boundary conditions in (10.14).

Since we are dealing with an ode, these solutions are infinitely differentiable. Since 0 is not an eigenvalue of $D_{\xi,R\xi}$, the Wronskian formed by these solutions,

$$\det\begin{pmatrix} v_1 & v_2 & v_3 & v_4 \\ v_1' & v_2' & v_3' & v_4' \\ v_1'' & v_2'' & v_3'' & v_4'' \\ v_1''' & v_2''' & v_3''' & v_4''' \end{pmatrix},$$

is nonzero and can be used to construct a Green's function $G$ for the problem. Then we express the unique solution of (10.15)–(10.16) as an integral:

$$v(x) = \int_{-1}^{1} G(x,y)\,\hat{g}(y)\,dy.$$

A bound for $v$ in terms of $\hat{g}$ is immediate as well, completing the proof that $D_{\xi,R\xi}$ is an isomorphism from $H^4(I,\mathcal{C}) \cap H_0^2(I,\mathcal{C})$ to $L^2(I,\mathcal{C})$, as desired.

## 10.5. On the spectrum of $D_{\xi,R\xi}$ for large $R$

We may apply ode techniques as well to the problem of the decay as $|x| \to \infty$ of small perturbations of the Poiseuille flow $\mathbf{u}_\Phi$ given by (10.5). It turns out that this decay is determined by the eigenvalues of the operator $D_{\xi,R\xi}$ discussed above. By "eigenvalues" we mean values of $\xi$ such that (10.14) has a nonzero solution. There have been a number of numerical studies to get further information, such as [**24**], but little analytical work. Here we give one result about the eigenvalues, in which we discuss a limiting result as $R \to \infty$.[8]

The mathematical problem is to determine, for a given $R$, all values of $\xi$ such that the boundary value problem (10.14) has a nonzero solution. In Theorem 10.2 it was shown that there are no real values of $\xi$ for which this is true. It is more common in the decay problem to consider the variable $\alpha = i\xi$, giving us the equation

$$v^{(iv)} + 2\alpha^2 v'' + \alpha^4 v + \alpha R\left[\left(1 - y^2\right)\left(v'' + \alpha^2 v\right) + 2v\right] = 0 \tag{10.21}$$

with boundary conditions

$$v(\pm 1) = v'(\pm 1) = 0. \tag{10.22}$$

We emphasize that this is for decay of a stationary flow. In the more general situation of a traveling wave solution, a further parameter, the wave speed, is involved.

Some analytical progress was made on this problem by S. Wilson [**254**], who considered two classes of eigenvalues, the first being where $R \to \infty$ but $\alpha \to 0$ in such a way that $\alpha R \to \lambda \neq 0$. The eigenvalues are then the values of $\lambda$. In the second case, $\alpha \nrightarrow 0$. Here we concentrate on the first case, where the limiting problem as $R \to \infty$ is

$$v^{(iv)} + \lambda\left[\left(1 - y^2\right) v'' + 2v\right] = 0 \tag{10.23}$$

with (10.22). The physical symmetry of the problem imposes symmetry conditions on the eigenfunctions, which turn out to be either even or odd. Numerical computations showed that all of the eigenvalues $\lambda$ are real, and Wilson used matched asymptotic expansions to show this for those corresponding to even eigenfunctions, the "even eigenvalues". He was, however, unable to show this for the odd eigenvalues. Proving this for all of the eigenvalues of the limiting problem is the advance made here. The reference [**24**] contains further discussion and numerical computations for this problem.

**Theorem 10.3.** *All of the eigenvalues* $\lambda$ *of* (10.23), (10.22) *are real and positive.*

[8] Numerical work becomes difficult for large $R$.

**Proof.** Starting with (10.23), suppose that $v$ is an eigenfunction for some $\lambda$, and as before, let $v = \left(1 - y^2\right) w$. As in the previous section, $w(\pm 1) = 0$.[9] Proceeding as in the proof of Theorem 10.2 up to (10.19), but with $\xi = 0$ and $ik = \lambda$, we obtain

$$(10.24) \qquad \left(1 - y^2\right) w''' - 6yw'' - 6w' + \lambda \left(1 - y^2\right)^2 w' = c.$$

First we consider the case of even eigenfunctions. Then $w'(0) = w'''(0) = 0$, so that from (10.24) $c = 0$. Setting $w' = \psi$ gives

$$(10.25) \qquad \left(1 - y^2\right) \psi'' - 6y\psi' + \left(\lambda \left(1 - y^2\right)^2 - 6\right) \psi = 0,$$

with

$$(10.26) \qquad \psi(0) = 0.$$

This initial condition defines a solution up to a multiplicative constant. To obtain an eigenvalue problem for $w$ we need a further boundary condition. Since $v = \left(1 - y^2\right) w$, we see from Exercise 10.2 that the boundary conditions (10.22) in $v$ are satisfied if and only if $w(\pm 1) = 0$ and $w'(\pm 1)$ are bounded, i.e. $\psi(\pm 1)$ are bounded.

To set this problem in more standard form, we write

$$(10.27) \qquad x = \int_0^y \sqrt{1 - t^2} dt = \frac{1}{2} y \sqrt{1 - y^2} + \frac{1}{2} \sin^{-1} y,$$

so that

$$x - \frac{\pi}{4} \sim -\frac{1}{3} \left(1 - y^2\right)^{3/2}$$

as $y \to 1$.[10] Also, set

$$(10.28) \qquad z(x) = \left(1 - y^2\right)^{7/4} \psi(y),$$

and it is a routine calculation to check that

$$(10.29) \qquad \frac{d^2 z}{dx^2} + \left\{ \lambda - \frac{\frac{5}{2} - \frac{3}{4} y^2}{\left(1 - y^2\right)^3} \right\} z = 0,$$

where $y$ is expressed in terms of $x$ through (10.27). As $y \to 1$ and $x \to \frac{\pi}{4}$,

$$\frac{\frac{5}{2} - \frac{3}{4} y^2}{\left(1 - y^2\right)^3} \sim \frac{\frac{7}{36}}{\left(\frac{\pi}{4} - x\right)^2},$$

with a similar behavior as $y \to -1$ and $x \to -\frac{\pi}{4}$. We consider the operator

$$T = -\frac{d^2}{dx^2} + \frac{\frac{5}{2} - \frac{3}{4} y^2}{\left(1 - y^2\right)^3},$$

[9] See Exercise 10.2.

[10] See footnote 4 in Chapter 3.

with domain those functions $z$ in $L^2\left(-\frac{1}{4}\pi, \frac{1}{4}\pi\right)$ such that $Tz$ is also in $L^2\left(-\frac{1}{4}\pi, \frac{1}{4}\pi\right)$. This operator is then selfadjoint, so its eigenvalues must be real. Note that the solutions of the Euler equation

$$\frac{d^2 z}{dx^2} - \frac{\frac{7}{36}}{\left(\frac{\pi}{4} - x\right)^2} z = 0$$

behave either as $\left(\frac{\pi}{4} - x\right)^{7/6}$ or as $\left(\frac{\pi}{4} - x\right)^{-1/6}$ as $x \to \frac{1}{4}\pi$. In view of (10.28), it is the first of these behaviors which corresponds to $\psi(1)$ being bounded. Then the equation (10.29) is just

$$Tz = \lambda z, \tag{10.30}$$

so that $\lambda$ must be real, as required. Note also that $T$ is a positive operator, since the potential function $\frac{\frac{5}{2} - \frac{3}{4}y^2}{(1-y^2)^3}$ is positive on $(-1, 1)$. Hence $\lambda$ is positive. This also follows from Sturm's oscillation theorem.

As usual, we let $\lambda_n$ denote the eigenvalue of $T$ corresponding to an eigenfunction with $n$ zeros in $\left(-\frac{1}{4}\pi, \frac{1}{4}\pi\right)$. Then the odd eigenfunctions found above correspond to the odd eigenvalues $\lambda_{2n+1}$ and have $2n+1$ zeros, or $n$ in the interval $\left(0, \frac{1}{4}\pi\right)$.

To deal with the case of odd eigenfunctions for $w$, or even eigenfunctions for $\psi$ or $z$, we have to solve

$$\left(1 - y^2\right)\psi'' - 6y\psi' + \left(\lambda\left(1 - y^2\right)^2 - 6\right)\psi = c, \tag{10.31}$$

with

$$\psi'(0) = 0, \tag{10.32}$$

where $c$ is no longer necessarily zero. (The precise value of $c \neq 0$ is irrelevant, since changing $c$ merely multiplies $\psi$ by a constant.) For $z$ we obtain

$$\frac{d^2 z}{dx^2} + \left\{\lambda - \frac{\frac{5}{2} - \frac{3}{4}y^2}{\left(1 - y^2\right)^3}\right\} z = \frac{c}{\left(1 - y^2\right)^{1/4}}, \tag{10.33}$$

with

$$z'(0) = 0. \tag{10.34}$$

Let $\lambda_{2n}$ be the even eigenvalues of $T$ in (10.30). Again, these are real, and the corresponding eigenfunction $z_{2n}$ has $2n$ zeros in $\left(-\frac{\pi}{4}, \frac{\pi}{4}\right)$. An even solution of (10.33)–(10.34) may be expressed as a series in the eigenfunctions $z_{2n}$. These functions satisfy

$$\frac{d^2 z_{2n}}{dx^2} + \left\{\lambda_{2n} - \frac{\frac{5}{2} - \frac{3}{4}y^2}{\left(1 - y^2\right)^3}\right\} z_{2n} = 0, \tag{10.35}$$

with

$$z_{2n}'(0) = 0.$$

We multiply (10.33) by $z_{2n}$ and (10.35) by $z$, subtract, and integrate over $\left(0, \frac{\pi}{4}\right)$ to obtain

$$(\lambda - \lambda_{2n}) \int_0^{\frac{\pi}{4}} z\, z_{2n} dx = c \int_0^{\frac{\pi}{4}} \frac{z_{2n}}{(1-y^2)^{1/4}} dx,$$

so that if the $z_{2n}$ are normalized then

$$z(x) = \sum_{n=0}^{\infty} \frac{c\, z_{2n}(x)}{\lambda - \lambda_{2n}} \int_0^{\frac{\pi}{4}} \frac{z_{2n}(t)}{(1-y^2)^{1/4}} dt. \tag{10.36}$$

We have said that $c$ is not necessarily zero. In fact, $c \neq 0$, for if $c = 0$ then from (10.33), $z$ becomes an even eigenfunction of $T$, and necessarily $\lambda = \lambda_{2n}$ for some $n$, and $z = z_{2n}$. Since

$$z = \left(1-y^2\right)^{7/4} \psi = \left(1-y^2\right)^{7/4} w',$$

this implies that

$$\int_0^1 \left(1-y^2\right)^{-7/4} z_{2n}\, dy = [w]_0^1 = 0, \tag{10.37}$$

since we are now considering $w$ to be an odd eigenfunction. Thus,

$$\int_0^{\pi/4} \left(1-y^2\right)^{-9/4} z_{2n}\, dx = 0,$$

and we shall prove to the contrary that

$$\int_0^{\pi/4} \left(1-y^2\right)^{-9/4} z_{2n}\, dx \neq 0. \tag{10.38}$$

To prove (10.38) we note that

$$\frac{d}{dy}\left(\frac{\frac{5}{2} - \frac{3}{4}y^2}{(1-y^2)^3}\right) = \frac{\frac{27}{2}y - 3y^3}{(1-y^2)^4} > 0$$

for $0 < y < 1$. Hence (10.35) is of the form

$$\frac{d^2Y}{dx^2} - Q(x)\, Y = 0, \tag{10.39}$$

where $Q$ is increasing.

The following lemma then implies (10.38).

**Lemma 10.4.** *If $Q$ is continuous and increasing on an interval $[x_1, x_3]$ and $Y$ is a solution of (10.39) which has successive zeros at $x_1 < x_2 < x_3$, then*

$$\left|\int_{x_1}^{x_2} fY\,dx\right| < \left|\int_{x_2}^{x_3} fY\,dx\right| \tag{10.40}$$

*for each positive increasing function $f$.*

**Proof.** Clearly, the worst case is $f$ constant, which we now consider. Without loss of generality, let us suppose that $Y > 0$ in $(x_2, x_3)$, $Y < 0$ in $(x_1, x_2)$. Then $Y$ must have a positive maximum in $(x_2, x_3)$, say at $x_2^*$, and from (10.39), $Q(x_2^*) \le 0$. Since $Q$ is increasing, $Q(x) < 0$ for $x < x_2^*$, which implies that there is a unique point $x_1^*$ in $(x_1, x_2)$ where $Y' = 0$. There is no point in $(x_2^*, x_3)$ where $Y' = 0$, for this would be a positive minimum, with $Q > 0$ thereafter, which prevents $Y(x_3) = 0$. Thus $Y' < 0$ in $(x_2^*, x_3)$, and in some interval to the right of $x_2^*$, $Q < 0$.

Now compare $\int Y$ in the two intervals $(x_2, x_2^*)$ and $(x_2^*, x_3)$. First consider $(x_2^*, x_3)$, in which $Y' < 0$. If $Y' = P_+(Y)$ in this interval then (10.39) may be written in the form

$$P_+ \frac{dP_+}{dY} = Q(x)\,Y.$$

Integrating forward we get

$$\frac{1}{2} P_+^2(Y) = \int_{Y(x_2^*)}^{Y} Q(x_+(Y))\,Y dY,$$

where $x_+$ is the inverse function of $Y$ taking values in $(x_2^*, x_3)$. (Note that $Y(x_2^*) > Y$.) This integral is positive for $Y$ in any interval below $Y(x_2^*)$ where $Y > 0$ and $Q < 0$ and so must be positive on $(0, Y(x_2^*))$, since $Y'$ is not zero there. Similarly, integrating backward and again recalling that $Y < Y(x_2^*)$,

$$\frac{1}{2} P_-^2(Y) = \int_{Y(x_2^*)}^{Y} Q(x_-(Y))\,Y dY.$$

Since $Q(x_+) > Q(x_-)$ for every $x_+ \in (x_2^*, x_3)$ and $x_- \in (x_2, x_2^*)$, we see that

$$P_+^2(Y) < P_-^2(Y). \tag{10.41}$$

Now,

$$\int_{x_2}^{x_2^*} Y dx = \int_0^{Y(x_2^*)} Y \frac{dx}{dY} dY < \int_{x_2^*}^{x_3} Y\,dx$$

from (10.41). Similar arguments show that

$$\left|\int_{x_1}^{x_1^*} Y\,dx\right| < \left|\int_{x_1^*}^{x_2} Y\,dx\right| < \int_{x_2}^{x_2^*} Y\,dx,$$

and from these we get (10.40) and so (10.38). □

We have thus established that $c \neq 0$, and so, without loss of generality, we may suppose that $c = 1$. We have also established from (10.40) that

$$\int_0^{\pi/4} \left(1-y^2\right)^{-9/4} z_{2n}\,dx \neq 0, \qquad \int_0^{\pi/4} \left(1-y^2\right)^{-1/4} z_{2n}\,dx \neq 0.$$

If we suppose without loss of generality that $z_{2n} > 0$ between its last zero and $\frac{\pi}{4}$, then we have

$$\int_0^{\pi/4} \left(1-y^2\right)^{-9/4} z_{2n}\,dx > 0, \qquad \int_0^{\pi/4} \left(1-y^2\right)^{-1/4} z_{2n}\,dx > 0.$$

Now multiply (10.36) by $\left(1-y^2\right)^{-9/4}$ and integrate. Since (10.37) holds when $z_{2n}$ is replaced by an even solution $z$ of (10.33) (corresponding to an odd eigenfunction of (10.23), (10.22)), we find that

(10.42)
$$0 = \sum_{n=0}^{\infty} \frac{1}{\lambda - \lambda_{2n}} \int_0^{\pi/4} \left(1-y^2\right)^{-9/4} z_{2n}\,dx \int_0^{\pi/4} \left(1-y^2\right)^{-1/4} z_{2n}\,dx.$$

As $\lambda$ increases from just above some $\lambda_{2N}$ to just below $\lambda_{2(N+1)}$, the expression on the right of (10.42) moves monotonically from $\infty$ to $-\infty$ and thus takes the value 0 once and only once. Exactly one even eigenvalue of (10.23), (10.22) lies between two successive even eigenvalues of $T$.

Further, there are no complex even eigenvalues, for if $\lambda = \mu + i\nu$ in (10.42), then

$$0 = \sum_{n=0}^{\infty} \frac{(\mu - \lambda_{2n}) - i\nu}{(\mu - \lambda_{2n})^2 + \nu^2} \int_0^{\pi/4} \left(1-y^2\right)^{-9/4} z_{2n}\,dx \int_0^{\pi/4} \left(1-y^2\right)^{-1/4} z_{2n}\,dx,$$

and the imaginary part does not vanish unless $\nu = 0$. This completes the proof of the theorem. □

## 10.6. Exercises

**Exercise 10.1.** Prove (10.18), by observing that $v$ is smooth at $\pm 1$ and using the boundary conditions.

**Exercise 10.2.** Show that in the proof of Theorem 10.3, $w'(\pm 1)$ are bounded.

*Chapter 11*

# Bending of a tapered rod; variational methods and shooting

## 11.1. Introduction

This chapter is inspired by work of Stuart, who developed and studied extensively a model for the planar buckling of a tapered column [**229**], [**230**], [**231**] and references therein, and [**246**], and whom we thank for introducing us to this topic. His model, like so many previous studies of buckling, is based on the original work of Euler. The novelty of Stuart's approach is that it allows for a horizontal cross section of the unbuckled column which varies in shape, as well as diameter.

The derivation starts with a reference configuration, in which the centroid of each cross section is on the $z$-axis, with the bottom, at $z = 0$, being clamped and the top, at $z = 1$, free. However the weight of the rod itself may be enough to cause it to bend. Depending on the amount of tapering which is assumed, the free end may even bend over and down, and in the stable position, point vertically downward.

A planar configuration of the rod is identified with a smooth curve in the $(x, z)$-plane formed by the centroids of the cross sections. The rod is assumed always to have unit length, and the curve of centroids is parametrized by the arc length $s$ from the free end. The physical variables used in the derivation are the area $S(z)$ of a horizontal cross section of the reference (vertical) configuration and the moment of inertia $I(z)$ of this cross section around the horizontal line in the $(y, z)$-plane through $(0, 0, z)$. The shape of the curve of centroids after bending is described by the angle $\theta(s)$ between the tangent to the curve and the vertical.

The final independent variable, $t$, in the equations is obtained from a nonlinear rescaling of $s$ based on the volume of the rod between $s$ and the free end, where the diameter is assumed to tend to zero. Hence $t = 0$ is the top of the rod, and (from the rescaling) $t = 1$ is the bottom. The dependent variable, $u$, is the angle $\theta$ in terms of $t$. The equations involve a function $A(t)$ defined by

$$A(t(s)) = I(1-s)\int_0^s S(1-\tau)\,d\tau,$$

called a "profile". See [**229**] for further details.

The problem to be studied here is

$$(Au')' + \mu \sin u = 0, \tag{11.1}$$

$$u(1) = 0, \tag{11.2}$$

$$\lim_{t\to 0+} A(t)\,u'(t) = 0, \tag{11.3}$$

where $\mu > 0$ is a parameter arising in the derivation of (11.1). Solutions will also be required to satisfy

$$\int_0^1 A(s)\,u'(s)^2\,ds < \infty, \tag{11.4}$$

thereby having finite energy in an appropriate sense. Of course, the solution $u \equiv 0$ satisfies all of these conditions. We shall refer to the problem of finding nontrivial solutions of (11.1)–(11.4) as "Problem P". The functions $A$ and $A'$ are assumed to be continuous on $(0,1]$, and $A$ satisfies the following additional conditions:

$$A > 0 \quad \text{on } (0,1], \tag{11.5}$$

and

$$\text{for some } p > 0,\ Q = \lim_{t\to 0+}\frac{A(t)}{t^p} \text{ exists and is positive.} \tag{11.6}$$

A further condition is needed for some of the results in [**229**]:

$$A' \geq 0 \quad \text{on } (0,1]. \tag{11.7}$$

Condition (11.5) is natural from the physical model. In the case of a rod of uniform density and circular cross section, condition (11.7) says that the diameter decreases monotonically from the base to the tip. Many of the results in [**229**] and its successors do not require this condition. We use it particularly in Section 11.3 and have not investigated whether it could be avoided in the arguments given there. The constant $p$ in (11.6) determines the degree of tapering.

The goal is to find all nontrivial solutions of the problem, classifying them according to the number of zeros of $u$ in $(0,1)$.[1] Also, one wants to know if a solution minimizes energy in some sense, which means it is physically stable. Stuart's analysis of this problem, some done with his Ph. D. student Vuillaume, was based on the calculus of variations. We will outline this method in the next section, though we will by no means cover all of their results. The techniques are heavily functional analytic, using bifurcation theory in a relevant Hilbert space. There are many results about bifurcation properties of the problem, some of which only make sense in the Hilbert space setting. We will show in Section 11.3 that elementary methods give results similar to theirs about the existence and qualitative behavior of solutions to the Problem P.[2] We also verify Stuart's conjecture that shooting can give details of the solutions which are not attainable by a standard variational method. In private correspondence, Stuart has described an extension of the variational method, based on work of Nehari, which does give more complete results, and we outline this method as well.

In this section we illustrate some of the interesting points by considering three special cases, namely

$$A(t) = t, \tag{11.8}$$

$$A(t) = t^2, \tag{11.9}$$

$$A(t) = t^3. \tag{11.10}$$

For these specific forms of $A$, at least, it is convenient to put the problem on the half-line $x \geq 0$, by letting $x = -\int_1^t \frac{1}{\sqrt{A(s)}}ds,\ u(t) = w(x)$. We obtain the following three differential equations for $w$:

$$w'' - \frac{1}{2-x}w' + \mu \sin w = 0, \tag{11.11}$$

$$w'' - w' + \mu \sin w = 0, \tag{11.12}$$

$$w'' - \frac{3}{x+2}w' + \mu \sin w = 0. \tag{11.13}$$

Equation (11.12) is the easiest to analyze because, as was pointed out in [**229**], it can be studied in the phase plane. Even without considering the boundary conditions on $w$, we see by linearizing (11.12) that $w$ oscillates when $|w|$ is small and $\mu > \frac{1}{4}$. Hence, a solution $u$ of Problem P with $\mu > \frac{1}{4}$ oscillates as long as it remains small. We will see, however, that $u(t)$ is not small if $t$ is sufficiently close to zero, and so $u$ has only a finite number of

[1] Recall that $u(t)$ is the angle between the vertical and the tangent to the rod at a distance (arc length) $s(t)$ from the tip.

[2] We do not address more sophisticated issues, such as the essential spectrum of the related linearized problem, where modern techniques are probably essential.

zeros. By choosing $u'(1)$ small, we obtain arbitrarily many zeros before the solution grows in amplitude.

The linearization around zero of equation (11.13) is a form of Bessel's equation. (See below for more detail on this.) Hence, for $w$ small, there are oscillations for all positive values of $\mu$. By contrast, equation (11.11) is only valid on $0 \leq x < 2$, and there will be few, if any, oscillations, because the coefficient $\frac{1}{2-x}$ becomes large.

This simple analysis shows that the behavior of solutions to Problem P may be quite different for $0 < p < 2$ and $p > 2$ and that the case $p = 2$ may be of particular interest. Stuart confirms that this is correct and that for many profiles $A$ there are interesting phenomena which do not occur in the three special cases above.

## 11.2. A calculus of variations approach in Hilbert space

Here we summarize just a few of the techniques and results in the basic papers of Stuart, [**229**] and [**230**]. Much of what is proved in these papers may be put into one of two categories: (I) results about the existence and qualitative behavior of solutions to Problem P and (II) results which explain the bifurcation aspects of the problem in certain relevant Hilbert spaces and characterize the solutions as critical points of an energy functional on these spaces. Many of the conclusions of type (I) are derived from the more abstract results of type (II). Our classical methods, by and large, may not be efficient in dealing with the problems of type (II) and instead give results of type (I) more directly, sometimes allowing greater detail in conclusions of that type, but not bringing out the general abstract functional analytic framework of the problem. Fundamentally, of course, all analysis is "classical", with the abstractions of Hilbert space rooted in more down-to-earth concepts. So it is possible in theory to derive all of Stuart's results by our methods, but there comes a point of diminishing returns if one goes too far in this direction, which is why abstraction is important.

We will consider only results of type (I) and, furthermore, restrict our attention to the main paper [**229**]. But we urge those interested to consult the other papers cited at the beginning of this chapter for many other details of the theory.

Among the results in [**229**] are the following, where it is convenient to consider $\mu \geq 0$.

(A) Under the assumptions (11.5) and (11.6) above, there is a number $\Lambda(A) \geq 0$ such that for $\mu \leq \Lambda(A)$, $u \equiv 0$ is the only solution to Problem P, while for $\mu > \Lambda(A)$ there are nontrivial solutions. In particular, for $\mu > \Lambda(A)$ there is a nontrivial energy minimizer.

(B) For $0 \leq p < 2$, $\lim_{t \to 0^+} u(t) \in (-\pi, \pi)$ for all nonzero solutions of Problem P. If $p \geq 2$, $\mu > \Lambda(A)$, and (11.7) holds, then $|u| < \pi$ on $(0,1]$ and $\lim_{t \to 0^+} u(t) = \pm\pi$.

(C) For $0 \leq p \leq 2$, $\Lambda(A) > 0$, whereas $\Lambda(A) = 0$ for $p > 2$.

(D) When $p = 2$, $\Lambda(A) \leq \frac{Q}{4}$, where $Q$ is the limit in (11.6).

(E) Assume (11.5), (11.6), and (11.7). If $p = 2$ and $\mu > \frac{Q}{4}$ then there are infinitely many solutions $\{u_k\}$ of Problem P with the property that $|u_k(t)| < \pi$ for all $t \in (0,1]$, and for each $k$, $\lim_{t \to 0^+} u_k(t) = \pm\pi$. Further, the number of zeros of $u_k$ tends to infinity as $k \to \infty$.

In the rest of this chapter, we will concentrate on parts of (B), (C), (D), and (E) and on some new results for $p > 2$.

**Theorem 11.1.** *If $p > 2$ then $\Lambda(A) = 0$.*

**Outline of variational proof.** The discussion here requires Lebesgue integration theory and some functional analysis, as well as calculus of variations. See, for example, [**228**] or [**62**]. Some of the required definitions are in Chapter 10. The first step in Stuart's approach is to define certain "energy spaces" $H_p$ and $H_A$, which are Hilbert spaces especially suited to the problem. Suppose that $u \in L^1_{\text{loc}}((0,1])$, meaning that $u$ is locally integrable in $(0,1]$. Suppose also that $u$ has a generalized, or weak, derivative $u'$,[3] and $\int_0^1 s^p u'(s)^2 \, ds < \infty$. It then follows that $u$ is continuous on $(0,1]$.[4] Let

$$H_p = \{u \in L^1_{\text{loc}}((0,1]) \mid u \text{ has a weak derivative } u' \text{ on } (0,1), \\ \textstyle\int_0^1 s^p u'(s)^2 \, ds < \infty, \text{ and } u(1) = 0\}.$$

(The boundary condition refers to the continuous representative of $u$ in $L^1_{\text{loc}}((0,1])$.) Then $H_p$ is a Hilbert space if we define the inner product by

$$\langle u, v \rangle_p = \int_0^1 s^p u'(s) \, v'(s) \, ds.$$

As usual, $||u||_p^2 = \langle u, u \rangle_p$, for $u \in H_p$.

A second inner product, involving the profile function $A(t)$ of the rod, is given by

$$\langle u, v \rangle_A = \int_0^1 A(s) \, u'(s) \, v'(s) \, ds.$$

For functions $u, v$ in $H_p$, the convergence of this integral depends on the behavior of the integrand near 0, and by using (11.6) it is seen that $\langle u, v \rangle_A$ defines a norm which is topologically equivalent to the norm $||\ ||_p$ ([**229**]).

[3]See Section 10.3.1 for a definition of weak derivatives.

[4]See [**25**, Theorem VIII.2]. Since we are using Lebesgue measure, this statement means that $u$ is equal almost everywhere to a continuous function.

We will denote the Hilbert space consisting of the set of functions in $H_p$ with this inner product by $H_A$.

Recall that with this definition, a sequence $\{u_n\}$ of functions in $H_A$ is said to converge weakly to a limit $u \in H_A$ if $\lim_{n\to\infty}\langle u_n, v\rangle_A = \langle u, v\rangle_A$ for every $v \in H_A$. As mentioned earlier in the case of $H_p$, functions in $H_A$ are continuous. If $\{u_n\}$ converges weakly to $u$ then $u_n \to u$ uniformly on $[\varepsilon, 1]$, for every $\varepsilon \in (0,1)$.

The calculus of variations technique uses an energy functional $J_\mu$ defined as follows:

$$J_\mu(u) = \int_0^1 \left\{ \frac{1}{2} A(s)\, u'(s)^2 - \mu\left[1 - \cos u(s)\right] \right\} ds. \tag{11.14}$$

The Euler-Lagrange equations for the problem of minimizing $J_\mu$ over the space $H_A$ are (11.1)–(11.3). Solutions to Problem P are stationary points of $J_\mu$ in $H_A$. Among these are all possible energy minimizers. An energy minimizer with $J_\mu(u) \neq 0$ is a nontrivial solution of Problem P.

Clearly $J_\mu(0) = 0$. No constraint on the functions in $H_A$ prevents 0 from being the minimizer. Therefore, we get a nontrivial solution by showing that the infimum $m(\mu)$ of $J_\mu$ in $H_A$ is negative and that there is a $u_\mu \in H_A$ with $J_\mu(u) = m(\mu)$.

The technique for proving the existence of a minimizing function $u_\mu$ is standard in the calculus of variations, and we refer to [**229**] for details. The nonstandard part is to prove that $m(\mu) < 0$, and this is accomplished by exhibiting a specific function $u$ in $H_A$ such that $J_\mu(u) < 0$.

The function used in [**229**] is

$$u_\alpha(s) = s^\alpha (1 - s), \tag{11.15}$$

for a suitably chosen $\alpha$. We find that

$$\begin{aligned} \int_0^1 s^p u_\alpha'(s)^2\, ds &= \int_0^1 s^p \left(\alpha s^{\alpha-1} - (\alpha+1) s^\alpha\right)^2 ds \\ &= \int_0^1 s^{2\alpha+p}\Big(1 + 2\alpha + \alpha^2 - \frac{2}{s}\alpha - \frac{2}{s}\alpha^2 + \frac{1}{s^2}\alpha^2\Big) ds, \end{aligned} \tag{11.16}$$

where this integral converges if $p + 2\alpha - 2 > -1$, or $\alpha > \frac{1-p}{2}$. Hence, under this condition $u_\alpha \in H_p = H_A$.

For $p > 2$, $-\frac{1}{2} > \frac{1-p}{2}$, and so we may use any $\alpha > -\frac{1}{2}$. Consider $\frac{J_\mu(\lambda u_\alpha)}{\lambda^2}$, for $\lambda > 0$ and $-\frac{1}{2} < \alpha < 0$. Since $1 - \cos\theta \geq \frac{1}{4}\theta^2$ for $0 \leq \theta \leq \frac{\pi}{2}$, we obtain

$$\frac{J_\mu(\lambda u_\alpha)}{\lambda^2} \leq \frac{1}{2\lambda^2}\left( \int_0^1 A(s)\, \lambda^2 u_\alpha'(s)^2\, ds - \frac{\mu}{2} \int_{S(\lambda,\alpha)}^1 \lambda^2 u_\alpha^2(s)\, ds \right),$$

where $S(\lambda, \alpha)$ is the largest $s \in (0,1)$ such that $\lambda u_a(s) = \frac{\pi}{2}$. Note that for each $\alpha \in \left(-\frac{1}{2}, 0\right)$,

$$\lim_{\lambda \to 0+} S(\lambda, \alpha) = 0. \tag{11.17}$$

Also, there is a $K_1 > 0$ such that $A(s) \leq K_1 s^p$ for $0 < s \leq 1$. Hence,

$$\frac{J_\mu(\lambda u_\alpha)}{\lambda^2} \leq \frac{1}{2}\left(\int_0^1 K_1 s^p u_\alpha'(s)^2\, ds - \frac{\mu}{2}\int_{S(\lambda,\alpha)}^1 u_\alpha^2(s)\, ds\right).$$

Because $-\frac{1}{2} > \frac{1-p}{2}$, (11.16) implies that $\int_0^1 K_1 s^p u_\alpha'(s)^2\, ds$ is bounded as $\alpha \to -\frac{1}{2}^+$, while (11.15) and (11.17) imply that

$$\lim_{\alpha \to -\frac{1}{2}^+} \lim_{\lambda \to 0+} \int_{S(\lambda,\alpha)}^1 u_\alpha^2(s)\, ds = \infty.$$

Hence, there exist $\alpha > -\frac{1}{2}$ and $\lambda > 0$ such that $J_\mu(\lambda u_a) < 0$, as desired. This completes our outline of the proof that $\Lambda(A) = 0$ if $p > 2$. We have also seen that in this case there is a nontrivial energy minimizer. □

In the next section we will prove further results for the case $p > 2$. These will include the existence of many solutions of Problem P, but the techniques used in that section do not imply the existence of a nontrivial energy minimizer. We also give a proof by shooting that if (11.7) holds and $p > 2$, then $\Lambda(A) > 0$.

**11.2.1. Results for $p = 2$.** We now turn to the results in (D) and (E) above, which are for the case $p = 2$. In particular, we outline the proof of (E), which implies (D). The following result includes most of (E) and adds a bit more information. Recall that $Q = \lim_{t \to 0+} \frac{A(t)}{t^p}$.

**Theorem 11.2.** *If $p = 2$ and $\mu > \frac{Q}{4}$ then there are infinitely many solutions $\{u_k\}$ of Problem P with the property that $|u_k(t)| < \pi$ for all $t \in (0,1]$. Further, $\lim_{k\to\infty} ||u_k||_2 = 0$, and the number of zeros of $u_k$ tends to infinity as $k \to \infty$.*

**Remark 11.3.** We have observed earlier that for the special case (11.9) a phase plane argument gives infinitely many solutions for $\mu > \frac{1}{4}$. It can also be used to show that in this case, for each $k$ there is a solution with exactly $k$ zeros. On the other hand, for more general $A$ with $p = 2$, the standard variational technique only proves that for each $k$ there is a solution with at least $k$ zeros. Later we will see that a shooting proof gives solutions with exactly $k$ zeros. Also, we will outline an extension of the variational method, based on work of Nehari [**178**] and developed for this problem by C. Stuart (private communication), which can be used to establish the same result.

**Outline of proof.** We break the proof into two parts:

(a) If $\mu > \frac{Q}{4}$ then there is an infinite sequence $\{u_k\}_{k=0,1,\dots}$ of nontrivial solutions of Problem P, and $\lim_{k\to\infty} ||u_k||_2 = 0$.

(b) For each $k$, $|u_k| < \pi$ on $(0,1]$. Also, the number of zeros of $u_k$ tends to infinity with $k$.

**Remark 11.4.** Under the hypotheses above, including the condition that $A' \geq 0$, it can be shown in addition that for every nontrivial solution of Problem P,

$$\lim_{t\to 0^+} u(t) = \pm\pi. \tag{11.18}$$

However the proof is somewhat complicated. (See [**229**].) In Section 11.3 we will see that there is an infinite sequence of solutions $u_k$ such that (11.18) holds, but we will not show that every solution of Problem P has this property.

### 11.2.2. Proof of (b) assuming (a).

**Proof.** Assuming that (a) is true, the proof of (b) is elementary. First consider the function

$$V(t) = \frac{1}{2}\left[A(t)\,u'(t)\right]^2 - \mu A(t)\,(1+\cos u(t)).$$

Then

$$V'(t) = -\mu A'(t)\,(1+\cos u(t)) \leq 0.$$

From (11.3) we see that $\lim_{t\to 0^+} V(t) \leq 0$, and so $V(t) \leq 0$ for $t \in (0,1]$. If $|u(t)| = \pi$ for some $t$ then $u'(t) = 0$, and so $u = \pm\pi$ on $(0,1]$, which contradicts (11.2).

To show that the number of zeros of $u_k$ tends to infinity with $k$, we use the Sturm comparison theorem. (See Chapter 8 of [**41**].) Because $||u_k||_2 \to 0$, solutions can be compared with those of a linear equation. Specifically, choose $\sigma \in \left(\frac{Q}{4}, \mu\right)$ and consider the equation

$$(Av')' + \sigma v = 0. \tag{11.19}$$

Choose $\varepsilon > 0$ such that $\frac{\sigma}{1+\varepsilon} > \frac{Q}{4}$. By (11.6) there is a $\delta > 0$ such that for $0 < s \leq \delta$,

$$Q(1-\varepsilon)\,s^2 < A(s) < Q(1+\varepsilon)\,s^2.$$

The equation

$$Q\left((1+\varepsilon)\,s^2 v'\right)' + \sigma v = 0$$

can be solved exactly to prove that with our choice of $\sigma$ and $\varepsilon$, all solutions have infinitely many zeros in $(0, \delta)$.[5] The Sturm comparison theorem then gives the same conclusion for all solutions to (11.19).

Now consider some interval $[t_0, 1] \subset (0, 1]$ with $t_0$ small, and the solutions $u_k$ of (11.1) known to exist from (a). Integrating (11.1) once and using (11.3), we see that $u_k'(1)$ is bounded independently of $k$. Since $A$ is bounded below by a positive number on $[t_0, 1]$, it follows that $u_k'$ is bounded on this interval independent of $k$. Because $\lim_{k\to\infty} ||u_k||_2 = 0$ (from (a)), it then follows that $|u_k| \to 0$ uniformly on $[t_0, 1]$. The solutions $u_k$ satisfy

$$(Au_k')' + \mu \frac{\sin u_k}{u_k} u_k = 0 \qquad (11.20)$$

wherever $u_k \neq 0$, and if $k$ is large enough then $\mu \frac{\sin u_k}{u_k} > \sigma$ wherever $u_k \neq 0$ in $[t_0, 1]$.

Now suppose that $N$ is a positive integer. Since solutions of (11.19) have infinitely many zeros in $(0, 1]$, we may choose $t_N \in (0, 1)$ such that each such solution has at least $N$ zeros in $(t_N, 1)$. A further use of the Sturm comparison theorem yields the same conclusion for $u_k$ if $k$ is sufficiently large. □

**Remark 11.5.** Because the Sturm theorem used in this way gives only a lower bound on the number of zeros of a solution, we cannot conclude that solutions have a finite number of zeros, or that for each $k$ there is a solution of Problem P with exactly $k$ zeros. These results will follow from the existence proof in Section 11.3.[6]

**11.2.3. Remarks on the proof of (a).** We turn now to the proof of (a) above. The proof in [**229**] uses considerably more sophisticated arguments than we have space for in this book, and only a brief suggestion of the method will be provided.

The basic tool is the Liusternik-Schnirelmann theory [**228**], in a version developed by Clark [**37**]. In this theory (as presented in [**229**]) one considers a $C^1$ functional on a real Hilbert space $H$. For our application, $H = H_A$. It is simpler to assume that $A' > 0$, in which case (it turns out) the functional to be considered is the energy $J_\mu$, defined in (11.14). Note that this functional is even: $J_\mu(u) = J_\mu(-u)$, and that $J_\mu(0) = 0$. To find critical points of $J_\mu$, and hence solutions of Problem P, we will use a "minimax" principle, as described in the next subsection. However before turning to this, we must consider a topological concept called the "genus" of a subset of $H_A$.

[5] Alternatively, the equation can be transformed into an autonomous system and analyzed in the phase plane.

[6] Another problem of this sort, in which a preliminary argument shows the existence of solutions with at least $k$ zeros to a certain boundary value problem but one wants to show that there are exactly $k$ zeros, is presented in the last section of Chapter 19.

Let

$$\Sigma = \{\Omega \subset H_A \mid \Omega \text{ is closed and } \Omega = -\Omega\}.$$

One may think particularly of unit spheres in finite dimensional subspaces of $H_A$. Then define $g : \Sigma \to N \cup \{0, \infty\}$, where $N$ denotes the positive integers, by setting

$$\begin{aligned} &g(\emptyset) = 0, \\ &g(\Omega) = k \text{ if there is an odd mapping } h \in C\left(\Omega, R^k \setminus \{\mathbf{0}\}\right) \\ &\qquad\qquad \text{and } k \text{ is the smallest integer with this property,} \\ &g(\Omega) = \infty \text{ if there is no } k \text{ with the property above.} \end{aligned}$$

Genus is defined in the same way for subsets of finite dimensional spaces, and this helps us gain intuition about the concept. For example, we have the following result ([**228**, p. 87]):

**Proposition 11.6.** *For each bounded symmetric neighborhood* $\Omega$ *of the origin in* $R^m$, $g(\partial\Omega) = m$, *where* $\partial\Omega$ *is the boundary of* $\Omega$.

**Proof.** To see that $g(\partial\Omega) \le m$ we choose $h$ to be the identity map, which is odd and maps $\partial\Omega$ into $R^m \setminus \{0\}$. The reverse inequality uses the concept of topological degree; see [**228**]. □

Another helpful result for understanding genus, applicable in finite or infinite dimensions and also found in [**228**], is:

**Proposition 11.7.** $g(\Omega) = k$ *if there is an odd homeomorphism from* $\Omega$ *onto the unit sphere in* $R^k$.

11.2.3.1. *Minimax principles.* The additional technique used in [**229**] is an appropriate minimax principle. The basic idea can be illustrated in $R^n$. Suppose that $L : R^n \to R^n$ is a symmetric linear transformation. Then $L$ has a set of eigenvalues $\lambda_1 \le \lambda_2 \le \cdots \le \lambda_n$ and a corresponding orthonormal set of eigenvectors $\{\mathbf{x}_1, \ldots, \mathbf{x}_n\}$. The minimax principle in this setting is the formula

$$\lambda_k = \min_{\substack{\text{subspaces } V \text{ of } R^n \\ \text{such that } \dim V = k}} \max_{\substack{\mathbf{x} \in V \\ \|\mathbf{x}\| = 1}} \langle L\mathbf{x}, \mathbf{x} \rangle, \tag{11.21}$$

where $\langle\ ,\ \rangle$ is the standard inner product on $R^n$ ([**142**, p. 89]). See Exercise 11.1.

To put this in the context of genus, using $H = R^n$ instead of $H_A$, we let $M = S^{n-1}$, the unit sphere in $R^n$. Then set

$$F_k = \{\Omega \in \Sigma \mid \Omega \subset M,\ g(\Omega) \ge k\}.$$

Each $F_k$ with $k \leq n-1$ is nonempty, because the unit sphere in every $k$-dimensional subspace of $H$ is contained in $F_k$. Further, let

$$\beta_k = \min_{\Omega \in F_k} \max_{\mathbf{x} \in \Omega} \langle L\mathbf{x}, \mathbf{x} \rangle. \tag{11.22}$$

Using (11.21) and Proposition 11.6, $\beta_k \leq \lambda_k$. On the other hand, the linearity of $L$ leads to the conclusion that $\beta_k \geq \lambda_k$ ([**228**, p. 80]).

Continuing with the finite dimensional case, we examine the relation between eigenvalues and critical points of a functional. If $L$ is a symmetric linear transformation on $R^n$, define the functional $E(\mathbf{x})$ on $M = S^{n-1}$ by $E(\mathbf{x}) = \langle L\mathbf{x}, \mathbf{x} \rangle$. Then the theory of Lagrange multipliers shows that critical points of $E$ are exactly the eigenvalues of $L$, and so exactly the numbers $\beta_k$ defined by (11.22).

It turns out that all of this generalizes to general Hilbert spaces, and even Banach spaces. Again, for the general theory we refer to [**228**, Chapter 2]. The specific application to the tapered rod problem is found in [**229**], and further details are beyond the scope of this book, our goal being only to give an idea of the techniques needed in the variational approach to Problem P. We now turn to the use of shooting.

## 11.3. Existence by shooting for $p > 2$

While shooting can be used to prove existence for every $p > 0$, we illustrate its use here by proving what we believe is a new result, for a case not included in the list (A)–(E) of the previous section. We do need an extra condition on $A$, (iv) below, which a "reasonable" smooth function satisfying (11.6) would satisfy.

**Theorem 11.8.** *Suppose that*

(i) $A > 0$,

(ii) $A' > 0$,

(iii) $\lim_{t \to 0^+} \frac{A(t)}{t^p} = Q$, *for some* $Q > 0$ *and* $p > 2$,

(iv) $\lim_{t \to 0^+} \frac{A'}{\sqrt{A}} = 0$, *and for some* $q > 0$ *and* $K > 0$, $\frac{A'(t)}{\sqrt{A(t)}} \geq Kt^q$ *for* $0 < t \leq 1$.

*Then Problem P has, for each* $\mu > 0$ *and each integer* $k \geq 0$, *a solution with exactly* $k$ *zeros such that* $\lim_{t \to 0^+} u(t) = \pm\pi$.

**Corollary:** (Does not require (iv)) If $p > 2$ then $\Lambda(A) > 0$.

**Proof.** We begin by proving that solutions of (11.1) which are small in an appropriate sense must oscillate.

**Lemma 11.9.** *There exists a $\rho > 0$ such that for every positive integer $N$ there is a $t_N \in (0,1)$ such that if $|u| < \rho$ on $(t_N, 1]$ then $u$ has at least $N$ zeros on $(t_N, 1]$.*

**Proof.** The proof is similar to our discussion above of (11.19) and (11.20). Choose $\rho$ so that for $0 < |u| < \rho$, $\frac{\sin u}{u} > \frac{1}{2}$, and assume that $A(t) < Kt^p$ for $t \in (0,1)$. We then apply the Sturm oscillation theorem ([**41**, Chapter 8]) to compare $u$ with nontrivial solutions of

$$(Kt^p v')' + \frac{1}{2}\mu v = 0. \tag{11.23}$$

If $|u| < \rho$ on an interval $[t_1, t_2] \subset (0,1]$ and $v$ vanishes at $t_1$ and $t_2$, then $u$ vanishes in $(t_1, t_2)$. In Exercise 11.2 you are asked to verify that $v$ can be transformed to a Bessel function and so has infinitely many zeros on $(0,1)$. Lemma 11.9 follows. □

We now start our shooting procedure. In this problem, the appropriate place to shoot from is $t = 1$, since initial value problems at that point have unique solutions. The shooting in the $t$ variable will be "backward", meaning to the left, towards $t = 0$. We are dealing with solutions which oscillate, and we wish to avoid confusion in the signs of derivatives as $t$ decreases. Hence, we find it convenient to change variables so that the shooting direction is "forward".

Let

$$x = \int_t^1 \frac{1}{\sqrt{A(s)}} ds, \tag{11.24}$$
$$u(t) = w(x).$$

Condition (iii) of the theorem implies that $x \to \infty$ as $t \to 0^+$. Problem P becomes

$$w'' - \frac{1}{2}\frac{A'(t(x))}{\sqrt{A(t(x))}} w' + \mu \sin w = 0,$$
$$w(0) = 0, \lim_{x\to\infty} \sqrt{A(t(x))} w'(x) = 0.$$

Let $f(x) = \frac{1}{2}\frac{A'(t(x))}{\sqrt{A(t(x))}}$. The assumptions on $A$ imply that $f > 0$ on $[0, \infty)$ and $\lim_{x\to\infty} f(x) = 0$.

The initial value problem for $w$ now takes the form

$$w'' - f(x) w' + \mu \sin w = 0, \tag{11.25}$$

with

$$w(0) = 0, \quad w'(0) = c. \tag{11.26}$$

We look for values of $c$ such that

$$\lim_{x\to\infty} \sqrt{A\left(t\left(x\right)\right)}w'\left(x\right) = 0. \tag{11.27}$$

We further want to find a solution with a prescribed number $k$ of zeros in $(0,\infty)$, and we wish to show that $\lim_{x\to\infty} w\left(x\right) = \pm\pi$.

**Lemma 11.10.** *There is a $c^* > 0$ such that the solution to (11.25)–(11.26) with $c = c^*$ satisfies $w' > 0$ on $[0,\infty)$ and $w\left(\infty\right) = \pi$. Further, (11.27) is satisfied.*

**Proof.** Consider two subsets of the semi-axis $c > 0$:

$$S_1 = \left\{c > 0 \mid \text{for some } x_1 > 0,\ w\left(x_1\right) > \pi \text{ and } w' > 0 \text{ on } \left[0, x_1\right]\right\},$$
$$S_2 = \left\{c > 0 \mid \text{for some } x_2 > 0,\ w'\left(x_2\right) < 0 \text{ and } w < \pi \text{ on } \left[0, x_2\right]\right\}.$$

These sets are clearly disjoint and open, and we wish to prove that each is nonempty. It is easy to check that large values of $c$ lie in $S_1$, and we leave this step to the reader. For $S_2$ we apply Lemma 11.9. Since $w \to 0$ as $c \to 0$, uniformly on compact sets $[0, X]$, it follows as in our discussion of (11.20) that for every positive integer $k$, $w$ has at least $k$ zeros for sufficiently small $c$. In particular, $w$ has at least one zero before crossing $\pi$, and therefore sufficiently small positive values of $c$ lie in $S_2$.

We note as well that no nonconstant solution is tangent to $w = \pi$, which is the nontangency condition for this problem. Hence, there is at least one $c^* > 0$ such that if $c = c^*$ then $w' > 0$ and $0 < w < \pi$ on $0 \le x < \infty$. If $\lim_{x\to\infty} w\left(x\right) = w_\infty \in (0,\pi)$ then a contradiction results because $\sin w_\infty \neq 0$, so $\lim_{x\to\infty} w\left(x\right) = \pi$.

Since $w''$ is bounded, $w'$ must tend to zero as $x \to \infty$, and since $A\left(t\left(x\right)\right) \to 0$, (11.27) is satisfied. This step completes the proof of Lemma 11.10. □

Hence the theorem is true for $k = 0$.[7] The set of possible values of $c^*$ in Lemma 11.10 is closed since $S_1$ and $S_2$ are open, and we assume that $c_0$ is the smallest such $c^*$. Then the interval $(0, c_0)$ is contained in $S_2$, and $c_0$ is on the boundary of $S_2$.

To prove the theorem for $k = 1$ we define three more sets:

$$S_3 = \{c \in (0, c_0) \mid w \text{ crosses } 0 \text{ at least twice before possibly crossing } -\pi\},$$
$$S_4 = \{c \in (0, c_0) \mid w \text{ crosses } -\pi \text{ after crossing } 0 \text{ only once}\},$$
$$\Sigma = \{c \in (0, c_0) \mid w \text{ crosses } 0 \text{ once and then tends monotonically to } -\pi\}.$$

Our goal is to prove that $\Sigma$ is nonempty. So we will assume that it is empty and look for a contradiction. This assumption implies that $(0, c_0) =$

[7]This proves the Corollary.

$S_3 \cup S_4$. It is clear that $S_3$ and $S_4$ are open and disjoint, and so the desired contradiction is obtained by showing that $S_3$ and $S_4$ are both nonempty.

Our previous observation that $w$ has more and more oscillations as $c \to 0$ implies that $S_3$ is nonempty. To obtain the desired solution for $k = 1$, we must prove that $S_4$ is nonempty.

Note that while a solution remains in $(-\pi, \pi)$, it does not have a positive minimum or a negative maximum. Also, the previous argument involving Bessel functions shows that no solution tends to zero without oscillating. Hence every solution with $c$ in $(0, c_0)$ must cross 0 at least once, with $w' < 0$ at the crossing point. We wish to prove that if $c$ is sufficiently close to $c_0$ and $c \notin \Sigma$, then $w$ crosses $-\pi$ before it has a local minimum. It is convenient now to let $w_c$ denote the solution to (11.25) satisfying (11.26).

**Lemma 11.11.** *Suppose that for* $c \in (0, c_0)$, *the first zero of* $w_c'$ *is at* $x_0(c)$. *Then*

(i) $w_c < w_{c_0}$ *on* $(0, x_0(c))$,

(ii) $\lim_{c \to c_0^-} x_0(c) = \infty$.

**Proof.** Since $w_c''(x_0(c)) < 0$, the implicit function theorem implies that $x_0(c)$ is continuous on $0 < c < c_0$. To prove (i), observe that it is true for small $c$, because comparison of $w$ with the solution of the linearized equation shows that as $c \to 0$, $x_0(c)$ is bounded and $w_c$ and $w_c'$ tend to zero uniformly in $(0, x_0(c))$. In particular, $w_c' < \min_{0 \le x \le x_0(c)} w_{c_0}'$.

If (i) is false then let

$$\hat{c} = \inf \{c < c_0 \mid w_c(x) \ge w_{c_0}(x) \text{ for some } x \in (0, x_0(c))\}.$$

Then $w_{\hat{c}} \le w_{c_0}$ on $[0, x_0(\hat{c})]$. Since $w_{c_0}' > 0$ on $[0, x_0(\hat{c})]$, while $w_{\hat{c}}'(x_0(\hat{c})) = 0$, we see that $w_{\hat{c}}(x_0(\hat{c})) < w_{c_0}(x_0(\hat{c}))$. Since $\hat{c} < c_0$, $w_{\hat{c}} < w_c$ on some interval $(0, \hat{x})$. It follows that for some $\zeta \in (\hat{x}, x_0(\hat{c}))$, $w_{\hat{c}}(\zeta) = w_{c_0}(\zeta)$ and $w_{\hat{c}}'(\zeta) = w_{c_0}'(\zeta)$. But since $w_{\hat{c}}$ and $w_{c_0}$ are distinct solutions of the same second order equation, this is impossible.

Assertion (ii) follows from the continuity of $w_c'$ with respect to $c$ on each compact interval $[0, X]$, completing the proof of Lemma 11.11. □

Suppose that $\varepsilon(c) = \pi - w_c(x_0(c))$. Suppressing the dependence of $x_0$ and $\varepsilon$ on $c$, we have that $x_0 \to \infty$ and $\varepsilon \to 0$ as $c \to c_0^-$. We wish to show that for $c$ sufficiently close to $c_0$, $w_c$ decreases monotonically after $x_0$ to a value below $-\pi$.

**Lemma 11.12.** *As* $c \to c_0$, $x_0(c) = O\left(\log \frac{1}{\varepsilon}\right)$.

**Proof.** Let $v = v_c = \pi - w_c$. Then

$$v'' - f(x)v' - \mu \frac{\sin v}{v} v = 0.$$

Further,

$$\begin{aligned} v(0) = \pi \quad &\text{and} \quad v'(0) = -c, \\ v > 0 \quad &\text{and} \quad v' < 0 \text{ on } [0, x_0), \\ v(x_0) = \varepsilon \quad &\text{and} \quad v'(x_0) = 0. \end{aligned}$$

Let

$$q(x) = \begin{cases} \frac{\sin v(x)}{v(x)} & \text{on } [0, x_0], \\ \frac{\sin v(x_0)}{v(x_0)} & \text{if } x > x_0. \end{cases}$$

(Note that $v(x) \neq 0$ on $[0, x_0]$.) Then $v_c$ satisfies

$$v'' - f(x)v' - \mu q(x) v = 0 \tag{11.28}$$

on $[0, x_0]$.

Further, (11.28) has two linearly independent solutions, $v_1$ and $v_2$, such that

$$\begin{aligned} v_1(0) = \pi, \qquad & v_1(\infty) = 0, \\ v_2(0) = 0, \qquad & v_2'(0) = 1. \end{aligned}$$

Then on $[0, x_0]$,

$$v_c = v_1 + \kappa v_2$$

for some $\kappa$. Also,

$$0 = v_c'(x_0) = v_1'(x_0) + \kappa v_2'(x_0),$$

and hence,

$$\kappa = -\frac{v_1'(x_0)}{v_2'(x_0)}.$$

On $[0, x_0]$,

$$v_c = v_1 - \frac{v_1'(x_0)}{v_2'(x_0)} v_2.$$

Therefore,

$$v_c(x_0) = \frac{W(x_0)}{v_2'(x_0)} \tag{11.29}$$

where

$$W = v_1 v_2' - v_2 v_1'$$

is the Wronskian of $\{v_1, v_2\}$. Note that $W(0) = \pi$ and

$$W' = f(x) W.$$

Thus,

$$W = \pi e^{F(x)},$$

where $F(x) = \int_0^x f(s)\,ds$. Recall that $f > 0, f(\infty) = 0$, and

$$x = \int_t^1 \frac{1}{\sqrt{A(s)}}ds,$$
$$f(x) = \frac{A'(t)}{2\sqrt{A(t)}}.$$

From these we find that

$$e^{F(x)} = \sqrt{\frac{A(1)}{A(t)}}.$$

For the case $A(t) = t^p$ this leads to

$$W = \pi\left(\frac{2+(p-2)x}{2}\right)^{\frac{p}{p-2}}.$$

For the general case, where

$$A(t) \geq ct^p$$

for some $c > 0$, a bound of the form

$$W(x) \leq K(1+x)^{\frac{p}{p-2}} \tag{11.30}$$

is obtained. For $x_0$ sufficiently large, $v_2 > 0, v_2' > 1$, and $v_2'' \geq \frac{\mu}{2}v_2$ on $\left[\frac{x_0}{2}, x_0\right]$. Equations (11.29)–(11.30) and a simple comparison of $v_2$ with the solution $\sqrt{\frac{2}{\mu}}\sinh\left(\sqrt{\frac{\mu}{2}}\left(x - \frac{x_0}{2}\right)\right)$ of $v'' = \frac{\mu}{2}v$ imply that there is a $\gamma > 0$ such that $\varepsilon = v_c(x_0) = O\left(e^{-\gamma\sqrt{\mu}x_0}\right)$ as $c \to c_0$. The lemma follows. □

We continue with the proof that if $c \in (0, c_0)$ and $c$ is sufficiently close to $c_0$, then $w_c'$ has exactly one zero before $w_c$ crosses $-\pi$. Suppose that $w' = w_c'$ has a second zero, at $x_1 > x_0$, before $w$ crosses $-\pi$. We wish to obtain a contradiction. Clearly, $w(x_1) \in (-\pi, 0)$. The following lemma shows the effect of the monotonicity assumption on $A$. A similar lemma will be used in Chapter 15.

**Lemma 11.13.** *With $w$, $x_0$, and $x_1$ as described,*

$$|w(x_1)| > |w(x_0)|.$$

**Proof.** Multiplying (11.25) by $w'$ and integrating gives

$$\left[\frac{1}{2}w'^2 - \mu\cos w\right]_{x_0}^{x_1} = \int_{x_0}^{x_1} f(x)\,w'(x)^2\,dx. \tag{11.31}$$

Since $w'(x_0) = w'(x_1) = 0$, $f > 0$, and $w(x_0) \in (0, \pi)$, $w(x_1) \in (-\pi, 0)$ and the lemma follows. □

However, we are not through with (11.31). Since $w(x_0) = \pi - \varepsilon$, Lemma 11.13 implies that $w(x_1) \in (-\pi, -\pi + \varepsilon)$. Therefore, from (11.31),

$$\int_{x_0}^{x_1} f(x)\, w'(x)^2\, dx = O\left(\varepsilon^2\right) \quad \text{as } c \to c_0^-. \tag{11.32}$$

A contradiction to our assumption that $w' = 0$ before $w = -\pi$ results from the following lemma, in which hypothesis (iv) of the theorem (Theorem 11.8) is used.

**Lemma 11.14.** *There exists a $K > 0$ such that for sufficiently small $\varepsilon$,*

$$\int_{x_0}^{x_1} f(x)\, w'(x)^2\, dx > \frac{K}{|\log \varepsilon|^{1+2q/(p-2)}}. \tag{11.33}$$

**Proof.** Let $x_2$ be the unique point in $(x_0, x_1)$ where $w = 0$. Then

$$\int_{x_0}^{x_1} f(x)\, w'(x)^2\, dx > \int_{x_0}^{x_2} f(x)\, w'(x)^2\, dx \geq \eta \int_{x_0}^{x_2} w'(x)^2\, dx, \tag{11.34}$$

where $\eta = \min_{x_0 \leq x \leq x_2} f(x)$.

From (11.24) and hypotheses (iv) of Theorem 11.8 it follows that for sufficiently large $x$, and appropriate positive $K$ independent of $\varepsilon$ at each step,

$$x < Kt^{\frac{2-p}{2}}$$

and

$$f(x) = \frac{A'(t)}{2\sqrt{A(t)}} > Kt^q > Kx^{\frac{2q}{2-p}}. \tag{11.35}$$

(In the case $A = t^p$, this becomes $f(x) > \frac{K}{x}$.) Hence in (11.34),

$$\eta > \frac{K}{x_2^{\frac{2q}{p-2}}}. \tag{11.36}$$

To obtain a contradiction we need an upper bound on $x_2 - x_0$. As in (11.31),

$$\frac{1}{2} w'(x)^2 > \mu\left(\cos w - \cos(\pi - \varepsilon)\right)$$

for $x_0 < x < x_2$, and since $w' < 0$ in this interval,

$$\frac{dw}{dx} < -\sqrt{2\mu}\sqrt{\cos w - \cos(\pi - \varepsilon)},$$

$$\frac{dx}{dw} > -\frac{1}{\sqrt{2\mu}\sqrt{\cos w - \cos(\pi - \varepsilon)}},$$

$$x_2 - x_0 < \frac{1}{\sqrt{2\mu}} \int_0^{\pi - \varepsilon} \frac{dw}{\sqrt{\cos w - \cos(\pi - \varepsilon)}}$$

$$(11.37) \qquad = O\left(\int_0^{\pi-\varepsilon} \frac{dy}{\sqrt{\varepsilon y + \frac{1}{2} y^2}}\right) = O(-\log \varepsilon) \quad \text{as } \varepsilon \to 0.$$

(Set $y = \pi - \varepsilon - w$.) From the definitions of $x_0$ and $x_2$ and from the Cauchy-Schwarz inequality,

$$(\pi - \varepsilon) = \int_{x_2}^{x_0} w' dx \leq \sqrt{\int_{x_0}^{x_2} w'^2 dx} \sqrt{\int_{x_0}^{x_2} 1 dx}.$$

Still adjusting $K$ as needed, we get

$$\int_{x_0}^{x_2} w'^2 dx \geq \frac{K}{\log \frac{1}{\varepsilon}}.$$

It then follows from (11.34), (11.36), and (11.37) that for some $K$ and $K_1$,

$$\int_{x_0}^{x_1} f(x)\, w'(x)^2\, dx > \frac{K}{\left(\log \frac{1}{\varepsilon}\right) x_2^{\frac{2q}{p-2}}} > \frac{K}{\log \frac{1}{\varepsilon} \left(x_0 + K_1 \left(\log \frac{1}{\varepsilon}\right)\right)^{\frac{2q}{p-2}}}.$$

Using Lemma 11.12 we get (11.33). □

This result contradicts (11.32), and hence our assumption that after $x_0$, $w' = 0$ before $w = -\pi$. Therefore, for $c \in (0, c_0)$ and sufficiently close to $c_0$, $w_c$ has a maximum near $\pi$ and then decreases to below $-\pi$. Thus, the set $S_4$ is nonempty, and so there is a $c_1 \in S_2$ which is not in $S_3$ or $S_4$.

The corresponding solution $u_{c_1}(t) = w_{c_1}(x)$ of (11.1) has one zero and satisfies (11.2). From the definition of $w$,

$$A(t)\, u'(t) = -\sqrt{A(t)} w'(x).$$

Also, for large $x$, $w'_{c_1}(x) < 0$ and $w_{c_1}(x) > -\pi$. It follows from (11.25) and the properties of $f$ that $\lim_{x\to\infty} w_{c_1}(x) = -\pi$ and $\lim_{x\to\infty} w'_{c_1}(x) = 0$. Hence $u_{c_1}$ satisfies (11.3) and $\lim_{t\to 0^+} u_{c_1}(t) = -\pi$. We must show that $u_{c_1}$ also satisfies (11.4). To see this, let $t_1$ denote the zero of $u_{c_1}$ in $(0,1)$. Then $u'_{c_1} > 0$ on $(0, t_1)$, and $\int_0^{t_1} u'_{c_1}(t)\, dt = \pi$. From (11.3) we see that $A(t)\, u'_{c_1}(t)^2 < u'_{c_1}(t)$ for sufficiently small positive $t$, and so (11.4) is satisfied. Thus, $u_{c_1}$ solves Problem P and has exactly one zero in $(0, 1]$.

We have thus proved Theorem 11.8 in the case $k = 1$. A straightforward induction process completes the proof of the theorem. □

## 11.4. Proof using Nehari's method

This section is based on a private communication from Professor Stuart. The method to be described was introduced in [**178**]. We will outline its use to show that for $p > 2$ and $\mu > 0$, Problem P has a solution with exactly one zero in the interval $(0, 1)$. The extension to solutions with additional zeros is more technical but involves no new ideas.

Our starting point is the variational proof outlined in Section 11.2.3, where we define the Hilbert space $H_A$, and in (11.14), the energy functional $J_\mu$. It is found that

$$m(\mu) = \min_{u \in H_A} J_\mu(u)$$

is negative. Fairly standard calculus of variation techniques then lead to the existence of a nontrivial $u_0 \in H_A$ with $J_\mu(u_0) = m(\mu)$, and it can be shown that $u_0$ is a solution of Problem P.

It should be remarked that there is a minimizing function $u_0$ with no zeros in $(0, 1)$.[8] To see this, observe that $J(u) = J(|u|)$, and so the minimum may be sought over functions which are nonnegative on this interval. Also, $u_0$ can be shown to be smooth, so that if it has a zero then $u_0 = u_0' = 0$ at some $t \in (0, 1)$. But since $u_0$ solves (11.1), this implies that $u_0 \equiv 0$, a contradiction.

To obtain a second solution, we define, for each $T \in (0, 1)$, subspaces $H_T^1$ and $H_T^2$ as follows:

$$H_T^1 = \{u \in H_A \mid u \equiv 0 \text{ on } T \le t \le 1\},$$
$$H_T^2 = \{u \in H_A \mid u \equiv 0 \text{ on } 0 \le t \le T\}.$$

Then let

$$\left\{ \begin{array}{l} E_T^1 = \inf_{u \in H_T^1} J_\mu(u) = \inf_{u \in H_T^1} \int_0^T \left(Au'^2 - \mu(1 - \cos u)\right) dt, \\ E_T^2 = \inf_{u \in H_T^2} J_\mu(u) = \inf_{u \in H_T^2} \int_T^1 \left(Au'^2 - \mu(1 - \cos u)\right) dt. \end{array} \right. \tag{11.38}$$

It can be shown that for each $T \in (0, 1)$, and for $i = 1$ and $i = 2$, there is a $u_T^i \in H_T^i$ such that $J_\mu(u_T^i) = E_T^i$. Then each of these functions satisfies (11.1) and the respective boundary conditions

$$\left(Au_T^{1\prime}\right)(0) = u_T^1(T) = 0,$$
$$u_T^2(T) = u_T^2(1) = 0.$$

Note the different nature of these two pairs of boundary conditions, with those for $u_T^1$ being essentially the same as those for Problem P, except on

[8] It is shown in [**229**] that the minimizer is unique up to sign.

the smaller interval $(0,T]$. We may assume that $u_T^1 > 0$ on $(0,T)$ and $u_T^2 > 0$ on $(T,1)$. Further, let

$$E_T = E_T^1 + E_T^2.$$

From (11.38), and because $u_T^i(T) = 0$ for $i = 1,2$, it follows (with some effort) that

$$\left\{ \begin{array}{l} \frac{d}{dT} E_T^1 = -\frac{1}{2} A(T)\, u_{T^-}^{1\prime}(T)^2, \\ \frac{d}{dT} E_T^2 = \frac{1}{2} A(T)\, u_{T^+}^{2\prime}(T)^2, \end{array} \right. \tag{11.39}$$

where $T^-$ and $T^+$ indicate left and right derivatives, respectively. Set

$$c_1 = \sup_{T\in(0,1)} E_T = \sup_{T\in(0,1)} \min_{\substack{u\in H_A \\ u(T)=0}} J(u).$$

We will outline how to prove that

$$0 > c_1 > \max\left\{ \lim_{T\to 1} E_T, \liminf_{T\to 0} E_T \right\} \tag{11.40}$$

and conclude that for some $T \in (0,1)$,

$$E_T = c_1 \quad \text{and} \quad \frac{d}{dT} E_T = 0.$$

We then deduce from (11.39) that $u_{T^-}^{1\prime}(T) = -u_{T^+}^{2\prime}(T)$. Hence, the solution formed by piecing together $u_T^1$ and $-u_T^2$ is smooth and nontrivial and has energy $c_1 > m(\mu)$, thus giving the desired solution to Problem P with exactly one zero in $(0,1)$.

To get an idea of why (11.40) is true, we observe from their definitions that $E_T^1$ is nonincreasing in $T$ while $E_T^2$ is nondecreasing, because, as $T$ increases, more functions lie in $H_T^1$, but fewer in $H_T^2$. These statements are also consistent with (11.39). Since in either case, $u = 0$ is in the relevant subspace, each is nonpositive. As noted above, the boundary conditions for $u_T^1$ closely resemble those of Problem P, so it is not surprising that for $T$ close to 1, $E_T^1$ is close to $m(\mu)$, and in particular, negative. Furthermore, it turns out that for $T$ close to 1, the minimizing function in $H_T^2$ is zero, so that $E_T^2 = 0$.[9] Thus, $\lim_{T\to 1^-} E_T = m(\mu)$.

More technicalities are needed in dealing with $E^T$ as $T \to 0^+$, because the type of boundary condition satisfied by $u_T^2$ on the left is different from that satisfied on the left by $u_T^1$. However we get insight into the behavior of $u_T^2$ by referring to the results of Section 11.3. While we carried out the analysis there in the variable $x$ defined by (11.24), it is easily interpreted in terms of $t$. Let $u_0$ be the nonnegative solution of Problem P which has, among all such solutions (if there is more than one), the maximum slope $-c$ at $t = 1$. A straightforward argument showed that for every small positive $T$, there are solutions $u$ with $u(1) = 0$, $u'(1) < 0$, $u(T) = 0$, and $u$ close

[9]See Exercise 11.3.

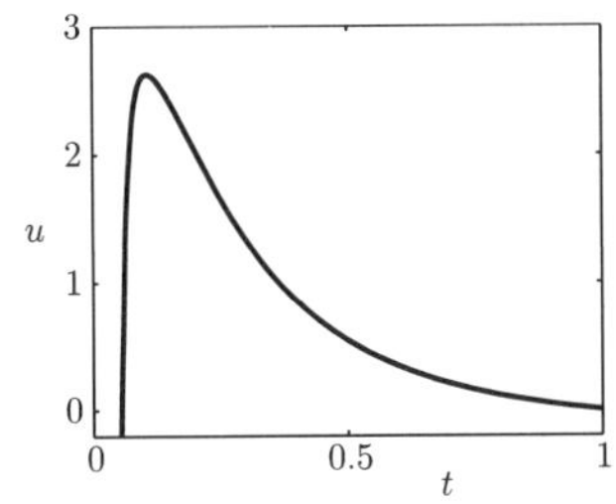


**Figure 11.1**

to $u_0$ on most of the interval $[T, 1]$. (See Figure 11.1. We do not need the more difficult conclusion that to the left of $T$, $u$ crosses $-\pi$.)

We have not shown here that $u_0$ is an energy minimizer. However this follows from Theorems 3.3 and 3.5 in [**229**], which imply the uniqueness of the positive solution. From these results we can prove that for the solutions $u$ described just above, $\int_T^1 \left(Au'^2 - \mu\left(1 - \cos u\right)\right) dt$ is close to $m\left(\mu\right)$, so that $\liminf_{T \to 0^+} E_T^2 \leq m\left(\mu\right)$. We have made no estimate of $E_T^1$, knowing only that it is nonpositive. Still, these observations lend credibility to (11.40).

Thus, the basic idea for proving Theorem 11.8 using variational methods is clear. However the technical details required are considerable, and we leave the subject here.

## 11.5. More about the case $p = 2$

Here we are concerned briefly with result (D) from Section 11.2. We wish to explain why it is possible that in the case $p = 2$, it may be that either $\Lambda\left(A\right) < \frac{Q}{4}$ or $\Lambda\left(A\right) = \frac{Q}{4}$. Two examples serve to illustrate these results, and the extension of our earlier methods to provide rigorous proofs requires no new ideas and is left to the reader. The first example is the case $A\left(t\right) = t^2$ considered before. As discussed earlier, in this case, $\Lambda\left(A\right) = \frac{1}{4}$. For the second example we take a number $\varepsilon \in (0, 1)$ and let

$$A\left(t\right) = \begin{cases} t^2 & \text{if } 0 \leq t \leq \varepsilon, \\ \varepsilon^2 & \text{if } \varepsilon < t \leq 1. \end{cases}$$

It is easy to show that for sufficiently large $\frac{\mu}{\varepsilon}$, even with $\mu < \frac{1}{4}$, $u\left(t\right)$ oscillates in $(\varepsilon, 1)$, but still $p = 2$ and there are no oscillations in $(0, \varepsilon)$.[10] A similar example appears in [**247**].

Many other results about the behavior of solutions when $p = 2$ are given in [**229**] and the other papers we have cited. The techniques used for many

[10] $A$ is not smooth, but can be altered slightly to be smooth, without affecting the conclusion that $\Lambda\left(A\right) < \frac{1}{4}$.

of these are classical, and we refer the interested reader to the original papers for details.

## 11.6. Exercises

**Exercise 11.1.** Verify the minimax principle (11.21) for the map $K(u_1, u_2) = (u_1 + 2u_2, 2u_1 - 2u_2)$. Hint: One method is to use Lagrange multipliers.

**Exercise 11.2.** (Computation) By writing (11.23) in the form

$$t^2 v'' + ptv' + \eta t^{2-p} v = 0,$$

with $p > 2$, show that if $v(t) = t^{-\frac{p-1}{2}} w\left(\frac{2\sqrt{\mu}}{p-2} t^{\frac{2-p}{2}}\right)$ then $w$ is a Bessel function, and so $v$ has infinitely many zeros on $(0, 1)$.

**Exercise 11.3.** In connection with the proof of (11.40), use Poincaré's inequality to show that there is a $T^*$ such that if $T^* \leq T < 1$ and $u \in C^1([T, 1])$ with $u(T) = u(1) = 0$, then $\int_T^1 \left(\frac{1}{2} A u'^2 - \mu(1 - \cos u)\right) dt \geq 0$. What does this imply about $m(\mu)$?

*Chapter 12*

# Uniqueness and multiplicity

## 12.1. Introduction

Many of the techniques we discuss in this book are used to prove the existence of a solution to a boundary value problem. But often one wants uniqueness or multiplicity results as well. Is there only one solution? If not, how many are there? How does the number of solutions depend on parameters in the problem? These questions lead into the subject of bifurcation diagrams, which we introduce, though only briefly, in Section 12.3. Uniqueness is often harder to prove than existence, particularly for systems of order three or higher. The methods for attacking uniqueness and multiplicity problems, particularly global problems (not near a bifurcation point) are often ad hoc, since frequently the available general theories do not apply.

The most common problems where a general uniqueness theory can be applied are those where the right-hand sides of the system obey some sort of monotonicity condition. Here is a simple result of this type.[1]

**Proposition 12.1.** *Suppose that $f(x,u,v)$ is continuous in the region $0 \leq x \leq 1$, $(u,v) \in R^2$. Suppose also that $\frac{\partial f}{\partial u}$ and $\frac{\partial f}{\partial v}$ exist and are continuous and $\frac{\partial f}{\partial u}$ is positive in this region. Then for each $p$ and $q$, the following boundary value problem has at most one solution:*

$$\begin{gathered} u'' = f(x,u,u'), \\ u(0) = p, \quad u(1) = q. \end{gathered} \tag{12.1}$$

[1]As usual, we have not tried to state the strongest possible result. Several other uniqueness results are scattered through the book, as indicated in the index.

**Proof 1.** Suppose that there are two solutions, say $u_1$ and $u_2$. We can assume that $u_1'(0) > u_2'(0)$. If there is a first point $x_1 \in (0,1)$ where $u_1'(x_1) = u_2'(x_1)$, then necessarily $u_1''(x_1) \leq u_2''(x_1)$. However, since $u_1' > u_2'$ on $[0, x_1)$, we see that $u_1(x_1) > u_2(x_1)$, and from the hypotheses, $u_1''(x_1) > u_2''(x_1)$, a contradiction. Hence $u_1' > u_2'$ on $[0,1]$, and since $u_1(0) = u_2(0) = p$, we have $u_1(1) > u_2(1)$, a further contradiction. □

**Proof 2.** Although the proof above is simple, it is frequently helpful to use monotonicity slightly differently, by considering the derivative of the solution with respect to one of the initial conditions. Thus, consider the initial value problem

$$u'' = f(x, u, u'),$$
$$u(0) = p, \quad u'(0) = \alpha,$$

with unique solution $u(t,\alpha)$, and attempt to analyze the function $v(t) = \frac{\partial u(t,\alpha)}{\partial \alpha}$. This function is easily seen to satisfy the "equation of first variation",

$$v'' = f_u(x, u, u')\, v + f_{u'}(x, u, u')\, v', \tag{12.2}$$

with initial conditions

$$v(0) = 0, \quad v'(0) = 1. \tag{12.3}$$

We claim that $v' > 0$ on $[0,1]$. If not, and $v'$ has its first zero at $x_0$, then $v''(x_0) \leq 0$. However, $v(x_0) > 0$, and so from (12.2) and the hypothesis that $f_u > 0$, we see that $v''(x_0) > 0$. This contradiction proves that $v' = \frac{\partial u'}{\partial \alpha} > 0$ on $[0,1]$. Hence, $v > 0$ on $(0,1]$, and in particular, $v(1) = \frac{\partial u(1,\alpha)}{\partial \alpha} > 0$. It follows that the equation $u(1,\alpha) = q$ has at most one solution, proving Proposition 12.1. □

To see what can occur if the inequality on $\frac{\partial f}{\partial u}$ goes the other way, observe that the simple linear problem

$$\begin{aligned} u'' &= -\pi^2 u, \\ u(0) &= u(1) = 0 \end{aligned} \tag{12.4}$$

is of the type (12.1), except that in this case $\frac{\partial f}{\partial u} < 0$, and $u(x) = c \sin \pi x$ solves this for every $c$.

### 12.1.1. An application of contraction mapping in a Banach space.

The contraction mapping theorem is an important general technique for both existence and uniqueness problems in analysis. We have commented in Chapter 1 on the use of this method for initial value problems, noting that it is equivalent to what we often prefer to call the method of successive approximations. Applications to boundary value problems are less common, but we present here an elegant result from the ode text [**248**] of Walter. It nicely complements Proposition 12.1. As one would expect for a contraction mapping proof, both existence and uniqueness are included. Here is an instance where use of Banach space terminology is convenient, particularly because a weighted norm is used.

**Theorem 12.2.** *Suppose that $f(x,u)$ is continuous on $[0,1]\times R^1$ and satisfies a Lipschitz condition*

$$|f(x,u)-f(x,v)|\le L|u-v|$$

*with $L<\pi^2$. Then the boundary value problem*

$$\begin{aligned} u'' &= f(x,u), \\ u(0) &= u(1)=0 \end{aligned} \tag{12.5}$$

*has a unique solution.*

**Proof.** We consider first the simple linear boundary value problem

$$\begin{aligned} u'' &= g(x), \\ u(0) &= u(1)=0. \end{aligned} \tag{12.6}$$

The Green's function $G(x,y)$ for this problem is readily found by simple integration. (See [**41**, Chapter 7].) For every continuous function $g$ the unique solution is

$$u(x)=\int_0^1 G(x,y)\,g(y)\,dy,$$

where

$$G(x,y)=\begin{cases} y(x-1) & \text{if } 0\le y\le x\le 1, \\ x(y-1) & \text{if } 0\le x\le y\le 1. \end{cases}$$

We then define an operator $T$ on the Banach space $C([0,1])$ of continuous real-valued functions on the interval $[0,1]$. Recall that the usual norm on this space is $\|u\|=\max_{0\le x\le 1}|u(x)|$. The operator $T$ is given by

$$(Tu)(x)=\int_0^1 G(x,y)\,f(y,u(y))\,dy, \tag{12.7}$$

where $f$ is the Lipschitz continuous function in (12.5). We are looking for a fixed point of the operator $T$, that is, a continuous function $u$ such that $Tu=u$. This is an integral equation, and every continuous solution must be twice differentiable, because of the form of $G$. In fact, every fixed point of $T$ satisfies (12.6), with $g(x)=f(x,u(x))$.

Recall that there is a unique fixed point if $T$ is a contraction on $C([0,1])$, meaning that for some $K<1$, $\|Tu-Tv\|\le K\|u-v\|$ for every $u,v$ in

$C([0,1])$.[2] From (12.7),

$$\begin{aligned}
&|(Tu)(x) - (Tv)(x)| \\
&\quad = \left| \int_0^1 G(x,y)\left(f(y,u(y)) - f(y,v(y))\right)\,dy \right| \\
&\quad \le L \int_0^1 |G(x,y)|\ |u(y) - v(y)|\ dy \le L\,||u-v|| \int_0^1 |G(x,y)|\,dy \\
&\quad = L\,||u-v||\,\frac{1}{2}x(1-x).
\end{aligned}$$

Hence $T$ is a contraction if $L < 8$.

However, Theorem 12.2 gives a better estimate, stating that $T$ has a unique fixed point if $L < \pi^2$. The proof in [**248**] is an interesting application of a weighted norm. The Banach space $C([0,1])$ is larger than is needed to study the problem (12.5), because only functions satisfying the homogeneous boundary conditions are eligible to be solutions. Hence, we consider

$$\mathcal{B} = \{u \in C([0,1]) \mid |u(x)| \le K \sin \pi x \text{ for some } K > 0\}.$$

Functions in $\mathcal{B}$ satisfy the boundary conditions $u(0) = u(1) = 0$. For $u \in \mathcal{B}$, let

$$||u||_{\mathcal{B}} = \sup_{0<x<1} \left| \frac{u(x)}{\sin \pi x} \right|.$$

It is easy to check that $||\ ||_{\mathcal{B}}$ is a norm and that with this norm, $\mathcal{B}$ is a Banach space. (See a functional analysis book, such as [**237**]. There is no requirement that $u$ be differentiable, or that the limits of $\frac{u(x)}{\sin \pi x}$ at $x = 0$ and 1 exist.)

Also,

$$||Tu - Tv||_{\mathcal{B}} = \sup_{0<x<1} \left| \frac{\int_0^1 G(x,y)\left(f(y,u(y)) - f(y,v(y))\right)dy}{\sin \pi x} \right|.$$

For each $x \in (0,1)$ and for all $u, v$ in $\mathcal{B}$,

$$\begin{aligned}
&\left| \int_0^1 G(x,y)\left(f(y,u(y)) - f(y,v(y))\right)dy \right| \\
&\quad = \left| \int_0^1 G(x,y) \sin \pi y\ \frac{f(y,u(y)) - f(y,v(y))}{\sin \pi y}\,dy \right| \\
&\quad \le L\,||u-v||_{\mathcal{B}} \left| \int_0^1 G(x,y) \sin \pi y\ dy \right|,
\end{aligned}$$

[2] Letting $u_0(x) = 0$, $u_{i+1} = Tu_i$, the sequence $\{u_i\}$ of successive approximations converges to a unique solution of $Tu = u$.

and

$$\left|\int_0^1 G(x,y)\sin \pi y \, dy\right| = \frac{1}{\pi^2}\sin \pi x.$$

It follows that $T$ is a contraction if $L < \pi^2$, proving Theorem 12.2. □

Example (12.4) shows that $L < \pi^2$ is the best possible estimate of this kind. This example is therefore of interest for both (12.1) and (12.5).

In the rest of this chapter we will discuss two further uniqueness problems which we find interesting. They illustrate another important technique which can be used on such problems. We also discuss examples where, instead of a unique solution, there are several. Often it is difficult to pin down the exact number of solutions. We will give some examples where this can be done.

## 12.2. Uniqueness for a third order problem

In Chapter 2 we discussed the following boundary value problem:

$$u''' = u^2 - 1, \tag{12.8}$$

$$u(-\infty) = -1, \quad u(\infty) = 1. \tag{12.9}$$

This problem is interesting because solutions must oscillate around $u = 1$ as they approach this limit. A shooting method based on an energy function was used to prove the existence of a solution. We will show how introducing a second energy function leads to a relatively simple uniqueness proof.[3] This presages the use in Section 12.6 below of five auxiliary functions in C. Coffman's proof of a landmark uniqueness result.

**Theorem 12.3.** *The boundary value problem* (12.8)–(12.9) *has exactly one solution, up to translation.*

**Proof.** The first part of the proof is easy:

**Lemma 12.4.** *There is a unique solution of* (12.8)–(12.9) *such that* $u(0) = 0$ *and* $u(t) \neq 0$ *for* $t \neq 0$.

**Proof.** The existence of at least one such solution was shown in Chapter 2. Suppose that there are two such solutions, say $u_1$ and $u_2$. Since (12.8) is autonomous, we can assume that $u_1(0) = u_2(0) = 0$ and both solutions are positive on $(0,\infty)$ and negative on $(-\infty, 0)$. Let

$$\phi = u_2 - u_1.$$

[3] Independent existence and uniqueness proofs using classical methods were given recently by Bona and Weissler [**21**].

Then

$$\phi''' = (u_1 + u_2)\,\phi,$$
$$\phi(0) = \phi(-\infty) = \phi(\infty) = 0.$$

We may also assume, by interchanging $u_1$ and $u_2$ if necessary, that

$$\phi''(0) \geq 0.$$

If $\phi'(0) \geq 0$ and at least one of $\phi'(0)$, $\phi''(0)$ is positive, then $\phi' > 0$ for all $t > 0$, since $u_1 + u_2 > 0$ on $(0,\infty)$. But this is impossible since we have assumed that $\phi(0) = 0$ and $\phi(\infty) = 0$.

If $\phi'(0) \leq 0$ and $\phi''(0) \geq 0$, with at least one of these nonzero, then we obtain the same contradiction by considering $-\phi(-t)$, since $-u_1(-t)$ and $-u_2(-t)$ also satisfy (12.8) and both are positive for $t > 0$. □

Suppose now that $u$ is a solution of (12.8)–(12.9) with at least two zeros. We can assume that the first zero is at $t = 0$. Let the first positive local maximum of $u$ be at $t_0 > 0$.

**Lemma 12.5.** *Then* $0 < u(t_0) \leq \sqrt{3}$.

**Proof.** Suppose for contradiction that $u(t_0) > \sqrt{3}$. We reintroduce, from Chapter 2, the energy function $Q(t) = u'u'' + u - \frac{1}{3}u^3$. Then $Q' = (u'')^2$, and so integrating from 0 to $t_0$ gives

$$-u'(0)\,u''(0) = \int_0^{t_0} (u'')^2\,dt + \frac{1}{3}u_0^3 - u_0 > 0, \tag{12.10}$$

where $u_0 = u(t_0)$. Hence $u'(0) > 0$ and $u''(0) < 0$.

Let $u_*(t) = -u(-t)$. Then $u_*$ satisfies (12.8), $u_* > 0$ on $(0,\infty)$, and $u_*(\infty) = 1$. Also

$$u_*'(0) > 0, \qquad u_*''(0) = -u''(0) > 0. \tag{12.11}$$

If $h = u_* - u$ then

$$h''' = (u_* + u)\,h,$$
$$h(0) = 0, \quad h'(0) = 0, \quad h''(0) = -2u''(0) > 0.$$

This implies that $h > 0$ as long as $u > 0$. Hence $u_*$ has a critical point at $t_0$ with $u_*(t_0) > u(t_0) > \sqrt{3}$.

Therefore $u_*$ also satisfies

$$-u_*'(0)\,u_*''(0) = \int_0^{t_0} (u_*'')^2\,dt + \frac{1}{3}u_*(t_0)^3 - u_*(t_0) > 0,$$

giving the contradictory conclusion that $u_*''(0) < 0$. This proves Lemma 12.5. □

We now assume that $u$ is a solution of (12.8)–(12.9) with at least two zeros. Let the first zero be at $t = 0$, and the first local maximum at $t_0 > 0$. Lemma 12.5 shows that $0 < u(t_0) \le \sqrt{3}$.

First suppose that $0 < u(t_0) < 1$. Then $u'''(t_0) < 0$, and so immediately after $t_0$, $u' < 0, u'' < 0, u''' < 0$, and $u < 1$, and these inequalities continue to hold until $u = -1$. Observe that $Q(t_0) = u_0 - \frac{1}{3}u_0^3 > 0$, where $u_0 = u(t_0)$. Also, if $0 < u < u_0$ then $u - \frac{1}{3}u^3 < u_0 - \frac{1}{3}u_0^3$. Since $Q$ is increasing, $u'u''$ does not vanish until after $u$ reaches a value below zero where $u - \frac{1}{3}u^3 = u_0 - \frac{1}{3}u_0^3$. In fact, such a value must be below $-\sqrt{3}$. Since $u(\infty) = 1$, $u$ has a minimum after $t_0$. Suppose that the first minimum after $t_0$ is at $t = t_1$. Since $u(t_1) < -\sqrt{3}$, $u'''$ is positive after $t_1$ until $u = -1$, and so $u' > 0$ and $u'' > 0$ at least until this point. Hence there is a $t_2 > t_1$ with $u(t_2) = -\sqrt{3}$ and $u'(t_2) > 0$, $u''(t_2) > 0$.

Now introduce the second functional mentioned earlier, by letting

$$F = \frac{1}{3}u'^3 - \frac{1}{12}u^4 + \frac{1}{2}u^2.$$

Then $F'(t) = u'Q(t)$, which is positive for $t > t_2$ as long as $u' > 0$. Hence, for as long after $t = t_2$ as $u' \ge 0$ we have

$$\frac{1}{3}u'^3 - \frac{1}{12}u^4 + \frac{1}{2}u^2 > \frac{1}{3}u'^3(t_2) - \frac{1}{12}u(t_2)^4 + \frac{1}{2}u(t_2)^2.$$

Since $u'(t_2) > 0$ this gives

$$\frac{1}{3}u'^3 > \frac{1}{12}u^4 - \frac{1}{2}u^2 - \left(\frac{1}{12}u(t_2)^4 - \frac{1}{2}u(t_2)^2\right).$$

But $\frac{1}{12}u^4 - \frac{1}{2}u^2$ has its global minima at $\pm\sqrt{3}$, and hence $u' > 0$ after $t = t_2$. Therefore $u(\infty) \ne 1$, for otherwise $u - 1$ has infinitely many zeros.

If $u(t_0) = 1$ and $u''(t_0) < 0$ then the same argument applies. If $u(t_0) = 1$ and $u'(t_0) = u''(t_0) = 0$, then $u \equiv 1$, which is impossible. Hence we have only to consider the case $1 < u(t_0) \le \sqrt{3}$. We are assuming that $u$ has a subsequent zero, at which point $Q(t) > Q(t_0) \ge 0$, so that $u'u'' > 0$. (This quantity might have been negative somewhere in between.) From there we proceed as in the case $0 < u_0 < 1$ to get a contradiction, completing the proof of Theorem 12.3. □

## 12.3. A problem with exactly two solutions

In the next three sections we consider problems of the form

$$\Delta u + \lambda f(u) = 0 \tag{12.12}$$

on the unit disk $D = \{\mathbf{x} \in R^n \mid |\mathbf{x}| \leq 1\}$ where $n = 1$, 2, or 3, with the boundary condition

$$u = 0 \text{ on the boundary } \partial D \text{ of } D. \tag{12.13}$$

Such problems arise in a variety of applied areas. For example, in [**113**] references are given to applications in thermal ignition of a chemically active mixture of gases, the theory of membrane buckling, and the gravitational equilibrium of polytropic stars, and a further citation there leads to other applications.

A famous and important result of Gidas, Ni, and Nirenberg [**77**] gives conditions under which every positive solution of this problem must be radially symmetric, and a function only of the distance $r$ from the origin. We will study only problems of this type, in which case we are led to an ode,

$$\left(r^{n-1}u'\right)' + \lambda r^{n-1} f(u) = 0 \quad \text{on } 0 < r < 1. \tag{12.14}$$

We assume that $f$ is continuously differentiable on $-\infty < u < \infty$. Further, we will consider only positive functions $f$ and positive values of $\lambda$, in which case it is clear that the solutions of (12.14) and (12.13) must satisfy

$$u'(0) = 0, \quad u(1) = 0, \tag{12.15}$$

and

$$\begin{aligned} u &> 0 \quad \text{on } [0,1), \\ u' &< 0 \quad \text{on } (0,1]. \end{aligned}$$

This suggests that solutions might be obtained by shooting, using the initial conditions

$$\begin{aligned} u(0) &= \alpha, \\ u'(0) &= 0. \end{aligned} \tag{12.16}$$

We wish to find $\alpha$ such that $u(1) = 0$.

For $n > 1$, the equation (12.14) is singular at $r = 0$, so that the standard existence and uniqueness theorem does not apply to (12.14) with initial condition (12.16). In Exercise 1.3 you were asked to use a successive approximations method to show that (12.14) has a unique solution $u(x, \alpha)$ satisfying (12.16) for every $\alpha > 0$ and that this solution depends continuously on $\alpha$. Further, because $\lambda > 0$ and $f(u) > 0$ for $u \geq 0$, $u(\cdot, \alpha)$ is decreasing and has a first zero at some $r_0 > 0$ which also depends continuously on $\lambda$.[4] For a given $\lambda$, the number of solutions of (12.14) and (12.16) such that $u(1) = 0$ can be found from the graph of $r_0$ as a function of $\alpha$, by determining how many intersections this curve has with the line $r_0 = 1$.

[4]See Exercise 12.9.

We therefore consider the graph

$$\Gamma = \{(\alpha, \lambda) \mid \text{the solution of (12.14) and (12.16) satisfies (12.15)}\}.$$

Such a graph, of a parameter versus some function of the solution (such as $u(0)$), is called a bifurcation diagram. For the graph $\Gamma$ given above we have the following observation, recalling that $f > 0$.

**Proposition 12.6.** *$\Gamma \cap \{\alpha \geq 0\}$ is the graph of a smooth function $\lambda = \lambda(\alpha)$, with $\lambda(0) = 0$, $\lambda(\alpha) > 0$ if $\alpha > 0$.*

The proof is left to the reader in Exercise 12.6. The nature of the bifurcation diagram for (12.14)–(12.15) depends on both $f$ and $n$. On the most basic level, we have the following relatively simple result.

**Proposition 12.7.** *Suppose that $f$ satisfies the conditions above (just after (12.14)). If $f$ is bounded then there is a solution to (12.14)–(12.15) for every $\lambda > 0$, so that the function $\lambda(\alpha)$ is unbounded.*

In Exercise 12.8 the reader is asked to prove this proposition and to demonstrate that the function $\lambda$ can be unbounded even if $f$ is unbounded. In the next section we will see an example where $\lambda(\alpha)$ is bounded, and in fact, $\lim_{\alpha\to\infty} \lambda(\alpha) = 0$.

**12.3.1. One-dimensional case; introduction of the time map.** The most straightforward case is $n = 1$, though we will see that even here the problem may be difficult. The equation is now

$$u'' + \lambda f(u) = 0, \tag{12.17}$$

with initial conditions (12.16). Multiplying by $u'$ and integrating gives

$$\frac{1}{2}u'^2 + \lambda F(u) = \lambda F(\alpha),$$

where

$$F(u) = \int_0^u f(v)\, dv.$$

Since the solution $u$ is monotonic, we can consider $u$ as the independent variable, with $t = T(u)$, and get

$$\frac{dT}{du} = \frac{1}{\sqrt{2\lambda}\sqrt{F(\alpha) - F(u)}} \tag{12.18}$$

on the interval $(0, T(\alpha))$ where $u > 0$. Hence,

$$T(\alpha) = \frac{1}{\sqrt{2\lambda}} \int_0^\alpha \frac{1}{\sqrt{F(\alpha) - F(u)}} du.$$

The function $T$ is called the "time map" for the problem, even if the independent variable for a particular model has nothing to do with time.

The solution $u_\alpha$ satisfies $u(1) = 0$ if

$$\sqrt{2\lambda} = \int_0^\alpha \frac{1}{\sqrt{F(\alpha) - F(u)}} du.$$

In the next two sections we analyze $T(\alpha)$, or more precisely, the integral function

$$H(\alpha) = \int_0^\alpha \frac{1}{\sqrt{F(\alpha) - F(u)}} du,$$

in two important special cases.

**12.3.2. The one-dimensional Gelfand equation.** One of the first problems of this kind to be studied was the case $f(u) = e^u$, which gives

$$u'' + \lambda e^u = 0, \tag{12.19}$$

with

$$u'(0) = 0, \quad u(1) = 0. \tag{12.20}$$

This boundary value problem was analyzed by I. M. Gelfand, in [**76**]. He observed that the equation (12.19) with initial conditions (12.16) can be solved exactly, with the solution

$$u(x) = \alpha + \log\left(\operatorname{sech}^2\left(\frac{\sqrt{2\lambda}}{2} x e^{\alpha/2}\right)\right).$$

From this he deduced that (12.19)–(12.20) has either two, one, or zero solutions, depending on $\lambda$. We can see this from the bifurcation diagram in Figure 12.1, which is simply the graph of $\lambda$ defined implicitly as a function of $\alpha$ by setting $u(1) = 0$ in the formula above. It is common to plot such diagrams with $\lambda$ as the horizontal axis.

We will proceed as though no exact solution is possible, partly in order to illustrate one kind of analysis that has been used on this type of problem more generally.[5] Also, this analysis will be helpful when we turn to a more difficult problem in subsequent sections. For the Gelfand case we obtain

$$H(\alpha) = \int_0^\alpha \frac{1}{\sqrt{e^\alpha - e^u}} du. \tag{12.21}$$

Since the Gelfand ode has an exact solution, it is not surprising that this integral can also be done exactly, but we will ignore this. Fixing $\lambda$, we try to determine how many solutions there are for the equation $H(\alpha) = \sqrt{2\lambda}$ by examining properties of $H$ and its derivatives. (Among the first to use this technique was Laetsch, in [**135**].) Note that the graph of $H$ is obtained from the bifurcation diagram in Figure 12.1 by reversing the axes and rescaling, giving Figure 12.2. This observation motivates the following result:

[5] The integral can indeed be evaluated, but the precise nature of the bifurcation curve may not be immediately obvious from the answer.

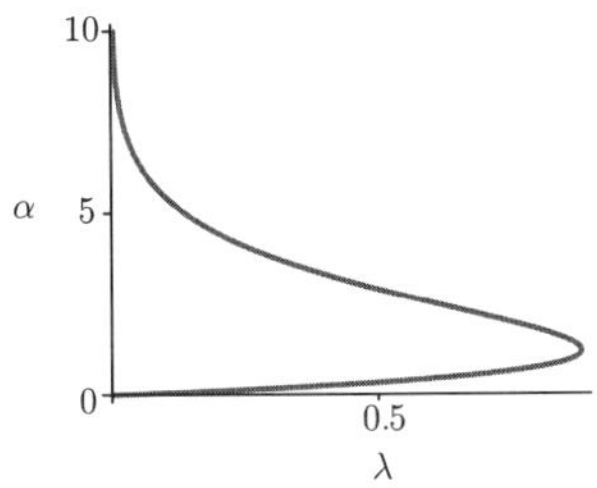


Figure 12.1

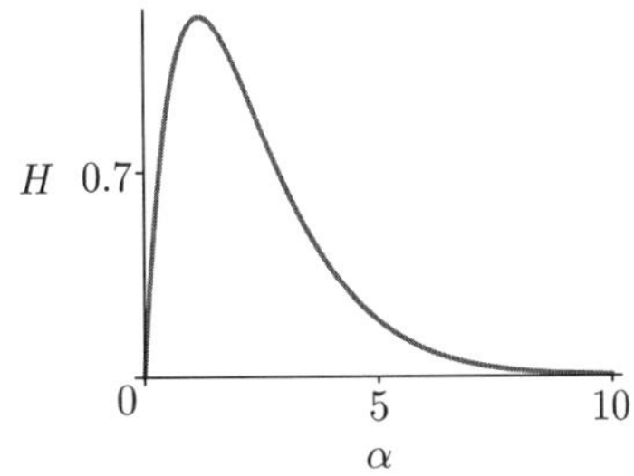


Figure 12.2

**Theorem 12.8.** *The function $H$ has the following properties:*

(i) $\lim_{\alpha\to 0^+} H(\alpha) = 0$, $\lim_{\alpha\to\infty} H(\alpha) = 0$, *and*

(ii) $H'$ *has exactly one zero in* $(0,\infty)$.

**Proof.** A number of techniques have been used for such analysis, as in [**135**], [**224**], [**249**], and other references in [**124**]. This example is fairly simple, even without using the exact solution. We start with an integration by parts, which is easily seen to be valid despite the singularity in the integrand and, in fact, removes the singularity:

$$H(\alpha) = \int_0^\alpha \frac{(-e^{-u})(-e^u)}{\sqrt{e^\alpha - e^u}}du = 2\sqrt{e^\alpha - 1} - 2\int_0^\alpha e^{-u}\sqrt{e^\alpha - e^u}du.$$

From this it is apparent that $\lim_{\alpha\to 0^+} H(\alpha) = 0$. Further, writing $H$ as

$$H(\alpha) = 2e^{\frac{\alpha}{2}}\left(\sqrt{1-e^{-\alpha}} - \int_0^\alpha e^{-u}\sqrt{1-e^{u-\alpha}}du\right) \tag{12.22}$$

leads to the second part of (i). (See Exercise 12.5.) From this we see that $H'$ has at least one zero in $(0,\infty)$.

Differentiating, followed by a second integration by parts, gives

$$H'(\alpha) = e^\alpha\left(\frac{3-2e^\alpha}{\sqrt{e^\alpha - 1}} + 4\int_0^\alpha e^{-2u}\sqrt{e^\alpha - e^u}du\right).$$

The first term in parentheses is decreasing in $\alpha$, while the second term is increasing. To analyze $H'$ more carefully, we write

$$H'(\alpha) = \frac{e^\alpha}{\sqrt{e^\alpha - 1}}\left(3 - 2e^\alpha + 4\sqrt{e^\alpha - 1}\int_0^\alpha e^{-2u}\sqrt{e^\alpha - e^u}du\right).$$

Let $P(\alpha) = \sqrt{e^\alpha - 1}H'(\alpha)$. Figure 12.3 is a plot of $P$, and we see that $H'$ probably has only one zero.

To prove this let

$$I(\alpha) = \int_0^\alpha e^{-2u}\sqrt{1-e^{u-\alpha}}du,$$

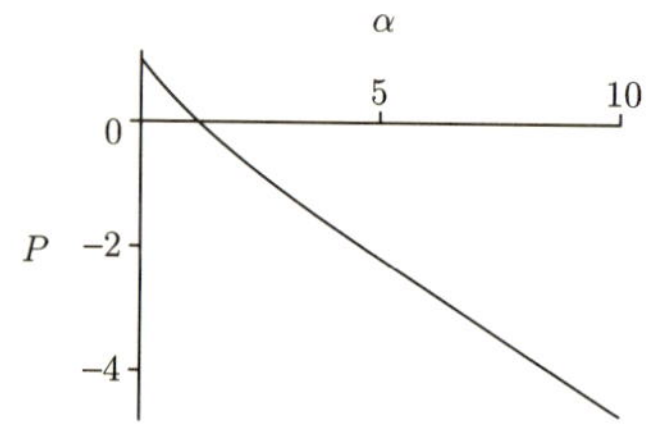


**Figure 12.3.** $P(\alpha)$.

so that

$$P(\alpha) = e^{\alpha}\left(3 - e^{\alpha}\left(2 - 4\sqrt{1-e^{-\alpha}}I(\alpha)\right)\right). \tag{12.23}$$

Then

$$I'(\alpha) = \sqrt{1-e^{-\alpha}} - 2I(\alpha). \tag{12.24}$$

Further, if $P(\alpha) = 0$ then from (12.23),

$$I(\alpha) = \frac{2-3e^{-\alpha}}{4\sqrt{1-e^{-\alpha}}}. \tag{12.25}$$

Differentiating $P$ and using (12.24) and (12.25) shows that when $P(\alpha) = 0$,

$$P'(\alpha) = -\frac{1}{2(1-e^{-\alpha})} < 0. \tag{12.26}$$

Hence $P$ has only one zero, implying that $H'$ also has exactly one zero in $(0,\infty)$ and proving Theorem 12.8. □

The point of this calculation is to give a relatively simple illustration of the kind of manipulations needed to study the function $H$. We will also need the result of this analysis below, for a more difficult problem where a closed form solution is not available.

## 12.4. A problem with exactly three solutions

We continue to study equations of the form (12.12), considering Dirichlet problems with radially symmetric solutions. Thus, we consider the problem (12.14)–(12.15). One of the earliest papers in this area was the landmark study of Joseph and Lundgren [**113**], who concentrated on the Gelfand case of $f(u) = e^u$, and on the case $f(u) = (1+\alpha u)^{\beta}$, where $\alpha$ and $\beta$ are positive. They found a fascinating array of solutions, for various values of $\alpha$, $\beta$, and the spatial dimension $n$. For example, they showed that in the Gelfand case, (12.14)–(12.15) can have

(i) either zero, one, or two solutions, depending on $\lambda$, if $n = 1$ or 2,

(ii) infinitely many solutions if $\lambda = 2(n-2)$ and $3 \leq n \leq 9$,

(iii) a large but finite number of solutions if $|\lambda - 2(n-2)|$ is small and $3 \leq n \leq 9$, or

(iv) a unique solution if $\lambda < 2(n-2)$ and $n \geq 10$.

**12.4.1. The perturbed Gelfand equation in one dimension.** There have been many papers where it has been shown that a problem of the form (12.14)–(12.15) has at least three solutions. In only a few of these is the exact number of solutions determined rigorously. When this number is three, a bifurcation curve, plotting some component of the solution against $\lambda$, will have a so-called "$S$" shape. The first of these studies, as far as we know, was by Smoller and Wasserman [**224**], who considered the case where $f(u)$ is a cubic polynomial. Another early example, less well known but both difficult and physically relevant, was studied by C. Lu [**150**], [**151**], who considered a biochemical model involving so-called Michaelis-Menton kinetics, with a nonlinear function of the form $f(u) = \frac{u}{1+au+bu^2}$ in which rescaling introduces a small parameter.

Further important work on $S$-shaped bifurcation curves includes papers by S.-H. Wang [**249**], K.-C. Hung and S.-H. Wang [**105**], Korman and Y. Li [**125**], and Y. Du and Y. Lou [**56**]. See Korman [**124**] for an excellent survey of much of this research, including many other references to the authors just cited here. These authors have done more recent work as well. Among the lessons one draws in studying this literature is that these problems are challenging and each problem is different, with analysis for one function $f$ rarely extending easily to other types of functions.

Here we emphasize the quadrature method described above, where results are obtained by analysis of the time map. This analysis is suitably classical, in the sense of this book. But quadrature is limited to one space dimension, although problems in two and three dimensions are of great interest. The most important technique for analyzing higher dimensional problems has been bifurcation. Bifurcation theory is largely neglected in this book, and we refer the reader to the book by Kuznetsov [**133**] for a well-regarded presentation.[6] The bifurcation theory needed for these problems is not classical, since it depends on results in infinite dimensional functional analysis. The basic tool used in much of this work is by Crandall and Rabinowitz [**47**]. In later subsections we will outline some of the results obtained on these problems by using bifurcation theory and refer the reader to relevant literature.

The equation to be studied in this subsection arises in chemical engineering, in the theory of combustion [**16**]. In one space dimension, which is

[6]One reviewer said that this book "underpromises (with its title) and overdelivers."

all we will study in detail, the boundary value problem of interest is

$$u'' + \lambda e^{\frac{u}{1+\frac{u}{\gamma}}} = 0 \qquad (12.27)$$
$$u(\pm 1) = 0.$$

As before we look for symmetric solutions and consider the boundary conditions

$$\begin{aligned} u'(0) &= 0, \\ u(1) &= 0. \end{aligned} \qquad (12.28)$$

The parameter $\gamma$ is taken as large, in accordance with physical evidence. Thus, when $u$ is small it behaves like a solution to the Gelfand equation. This observation leads us to believe that for large $\gamma$ the two solutions of (12.19)–(12.20) which we found earlier in a range of $\lambda$ will be perturbed slightly to two solutions of (12.27)–(12.28).

On the other hand, note that for every $\gamma > 0$ the function

$$f(u) = e^{\frac{\gamma u}{\gamma + u}} \qquad (12.29)$$

is bounded. By Proposition 12.7, the function $\lambda(\cdot)$ defining the bifurcation curve $\Gamma$ is unbounded. There is a solution for every $\lambda \geq 0$, in contrast to the situation depicted in Figure 12.1. Hence there must be some major changes in the bifurcation diagram for larger values of $u(0)$.

Consider the initial value problem

$$u'' + \lambda f(u) = 0,$$
$$u(0) = \alpha, \quad u'(0) = 0.$$

As in Section 12.3.1, we are led to a time map, which again can be written as

$$T_1(\alpha, \lambda) = \frac{1}{\sqrt{2\lambda}} \int_0^\alpha \frac{1}{\sqrt{F(\alpha) - F(v)}} dv,$$

where $F(u) = \int_0^u f(x)\, dx$. Again we need to analyze the integral, so we let

$$H_1(\alpha) = \int_0^\alpha \frac{1}{\sqrt{F(\alpha) - F(v)}} dv. \qquad (12.30)$$

We again use the integration by parts technique, to write

$$\begin{aligned} H_1(\alpha) &= -\int_0^\alpha \frac{1}{f(v)} \frac{-f(v)}{\sqrt{F(\alpha) - F(v)}} dv \\ &= \frac{2\sqrt{F(\alpha) - F(0)}}{f(0)} - 2\int_0^\alpha \frac{f'(v)}{f(v)^2} \sqrt{F(\alpha) - F(v)} dv. \end{aligned}$$

It is then clear from the previous analysis that for $\alpha$ in a bounded interval, and $\gamma$ large, $H_1$ is similar to $H$, which was defined in (12.21) and plotted in Figure 12.2. In particular, as $\alpha$ increases, $H_1(\alpha)$ increases to a maximum

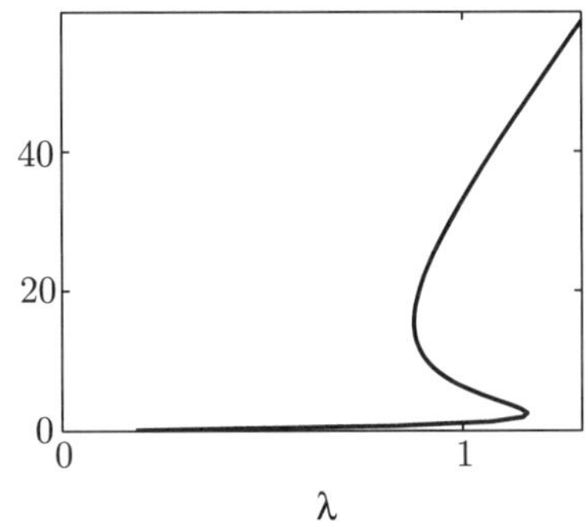


**Figure 12.4.** $u(0)$ vs $\lambda$ for (12.27).

and then decreases and becomes small, all this occurring while $\frac{\alpha}{\gamma}$ is still small. (We will make this precise below.)

Next we consider the behavior of $H_1(\alpha)$ when $\frac{\alpha}{\gamma}$ is not small. We saw previously that $\lim_{\alpha\to\infty} H(\alpha) = 0$. But this was based on the exponential growth of $e^\alpha$. In this case,

$$F(\alpha) = \int_0^\alpha e^{\frac{\gamma v}{\gamma+v}}\, dv, \tag{12.31}$$

and while $F$ initially increases exponentially as a function of $\alpha$, it later slows to linear growth, as the integrand approaches its limit of $e^\gamma$. It follows that $\lim_{\alpha\to\infty} H_1(\alpha) = \infty$, meaning that for large $\gamma$, $H_1'$ has at least two zeros, and so for a range of $\lambda$, the problem (12.27)–(12.28) has at least three solutions. (See Figure 12.4.)

**Theorem 12.9.** *For sufficiently large $\gamma$, there are $\lambda_2 > \lambda_1 > 0$ such that problem* (12.27)–(12.28) *has exactly one solution for $0 < \lambda < \lambda_1$, three solutions for $\lambda_1 < \lambda < \lambda_2$, one solution for $\lambda > \lambda_2$, and two solutions for $\lambda = \lambda_1$ and $\lambda = \lambda_2$.*

**Proof.** We must show that for sufficiently large $\gamma$, there are no more than three solutions for any $\lambda$, which follows if we show that $H_1'$ has exactly two zeros. The function $H_1$ depends smoothly on $\gamma$, and thus we can assert that as $\gamma \to \infty$, $H_1 \to H$, $H_1' \to H'$, and $H_1'' \to H''$, uniformly in each compact interval $0 \le \alpha \le Y$. Choosing $Y$ larger than the only zero of $H'$, it follows that for sufficiently large $\gamma$, $H_1'$ has exactly one zero in $[0, Y]$. We now have to show that it has only one zero in $[Y, \infty)$.

Our proof is adapted from one by S.-H.Wang [**249**], who uses some techniques from [**224**]. Wang proves considerably more, showing that there are exactly two zeros of $H_1'$ if $\gamma > 4.4967$. This number was lowered further in [**125**], and even further in [**105**].[7]

It is clear that for large $\gamma$, $H_1'$ has at least one zero in $(Y, \infty)$, since $H'(Y) < 0$ and so $H_1'(Y) < 0$ for sufficiently large $\gamma$, while $H_1(\infty) = \infty$. Changing variables in (12.30) gives

$$H_1(\alpha) = \int_0^1 \frac{\alpha}{\sqrt{F(\alpha) - F(\alpha u)}} du,$$

and differentiating this leads to

$$H_1'(\alpha) = \frac{1}{2} \int_0^1 \frac{\theta(\alpha) - \theta(\alpha u)}{(F(\alpha) - F(\alpha u))^{3/2}} du \tag{12.32}$$

$$= \frac{1}{2} \int_0^\alpha \frac{\theta(\alpha) - \theta(v)}{\alpha (F(\alpha) - F(v))^{3/2}} dv \tag{12.33}$$

where

$$\theta(u) = 2F(u) - uf(u). \tag{12.34}$$

The zeros in the top and bottom are found to cancel so that this integral converges.

Let $\Delta F(s) = F(\alpha) - F(s)$. Then differentiating (12.32) gives

$$H_1''(\alpha)$$
$$= \int_0^1 \frac{\Delta F(\alpha u)(\theta'(\alpha) - u\theta'(\alpha u)) - \frac{3}{2}(\theta(\alpha) - \theta(\alpha u))(f(\alpha) - uf(\alpha u))}{2\Delta F(\alpha u)^{5/2}} du$$
$$= \int_0^\alpha \frac{\Delta F(v)(\alpha\theta'(\alpha) - v\theta'(v)) - \frac{3}{2}(\theta(\alpha) - \theta(v))(\alpha f(\alpha) - vf(v))}{2\alpha^2 \Delta F(v)^{5/2}} dv,$$

where again the integrals converge. As in equation (1.11) of [**249**], a computation[8] using (12.34) and (12.33) shows that

$$H_1''(\alpha) + \frac{2}{\alpha} H_1'(\alpha) > \frac{1}{2\alpha^2} \int_0^\alpha \frac{\phi(\alpha) - \phi(v)}{(F(\alpha) - F(v))^{3/2}} dv, \tag{12.35}$$

where

$$\phi(u) = u\theta'(u) - \theta(u).$$

The key lemma in the proof turns out to be about the behavior of $\theta$, as follows:

---

[7] As noted, Wang established his bound for the critical $\gamma$ using classical analysis. Then, Korman and Li reduced the number to about 4.35 by using bifurcation theory in infinite dimensions. More recently, Wang and his student K-C. Hung returned to classical methods and got the bound down to 4.166. Perhaps by the time this book appears, bifurcation theory will again top the leader board.

[8] See Exercise 12.10. As pointed out in [**105**], a similar computation was done in [**224**].

**Lemma 12.10.** *For sufficiently large $\gamma$, $\theta(0) = 0$, $\theta'(0) = 1$, $\lim_{u\to\infty}\theta(u) = \infty$, and $\theta$ has a unique inflection point in $(0,\infty)$, at $u = C(\gamma) = \frac{1}{2}\gamma(\gamma-2)$. Moreover,*

$$\begin{aligned} \theta'' &< 0 \quad \text{in } (0, C(\gamma)), \\ \theta''(C(\gamma)) &= 0, \\ \theta'' &> 0 \quad \text{in } (C(\gamma), \infty), \end{aligned}$$

*and*

$$\theta(C(\gamma)) < 0.$$

**Proof.** The properties in the first sentence follow routinely from (12.29), (12.31), and (12.34). Differentiating (12.34) twice gives $\theta''(u) = -uf''(u)$, which is easily shown to be negative for small positive $u$, positive for large $u$, and to have a unique zero at $C(\gamma) = \frac{1}{2}\gamma(\gamma-2)$. To prove the lemma we must show that for sufficiently large $\gamma$, $\theta(C(\gamma)) < 0$. From (12.34), we need to show that if $C = C(\gamma)$ then

$$\frac{2F(C)}{Cf(C)} < 1. \tag{12.36}$$

An easy computation shows that

$$\frac{f\left(\frac{1}{2}C\right)}{f(C)} = e^{-2\frac{\gamma-2}{\gamma+2}},$$

which implies that for large $\gamma$,

$$f\left(\frac{1}{2}C\right) < \frac{1}{e}f(C). \tag{12.37}$$

Consider the graph of $f$ on the interval $[0, C]$, together with the chord $L$ connecting $(0,1)$ with $(C, f(C))$. Thus, the chord lies on the line $g(u) = 1 + \frac{f(C)-1}{C}u$. Let $x = x_1$ be the unique solution to $f\left(\frac{C}{2}\right) = g(x)$. Then it is seen from the graph of $f$ that

$$F(C) < C + \frac{1}{2}C(f(C)-1) - \frac{1}{2}\left(\frac{C}{2} - x_1\right)\left(\frac{f(C)}{2} - f\left(\frac{C}{2}\right)\right), \tag{12.38}$$

where the first two terms are the area under the chord $L$ and the last term is the area of the triangle containing the points $\left(\frac{C}{2}, f\left(\frac{C}{2}\right)\right)$, $\left(\frac{C}{2}, \frac{1}{2}f(C)\right)$, and $\left(x_1, f\left(\frac{C}{2}\right)\right)$. See Figure 12.5, noting that the upper vertex of the triangle is just below the line, since $g\left(\frac{C}{2}\right) = \frac{1}{2} + \frac{f(C)}{2}$.

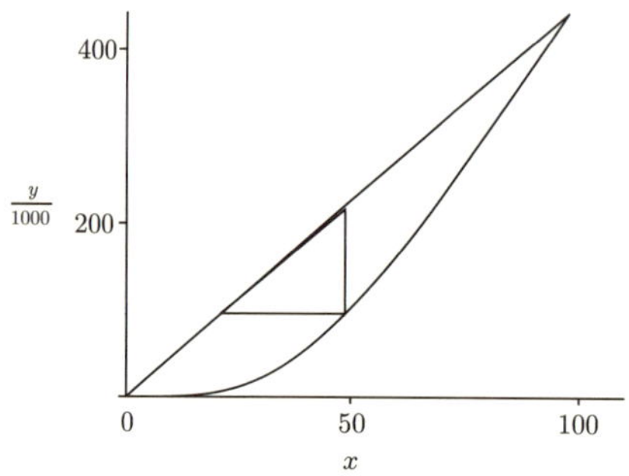


**Figure 12.5.** Graph of $f$ and chord $L$.

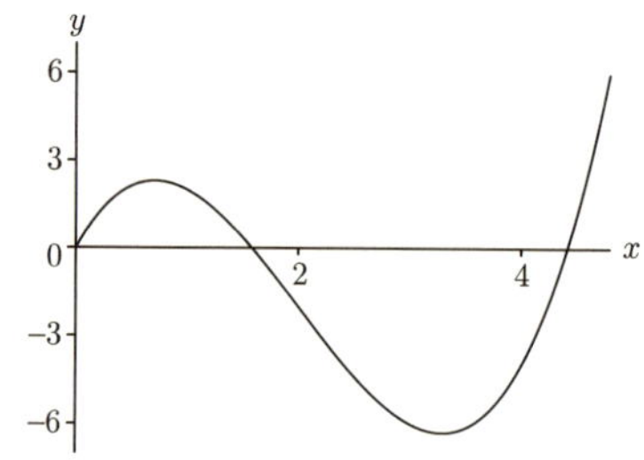


**Figure 12.6.** Graph of $\theta$.

We find that $x_1 = \frac{f\left(\frac{C}{2}\right)-1}{f(C)-1}C < 2e^{-2}C$ for large $\gamma$, and using (12.37) and (12.38) leads to

$$\frac{2F(C)}{Cf(C)} - 1 < \frac{1}{f(C)} - \left(\frac{1}{2} - \frac{2}{e^2}\right)\left(\frac{1}{2} - \frac{1}{e}\right) < 0,$$

which implies (12.36) and the lemma. □

The proof of Theorem 12.9 now mostly follows that in [**249**], with simplifications because we are aiming for a weaker result (by taking $\gamma$ arbitrarily large). From Lemma 12.10, the graph of $\theta$ is seen to be as in Figure 12.6.

Let $A$ denote the unique local maximum of $\theta$, $B$ the first zero of $\theta$, $C$ the unique zero of $\theta''$, $D$ the unique minimum of $\theta$, and $E$ the unique point beyond $D$ where $\theta(u) = \theta(A)$. All these exist because of Lemma 12.10, which also tells us that $\theta(C) < 0$.

From (12.33) we further see that

$$\begin{aligned} H_1' &> 0 \quad \text{in } (0, A], \\ H_1' &< 0 \quad \text{in } [B, D], \\ H_1' &> 0 \quad \text{in } [E, \infty). \end{aligned}$$

From (12.34) it follows that $\theta(2) < 0$, so that $B < 2$ for sufficiently large $\gamma$. However $C = \frac{1}{2}\gamma(\gamma - 2)$, and so we can draw no immediate conclusion for large $\gamma$ on $[D, E]$. To complete the proof we must consider the function $\phi$. We find that $\phi'(u) = -u^2 f''(u) = u\theta''(u)$, so that $\phi' < 0$ on $(0, C)$ and $\phi' > 0$ on $(C, \infty)$. Also, $\phi(D) = -\theta(D) > 0$. Hence, $\phi(\alpha) - \phi(u) > 0$ for $D < \alpha < E$ and $0 < u < \alpha$. Then (12.35) implies that $H_1'' + \frac{2}{\alpha}H_1' > 0$ in $(D, E)$. Hence $H_1'$ has only one zero in this interval, completing the proof of Theorem 12.9. □

**12.4.2. The two-dimensional case.** One of the first to obtain results for the perturbed Gelfand equation in more than one dimension was Dancer, in [**48**]. He considered a broader class of equations which, as he remarks, includes the problem above. He extended the result of Gelfand about (12.12) when $f(u) = e^u$ to problems which include the perturbed Gelfand equation, as studied above in one dimension. His methods depend crucially on degree theory and so do not come under the classical designation we have stressed here. In line with our earlier remarks, he uses the famous work on bifurcation theory of Crandall and Rabinowitz [**47**], as well as the theory of topological index. Thus, the techniques are beyond what we can describe here. He comes close to showing that the bifurcation curve in two dimensions is $S$-shaped in the same way as was proved above in one dimension, for large $\gamma$. Only for a narrow region of the parameter $\lambda$ does he fail to show that, for large $\gamma$, there are no more than three solutions. He also obtained results for three dimensions, as we will outline briefly below.

It took more than 20 years after the work of Dancer for the bifurcation curve of the two-dimensional perturbed Gelfand equation to be shown to be $S$-shaped, thus filling in the small gap in $\lambda$. This proof, by Du and Lou, is in [**56**], which also contains references to other interesting work on related problems. Like Dancer, they make important use of degree theory and bifurcation methods, particularly the Crandall-Rabinowitz result. They also use other nonclassical techniques, such as Sobolev inequalities. They point out at several spots that they are using techniques from a slightly earlier paper by Korman, Li, and T. Ouyang [**126**]. We do not know if these results are obtainable by classical ode methods.

## 12.5. The Gelfand and perturbed Gelfand equations in three dimensions

**12.5.1. Gelfand equation in three dimensions.** Next we consider problem (12.14)–(12.15) with $n = 3$, and the Gelfand nonlinearity $f(u) = e^u$. In this case a change of variables simplifies the problem. With $n = 3$, let $x = \log r$ and $w(x) = u(r) + 2x$. Then (12.14) become

$$w'' + w' - 2 + \lambda e^w = 0.$$

This autonomous equation can be studied in the phase plane. The boundary conditions $u'(0) = 0$ and $u(1) = 0$ correspond to

$$w'(-\infty) = 2, \quad w(0) = 0. \tag{12.39}$$

We leave it to the reader to show that there is a unique trajectory such that $w' < 2$ on $(-\infty, \infty)$, $\lim_{x\to-\infty} w'(x) = 2$, and $(w, w')$ spirals into the point $\left(\log \frac{2}{\lambda}, 0\right)$ as $x \to +\infty$. This trajectory is shown in Figure 12.7 for the case $\lambda = 1.9$. Figure 12.7 and its enlargement in Figure 12.8 suggest that this trajectory intersects $w = 0$ three times. Each intersection, say at a

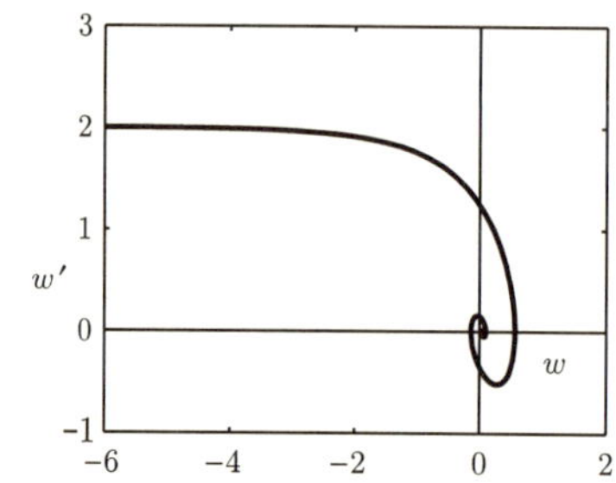


**Figure 12.7.** Three-dimensional Gelfand equation phase plane.

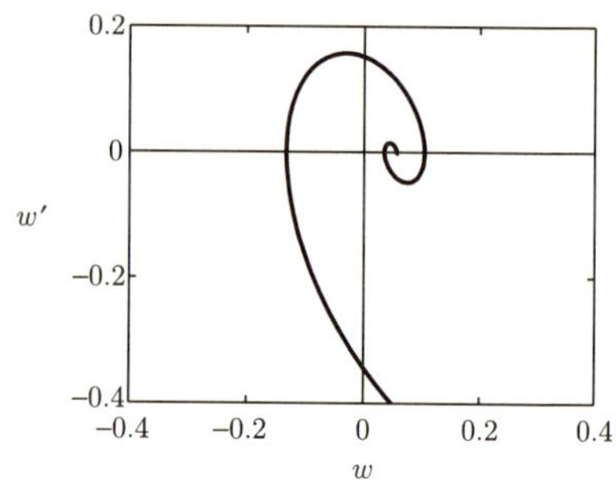


**Figure 12.8.** Enlargement.

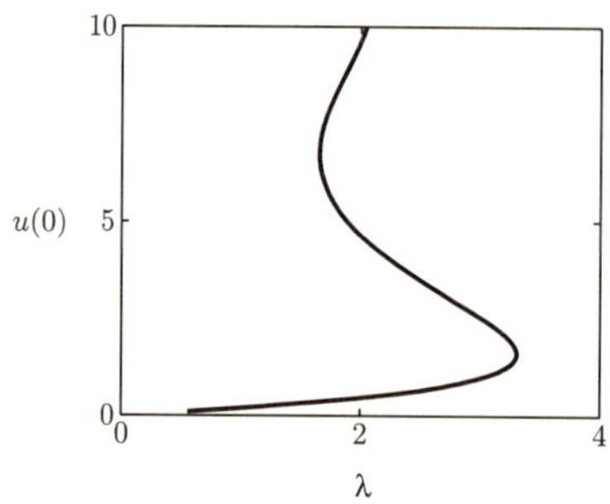


**Figure 12.9.** $u(0)$ vs $\lambda$.

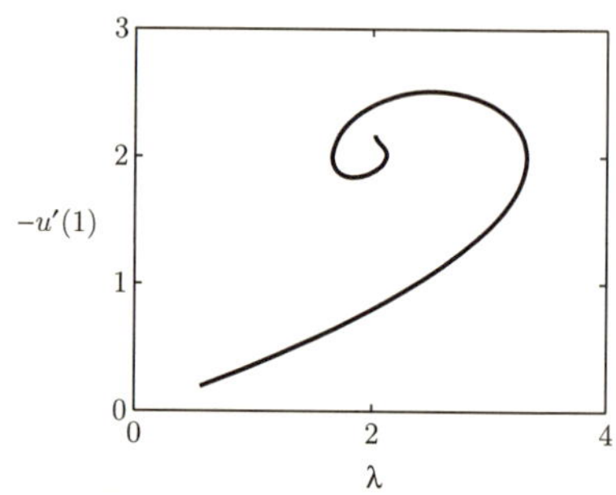


**Figure 12.10.** $-u'(1)$ vs $\lambda$.

point $(0, w'_0)$, gives a solution of (12.14)–(12.15), defined by setting $u\,(1) = 0$, $u'\,(1) = w'_0 - 2$.

More generally, if $\lambda$ is positive and unequal to 2 then there are finitely many solutions, while if $\lambda = 2$ then the equilibrium point is $(0, 0)$ and there are infinitely many. If $\lambda$ is large then the equilibrium point is far enough into the left half-plane that there are no solutions.

There are two common ways of depicting this in a bifurcation diagram. In one, we plot all points $(u\,(0)\,, \lambda)$ such that $u' < 0$ on $(0, 1]$ and $u\,(1) = 0$, while in the other, we plot $-u'\,(1)$ vs $\lambda$. The phase plane argument shows that Figures 12.9 and 12.10 are obtained. (Numerical challenges prevent us from finding more than a few solutions, even for $\lambda = 2$, where the phase plane argument shows that there are infinitely many.)

**12.5.2. Perturbed Gelfand equation in three dimensions.** We now consider Figure 12.10. The phase plane argument given above shows that the function $\lambda\,(\cdot)$ defining the bifurcation curve is bounded. Hence there is no solution for $\lambda$ above the right-most turning point of this curve. Yet when we switch to the perturbed Gelfand equation, with a bounded function $f$,

Proposition 12.7 shows that there is a solution for every $\lambda$. So there must be some radical change in the bifurcation diagram. The major work here is by Dancer [**48**], in the paper cited earlier in our discussion of the one- and two-dimensional cases. His results can be seen most easily by consulting a set of bifurcation diagrams which appear near the beginning of his paper. From these it can be seen that if $\gamma$ is large then initially (for low $u(0)$), the curve $(\lambda, -u'(1))$ spirals in as in Figure 12.10, but as $u(0)$ increases, the spiral unwinds, eventually moving near to the axis $\lambda = 0$, moving up this axis, and then tending to the right, in the manner of the $S$-shaped curves found when $n = 1$ or 2. Some of the interesting features described above were found earlier by formal asymptotic methods (but not rigorously) in [**117**].

## 12.6. Uniqueness of the ground state for $\Delta u - u + u^3 = 0$

This section, in addition to illustrating an important technique, serves as an introduction to one of the major open problems in our subject. This problem is discussed further in Chapter 19, where we give some background and discuss some important generalizations.

The partial differential equation

$$\Delta u - u + u^3 = 0 \tag{12.40}$$

is considered here in $R^3$, with radial symmetry around the origin, and a solution is sought which, in an appropriate sense, vanishes at infinity. The symmetry assumption means that we can consider the ode

$$u'' + \frac{2}{r}u' - u + u^3 = 0. \tag{12.41}$$

To obtain a solution of (12.40) which is smooth on $R^3$, we consider only solutions to the ode (12.41) which satisfy

$$u'(0) = 0. \tag{12.42}$$

Exercise 1.3 implies that for each $\alpha > 0$ there is a unique solution to (12.41)–(12.42) such that $u(0) = \alpha$, and this solution depends continuously on $\alpha$.

The boundary condition for (12.41) at $\infty$ is

$$u(\infty) = 0. \tag{12.43}$$

More generally, one could discuss the problem in $R^n$, but we defer that extension to Chapter 19.

The so-called "ground state" is a solution to (12.41)–(12.43) which is positive on $[0, \infty)$. It is the state with the least possible "energy". The existence of such a solution is proved in Chapter 19.

### 12.6.1. Coffman's uniqueness proof.

**Theorem 12.11.** *The problem* (12.41)–(12.43) *has no more than one positive solution.*

**Proof.** We repeat the ingenious proof of Coffman in [**42**], with comments.

Let $w(r) = ru(r)$. Assuming that $u$ is a solution of (12.41)–(12.43), we find that

$$w'' - w + r^{-2}w^3 = 0. \tag{12.44}$$

It is enough to prove that there is at most one positive solution $w$ satisfying

$$\begin{gathered} \lim_{r\to 0^+} \tfrac{1}{r} w(r) < \infty, \\ \lim_{r\to\infty} w(r) = 0. \end{gathered} \tag{12.45}$$

This problem has a slightly different sort of singularity at 0 from the standard radial Laplacian as in (12.41). The appropriate initial value problem can easily be shown to be (12.44) with

$$\lim_{r\to 0} \frac{w(r)}{r} = \alpha,$$

which we consider for $\alpha > 0$. It can be shown that this has a unique solution $w(r,\alpha)$ and that $w$ and $\frac{\partial w}{\partial \alpha}$ are continuous as long as the solution exists. Further, if $\delta = \frac{\partial w}{\partial \alpha}$ then $\delta$ satisfies the linearized equation

$$\delta'' - \delta + 3r^{-2}w^2\delta = 0, \tag{12.46}$$

with

$$\delta(0) = 0, \quad \delta'(0) = 1. \tag{12.47}$$

**Remark 12.12.** It is interesting to note here that Coffman is considering a nonmonotonic $w$, since $w$ is positive, $w(0) = 0$, and $w(\infty) = 0$. Uniqueness results are usually easier for monotonic solutions, and the original positive solution $u$, to (12.41)–(12.43), is decreasing. Nevertheless, he finds it more convenient to work with $w$, which removes a derivative term from the equation.

**Lemma 12.13.** (i) *If* $\alpha > 0$ *and* $w(r,\alpha) > 0$ *on* $(0, z_1)$, *with* $w(z_1,\alpha) = 0$, *then* $\delta(z_1,\alpha) < 0$.

(ii) *If* $\alpha > 0$ *and* $w$ *is a positive solution to* (12.44)–(12.45), *then*

$$\lim_{r\to\infty} \delta(r,\alpha) = -\infty.$$

**Remark 12.14** (Remark on notation). In the proof, we follow Coffman by referring both to $w(r)$ and $w(r,\alpha)$, using the first when dependence on $\alpha$ is not immediately relevant and the second when it is. At all times, $w' = \frac{\partial w}{\partial r}$. Similar notation is used for $\delta$.

**Proof of Theorem 12.11 assuming Lemma 12.13.** Suppose that (12.44) has at least one solution satisfying (12.45). Let

$$A = \{\alpha > 0 \mid w(r,\alpha) \text{ has at least one zero in } (0,\infty)\}.$$

By the implicit function theorem, if $w(r,\tilde{a})$ has an initial zero at $z_1(\tilde{\alpha})$, then it has a zero for nearby $\alpha$, and by Lemma 12.13, $\frac{dz_1}{d\alpha} = -\frac{\delta(z_1,\alpha)}{w'(z_1,\alpha)} < 0$. Thus, if $A$ is nonempty then the first zero of $w$ moves to the left as $\alpha$ increases. This zero cannot disappear by a tangency with $w = 0$, and so the set $A$ must be an open interval of the form $(\alpha_1, \infty)$. Since (12.44)–(12.45) is assumed to have at least one positive solution, either $A$ is empty or $0 < \alpha_1 < \infty$. We wish to show that $A$ is nonempty and that $w(r,\alpha_1)$ is the only positive solution of (12.44)–(12.45).

Suppose that $w(r,\alpha_2)$ is a positive solution of (12.44)–(12.45), and let $\alpha_3 > \alpha_2$. We will show that $\alpha_3 \in A$, which will imply that $\alpha_2 = \alpha_1$ and complete the proof of the theorem.

It is sufficient to prove that $\alpha_3 \in A$ if $\alpha_3 - \alpha_2$ is small. By part (ii) of Lemma 12.13 there is an $r_1$ so large that $\delta(r_1,\alpha_2) < 0$ and $\delta'(r_1,\alpha_2) < 0$. We can then pick $\alpha_3$ so close to $\alpha_2$ that $w(r,\alpha_3) < w(r,\alpha_2)$ and $w'(r,\alpha_3) < w'(r,\alpha_2)$, both inequalities holding on some interval $[r_1, r_2]$. Putting $v = w(r,\alpha_3) - w(r,\alpha_2)$, we see that $v < 0$ and $v' < 0$ in $[r_1, r_2]$, and

$$v'' = v\left(1 - r^{-2}\left(w_2^2 + w_2 w_3 + w_3^2\right)\right), \tag{12.48}$$

where $w_i(r) = w(r,\alpha_i)$. By choosing $\alpha_3$ sufficiently close to $\alpha_2$ we can choose $r_1$ large enough to ensure that $v'' < 0$ on $[r_1, r_2]$, and this guarantees that either $w_3$ has a zero or $0 < w_3 < w_2$ on $[r_1, \infty)$. But in the latter case, $w_2^2 + w_2 w_3 + w_3^2 \to 0$, and (12.48) implies that $v \to -\infty$. Hence $w_3$ has a zero after all, which proves Theorem 12.11. There remains to prove Lemma 12.13.

The proof of this lemma depends on a set of ingenious Wronskian type identities discovered by Coffman which connect $\delta$ and $w$. They are straightforward to verify, and we leave that to the reader. What is not straightforward is finding them in the first place. The identities are

$$\left(w'\delta' - w''\delta\right)' = -2r^{-3}w^3\delta, \tag{12.49}$$

$$\left(r\left(w'\delta' - w''\delta\right) - w'\delta\right)' = -2w\delta, \tag{12.50}$$

$$\left(\left(w' - 1\right)\delta' - w''\delta\right)' = -r^{-3}\delta\left(w - r\right)^2\left(2w + r\right), \tag{12.51}$$

$$\left[r\left(\left(w' - 1\right)\delta' - w''\delta\right) - \left(w' - 1\right)\delta\right]' = r^{-1}\delta\left(w - r\right)\left(3w + r\right). \tag{12.52}$$

In addition, use is made of the standard energy function

$$W = \frac{1}{2}u'^2 + \frac{1}{4}u^4 - \frac{1}{2}u^2,$$

where $u$ satisfies (12.41). Using $W$ we obtain the following result.

**Lemma 12.15.** *If $\alpha > 0$ and $w(r,\alpha)$ either has a zero in $(0,\infty)$ or satisfies $w(\infty) = 0$, then*

(a) $\alpha > \sqrt{2}$,

(b) $w(r) = r$ *for precisely one value* $r = r_0 > 0$,

(c) $w'(r_0) < 0$.

**Proof.** (a) This inequality follows because $W$ is decreasing. If $\alpha \le \sqrt{2}$ then $W < 0$ for $r > 0$ and $u = \frac{w}{r}$ does not vanish.

(b) Since $w(0) \ge \sqrt{2}$ and either $w(r) = 0$ for some $r$ or $w(\infty) = 0$, there is at least one point, say $r_0$, where $w = r$. If there is a second such point, at $r_1$, then $u(r_0) = u(r_1) = 1$, and so $u'(r_3) = 0$ for some $r_3 \in (r_0, r_1)$, and moreover, $u(r_3) < 1$. In that case $W(r_3) < 0$, and as in (a) this means that $w$ does not vanish or tend to zero at infinity, which is a contradiction.

(c) Note that $w'' > 0$ if $0 < w(r) < r$. If $w'(r_0) \ge 0$ then $w' > 0$ as long as $w < r$, and so to enable $w(\infty) = 0$ there must be a second point where $w(r) = r$, contradicting (b) and proving Lemma 12.15. □

It follows that for $w$ as in Lemma 12.13 there are positive numbers $\rho$, $\sigma$, and $\tau$ such that

$$w'(\rho) = 1, \quad w'(\sigma) = 0, \quad w(\tau) = \tau,$$

and these are each the first positive numbers satisfying the given equation. Further, Lemma 12.15 implies that

$$0 < \rho < \sigma < \tau < z_1,$$

where either (case (i) of Lemma 12.13) $z_1$ is the first zero of $w$ or (case (ii)) $w$ remains positive and $z_1 = \infty$. Further, $w'' < 0$ on $(0,\tau)$ and $w'' > 0$ on $(\tau, z_1)$.

**Lemma 12.16.** *Let $y_1$ denote the smallest positive zero of $\delta(r,\alpha)$. Then $\rho < y_1 < \sigma$.*

**Proof.** Assume that $y_1 \le \rho$, and integrate (12.52) from 0 to $y_1$. The right-hand side of this equation is positive on $(0,\tau)$, which implies that

$$(w'(y_1) - 1)\,\delta'(y_1) > 0.$$

However, the definition of $\rho$ implies that $w'(y_1) \ge 1$, while $\delta'(y_1) < 0$, giving a contradiction.

Now suppose that $y_1 \ge \sigma$, and integrate (12.50) between 0 and $\sigma$. We find that $-w''(\sigma)\,\delta(\sigma) < 0$. However our assumption implies that $\delta(\sigma) \ge 0$, and clearly $w''(\sigma) < 0$, another contradiction and the lemma is proved. □

Continuing with the proof of Lemma 12.13, $w'' < 0$ on $(0, \tau)$, and so $w' < 1$ on $(\rho, \tau]$. In particular,

$$w'(y_1) < 1. \tag{12.53}$$

Suppose first that we are in case (i) of Lemma 12.13. Then $w'(z_1) < 0$. Also, $w < r$ on $(\tau, z_1)$, so that $w'' > 0$ in this interval. Hence, $w' < 0$ on $(\tau, z_1]$.

Suppose that $\delta$ has a zero, say at $y_2$, in $(y_1, z_1]$. Then integrating (12.51) from $y_1$ to $y_2$ shows that

$$\begin{aligned}(w'(y_2) - 1)\,\delta'(y_2)& \\ &= (w'(y_1) - 1)\,\delta'(y_1) - \int_{y_1}^{y_2} r^{-3}\delta(w-r)^2(2w+r)\;dr.\end{aligned} \tag{12.54}$$

By taking $y_2$ as the first zero of $\delta$ after $y_1$, we obtain $\delta < 0$ in $(y_1, y_2)$. Also, $\delta'(y_1) < 0$ and by (12.53), $w'(y_1) - 1 < 0$. Hence, the right side of (12.54) is positive. But on the left, $w'(y_2) - 1 < 0$ and $\delta'(y_2) > 0$, giving a contradiction. Therefore, $\delta < 0$ in $(y_1, z_1]$, proving (i).

Now suppose that case (ii) holds, so that $w > 0$ on $(0, \infty)$ and $w(\infty) = 0$. Then $w' < 0$ on $(\sigma, \infty)$ and by the previous argument, $\delta < 0$ in $(y_1, \infty)$. Integrating (12.51) from $y_1$ to $\tau$ gives $\delta'(\tau) < 0$, and integration of (12.49) from $\tau$ to $r$ gives

$$\lim_{r\to\infty}\left(w'(r)\,\delta'(r) - w''(r)\,\delta(r)\right) > 0.$$

Hence, there is an $R > 0$ such that for $r \geq R$,

$$\frac{\delta'(r)}{\delta(r)} > \frac{w''(r)}{w'(r)}.$$

Since $w(\infty) = 0$ and $w' < 0, w'' > 0$ for $r > R$, $w'(\infty) = 0$ and it follows that $\delta$ tends to $-\infty$. This result completes the proof of Lemma 12.13. □

Hence Theorem 12.11 is proved. □

## 12.7. Exercises

**Exercise 12.1.** (Routine) Prove that a solution $u$ of (12.8) satisfying $u(0) = 0$ is odd.

**Exercise 12.2.** (Easy) Prove that the assumptions of Proposition 12.1 also imply the uniqueness of a solution to the problem with boundary conditions $u'(0) = 0$, $u'(1) = 0$. (This is called a "Neumann" problem.)

**Exercise 12.3.** Do the hypotheses of Proposition 12.1 also imply the existence of a solution to the problem (12.1)?

**Exercise 12.4.** Do the hypotheses of Proposition 12.1 also imply the existence of a solution to the Neumann problem of Exercise 12.2?

**Exercise 12.5.** Use (12.22) to verify that $\lim_{\alpha\to\infty} H(\alpha) = 0$. Hint: Break the interval of integration into $\left[0, \frac{3\alpha}{4}\right]$ and $\left[\frac{3\alpha}{4}, \alpha\right]$.

**Exercise 12.6.** Prove Proposition 12.6, by showing first that for each $\alpha > 0$ there is a unique $\lambda > 0$ such that the solution to (12.14)–(12.16) satisfies $u(1) = 0$. Hint: Rescale the independent variable to remove $\lambda$ from the differential equation. See [**48**].

**Exercise 12.7.** Determine the bifurcation diagrams ($u(0)$ plotted against $\lambda$) for the boundary value problems

$$\left(r^{n-1}u'\right)' + \lambda r^{n-1}\left(|u| + 1\right) = 0,$$
$$u'(0) = 0, \quad u(1) = 0$$

in the region $\lambda \geq 0, u(0) \geq 0$, for $n = 1, 2, 3$. Hint: Look for special functions.

**Exercise 12.8.** Prove Proposition 12.7. Show that if $f(u) = \sqrt{u}$ and $n = 1$, then (12.14)–(12.15) has a unique solution for every $\lambda \geq 0$, so that the function $\lambda(\alpha)$ is monotonic and unbounded.

**Exercise 12.9.** Suppose that $u$ satisfies (12.14) and (12.16) for some $\alpha > 0$, with $f(u) > 0$ if $u \geq 0$. Prove that $u(r_0) = 0$ for some $r_0 > 0$. Is it necessarily true that $u''(r) < 0$ if $r > 0$?

**Exercise 12.10.** (Straightforward) Prove (12.35).

Chapter 13

# Shooting with more parameters

## 13.1. A problem from the theory of compressible flow

In Chapter 2 we gave an example of shooting in which two initial conditions had to be adjusted to satisfy the boundary conditions. In this chapter we discuss a more difficult example of shooting with two parameters than the model problem treated earlier. The setting is flow of a compressible fluid past a fixed boundary surface. The differential equations in the fluid boundary layer were derived by Stewartson [**227**]. They are

$$\begin{gathered} f''' + ff'' + \lambda\left(h - f'^2\right) = 0, \\ h'' + fh' = 0, \end{gathered} \tag{13.1}$$

where $f'$ represents the velocity tangential to the surface and $h$ is proportional to the total energy of the system.

The boundary conditions in the compressible flow case (13.1) involve a constant $h_0$, which we will assume satisfies

$$0 < h_0 < 1. \tag{13.2}$$

Then we want

$$\begin{gathered} f(0) = f'(0) = 0, \quad f'(\infty) = 1, \\ h(0) = h_0, \quad h(\infty) = 1. \end{gathered} \tag{13.3}$$

In this section we prove the existence of a solution to (13.1) and (13.3) when (13.2) holds. The proof is based on one given by McLeod and Serrin in [**166**], where the physical background is discussed and further references given. McLeod and Serrin treat a more general problem, but the essence of the shooting method they use is seen in the problem (13.1)–(13.3) with condition (13.2). A uniqueness result for this problem is given in [**85**]. Existence when $h_0 > 1$ is also proved in [**166**].

Note that we obtain a family of solutions to (13.1) by setting $h = 1$ and letting $f$ be a solution of the Falkner-Skan equation, which was discussed in Chapter 9. However, these do not satisfy the boundary conditions (13.3).

**13.1.1. Existence of a solution.** The following proof differs in some details from the one given in [**166**]. We take advantage of the relative simplicity of the problem we are treating, and also of insights obtained from the efficient and easy to use numerical ode packages now available. We used the program XPP [**55**], of G. B. Ermentrout, which we find particularly convenient for shooting experiments, but other programs could be used instead.

**Theorem 13.1.** *The equation* (13.1) *has a solution satisfying* (13.3) *for each* $h_0 \in (0, 1)$.

**Proof.** As expected in a shooting proof, we consider the initial value problem consisting of (13.1) and

$$\begin{aligned} f(0) = 0, \quad f'(0) = 0, \quad f''(0) = \alpha, \\ h(0) = h_0, \quad h'(0) = \beta. \end{aligned} \tag{13.4}$$

There are two shooting parameters, $\alpha$ and $\beta$, and two conditions to be satisfied at infinity. We use the following result from plane topology. This is similar to Lemma 2.8.

**Lemma 13.2.** *Let $J$ be a closed rectangle $a \le x \le b, c \le y \le d$, and let $S^+$ and $S^-$ be disjoint open subsets of $R^2$ such that $S^-$ contains the left edge $x = a$ of $J$ and $S^+$ contains the right edge $x = b$. Then the complement $D$ in $J$ of $(S^+ \cup S^-) \cap J$ contains a continuum $\gamma$ connecting the lines $y = c$ and $y = d$ in $J$.*

**Proof.** A short elementary proof is given in [**166**], and a similar result is proved in [**180**]. Note that $S^-$ and $S^+$ need not be connected sets. They only need to have connected components which contain the left and right sides of the rectangle. The theorem proved by degree theory in the appendix to this chapter includes this result. □

We will take a moment to discuss the use of computations to help find appropriate sets $S^+$ and $S^-$. In panels $A$–$F$ of Figure 13.1 we plot $h$ and $f'^2$ for a variety of initial conditions. In each case $h(0) = \frac{1}{2}$. These graphs guided us to the choices we made of $S^-$ and $S^+$, which are different from the sets used in [**166**]. At the time of the original proof, the available numerical packages were much more cumbersome than what is now available, making it more time consuming to search in this way, especially when two shooting

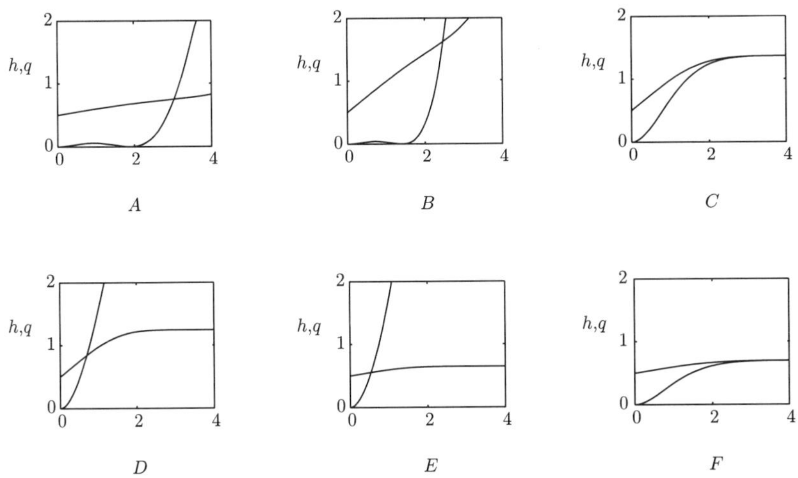


**Figure 13.1.** Graphs of $h$ and $q = (f')^2$; note that $q(0) = 0$.

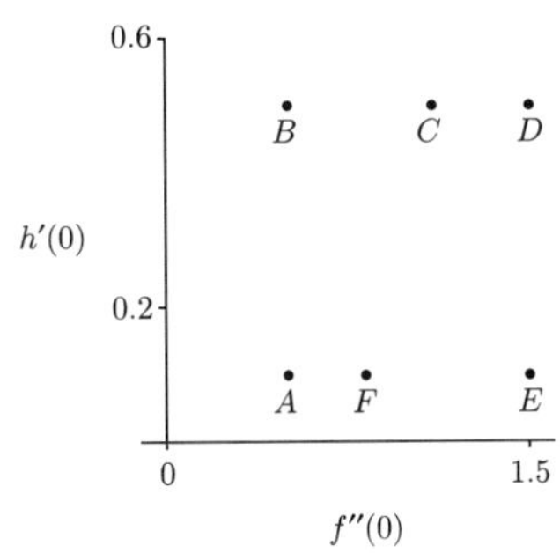


**Figure 13.2.** Initial conditions for solutions above.

parameters were involved. The locations in the $(\alpha, \beta)$-plane of the initial conditions which gave each of these pairs of graphs is shown in Figure 13.2.

We now define $S^+$ and $S^-$. In these definitions, $(f, h)$ denotes the unique solution to (13.1) and (13.4), with the dependence on $(\alpha, \beta)$ understood. From (13.4) we see that if $\alpha > 0$ and $\beta > 0$ then on some initial interval $(0, \delta)$, $f'' > 0$ and $h - f'^2 > 0$. Let $R_+^2$ denote the closed quadrant $\alpha \geq 0, \beta \geq 0$. Then set

$$S^- = \{(\alpha, \beta) \in R_+^2 \mid \text{there is a } t_1 > 0 \text{ such that } f''(t_1) < 0 \\ \text{and } h - f'^2 > 0 \text{ on } [0, t_1]\},$$

$$S^+ = \Big\{(\alpha, \beta) \in R_+^2 \mid \text{there is a } t_2 > 0 \text{ such that } h(t_2) - f'(t_2)^2 < 0 \\ \text{and } f'' > 0 \text{ on } [0, t_2]\Big\}.$$

By referring to Figure 13.1, with panels labeled $A$ through $F$, it can be seen that the points labeled $A$ and $B$ in Figure 13.2 are initial conditions in $S^-$, while $D$ and $E$ are in $S^+$. Also, $C$ and $F$ are not in $S^+ \cup S^-$, and so these two points may be in $\gamma$.

Since $h_0 > 0$, we see that if $\alpha = 0$ then $f'''(0) < 0$, and so $(0, \beta) \in S^-$ for every $\beta \geq 0$. The sets $S^+$ and $S^-$ are clearly open, because the solutions of (13.1) are continuous with respect to the initial conditions. Also, they are disjoint from their definitions. The following lemma restricts the behavior of $(f, h)$ when $(\alpha, \beta) \notin S^- \cup S^+$.

**Lemma 13.3.** *None of the following are possible for a solution of* (13.1) *and* (13.4) *with* $\alpha \geq 0$, $\beta \geq 0$, *and* $(\alpha, \beta) \notin S^+ \cup S^-$.

(i) $f'' > 0$ *on* $[0, \tau)$, $f''(\tau) = 0$, $h - f'^2 > 0$ *in* $[0, \tau]$, *and* $f'''(\tau) = 0$.

(ii) $f'' > 0$ *on* $[0, \tau]$, $h - f'^2 > 0$ *in* $[0, \tau)$, $h(\tau) = f'(\tau)^2$ *and* $\left(h - f'^2\right)' = 0$ *at* $\tau$.

(iii) $f'' > 0$ *on* $[0, \tau)$, $h - f'^2 > 0$ *on* $[0, \tau)$, *and* $f''(\tau) = h(\tau) - f'(\tau)^2 = 0$.

**Proof.** (i) If $f'' = 0$ then $f''' = -\lambda\left(h - f'^2\right)$, which is negative at $\tau$. (Thus, $f'' < 0$ just after $\tau$, and $(\alpha, \beta) \in S^-$.)

(ii) An easy computation shows that at $\tau$,

$$\left(h - f'^2\right)'' = -2f''(\tau)^2 < 0,$$

and so $h - f'^2$ must be negative somewhere in $(0, \tau)$, a contradiction.

(iii) Note that $h'(t) = \beta e^{-\int_0^t f(s)ds}$. Under the given conditions,

$$\left(h - f'^2\right)'(\tau) = h'(\tau) \geq 0.$$

If $\beta > 0$, then $h'(\tau) > 0$ and $h - f'^2$ is negative somewhere in $(0, \tau)$, again a contradiction. If $\beta = 0$, then $h \equiv h_0$, $f' \equiv \sqrt{h_0}$, and the solution does not satisfy $f'' > 0$ on $[0, \tau)$. □

**Corollary 13.4.** *If* $\alpha \geq 0$, $\beta \geq 0$ *and* $(\alpha, \beta) \notin S^- \cup S^+$, *then the solution exists on* $[0, \infty)$, *with* $f'' > 0$ *and* $h - f'^2 > 0$. *Further,* $\lim_{x \to \infty} h(x)$ *exists.*

**Proof.** If $\alpha = 0$ and $\beta \geq 0$, then $f'''(0) < 0$ and $(\alpha, \beta) \in S^-$. Therefore, the lemma implies that if $\alpha \geq 0$, $\beta > 0$ and $(\alpha, \beta) \notin S^- \cup S^+$, then $f'' > 0$ and $h - f'^2 > 0$ as long as the solution exists. Thus, $f'$ is increasing and $f$ is unbounded. Integrating the equation for $h$ in (13.1) shows that $h$ is bounded, and since $f'^2 < h$, $f'$ is also bounded. Thus the solution exists on $[0, \infty)$ and $h$, which is increasing, approaches a limit $h(\infty)$. □

We will choose $J$ to be a rectangle of the form

$$J = \{(\alpha, \beta) \mid 0 \le \alpha \le \alpha_2, \beta_1 \le \beta \le \beta_2\},$$

for some $\alpha_2 > 0$ and $\beta_2 > \beta_1 > 0$.

**Lemma 13.5.** *The numbers $\alpha_2, \beta_1$, and $\beta_2$ can be chosen so that $S^-$ contains the left side of $J$ and $S^+$ contains the right side of $J$ and so that if $(\alpha, \beta)$ is on the top edge $\beta = \beta_2$ of $J$ and not in $S^+ \cup S^-$, then $h(t) > 1$ for some $t > 0$, while if $(\alpha, \beta)$ is on the bottom edge $\beta = \beta_1$ of $J$ and not in $S^+ \cup S^-$, then $\sup_{t \ge 0} h(t) < 1$.*

**Proof.** We identify several steps.

(a) Recall that $(0, \beta) \in S^-$ for every $\beta \ge 0$ and that $S^-$ is open. It follows that there is a continuous positive function $\alpha(\beta)$ defined for $\beta \ge 0$ such that if $0 \le \alpha < \alpha(\beta)$ then $(\alpha, \beta) \in S^-$.

(b) We use the following result.

**Lemma 13.6.** *If $\beta = \sqrt{\frac{2}{\pi}}$ and $(\alpha, \beta) \notin S^+ \cup S^-$, then $h(\infty) > 1$.*

**Proof.** We saw earlier that $h(\infty)$ exists. Suppose that $h(\infty) \le 1$. Since $f'^2 < h$ on $(0, \infty)$, $f' < 1$ and $f < x$. Hence,

$$h'(x) = \beta e^{-\int_0^x f ds} \ge \beta e^{-\frac{x^2}{2}}, \tag{13.5}$$

$$h(x) \ge h_0 + \beta \int_0^x e^{-\frac{s^2}{2}} ds,$$

and a contradiction results if $\beta = \sqrt{\frac{2}{\pi}}$. □

Hence, we take $\beta_2 = \sqrt{\frac{2}{\pi}}$.

(c) If $\beta \le \beta_2$ then $h \le h_0 + \sqrt{\frac{2}{\pi}} x$. Then an argument used in Chapter 9, in the proof of Theorem 9.1, shows that $\alpha_2$ can be chosen independently of $\beta$ so that if $\alpha = \alpha_2$ then $f'(x)^2 > h(x)$ for some $x \in (0, 1]$.

(d) This step requires the most thought. Choosing $\alpha_2$ from part (c), we have to find a positive number $\beta_1$ such that if $0 < \alpha \le \alpha_2$ and $(\alpha, \beta_1) \notin S^+ \cup S^-$, then $h(\infty) < 1$. It was shown above that $(0, 0) \in S^-$. Hence there is a $\rho > 0$ such that the square $\sigma_\rho$ of points $(\alpha, \beta)$ such that $0 \le \alpha \le \rho$ and $0 \le \beta \le \rho$ is contained in $S^-$. We will choose $\beta_1 \in (0, \rho)$.

To give further conditions on $\beta_1$ we assume that $(h_0, \beta_1) \in J \backslash (S^- \cup S^+)$, which implies that $(\alpha, \beta_1) \notin \sigma_\rho$, and so $\rho < \alpha \le \alpha_2$. Initially, therefore, $f'' > h'$.

**Lemma 13.7.** *We can choose $\beta_1$ so small that if $\beta = \beta_1$ and $f'' > h'$ on an interval $[0, T)$, then on this interval, $h' \le \rho$ and $h \le \frac{1+h_0}{2}$.*

**Proof.** Choosing $\beta_1 < \rho$ ensures that $h' < \rho$ on $[0,\infty)$, since $h'' \le 0$. In the interval $[0,T)$,

$$f' > h - h_0,$$

$$f(x) > \int_0^x (h(s) - h_0)\, ds,$$

$$\int_0^x f(s)\, ds > \int_0^x (x-s)(h(s) - h_0)\, ds.$$

Hence, $h'(x) = \beta_1 e^{-\int_0^x f(s)ds} \le \beta_1 e^{-\int_0^x (x-s)(h(s)-h_0)ds} ds.$

If $h \le \frac{2h_0+1}{3}$ on $[0,T)$ then (13.2) implies the result. Suppose that $h(\tau) = \frac{2h_0+1}{3}$ for some $\tau < T$. Then on $[\tau, T]$,

$$h'(t) \le \beta_1 e^{-\int_\tau^t (t-s)\frac{1-h_0}{3} ds} \le \beta_1 e^{-\frac{1-h_0}{3}\frac{(t-\tau)^2}{2}}.$$

Integrating from $\tau$ to $\infty$, we get

$$h(t) \le \frac{2h_0+1}{3} + \sqrt{\frac{3}{2}}\beta_1 \frac{\sqrt{\pi}}{\sqrt{1-h_0}}.$$

We can thus choose $\beta_1$ so small (and independent of $T$) that $h(t) \le \frac{1+h_0}{2}$ on $[0,T)$. This proves the lemma. □

Choose $\beta_1$ as in Lemma 13.7, and assume that $(\alpha, \beta_1) \notin S^- \cup S^+$. Further, suppose that $f''(T) = h'(T)$, with $f'' > h'$ on $[0,T)$. Then $h(T) \le \frac{1+h_0}{2}$ and $f''(T) = h'(T) \le \beta_1$. Recalling the earlier stage of the argument where we considered $\alpha = \beta = 0$, we see that if $\beta_1$ is sufficiently small, independent of $T$, then $f''$ turns negative, contradicting our assumption that $(\alpha, \beta_1) \notin S^- \cup S^+$. Hence, if $(\alpha, \beta_1) \in \gamma$ then $f'' > h'$ on $[0,\infty)$ and $h(\infty) < 1$, as desired. This completes the proof of Lemma 13.5. □

Lemmas 13.2 and 13.5 imply that there is a continuum $\gamma$ connecting $\beta = \beta_2$ and $\beta = \beta_1$ and lying in $J$ such that $\gamma \cap (S^- \cup S^+)$ is empty. Suppose that $(\alpha, \beta) \in \gamma$. Then $\alpha > 0$. Using Lemma 13.3, we see that $f'' \ge 0$ and $f'^2 \le h$ as long as the solution exists. Also, if $x > 0$ and $f(x)$ exists, then $f'(x) > 0$ and $f(x) > 0$. Since $\beta > 0$, it follows from (13.1) that $h' > 0, h'' < 0$, and so $h' \le \beta$, $h \le h_0 + \beta x$. Thus, $f' \le \sqrt{h_0 + \beta x}$, from which it follows that the solution exists in $[0,\infty)$.

We also see that for $x \ge 1$, $f' \ge f'(1) > 0$, and so

$$h'(x) = \beta e^{-\int_0^x f} \le c e^{-\mu x}$$

for some $\mu > 0, c > 0$. Thus, $\lim_{x\to\infty} h(x) = h_\infty$ exists. Also, $f'^2 < h_\infty$, and $f'' > 0$, so that $f'(\infty)$ exists and is positive. If $p = h(\infty) - f'(\infty)^2 > 0$ then for large $x$,

$$f''' + ff'' \le -\lambda \frac{p}{2}.$$

Since $f>0, f''>0$, it follows that for large $x$, $f'''(x) \leq -\lambda\frac{p}{2}$, which implies that $f''$ becomes negative, a contradiction. Hence,

$$\lim_{x\to\infty}\left(h(x)-f'(x)^2\right)=0.$$

We must now show that $h(\infty)$ is a continuous function of $(\alpha,\beta)$ on $\gamma$. In other words, suppose $(\bar{\alpha},\bar{\beta})\in\gamma$ and $\varepsilon>0$. Then we must show that there is a $\delta>0$ such that if $(\alpha,\beta)\in\gamma$ and

$$|\alpha-\bar{\alpha}|+\left|\beta-\bar{\beta}\right|<\delta, \tag{13.6}$$

then

$$\left|h(\infty)-\bar{h}(\infty)\right|<\varepsilon, \tag{13.7}$$

where $(\bar{f},\bar{h})$ is the solution with initial conditions $(\bar{\alpha},\bar{\beta})$. Since $\bar{\beta}\geq\beta_1>0$, this is a fairly routine step, which is mostly left to the reader in Exercise 13.1. Given $\varepsilon$, we choose $T>0$ such that $\int_T^\infty \bar{\beta}e^{-\int_0^s \bar{f}(s)ds}<\frac{\varepsilon}{4}$, for example. We can then choose $\delta$ so small that if (13.6) holds and $(\alpha,\beta)\in\gamma$, then $\left|h(T)-\bar{h}(T)\right|<\frac{\varepsilon}{2}$ and $\beta\int_T^\infty e^{-\int_0^s f(s)ds}<\frac{\varepsilon}{2}$, implying (13.7) and proving the continuity of $h(\infty)$. The existence of an $(\alpha,\beta)\in\gamma$ such that $h(\infty)=1$ then follows from Lemma 13.5. □

## 13.2. A result of Y.-H. Wan

The McLeod-Serrin two-parameter shooting principle, and similar topological results such as Lemma 2.8, can be extended to higher dimensions. In this section we state such a generalization. It is an intriguing result which one feels should be useful in some circumstances, and it was helpful in one three-dimensional setting [**88**]. However we have never encountered another problem where it could be applied. We state this result in the hope that a reader may find it of interest. It is due to our former colleague, Y.-H. Wan, whom we thank for allowing us to include it here. Only the three-dimensional version was published previously, for lack of other applications. We should also mention a difficult four-dimensional shooting proof by Dunbar [**53**]. We have not investigated whether Wan's result would simplify Dunbar's proof.

**Theorem 13.8.** *Let $B=[-1,1]^n\times[-1,1]$. For $i=1,\ldots,n$ let $\Omega_i^\pm$ be open sets in $B$. Further, let $\Delta_1=B$, and for $j=2,\ldots,n+1$, let*

$$\Delta_j=B\backslash\left\{\bigcup_{i=1}^{j-1}\left(\Omega_i^+\cup\Omega_i^-\right)\right\}. \tag{13.8}$$

*Suppose that*

(i) *if* $x_i = \pm 1$ *and* $x \in \Delta_i$, *then* $x \in \Omega_i^{\pm}$, *for* $i = 1, \ldots, n$,

(ii) $\Delta_j \cap \Omega_j^+ \cap \Omega_j^-$ *is empty, for* $j = 1, \ldots, n$.

*Then,* $\Delta_{n+1}$ *contains a continuum* $\gamma$ *connecting* $x_{n+1} = -1$ *to* $x_{n+1} = 1$.

**Remark 13.9.** The case $n = 1$ is the McLeod-Serrin result.

**Corollary 13.10.** *The continuum* $\gamma$ *can be chosen so that*

$$\gamma \subset \partial\Omega_1^+ \cap \cdots \cap \partial\Omega_n^+ \cap \Delta_{n+1}.$$

**Proof of Corollary 13.10.** Define

$$\begin{aligned}\tilde{\Omega}_i^+ &= \Omega_i^+, \\ \tilde{\Omega}_i^- &= \Omega_i^- \cup \{\text{exterior of } \Omega_i^+\}.\end{aligned}$$

Also, define $\tilde{\Delta}_i$ by replacing $\Omega_i^{\pm}$ with $\tilde{\Omega}_i^{\pm}$ in (13.8), for $i = 1, \ldots, n$. Since $\tilde{\Omega}_i^- \supseteq \Omega_i^-$, we see that $\tilde{\Delta}_j \subseteq \Delta_j$. The theorem still applies to $\tilde{\Omega}_i^{\pm}$. Hence, $\gamma$ can be taken in $\tilde{\Delta}_{n+1}$, implying that

$$\gamma \cap \left( \bigcup_{i=1}^{n} \left( \Omega_i^+ \cup \text{ext } \Omega_i^+ \cup \Omega_i^- \right) \right) = \phi.$$

Hence,

$$\gamma \subset \partial\Omega_1^+ \cap \cdots \cap \partial\Omega_n^+ \cap \Delta_{n+1}.$$

□

The theorem will be proved in the appendix to this chapter and makes use of topological degree theory.

## 13.3. Exercise

**Exercise 13.1.** (Routine) Complete the proof of existence of a solution to (13.1) and (13.3) by giving the details of the argument outlined above that $h(\infty)$ is a continuous function of $(\alpha, \beta)$ in $\gamma$.

## 13.4. Appendix: Proof of Wan's theorem

**Proof.** Note that condition (i) is equivalent to the following:

(i)$'$ For $i = 1, \ldots, n$, if $x_i = \pm 1$ then $x \in \Omega_1^+ \cup \Omega_1^- \cup \cdots \cup \Omega_{i-1}^+ \cup \Omega_{i-1}^- \cup \Omega_i^{\pm}$.

We will shrink each $\Omega_i^{\pm}$ by a small $\delta_i > 0$. Specifically, suppose that $\delta_1, \ldots, \delta_n$ are given positive numbers, and set $\delta = (\delta_1, \ldots, \delta_n)$. Then for each $i$ with $1 \le i \le n$ let $\Omega_{i,\delta}^{\pm} = \{p \in B \mid d(p, (\Omega_i^{\pm})^c) > \delta_i\}$. Further, let

$$\Delta_{i,\delta} = B \backslash \left( \Omega_{1,\delta}^+ \cup \Omega_{1,\delta}^- \cup \cdots \cup \Omega_{i-1,\delta}^+ \cup \Omega_{i-1,\delta}^- \right), \text{ for } i = 1, \ldots, n+1.$$

Note that $\Delta_i \subset \Delta_{i,\delta}$. The shrinkage $\delta = (\delta_i)$ is said to be *admissible* if the following hold:

(i)$_\delta$ For each $i$ with $1 \le i \le n$, if $x_i = \pm 1$ and $x \in \Delta_{i,\delta}$ then $x \in \Omega^{\pm}_{i,\delta}$.

(ii)$_\delta$ $\overline{\Delta_{i,\delta} \cap \Omega^{+}_{i,\delta}} \cap \overline{\Delta_{i,\delta} \cap \Omega^{-}_{i,\delta}}$ is empty for each $i$, $1 \le i \le n$.

Note that (i)$_\delta$ is equivalent to the following:

(i)$'_\delta$ For $i = 1, \ldots, n$, if $x_i = \pm 1$ then $x \in \Omega^{+}_{1,\delta} \cup \Omega^{-}_{1,\delta} \cup \cdots \cup \Omega^{+}_{i-1,\delta} \cup \Omega^{-}_{i-1,\delta} \cup \Omega^{\pm}_{i,\delta}$.

It is also equivalent to the following:

(i)$''_\delta$ For $i = 1, \ldots, n$, if $x_i = \pm 1$ then $x \in \left(\Delta_{k,\delta} \cap \Omega^{+}_{k,\delta}\right) \cup \left(\Delta_{k,\delta} \cap \Omega^{-}_{k,\delta}\right)$ for some $k$ with $1 \le k \le i$, or $x \in \Delta_{i,\delta} \cap \Omega^{\pm}_{i,\delta}$.

We claim that for each $\bar{b} > 0$ there exists an admissible shrinkage $\delta = (\delta_1, \ldots, \delta_n)$ with $\delta_i \le \bar{b}$ for $i = 1, \ldots, n$.

To show this for a given $\bar{b} > 0$, we will use backward induction to define $\bar{\delta} = (\bar{b}_1, \ldots, \bar{b}_n)$ with $0 < \bar{b}_1, \ldots, \bar{b}_n \le \bar{b}$. Set $\bar{b}_n = \bar{b}$. Assuming that $\bar{b}_j, \ldots, \bar{b}_n$ have been defined, choose $\bar{b}_{j-1} \in (0, \bar{b}]$ so that for every choice of $b_1, \ldots, b_{j-1}$ with $0 < b_1, \ldots, b_{j-1} \le \bar{b}_{j-1}$, $d(p, \Delta_j) \le \frac{\bar{b}_j}{4}$ when $p \in \Delta_{j,\delta}$, where $\delta = \left(b_1, \ldots, b_{j-1}, \bar{b}_j, \ldots, \bar{b}_n\right)$. We demonstrate that for each $j$, $\overline{\Delta_{j,\delta} \cap \Omega^{+}_{j,\delta}} \cap \overline{\Delta_{j,\delta} \cap \Omega^{-}_{j,\delta}}$ is empty, by showing that $d\left(\Delta_{j,\delta} \cap \Omega^{+}_{j,\delta}, \Delta_{j,\delta} \cap \Omega^{-}_{j,\delta}\right) \ge \frac{\bar{b}_j}{4}$.

To see this, suppose that $p \in \Delta_{j,\delta} \cap \Omega^{+}_{j,\delta}$ and $q \in \Delta_{j,\delta} \cap \Omega^{-}_{j,\delta}$. Then there exists $p' \in \Delta_j, q' \in \Delta_j$ such that $d(p, p') \le \frac{\bar{b}_j}{4}$, $d(q, q') \le \frac{\bar{b}_j}{4}$. Thus, $d\left(p', \left(\Omega^{+}_j\right)^c\right) \ge \frac{3\bar{b}_j}{4}$, and $d\left(q', \left(\Omega^{-}_j\right)^c\right) \ge \frac{3\bar{b}_j}{4}$. Hence, if $q' \in \Omega^{-}_j \cap \Delta_j$ then $q' \in \left(\Omega^{+}_j\right)^c$ (by (ii)) and $d(p', q') \ge \frac{3\bar{b}_j}{4}$. Therefore $d(p, q) \ge d(p', q') - d(p, p') - d(q, q') \ge \frac{\bar{b}_j}{4}$. Our choice of $\bar{\delta} = (b_1, \ldots, b_n)$ satisfies condition (ii)$_{\bar{\delta}}$. By taking $\bar{b}$ even smaller if necessary, condition (i)$'_{\bar{\delta}}$ will be satisfied. Thus, $\bar{\delta}$ is admissible and each $\bar{b}_j \le \bar{b}$.

After shrinking each $\Omega^{\pm}_i$ by an admissible $\delta$, one can define smooth functions $f_{i,\delta} : B \to [-1, 1]$ such that

$$f_{i,\delta}\left(\Delta_{i,\delta} \cap \Omega^{+}_{i,\delta}\right) = 1,$$
$$f_{i,\delta}\left(\Delta_{i,\delta} \cap \Omega^{-}_{i,\delta}\right) = -1,$$

for each $i$, $1 \le i \le n$.

Next, let

$$I_j = [-1, 1]^j = \{(x_1, \ldots, x_j) \mid -1 \le x_i \le 1 \text{ for } i = 1, \ldots, j\}.$$

We embed $I_j$ into $B$ by the mapping $(x_1, \ldots, x_j) \to (x_1, \ldots, x_j, 0, \ldots, 0)$. Identifying $I_j$ with its embedded image, one now has $I_1 \subset I_2 \subset \cdots \subset I_n \subset B$. Define $P : I_n \times [-1, 1] \to I_n$ by $P = (f_{1,\delta}, \ldots, f_{n,\delta})$ for each admissible $\delta$.

For each $j \in \{1, \ldots, n\}$ and each $x_{j+1}, \ldots, x_{n+1}$, define $P_{x_{j+1},\ldots,x_{n+1}} : I_j \to I_j$ by

$$P_{x_{j+1},\ldots,x_{n+1}} (x_1, \ldots, x_j) = (f_{1,\delta}(x), \ldots, f_{j,\delta}(x))$$

where $x = (x_1, \ldots, x_j, x_{j+1}, \ldots, x_{n+1})$. From $(\mathrm{i})''_\delta$ one sees that

$$P_{x_{j+1},\ldots,x_{n+1}} (\partial I_j) \subseteq \partial I_j.$$

**Lemma 13.11.** *For admissible* $\delta$, $\deg P_{x_{j+1},\ldots,x_{n+1}} (\partial I_j) = 1$, $1 \le j \le n$.

Recall from [**173**] that for a smooth function $F : I_k \to I_k$, with $F(\partial I_k) \subset \partial I_k$, we have the formula $\deg (F|\partial I_k) = \sum_{x \in F^{-1}(y)} \mathrm{sign}\big(\det\big(\frac{\partial F}{\partial x}\big)\big)$, where $y \in I_k \backslash \partial I_k$ is a regular value of $F$. Note here that $\frac{\partial F}{\partial x}$ stands for the Jacobian and det for its determinant, evaluated at $x$.

**Proof.** Use induction. The case $j = 1$ follows from the degree formula in [**173**]. Assuming that $2 \le j \le n$ and $\deg P_{x_j,\ldots,x_{n+1}} (\partial I_{j-1}) = 1$, define $P^*_{x_{j+1},\ldots,x_{n+1}} : I_j \to I_j$ by $(x_1, \ldots, x_j) \to (f_{1,\delta}(x), \ldots, f_{j-1,\delta}(x), x_j)$. Using $(\mathrm{i})''_\delta$ one readily sees that $P^*_{x_{j+1},\ldots,x_{n+1}} (\partial I_j) \subseteq \partial I_j$, and $P_{x_{j+1},\ldots,x_{n+1}}|\partial I_j$ is homotopic to $P^*_{x_{j+1},\ldots,x_{n+1}}|\partial I_j$ in $\partial I_j$, via $(1-t) P_{x_{j+1},\ldots,x_{n+1}} + tP^*_{x_{j+1},\ldots,x_{n+1}}$, $0 \le t \le 1$. Thus, $\deg \left(P_{x_{j+1},\ldots,x_{n+1}}|\partial I_j\right) = \deg \left(P^*_{x_{j+1},\ldots,x_{n+1}}|\partial I_j\right)$. From the degree formula we have

$$\begin{aligned}\deg \left(P^*_{x_{j+1},\ldots,x_{n+1}}|\partial I_j\right) &= \deg \left(P^*_{x_{j+1},\ldots,x_{n+1}}|\partial I_{j-1}\right)\\ &= \deg \left(P_{x_j,\ldots,x_{n+1}}|\partial I_{j-1}\right) \text{ with } x_j = 0 \text{ (by definition)}\\ &= 1 \text{ (by the induction hypothesis).}\end{aligned}$$

Thus, our lemma is proved. □

Now choose a point $\beta$ near $(0, \ldots, 0) \in I_n$ which is a common regular value of $P_{-1}, P_1$, and $P$. From arguments as in smooth degree theory [**173**], [**100**], [**158**], one has that:

(a) $P_{-1}^{-1}(\beta)$ contains an odd number of points, because $P_{-1}|\partial I_n$ is of degree one by the lemma above and the degree formula.

(b) $P^{-1}(\beta)$ is a one-dimensional manifold (by the implicit function theorem), which consists of finitely many circles inside $I_{n+1}$ and finitely many arcs with endpoints in $x_{n+1} = \pm 1$.

Statements (a) and (b) imply that there exists an arc in $P^{-1}(\beta)$ connecting $x_{n+1} = -1$ to $x_{n+1} = 1$ in $B$. Thus, $\Delta_{n+1,\delta}$ contains an arc connecting

$x_{n+1} = -1$ to $x_{n+1} = 1$, where $\Delta_{n+1,\delta} = B \backslash \left( \Omega^+_{1,\delta} \cup \Omega^-_{1,\delta} \cup \cdots \cup \Omega^+_{n,\delta} \cup \Omega^+_{n,\delta} \right)$. Observe that $\Delta_{n+1,\delta} \supset \Delta_{n+1}$ and $\Delta_{n+1,\delta} \searrow \Delta_{n+1}$ as $\delta \searrow 0$ with $\delta$ admissible.

For each $\varepsilon > 0$ let $\{K_\varepsilon\}$ be the nonempty collection of connected components of the closed $\varepsilon$-neighborhood of $\Delta_{n+1}$ in $B$ which meet $x_{n+1} = -1$ and $x_{n+1} = 1$. Then $\{K_\varepsilon\}$ must have a finite number of elements. We can choose suitable connected components $K_{\frac{1}{m}}$ so that $K_1 \supset K_{1/2} \supset \cdots$. Let $K = \bigcap_{m=1}^\infty K_{1/m}$. Then $K$ is connected, contained in $\Delta_{n+1}$, and meets $x_{n+1} = -1$ and $x_{n+1} = 1$. This completes the proof of Wan's theorem. $\square$

Chapter 14

# Some problems of A. C. Lazer

## 14.1. Introduction

In 1970, Landesman and Lazer considered a class of elliptic boundary value problems

$$Lu + \alpha u + g(u) = h(x), \quad u = 0 \text{ on } \partial\Omega, \tag{14.1}$$

in an appropriate bounded region $\Omega$ in $R^n$ with boundary $\partial\Omega$ [**139**]. Their interest was in the case where the linear problem

$$\begin{gathered} Lu + \alpha u = 0, \\ u = 0 \text{ on } \partial\Omega \end{gathered} \tag{14.2}$$

has a one-dimensional space, say span $\{w\}$, of weak solutions. Under the assumptions that $h \in L^2(\Omega)$,

$$\lim_{u \to \infty} g(u) \text{ and } \lim_{u \to -\infty} g(u) \text{ exist,} \tag{14.3}$$

and

$$g(-\infty) < g(u) < g(\infty) \quad \text{for all } u, \tag{14.4}$$

a condition was given relating $h$, $g$, and $w$ which is necessary and sufficient for the problem (14.1) to have a solution.

This paper has attracted much attention over the years. It was preceded by a similar result for ode's, by Lazer and Leach [**144**], which is also fairly well known and which set the stage for [**139**]. Several new proofs have been found for each result, and a number of extensions have been found. As evidence for the continued influence of this work, we mention that at least fifty papers were published in 2009 citing [**139**], and ten citing [**144**].

In both cases, the years since the original paper have seen many advances inspired by this work. Fortunately, there is a recent survey by Mawhin [**159**] of the ode aspects, following on from [**144**], and we refer the reader to this work for updates. We know of no comparable reference for the theorem of Landesman and Lazer, but its influence has been widespread and continuing. We do not attempt to discuss recent developments here.[1]

A year later, Landesman and Lazer considered another class of elliptic boundary value problems:

$$\Delta u + p(x,u,\nabla u)\,u = h(x,u,\nabla u) \quad \text{in } \Omega,$$
$$u = 0 \quad \text{on } \partial\Omega$$

for suitable bounded regions $\Omega$ in $R^n$. Their key assumption was that for some $\delta > 0$, and some $k$,

$$\lambda_k + \delta \le p(x,u,\nabla u) \le \lambda_{k+1} - \delta,$$

where $\lambda_k$ and $\lambda_{k+1}$ are successive eigenvalues of the problem

$$\Delta u + \lambda u = 0,$$
$$u = 0 \quad \text{on } \partial\Omega.$$

Again conditions were given which ensure that there is a solution [**140**]. And again, this paper was preceded by an ode version with Leach [**145**]. As in [**144**], their result was proved using the Schauder fixed point theorem (see Chapter 5). In this chapter we will do the following:

(1) Give the original proof of Lazer and Leach for the ode result in [**144**], which is an elegant application of the Schauder fixed point theorem.

(2) Give a short new proof of this result which does not use functional analysis. This and item (3) come from [**92**].

(3) Give a shortened version of the proof of Landesman and Lazer in [**139**].

(4) Give a short shooting proof of the result in [**145**].

(5) Give a short proof of the result in [**140**]. As in item (3), this is basically the original proof, but done a little more efficiently.

(6) Briefly introduce a class of ode problems due to Littlewood bearing a superficial resemblance to the problem in [**144**], but where, so far, we have not found a classical proof. The only technique we know

[1] In [**143**] Lazer emphasizes that the paper [**72**] was the first place where "Landesman-Lazer" conditions appeared, giving significant credit to P. O. Fredrickson.

of for attacking them is what is known as the Moser twist theorem, which we do not have space to discuss here in any detail.

## 14.2. First Lazer-Leach problem

We start by recalling a property of the linear equation

$$y'' + y = p(t),$$

where $p$ is periodic of period $2\pi$. From the variation of parameters formula, every solution is of the form

$$c_1 \sin t + c_2 \cos t + \int_0^t (\sin t \cos s - \cos t \sin s)\, p(s)\, ds.$$

It follows that either no solution is $2\pi$-periodic or every solution is, and the latter holds if and only if

$$\int_0^{2\pi} p(s) \sin s = \int_0^{2\pi} p(s) \cos ds = 0.$$

Further, if either of these integrals is not zero then all solutions are unbounded. This is the familiar phenomenon of resonance.

In [**144**], Lazer and Leach consider a nonlinear equation

$$y'' + n^2 y + g(y) = p(t), \tag{14.5}$$

again with $p$ of period $2\pi$. It is assumed that $g$ is continuous and satisfies the conditions (14.3) and (14.4).

The case $n = 1$ is illustrative. The interest is in determining the existence of $2\pi$-periodic solutions. If $y$ is such a solution then we can write the $n = 1$ case as

$$y'' + y = p(t) - g(y), \tag{14.6}$$

and our previous argument shows that

$$\int_0^{2\pi} (p(t) - g(y(t))) \sin t\, dt = \int_0^{2\pi} (p(t) - g(y(t))) \cos t\, dt = 0. \tag{14.7}$$

Further, for each $\alpha_0$, $v(t) = y(t - \alpha_0)$ is a $2\pi$-periodic solution of

$$v'' + v + g(v) = p(t - \alpha_0),$$

and in this case,

$$\int_0^{2\pi} (p(t - \alpha_0) - g(v(t))) \sin t\, dt = 0. \tag{14.8}$$

Using the change of variables $s = t - \alpha_0$ and the periodicity of $p$ and of the sine function, we get

$$
\begin{aligned}
\int_0^{2\pi} p(t-\alpha_0)\sin t\, dt &= \int_{-\alpha_0}^{2\pi-\alpha_0} p(s)\sin(s+\alpha_0)\ ds \\
&= \int_0^{2\pi} p(s)\sin(s+\alpha_0)\,ds = A\cos\alpha_0 + B\sin\alpha_0,
\end{aligned}
\tag{14.9}
$$

where

$$
\begin{aligned}
A &= \textstyle\int_0^{2\pi} p(s)\sin s\ ds, \\
B &= \textstyle\int_0^{2\pi} p(s)\cos s\ ds.
\end{aligned}
\tag{14.10}
$$

Suppose that $A^2 + B^2 > 0$. If we choose $\alpha_0$ so that

$$
\cos\alpha_0 = \frac{A}{\sqrt{A^2+B^2}}, \qquad \sin\alpha_0 = \frac{B}{\sqrt{A^2+B^2}},
\tag{14.11}
$$

then

$$
\int_0^{2\pi} p(t-\alpha_0)\sin t\ dt = \sqrt{A^2+B^2}.
$$

Therefore, from (14.8),

$$
\int_0^{2\pi} g(v(t))\sin t\ dt = \sqrt{A^2+B^2}.
$$

From (14.3) and (14.4) it is seen that

$$
\begin{aligned}
\int_0^{2\pi} g(v(t))\sin t\ dt &= \int_0^{\pi} g(v(t))\sin t\ dt + \int_{\pi}^{2\pi} g(v(t))\sin t\ dt \\
&< 2g(\infty) - 2g(-\infty).
\end{aligned}
\tag{14.12}
$$

A similar argument for general integers $n$ proves the following result.

**Theorem 14.1.** *If, in (14.5), $g$ is continuous and satisfies (14.3) and (14.4), $p$ is continuous and $2\pi$-periodic, and for some integer $n \geq 1$ there is a solution with period $2\pi$, then*

$$
\sqrt{A^2+B^2} < 2(g(\infty) - g(-\infty)),
\tag{14.13}
$$

*where*

$$
A = \int_0^{2\pi} p(s)\sin ns\ ds, \qquad B = \int_0^{2\pi} p(s)\cos ns\ ds.
\tag{14.14}
$$

(Note that (14.13) holds trivially if $A = B = 0$.) Lazer and Leach were able to show much more however:

**Theorem 14.2.** *Assume the same conditions on $g$ and $p$. If $n$ is a positive integer and $A$ and $B$ are defined by* (14.14), *then the condition* (14.13) *is both necessary and sufficient for the existence of a $2\pi$-periodic solution to* (14.5). *This is true even with $\leq$ in place of $<$ in* (14.4), *so long as $g(\infty) > g(-\infty)$.*

We will give two proofs. The first is the original one of Lazer and Leach, which is a nice application of the Schauder fixed point theorem. For simplicity we give some details of this proof only for the case where $g$ is monotone. The second proof below appears to be quite new [**92**]. It is a simple example of a winding number argument, familiar from the field of complex analysis. While this may be called an application of degree theory, which was also used for this problem in [**73**], it is an elementary example of this technique, with no prior background needed beyond elementary analysis and ode theory.

### 14.2.1. Proof of Theorem 14.2 using the Schauder fixed point theorem.

**Proof.** Necessity has been proved. The proof of sufficiency by Lazer and Leach is an application of the so-called Lyapunov-Schmidt procedure for solving nonlinear equations. This method involves a decomposition of the space of periodic functions into the subspace of constant functions and the subspace of functions with mean zero and appears in many books, e.g. [**216**], or see [**226**]. The application given in this section does not require any background beyond Schauder's fixed point theorem (Chapter 5).

To apply Schauder's theorem we first need to define a specific normed linear space. We will again discuss only the case $n = 1$ and begin by writing (14.5) as a first order system,

$$\begin{aligned} x_1' &= x_2, \\ x_2' &= -x_1 - g(x_1) + p(t). \end{aligned}$$

Make the transformation

$$\begin{aligned} x_1 &= z_1 \cos t + z_2 \sin t, \\ x_2 &= -z_1 \sin t + z_2 \cos t. \end{aligned}$$

This substitution gives the transformed system

$$\begin{aligned} z_1' &= (g(z) - p(t)) \sin t, \\ z_2' &= (p(t) - g(z)) \cos t, \end{aligned} \tag{14.15}$$

where $g(z)$ denotes $g(z_1 \cos t + z_2 \sin t)$. The relevance of this system to finding periodic solutions is obvious from (14.7).

We will denote a solution to (14.15) by $(\phi, \omega)$. In terms of these new variables, (14.7) becomes

$$\begin{aligned} &\textstyle\int_0^{2\pi} (p(s) - g(\phi(s)\cos s + \omega(s)\sin s))\sin s\; ds = 0, \\ &\textstyle\int_0^{2\pi} (p(s) - g(\phi(s)\cos s + \omega(s)\sin s))\cos s\; ds = 0. \end{aligned} \tag{14.16}$$

One could try a shooting approach starting from (14.15), but this would be somewhat complicated. Lazer and Leach start by turning (14.15) into a system of integral equations in the usual way:

$$\begin{aligned} \phi(t) &= a + \textstyle\int_0^t (g(\phi(s)\cos s + \omega(s)\sin s) - p(s))\sin s\; ds, \\ \omega(t) &= b + \textstyle\int_0^t (p(s) - g(\phi(s)\cos s + \omega(s)\sin s))\cos s\; ds. \end{aligned} \tag{14.17}$$

They wish to ensure that the solutions under consideration satisfy (14.16). Observe that (14.16) implies that the averages of the integrands in the two integrals in (14.17) are both zero. Hence, they subtract these averages when defining their function space mapping, as we will now explain in more detail.

First we define the Banach space to be used. Let

$$V = \left\{(\theta, \xi) \in C\left(R^1, R^2\right) \times R^2 \mid \theta(t) = \theta(t + 2\pi) \text{ for } t \in R^1\right\}.$$

Think of $\theta$ as the pair $(\phi, \omega)$ and $\xi$ as the initial condition $\theta(0) = (a, b)$. Let

$$\begin{aligned} ||\theta|| &= \max_t \sqrt{\phi(t)^2 + \omega(t)^2}, \\ |\xi| &= \sqrt{a^2 + b^2} \end{aligned}$$

and set

$$||(\theta, \xi)|| = ||\theta|| + |\xi|.$$

Also, define addition and scalar multiplication componentwise:

$$c_1(\theta_1, \xi_1) + c_2(\theta_2, \xi_2) = (c_1\theta_1 + c_2\theta_2, c_1\xi_1 + c_2\xi_2),$$

where we are using the standard addition and scalar multiplication in $R^2$ and $C\left(R^1, R^2\right)$. With these definitions, $V$ is a Banach space.

We must now define a mapping $F : V \to V$. This mapping is designed so that if $((\phi, \omega), (a, b))$ is a fixed point of $F$ then $(\phi, \omega)$ satisfies (14.16). To define $F$, Lazer and Leach consider an element $((\phi, \omega), (a, b))$ of $V$ and let

$$\begin{aligned} M(\phi, \omega) &= \frac{1}{2\pi}\int_0^{2\pi} (g(\phi(s)\cos s + \omega(s)\sin s) - p(s))\sin s\; ds, \\ N(\phi, \omega) &= \frac{1}{2\pi}\int_0^{2\pi} (p(s) - g(\phi(s)\cos s + \omega(s)\sin s))\cos s\; ds. \end{aligned}$$

They set

(14.18)
$$\phi^*(t) = a + \int_0^t \left( \left( g\left(\phi(s)\cos s + \omega(s)\sin s\right) - p(s) \right) \sin s - M(\phi,\omega) \right) \, ds,$$
$$\omega^*(t) = b + \int_0^t \left( \left( p(s) - g\left(\phi(s)\cos s + \omega(s)\sin s\right) \right) \cos s - N(\phi,\omega) \right) \, ds,$$

and then let

$$\begin{aligned} a^* &= a + N(\phi^*,\omega^*), \\ b^* &= b - M(\phi^*,\omega^*). \end{aligned} \tag{14.19}$$

This choice for $(a^*, b^*)$ will be seen to be just what is needed below. The map $F$ is given by

$$F\left(\left(\left(\phi, w\right), \left(a, b\right)\right)\right) = \left(\left(\phi^*, \omega^*\right), \left(a^*, b^*\right)\right).$$

Observe that if $a^* = a$ and $b^* = b$ then $N(\phi^*,\omega^*) = M(\phi^*,\omega^*) = 0$. If, in addition, $\phi^* = \phi$ and $\omega^* = \omega$, then the definitions of $M$ and $N$, and (14.18), imply that $(\phi^*,\omega^*)$ is a periodic solution of (14.15).

To show that $F$ has a fixed point using Schauder's theorem, we look for a closed, bounded, and convex set $K \subset V$ such that $F(K)$ is a relatively compact set in $V$ and $F(K) \subset K$. This set will be of the form

$$\{(\theta,\xi) \mid ||\theta|| \le r_1, \ |\xi| \le r_2\},$$

for some $r_1$ and $r_2$. Such a set is obviously closed, bounded, and convex.

To choose $r_1$ and $r_2$ so that $F(K) \subset K$, we consider some $(\theta,\xi) = ((\phi, w), (a, b)) \in V$. From the definition of $(a^*, b^*)$ in (14.18) and (14.19),

(14.20)
$$|\xi^*|^2 = |\xi|^2 + 2\left(aN(\phi^*,\omega^*) - bM(\phi^*,\omega^*)\right) + N(\phi^*,\omega^*)^2 + M(\phi^*,\omega^*)^2.$$

We wish to find an $r_2 > 0$ such that if $|\xi| \le r_2$ then $|\xi^*| \le r_2$.

Let

$$H = \sup_{y,t} \left( |g(y)| + |p(t)| \right),$$

which is independent of $(\theta,\xi)$. Then by the definitions of $N$ and $M$,

$$N(\phi^*,\omega^*)^2 + M(\phi^*,\omega^*)^2 \le 2H^2. \tag{14.21}$$

The proof now hinges on showing that the second term on the right side of (14.20) can be made sufficiently negative to offset the last two terms if $|\xi|$ is large. Showing this turns out to involve steps much like those used in the proof of Theorem 14.1.

By the definitions of $N$ and $M$,

(14.22)
$$aN(\phi^*,\omega^*) - bM(\phi^*,\omega^*)$$
$$= \frac{1}{2\pi} \int_0^{2\pi} \left( p(s) - g\left(\phi^*(s)\cos s + \omega^*(s)\sin s\right) \right) \left( a\cos s + b\sin s \right) \, ds.$$

Further, $\phi^*(s) = a + \alpha(s)$ and $\omega^*(s) = b + \beta(s)$, where $\alpha$ and $\beta$ are continuous and $2\pi$-periodic. Hence the argument of $g$ in the integrand above can be written as

$$a\cos s + b\sin s + \gamma(s),$$

where $\gamma$ is continuous and $2\pi$-periodic.

Letting $s_0 = \arctan\left(\frac{a}{b}\right)$, we can write

$$a\cos s + b\sin s = |\xi|\sin(s + s_0).$$

Substituting this into (14.22) gives

$$\begin{aligned} &aN(\phi^*, \omega^*) - bM(\phi^*, \omega^*) \\ &\qquad = \frac{1}{2\pi}\int_0^{2\pi} \left(p(s) - g(|\xi|\sin(s+s_0) + \gamma(s))\right)\left(|\xi|\sin(s+s_0)\right)\, ds. \end{aligned}$$

Letting $\sigma = s + s_0$ and using the periodicity of $p$, $\gamma$, and sine gives

(14.23)

$$\begin{aligned} &aN(\phi^*, \omega^*) - bM(\phi^*, \omega^*) \\ &= \frac{1}{2\pi}\int_{s_0}^{2\pi+s_0} \left(p(\sigma - s_0) - g(|\xi|\sin\sigma + \gamma(\sigma - s_0))\right)\left(|\xi|\sin\sigma\right)\, d\sigma \\ &= \frac{1}{2\pi}|\xi|\left(\int_0^{2\pi} p(\sigma - s_0)\sin\sigma\, d\sigma - \int_0^{2\pi} g(|\xi|\sin\sigma + \gamma(\sigma - s_0))\sin\sigma\, d\sigma\right). \end{aligned}$$

Again using periodicity and the definitions of $A$ and $B$,

$$\begin{aligned} \int_0^{2\pi} p(\sigma - s_0)\sin\sigma\, d\sigma &= \int_0^{2\pi} p(\sigma)\sin(\sigma + s_0)\, d\sigma \\ &= A\cos s_0 + B\sin s_0 \le \sqrt{A^2 + B^2}. \end{aligned} \tag{14.24}$$

Write the second integral in the last line of (14.23) as

$$\begin{aligned} -\frac{1}{2\pi}|\xi|\int_0^{\pi} g(|\xi|\sin\sigma + \gamma(\sigma - s_0))\sin\sigma\, d\sigma& \\ -\frac{1}{2\pi}|\xi|\int_{\pi}^{2\pi} g(|\xi|\sin\sigma + \gamma(\sigma - s_0))\sin\sigma\, d\sigma&. \end{aligned}$$

Recall our assumption (made for simplification) that $g$ is increasing. For $\delta \in \left(0, \frac{\pi}{2}\right)$ and $\sigma \in [\delta, \pi - \delta]$,

$$g(|\xi|\sin\sigma + \gamma(\sigma - s_0)) \ge g(|\xi|\sin\delta - \gamma_1), \tag{14.25}$$

where $\gamma_1 = \max_{0\le t\le 2\pi}(|\gamma(t)|)$, while for the same range of $\delta$ and for $\sigma \in [\pi + \delta, 2\pi - \delta]$,

$$g(|\xi|\sin\sigma + \gamma(\sigma - s_0)) \le g(-|\xi|\sin\delta + \gamma_1). \tag{14.26}$$

Hence by choosing $\delta$ sufficiently small, and then $|\xi|$ sufficiently large, both $\int_0^\pi g(|\xi|\sin\sigma+\gamma(\sigma-s_0))\sin\sigma\,d\sigma$ and $\int_\pi^{2\pi} g(|\xi|\sin\sigma+\gamma(\sigma-s_0))\sin\sigma\,d\sigma$ can be made as close as desired to, respectively, $2g(\infty)$ and $-2g(-\infty)$.[2]

Our hypothesis is that

$$2(g(\infty)-g(-\infty)) > \sqrt{A^2+B^2}.$$

From (14.23), (14.24), (14.25), and (14.26) it now follows that there are $\mu>0$ and $m_1>0$ such that if $|\xi|\geq m_1$ then

$$2(aN(\phi^*,\omega^*)-bM(\phi^*,\omega^*))\leq-\mu|\xi|=-\mu\sqrt{a^2+b^2}. \tag{14.27}$$

This is the key inequality in the proof of Theorem 14.2.

From (14.20), (14.21), (14.23), and (14.27) we obtain that if $|\xi|\geq m_1$ then

$$|\xi^*|^2\leq|\xi|^2-\mu|\xi|+2H^2$$

and so, choosing $m_1$ larger if necessary, $|\xi^*|\leq|\xi|$.

There is also an $m_2$ such that if $|\xi|\leq m_1$ then $|\xi^*|\leq m_2$. Let $r_2=m_1+m_2$. Then the inequalities above prove that if $|\xi|\leq r_2$ then $|\xi^*|\leq r_2$. Referring now to (14.18), we see that since $g$ and $p$ are bounded, there is an $r_1$ such that if $|\xi|\leq r_2$ then $|(\phi^*,\omega^*)|\leq r_1$. Combining these two results, we let

$$K=\{(\theta,\xi)\mid \|\theta\|\leq r_1 \text{ and } |\xi|\leq r_2\},$$

and conclude that $F:K\to K$, as desired.

The final property needed to apply the Schauder fixed point theorem to $F$ is that $F(K)$ is relatively compact in $V$. For this we consider a sequence $\{(\theta_i,\xi_i)\}\subset F(K)$, and we wish to show that some subsequence converges to an element of $K$. There is no problem with convergence of a subsequence of $\{\xi_i\}$ in $R^2$, since in $K$, $|\xi|\leq r_2$. Also, the $||\theta_i||$ are uniformly bounded by $r_1$. We wish to apply the Ascoli-Arzela theorem and therefore need to show that the $\theta_i$ are equicontinuous. But each $\theta_i$ is actually a $\theta^*$, defined by (14.18) for some $(\phi,\omega)\in K$. Hence,

$$\phi_i'=(g(\phi\cos s+\omega\cos s)-p(s))\sin s-M(\phi,\omega)$$

for some $(\phi,\omega)\subset K$, and the boundedness of $g$ and $p$ gives a uniform bound (i.e. independent of $i$) for $\phi_i'$. Similarly, $\omega_i'$ has a uniform bound, and so the sequence $\{\theta_i\}$ is equicontinuous and uniformly bounded. This argument proves that some subsequence of $\{(\theta_i,\xi_i)\}$ has a limit in $K$ and completes the proof of Theorem 14.2 in the case where $g$ is increasing. $\square$

[2]For each $\varepsilon>0$ there are numbers $\tilde\delta>0$ and $\tilde M>0$ such that if $\tilde\delta\leq u\leq\pi-\tilde\delta$ and $|\xi|\geq\tilde M$, then

$$g\left(|\xi|\sin\tilde\delta-\gamma\right)\geq g(\infty)-\varepsilon.$$

**14.2.2. Proof using winding number.** As above, we give the details only for $n = 1$ and ask the reader to make the changes necessary for larger $n$. (See Exercise 14.1.) We do not assume that $g$ is monotonic. Begin by assuming that $g$ satisfies a local Lipschitz condition. More precisely, we assume

**Condition I:** *For each $y_0$ there is an $\varepsilon > 0$ and an $L > 0$ such that if $y_1$ and $y_2$ are in the interval $[y_0 - \varepsilon, y_0 + \varepsilon]$, then $|g(y_1) - g(y_2)| \le L|y_1 - y_2|$.*

This condition implies that solutions to initial value problems for (14.5) are unique and depend continuously on the initial conditions. We then have the following lemma.

**Lemma 14.3.** *Suppose that $g$ is not constant and satisfies* (14.3), (14.4) *with $\le$ instead of $<$, and Condition I. For given $r > 0$ and $\beta$, consider the unique solution of* (14.5) *with $n = 1$ such that*

$$y(0) = r\sin\beta, \qquad y'(0) = r\cos\beta.$$

*If $r$ is sufficiently large then for every $\beta$ this solution satisfies*

$$(y(2\pi) - y(0))\cos\beta - \left(y'(2\pi) - y'(0)\right)\sin\beta > 0. \tag{14.28}$$

We will prove this lemma below. To apply it, take $r$ large enough that (14.28) applies for every $\beta$. Thus, it applies for all initial conditions $(y(0), y'(0))$ on a circle $C_{r_0}$ of sufficiently large radius $r_0$. For every $\beta$, there are $R > 0$ and $\gamma$, both depending continuously on $\beta$, such that

$$y(2\pi) - y(0) = R\sin\gamma, \qquad y'(2\pi) - y'(0) = R\cos\gamma, \tag{14.29}$$

and from this and (14.28) we get

$$\sin(\gamma - \beta) > 0.$$

It follows that as $\beta$ increases from 0 to $2\pi$, $\gamma$ must also increase by $2\pi$, since the initial conditions at $\beta = 0$ and $\beta = 2\pi$ are the same.

Now assume that there is no periodic solution. In this case, as $r$ is decreased from $r_0$, $R$ remains positive and $\gamma$ continues to be well defined, by (14.29) and the requirement that $\gamma$ vary continuously with $(r, \beta)$, for all $\beta \in [0, 2\pi]$ and $r > 0$. As $r$ decreases, $\gamma$ must continue to increase by $2\pi$ as $\beta$ goes from 0 to $2\pi$. However, for small $r$, the total change in $v(r,\beta) = (y(2\pi) - y(0), y'(2\pi) - y'(0))$ as $\beta$ goes around the circle is small, because the solution depends continuously on its initial conditions. But also, as $r \to 0$, $R$ in (14.29) is bounded away from zero, because we are assuming that the solution with $r = 0$ is not periodic. Therefore $\gamma$ cannot increase continuously by a total amount of $2\pi$. This contradiction implies that under Condition I there must be a periodic solution. We will remove Condition I as a requirement after proving Lemma 14.3.

14.2.2.1. *Proof of Lemma* 14.3.

**Proof.** First, multiply (14.5) by $\sin(t+\beta)$ and integrate from 0 to $2\pi$. With two integrations by parts this gives

(14.30)
$$\left(y'(2\pi)-y'(0)\right)\sin\beta-\left(y(2\pi)-y(0)\right)\cos\beta+\int_0^{2\pi}g(y)\sin(t+\beta)\,dt$$
$$=A\cos\beta+B\sin\beta,$$

where $A$ and $B$ are given by (14.14). Also,

$$\int_0^{2\pi}g(y(t))\sin(t+\beta)\,dt=\int_\beta^{2\pi+\beta}g(y(s-\beta))\sin s\;ds. \tag{14.31}$$

The idea now is that for large $r$, $y(s-\beta)$ is large and positive for most of each interval $(2k\pi,(2k+1)\pi)$ where $\sin s>0$, and large and negative for most of each interval where $\sin s<0$. To make this precise, use the variation of parameters formula to give

(14.32)
$$y(s)=r\sin\beta\cos s+r\cos\beta\sin s+\textstyle\int_0^s\sin(s-\tau)(p(\tau)-g(y(\tau)))\,d\tau,$$
$$y(s-\beta)=$$
$$r(\sin\beta(\cos s\cos\beta+\sin s\sin\beta)+\cos\beta(\sin s\cos\beta-\sin\beta\cos s))+f(s)$$
$$=r\sin s+f(s),$$

where $f$ is bounded independently of $s$, $r$, and $\beta$. Hence $\lim_{r\to\infty}g(y(s-\beta))=g(\infty)$ uniformly on $\{s\mid\sin s\geq\delta\}$ and $\lim_{r\to\infty}g(y(s-\beta))=g(-\infty)$ uniformly on $\{s\mid\sin s\leq-\delta\}$, for every $\delta\in(0,1]$. It follows that

$$\lim_{r\to\infty}\int_0^{2\pi}g(y(t))\sin(t+\beta)\,dt=2(g(\infty)-g(-\infty)).$$

Then, (14.30) and (14.13) imply (14.28) for sufficiently large $r$, completing the proof of Lemma 14.3 and Theorem 14.2 when $g$ is Lipschitz continuous.

If Condition I does not hold and $g$ is only continuous, then we note that the bound on $f(s)$ in the proof of Lemma 14.3 depends only on the bounds on $p$ and $g$. Hence, $r=\bar{r}$ can be chosen so that the conclusion of Lemma 14.3 holds for every continuous function $g_1$ such that

$$g(-\infty)\leq g_1\leq g(\infty) \tag{14.33}$$

on $(-\infty,\infty)$ and $g_1(\pm\infty)=g(\pm\infty)$.

We can then find a sequence of functions $g_n$ which all satisfy (14.33) such that each $g_n$ satisfies a local Lipschitz condition, $g_n(\pm\infty)=g(\pm\infty)$ and $g_n$ converges to $g$ uniformly on $[-\bar{r},\bar{r}]$. Corresponding to each of these there is a periodic solution $y_n$ of (14.6) (with $g_n$ instead of $g$). These all take values in $[-\bar{r},\bar{r}]$, so that they are uniformly bounded, and they have bounded derivatives since $y_n''$ is uniformly bounded. Hence some subsequence of $\{y_n\}$ converges uniformly on $[0, 2\pi]$ to a $2\pi$-periodic function $y$, and by using an equivalent integral equation, this function is seen to be a solution of (14.6). □

## 14.3. The pde result of Landesman and Lazer

This section and Section 14.5 are our only forays into partial differential equations. As mentioned earlier, the result can be extended to general elliptic operators, but for simplicity we will consider only the equation

$$\Delta u + \alpha u + g(u) = h(x) \tag{14.34}$$

in a bounded domain $\Omega$ in $R^n$.[3] It is assumed that $g$ and $h$ are continuous in $R^1$ and in $\bar{\Omega}$, respectively, and that for every $s \in (-\infty, \infty)$,

$$\lim_{u\to\infty} g(u) > g(s) > \lim_{u\to-\infty} g(u). \tag{14.35}$$

To state the result of Landesman and Lazer for (14.34), suppose that $w$ is a nonzero weak solution of the homogeneous boundary value problem

$$\Delta u + \alpha u = 0, \quad u = 0 \text{ on } \partial\Omega, \tag{14.36}$$

and let

$$\Omega^+ = \{x \in \Omega \mid w(x) > 0\}, \qquad \Omega^- = \{x \in \Omega \mid w(x) < 0\}.$$

**Theorem 14.4.** *Suppose that all weak solutions of* (14.36) *are of the form* $cw$ *for some constant* $c$. *Assume also that* $g$ *is continuous and satisfies* (14.35) *and that* $h \in L^2(\Omega)$. *Then the problem* (14.34) *has a weak solution if and only if*

$$\begin{aligned} g(-\infty)\int_{\Omega^+} |w|\, dx - g(\infty)\int_{\Omega^-} |w|\, dx \\ < \int_\Omega hw\, dx < g(\infty)\int_{\Omega^+} |w|\, dx - g(-\infty)\int_{\Omega^-} |w|\, dx. \end{aligned} \tag{14.37}$$

**Remark 14.5.** A short proof was given by Hess [**98**] in 1974. The proof here is a little shorter, and more elementary. Unlike the ode case, the proofs we know require the strict inequalities in (14.35).

**Proof.** First suppose that (14.34) has a weak solution $u$. Then the linear theory tells us that

$$\int_\Omega (h - g)\, w\, dx = 0.$$

[3] The original paper, [**139**], has a good outline of the needed background in pde theory, such as the definition of weak solutions and of the function spaces $H_0(\Omega)$ and $H_0^1(\Omega)$ needed for this definition. (Similar definitions are also given in our Chapter 10.) While there are many such expositions, the one in [**139**] is particularly tailored to this problem. See also [**62**]. Throughout this section $\partial\Omega$ is assumed sufficiently regular for the basic theory to hold.

So using (14.35),

$$\int_{\Omega^+} g(-\infty)|w|\ dx - \int_{\Omega^-} g(\infty)|w|\ dx \le \int_{\Omega^+} gw\ dx + \int_{\Omega^-} gw\ dx$$
$$= \int_\Omega hw\ dx \le \int_{\Omega^+} g(\infty)|w|\ dx - \int_{\Omega^-} g(-\infty)|w|\ dx.$$

Since $w$ is a nontrivial weak solution of (14.36), at least one of the sets $\Omega^+$ and $\Omega^-$ must have positive measure, leading to the strict inequalities of (14.37).

Now assume that (14.36) and (14.37) hold for some nontrivial $w$. We will assume that $||w|| = 1$ in $L^2(\Omega)$. Given $v_1 \in w^\perp$ (the subspace of $L^2(\Omega)$ orthogonal to $w$) and a real number $c_1$, observe first that the function

$$q = h - g(v_1 + c_1 w) - w\int_\Omega [h - g(v_1 + c_1 w)]\, w\ dx$$

is in $w^\perp$. It follows as in Section 9.2.2 of [**62**] that we can define a mapping $T : w^\perp \times R^1 \to w^\perp \times R^1$ by letting $v_2$ be the unique solution to

$$\begin{aligned} \Delta v + \alpha v &= q \quad \text{in } \Omega, \\ v &= 0 \quad \text{on } \partial\Omega \end{aligned}$$

and setting

$$c_2 = c_1 + \int_\Omega [h - g(v_1 + c_1 w)]\, w\ dx.$$

Then let $T(v_1, c_1) = (v_2, c_2)$. The mapping $T$ is compact, and because $h \in L^2(\Omega)$ and $g$ is bounded, there is an $M$ independent of $v_1$ and $c_1$ such that $||v_2|| \le M$.

Further, (14.35) implies that

$$\lim_{c_1 \to \infty} \int_\Omega g(v_1 + c_1 w)\, w\ dx = g(\infty)\int_{\Omega_+} |w|\ dx - g(-\infty)\int_{\Omega_-} |w|\ dx,$$
$$\lim_{c_1 \to -\infty} \int_\Omega g(v_1 + c_1 w)\, w\ dx = g(-\infty)\int_{\Omega_+} |w|\ dx - g(\infty)\int_{\Omega_-} |w|\ dx.$$

As a result, (14.37) implies that if $c_1$ is sufficiently large and positive then $c_2 < c_1$, while if $c_1$ is sufficiently large and negative then $c_2 > c_1$. In addition, $g$ is bounded, so that there is a $\Lambda > 0$ such that if $c_1 \in [-\Lambda, \Lambda]$ then $c_2 \in [-\Lambda, \Lambda]$. The mapping $T$ is continuous and maps the closed convex set $w^\perp \times [-\Lambda, \Lambda]$ into a subset of itself with compact closure. By Schauder's fixed point theorem, $T$ has a fixed point.

If $c_2 = c_1$, then

$$\int_\Omega [h - g(v_1 + c_1 w_1)]\, w\ dx = 0,$$

and so $v_1 = v_2$ implies that

$$\Delta v_1 + \alpha v_1 = h - g\,(v_1 + c_1 w_1).$$

From the definition of $w$ we obtain

$$(\Delta + \alpha)\,(v_1 + c_1 w) = h - g\,(v_1 + c_1 w),$$

which completes the proof of Theorem 14.4. □

## 14.4. Second Lazer-Leach problem

A second problem considered by Lazer and Leach around 1970, [**145**], has also attracted attention in the intervening years and was also generalized by Landesman and Lazer to the pde case [**140**]. This is a boundary value problem of the form

$$\begin{gathered} y'' + g\,(t, y, y')\,y = f\,(t, y, y')\,, \\ y\,(0) = y\,(\pi) = 0, \end{gathered} \tag{14.38}$$

where $f$ and $g$ are continuous and satisfy the following two conditions:

(i) $f$ is bounded on $[0, \pi] \times R^1 \times R^1$.

(ii) There is a nonnegative integer $n$ and a positive number $d$ such that

$$n^2 + d \leq g\left(t, y, y'\right) \leq (n+1)^2 - d \tag{14.39}$$

for $0 \leq t \leq \pi$ and all $y, y'$.

**Theorem 14.6.** *Under the hypotheses* (i) *and* (ii) *above, the problem* (14.38) *has at least one solution.*

**Proof.** As noted above, Lazer and Leach gave a proof based on the Schauder fixed point theorem. Here we show that a shooting technique also works for this problem, giving a relatively short proof. Therefore we consider the initial value problem

$$\begin{gathered} y'' + g\,(t, y, y')\,y = f\,(t, y, y'), \\ y\,(0) = 0, \quad y'\,(0) = \alpha \end{gathered} \tag{14.40}$$

and show that there is an $\alpha^*$ such that the solution $y_{\alpha^*}$ to (14.40) satisfies (14.38).

For every $\alpha$,

$$y_\alpha = \alpha y_1 + y_p, \tag{14.41}$$

where $y_1$ satisfies

$$\begin{gathered} y'' + g\,(t, y_\alpha, y_\alpha')\,y = 0, \\ y\,(0) = 0, \quad y'\,(0) = 1 \end{gathered} \tag{14.42}$$

and $y_p$ satisfies

$$\begin{gathered} y''+g\left(t,y_\alpha,y_\alpha'\right)y=f\left(t,y_\alpha,y_\alpha'\right),\\ y\left(0\right)=0,\quad y'\left(0\right)=0. \end{gathered}$$

We will need to consider some solutions for $t>\pi$, and so we extend $f$ and $g$ periodically in $t$. Both $y_1$ and $y_p$ depend on $\alpha$. However, because $f$ and $g$ are bounded, there is an $M>0$ independent of $\alpha$ such that $\left|y_p\left(t\right)\right|\leq M$ and $\left|y_p'\left(t\right)\right|\leq M$, for $0\leq t\leq T=\max\left\{2\pi,2\frac{\pi}{d+n^2}\right\}$, where $d$ appears in (14.39).[4]

On the other hand, the condition (14.39) on $g$ implies, by the Sturm comparison theorem, that $y_1$ has exactly $n$ zeros in $(0,\pi)$ and $y_1\left(\pi\right)\neq 0$. We further have the following lemma:

**Lemma 14.7.** *There is a $\mu>0$ such that for all $\alpha$, $\left|y_1\left(\pi\right)\right|>\mu$.*

**Proof.** The result is easy to prove if $n=0$. If $n>0$ let

$$\begin{aligned} P&=\frac{1}{2}\left(y_1'^2+n^2y_1^2\right),\\ Q&=\frac{1}{2}\left(y_1'^2+\left(n+1\right)^2y_1^2\right). \end{aligned}$$

Then

$$\begin{aligned} P'&=y_1'\left(y_1''+gy_1+\left(n^2-g\right)y_1\right),\\ Q'&=y_1'\left(y_1''+gy_1+\left(\left(n+1\right)^2-g\right)y_1\right). \end{aligned}$$

Hence, from (14.39) and (14.42),

$$\begin{gathered} P'\left(t\right)>0\quad\text{if } y_1\left(t\right)y_1'\left(t\right)<0,\\ Q'\left(t\right)>0\quad\text{if } y_1\left(t\right)y_1'\left(t\right)>0. \end{gathered}$$

Suppose that $y_1$ has successive zeros at $t_1,\ldots,t_{n+1}$, where $0<t_1<\cdots<t_n<t_{n+1}\leq T$ and $t_n<\pi<t_{n+1}$. Let the intervening zeros of $y_1'$ in $(0,\pi)$ be at $\tau_1,\ldots,\tau_n$ with $0<\tau_1<t_1<\tau_2<\cdots<\tau_n<t_n$, and possibly at $\tau_{n+1}\in\left(t_n,\pi\right)$. The Sturm oscillation theorem implies that

$$t_n\leq\frac{n\pi}{\sqrt{n^2+d}}<\pi<\frac{\left(n+1\right)\pi}{\sqrt{\left(n+1\right)^2-d}}\leq t_{n+1}.$$

Hence there is a $\nu>0$, depending on $d$ but not on $\alpha$, such that

$$\begin{aligned} \pi-t_n&>\nu,\\ t_{n+1}-\pi&>\nu. \end{aligned}$$

[4]This allows for the case $n=0$.

Let

$$H(t) = \begin{cases} Q(t) & \text{if } Q' \geq 0, \\ P(t) & \text{otherwise.} \end{cases}$$

Then $H(0) = \frac{1}{2}$, and $H'(t) > 0$ in the intervals between adjacent $t_i$ and $\tau_j$. In particular, $H(\tau_1) > \frac{1}{2}$, or if $n = 0$ and $\tau_1$ is not defined, then $H(\pi) > \frac{1}{2}$.

Since $P(t_i) = Q(t_i)$ for each $i = 1, \ldots, n+1$, $H$ is continuous in $[0, t_{n+1}]$ except at the zeros $\tau_i$ of $y_1'$, where $H$ has a jump discontinuity. For example, if $n \geq 1$ then $H(\tau_1^-) = Q(\tau_1^-) > \frac{1}{2}$, while $H(\tau_1^+) = P(\tau_1^+) = \lambda H(\tau_1^-)$, where $\lambda = \frac{n^2}{(n+1)^2}$. Hence, $H > \frac{1}{2}\lambda$ for $\tau_1 < t < \tau_2$. Continuing in this way, we find that $\frac{1}{2}y_1'(t_n)^2 = H(t_n) > \frac{1}{2}\lambda^n$ and $\frac{1}{2}y_1'(t_{n+1})^2 > \frac{1}{2}\lambda^{n+1}$.

Since $y_1''$ is bounded independent of $\alpha$, and since the distance from $\pi$ to either $t_n$ or $t_{n+1}$ is bounded below, it follows that there is a $\mu > 0$ independent of $\alpha$ such that $|y_1(t)| = \mu$ for a value of $t \in [t_n, \pi]$ and another value of $t \in [\pi, t_{n+1}]$. Since $y_1 y_1'' < 0$ on $(t_n, t_{n+1})$, Lemma 14.7 follows. □

From here, the shooting proof is easy. The sign of $y_1(\pi)$ depends on $n$. But no matter what this sign is, Lemma 14.7 and (14.41), where $y_p$ is bounded independent of $\alpha$, show that the sign of $y_\alpha(\pi)$ changes as $\alpha$ varies from large and negative to large and positive, and so for some intermediate $\alpha$, $y_\alpha(\pi) = 0$. This proves Theorem 14.6. □

## 14.5. Second Landesman-Lazer problem

Again there is a pde version, and we give a short proof. In this case, the ode theorem is simply the $n = 1$ case of the pde theorem, and the pde proof covers both. However, it is perforce less elementary, requiring once again standard results from functional analysis. The statement and proof below are for the Laplacian in two dimensions but can be easily extended to strongly elliptic selfadjoint operators on $R^n$.

**Theorem 14.8.** *Suppose that $p(x, y, u_x, u_y)$ and $h(x, y, u_x, u_y)$ are continuous for $(x, y)$ in the closure of a bounded connected domain $\Omega$ and all $u_x, u_y$. Further, suppose that $\lambda_k$ and $\lambda_{k+1}$ are successive eigenvalues of*

$$\begin{aligned} \Delta u + \lambda u &= 0 \quad \text{in } \Omega, \\ u &= 0 \quad \text{on } \partial\Omega. \end{aligned}$$

*Assume that $h$ is bounded and for some $d > 0$, $\lambda_k + d \leq p \leq \lambda_{k+1} - d$ in $\Omega \times R^2$. Then the problem*

$$\begin{aligned} \Delta u + p(x, y, u_x, u_y)\, u &= h(x, y, u_x, u_y) \quad \text{in } \Omega, \\ u &= 0 \quad \text{in } \partial\Omega \end{aligned} \tag{14.43}$$

*has at least one solution.*

**Proof.** The proof depends on the following lemma, which is of interest in itself.

**Lemma 14.9.** *If $p$ and $h$ are dependent only on $x$ and $y$ then the problem has a unique solution in $H_0^1(\Omega)$.*[5]

**Proof.** Write the problem in the form

$$\Delta u + \frac{1}{2}(\lambda_k + \lambda_{k+1})\, u = \left\{\frac{1}{2}(\lambda_k + \lambda_{k+1}) - p\right\} u + h. \tag{14.44}$$

Since $\frac{1}{2}(\lambda_k + \lambda_{k+1})$ is not an eigenvalue, the operator

$$\left(\Delta + \frac{1}{2}(\lambda_k + \lambda_{k+1})\right)^{-1}$$

is a bounded compact selfadjoint linear operator on $L^2(\Omega)$, and for each $u \in L^2(\Omega)$,

$$\left|\left(\Delta + \frac{1}{2}(\lambda_k + \lambda_{k+1})\right)^{-1} u\right| \leq \frac{1}{\frac{1}{2}(\lambda_{k+1} - \lambda_k)}\, ||u||.$$

This follows from spectral theory for selfadjoint operators. See Chapters 5 and 6 of [**237**] and also Chapter 6 of [**62**].

The denominator above is the distance from $\frac{1}{2}(\lambda_k + \lambda_{k+1})$ to the nearest eigenvalue. Then (14.43) can be written as

$$u = \left(\Delta + \frac{1}{2}(\lambda_k + \lambda_{k+1})\right)^{-1} \left[\left\{\frac{1}{2}(\lambda_k + \lambda_{k+1}) - p\right\} u + h\right], \tag{14.45}$$

and since

$$\left|\frac{1}{2}(\lambda_k + \lambda_{k+1}) - p\right| \leq \frac{1}{2}(\lambda_{k+1} - \lambda_k) - d,$$

we see that (14.44) can be written in the form

$$u = Tu + \left\{\Delta + \frac{1}{2}(\lambda_k + \lambda_{k+1})\right\}^{-1} h,$$

where

$$||T|| \leq \frac{\frac{1}{2}(\lambda_{k+1} - \lambda_k) - d}{\frac{1}{2}(\lambda_{k+1} - \lambda_k)} < 1.$$

Hence (14.45) has a unique solution.

Further, this solution satisfies

$$||u|| \leq \frac{1}{d}\, ||h||, \tag{14.46}$$

and by integrating (14.43) by parts (the divergence theorem), we obtain also

[5] See Chapter 10.

that

$$||\nabla u|| \le K, \tag{14.47}$$

where $K$ depends on $d$ and the bounds on $p$ and $h$. □

To prove the theorem, we set up a mapping as follows. Given $v \in H_0^1(\Omega)$, define $u = Qv$ by

$$\Delta u + p\left(x, y, v_x, v_y\right) u = h\left(x, y, v_x, v_y\right),$$
$$u = 0 \quad \text{on } \Omega,$$

which can be solved uniquely using the lemma. By (14.46) and (14.47), the image set $S = \left\{Qv \ : v \in H_0^1(\Omega)\right\}$ is contained in a bounded subset of $H_0^1(\Omega)$. Standard regularity theory for pde's, as given for example in Section 6.3 of [**62**], shows that in addition, $u \in H^2(\Omega)$, with a bound of the form

$$||u||_{H^2(\Omega)} \le C\,||v||_{H_0^1(\Omega)}.$$

(See as well [**62**, Example 2, pg. 505].) This inequality implies that the image set $S$ has compact closure in $H_0^1(\Omega)$. To apply Schauder's fixed point theorem (Chapter 5), we have left to show that $Q$ is continuous. But if $v_1, v_2 \in H_0^1(\Omega)$ and $u_i = Qv_i$, then

$$\begin{aligned} &\Delta\left(u_2 - u_1\right) + p\left(x, y, v_{1_x}, v_{1_y}\right)\left(u_2 - u_1\right) \\ &\quad = h\left(x, y, v_{2x}, v_{2_y}\right) - h\left(x, y, v_{1_x}, v_{1_y}\right) \\ &\quad - u_2\left(p\left(x, y, v_{2_x}, v_{2_y}\right) - p\left(x, y, v_{1_x}, v_{1\,y}\right)\right). \end{aligned} \tag{14.48}$$

From the strong maximal principle, $\lambda_k > 0$, and so $p > d$. Multiplying by $(u_2 - u_1)$ and integrating over $\Omega$ gives

$$\begin{aligned} -\int_\Omega \left|\nabla\left(u_2 - u_1\right)\right|^2 &= -\int_\Omega p\left(x, y, v_{1_x}, v_{1_y}\right)\left(u_2 - u_1\right)^2 \\ &+ \int_\Omega \left[h\left(x, y, v_{2_x}, v_{2_y}\right) - h\left(x, y, v_{1_x}, v_{1_y}\right)\right]\left(u_2 - u_1\right) \\ &+ \int_\Omega u_2\left(u_2 - u_1\right)\left[p\left(x, y, v_{1_x}, v_{1_y}\right) - p\left(x, y, v_{2_x}, v_{2_y}\right)\right]. \end{aligned}$$

Applying Lemma 14.9 and (14.46) to (14.48) and using the continuity of $h$ and $p$ and the fact that $u_2 \in S$, we see that $\int \left(u_2 - u_1\right)^2$ tends to zero with $||v_2 - v_1||_{H_0^1(\Omega)}$. Since $p$ is bounded, we obtain

$$\begin{aligned} &\int_\Omega \left|\nabla\left(u_2 - u_1\right)\right|^2 \\ &\le K \int_\Omega \Big\{\left(u_2 - u_1\right)^2 + \left|\left[h\left(x, y, v_{2_x}, v_{2_y}\right) - h\left(x, y, v_{1_x}, v_{1_y}\right)\right]\left(u_2 - u_1\right)\right| \\ &\qquad + \left|u_2\left(u_2 - u_1\right)\left[p\left(x, y, v_{1_x}, v_{1_y}\right) - p\left(x, y, v_{2_x}, v_{2_y}\right)\right]\right|\Big\} \end{aligned}$$

for some $K$. Assuming that $||v_2 - v_1||_{H_0^1(\Omega)}$ is small, we conclude that

$$\left|v_{2_x} - v_{1_x}\right| + \left|v_{2_y} - v_{1_y}\right|$$

is small, say less than $\varepsilon$, except possibly in a subset $\Omega_\varepsilon$ of $\Omega$ of small measure, in which $h$ and $p$ are bounded. Again using the continuity of $h$ and $p$ and the boundedness of $S$, and the Cauchy-Schwarz inequality, we obtain that

$$\begin{aligned}\int_\Omega (u_2 - u_1)^2 + \int_\Omega \nabla (u_2 - u_1)^2 &\leq K\varepsilon + K \int_{\Omega_\varepsilon} |u_2 - u_1| \\ &\leq K\varepsilon + K \left|\Omega_\varepsilon\right|^{1/2} \left[\int_\Omega (u_2 - u_1)^2\right]^{1/2},\end{aligned}$$

where $K$ may change from before but is independent of $\varepsilon$. Since $u_2$ and $u_1$ lie in a bounded subset of $H_0^1(\Omega)$, the continuity of $Q$ follows, completing the proof of the theorem. □

## 14.6. A problem of Littlewood, and the Moser twist technique

Return again to the simple linear problem

$$u'' + u = p(t)$$

where $p$ is periodic of period $2\pi$. As stated earlier, there are two possibilities: either every solution is periodic or every solution is unbounded. Contrast this with the equation

$$u'' + u = u^3 + \cos t. \tag{14.49}$$

In Exercise 2.4 it is shown that this equation has both a periodic solution and unbounded solutions. A simple shooting technique may be used.[6]

We can write the last equation as

$$u'' + h(u) = p(t), \tag{14.50}$$

where $h(u) = u - u^3$. A key property of $h$ allowing shooting to work is that $\lim_{|u|\to\infty} h(u)\text{sign}(u) = -\infty$.

On the other hand, J. E. Littlewood considered (14.50) in the case where $\lim_{|u|\to\infty} h(u)\text{sign}(u) = +\infty$. If $p(t) = 0$ then a simple energy argument shows that all solutions are bounded, and Littlewood asked if this is true also when $p(t)$ is periodic. The case $h(u) = u$ shows that some further

[6]This is related to a result of Massera [**157**]: *Suppose that $h(t, u, u')$ is smooth in all variables and that $h(t + 2\pi, u, u') = h(t, u, u')$ for all $t, u, u'$ and that*

$$u'' = h(t, u, u')$$

*has a solution which is bounded on $[0, \infty)$. Then this equation has a $2\pi$-periodic solution.* See Exercise 14.2.

condition on either $p$ or $h$ is needed, and he discussed the case where $h$ is "superlinear", meaning that

$$\lim_{|u|\to\infty} \frac{h(u)}{u} = \infty. \tag{14.51}$$

(A problem studied earlier in this chapter, (14.5), is of the form (14.50), but in that case, $\frac{h(u)}{u}$ is bounded in $|u| \geq 1$.) The first positive result was by Morris, in [**174**], and Levi settled the general question in [**148**], showing that if $p$ is periodic and (14.51) holds then all solutions of (14.50) are bounded.

There have been a number of other similar results. As far as we know, all make use of an important technique, the "Moser twist" theorem, which is more sophisticated than the methods we emphasize in this book [**175**]. We will not attempt to give an exposition here but refer the reader, for example, to work of Levi, B. Liu, or Ortega, three of the main authors who have developed this method [**184**], [**185**], [**148**], [**149**]. Put briefly, in the context of (14.50)–(14.51), the Moser twist theorem allows one to construct an unbounded sequence of concentric annuli in the $(x, x')$ phase space which are positively invariant under the Poincaré map which takes an initial point $(x(0), x'(0))$ into the image point $(x(2\pi), x'(2\pi))$. This implies that all solutions are bounded.[7] Equations for which the twist theorem has been effective remain a challenge for our classical techniques.

## 14.7. Exercises

**Exercise 14.1.** (Routine) Complete the proof of Theorem 14.2 by extending the arguments above to all values of $n$.

**Exercise 14.2.** It was shown in Chapter 2 that (14.49) has a $2\pi$-periodic solution. Use shooting to prove that if $\cos t$ in this equation is replaced by any bounded continuous function, then there is still a solution which is bounded on $[0, \infty)$.

**Exercise 14.3.** Work out the pde proof of Theorem 14.8 in the $n = 1$ (ode) case, letting $\Omega = [0, \pi]$ and $k = 1$, by expressing the operator $\left(\Delta + \frac{1}{2}(\lambda_1 + \lambda_2)\right)^{-1}$ in terms of the Green's function. In this way, verify the estimates needed, ending with (14.46) and (14.47).

[7] See Chapter 15 for a discussion of Poincaré maps.

Chapter 15

# Chaotic motion of a pendulum

## 15.1. Introduction

The main results in this chapter are about the equation

$$x'' + \varepsilon\delta x' + (1 + \gamma \sin \varepsilon t) \sin x = 0. \tag{15.1}$$

This is sometimes called the equation of a "parametrically excited pendulum", with $x$ being the angle between the pendulum and the vertical. It has received considerable attention, for instance in work of Bishop and Clifford [**20**], [**40**]. The pendulum turns out to be a good approximation to many physical and electrical systems. For examples, see [**40**] for references to work on the behavior of offshore structures, crane barges, and the response of Josephson junctions. The particular forcing used in (15.1) has the property that the vertical upward or downward positions of the pendulum, $x = k\pi$, are constant solutions. This equation represents the motion of a pendulum which is subjected to a vertical up and down motion of its support. We will discuss both the undamped ($\delta = 0$) and damped cases, for small values of $\varepsilon$.[1] However before getting to this problem, we will give an introduction to some background material on dynamical systems and chaos, including the important method of Melnikov.

[1]In [**82**] there is a section on a "damped excited pendulum with oscillating support." But the phenomena treated there are completely different from those we consider. An example of the type of equation discussed in [**82**] is $u'' + cu' + \left(1 + \varepsilon \sin \frac{1}{\varepsilon} t\right) \sin u = F \cos \omega t$. There are two forcing terms, and the one representing the oscillation of the support is of small amplitude and high frequency. The method of averaging is used to show, for example, that this pendulum can show stable oscillations around the vertical upward position.

Concerning the term "chaotic" used in the chapter title, we make no attempt in this book to give a careful discussion of chaos and its associated phenomena, such as "strange attractors". For this we can refer the reader, for example, to the classic monograph of Guckenheimer and Holmes [**81**], or the less comprehensive but influential works of Devaney and of Yorke, e.g. [**51**], [**7**]. We use the word chaos informally. Our theorems will include precise statements about the behavior of models which may loosely be described as chaotic. We will make some further comments on the type of chaos which our results imply later in this chapter.

## 15.2. Dynamical systems

Among the few analytical tools for studying chaos, that deriving from work of Melnikov in [**171**] has probably received the most attention. What has come to be called Melnikov's method depends crucially on some concepts from dynamical systems, and so we will start with a brief introduction to this important topic. We will then illustrate the use of Melnikov's method with a standard example.

**15.2.1. Continuous and discrete dynamical systems.** Although previously we have referred several times to "dynamical systems methods", we have not used this term in a precise way. Since much of ode theory is now often considered as part of dynamical systems, the reader may be surprised that we have waited so long before introducing this subject. The mathematical concept of chaos can hardly be said to exist outside of dynamical systems theory, and so we must now bring this material into the discussion.

We start with one-dimensional ode's and initial value problems of the form

$$x' = f(x), \tag{15.2}$$

$$x(0) = x_0. \tag{15.3}$$

It is assumed for simplicity that $f$ and $f'$ are continuous on $R^1$, and so for each $x_0$ there is a unique solution of (15.2)–(15.3) which exists on some maximal interval of existence $(\alpha, \omega)$ containing 0. If we denote this solution by $\phi(t, x_0)$ then this uniqueness implies the "semigroup property"

$$\phi(t+\tau, x_0) = \phi(t, \phi(\tau, x_0)). \tag{15.4}$$

This can be taken as the defining property of a dynamical system, in much more general settings. We observe that it does not hold for nonautonomous initial value problems.

The behavior of a dynamical system of this type is simple, because for each $x_0$, $\phi(t, x_0)$ is a monotonic function of $t$ in its entire interval of existence. The flow can be studied on the "phase line" $x \in R^1$. There are no periodic solutions, and for each $x_0$, either $\lim_{t\to\omega}\phi(t,x_0)$ exists or $\phi(t,x_0)$ tends monotonically to $\infty$ or $-\infty$. If the limit, say $x_\infty$, exists, then $\omega=\infty$, and $x_\infty$ is an equilibrium point of (15.2).

Turning to the $n$-dimensional case, we assume that each component $f_i$ of $f: R^n \to R^n$ is continuously differentiable, so that solutions of

$$\begin{aligned} x' &= f(x), \\ x(0) &= x_0 \end{aligned} \tag{15.5}$$

are unique and vary continuously with $x_0$.[2] The continuous map $x_0 \to \phi(\cdot, x_0)$ satisfies (15.4) and defines a "continuous dynamical system" on $R^n$. The set of points $\Gamma(x_0) = \{(t, x(t)) \mid \alpha < t < \omega\}$ is called an "orbit", or "trajectory" of the dynamical system.

We can also define a "discrete dynamical system" on $R^n$. This can be considered simply as a 1:1 map $F: R^n \to R^n$. The "positive semi-orbit" of a point $x_0$ is $\Gamma^+(x_0) = \{F^j(x_0) \mid j = 1,2,3,\dots\}$, where $F^1 = F$, $F^{j+1}(x_0) = F(F^j(x_0))$.

The inverse map $F^{-1}$ is defined, and also all the iterations of this, $F^{-n}$. We then have $\Gamma(x_0) = \{F^j(x_0) \mid -\infty < j < \infty\}$. Note that the semigroup property becomes $F^{j+k}(x_0) = F^j(F^k(x_0))$.

**15.2.2. Poincaré maps.** The standard theory of chaos in ordinary differential equations proceeds by way of certain maps which are defined by solutions of the differential equation, called Poincaré maps. There are two notions of Poincaré map which are used. The first applies to autonomous systems. Readers familiar with the famous Poincaré-Bendixson theorem will probably have encountered this concept in the proof (Chapter 16 of [**41**] for example), where one considers (15.5) in $R^2$. For simplicity, assume that the partial derivatives $\frac{\partial f_i}{\partial x_j}$ are all continuous on $R^2$, so that solutions to (15.5) are unique and depend continuously on $x_0$. Further, define a "transversal" as a line segment $l$ such that for every $x \in l$, $f(x)$ is not 0 and not parallel to $l$.

We will now denote the solution to (15.5) by $x(t, x_0)$. Suppose that $x_0$ is a point in the interior of $l$, and assume that the solution $x(t,x_0)$ intersects $l$ at a subsequent point $x(t_0, x_0)$. Assume also that this is the first intersection of $x(t)$ and $l$ with $t > 0$ and that $x(t_0,x_0)$ is also in the interior of $l$. Then for $x_1 \in l$, and $|x_1 - x_0|$ sufficiently small, the solution $x(t, x_1)$ will also intersect $l$ again, at some first point $x(t_1, x_1)$ which is close to $x(t_0, x_0)$, and with $|t_1 - t_0|$ small. Of course, $t_1$ depends on $x_1$. This defines a map $P(x_1) = x(t_1, x_1)$, defined for all $x_1$ on $l$ which are sufficiently close to

[2] Again this is for simplicity. As shown in Section 2.4 of [**41**], it is sufficient to assume that $f$ is continuous and that solutions of initial value problem (15.5) are unique.

$x_0$. This is called a Poincaré map for this system. Such maps are easily defined for systems in $R^n$ with $n > 2$, where the transversal is a portion of an $(n-1)$-dimensional hyperplane which is transverse to the flow.

There is a second concept of Poincaré map, which is used for nonautonomous periodic systems of the form

$$x' = f(t, x) \tag{15.6}$$

where for some fixed $T > 0$, $f(t+T, x) = f(t, x)$ for all $(t, x) \in R^1 \times R^n$. Assume that solutions to initial value problems for (15.6) are unique, and consider the map $P(x_0) = x(T, x_0)$. Assume also that all solutions exist on $(-\infty, \infty)$. Then in contrast to the previous example, this map is defined for all $x_0 \in R^n$. Further, it is 1:1 and so invertible, and we can consider its iterates as discussed above. If (15.6) has a periodic solution $p(t)$ of period $T$, then $p(0)$ is a fixed point for the Poincaré map for (15.6).[3]

**15.2.3. Horseshoe maps.** One of the key tools in analyzing chaos in ode's is that of the Smale horseshoe map [**219**]. There are many presentations of this, e.g. [**81**], [**78**], and we will not give a full description here. We simply summarize some of the important properties of a map of this type. We base this summary mostly on [**81**].

Let $S$ be the unit square $0 \le x \le 1$, $0 \le y \le 1$ on $R^2$. The Smale horseshoe map is a smooth 1:1 map $Q : S \to R^2$ which has the two properties given below. The first is stated in terms of its image $Q(S)$ and the image of its inverse $Q^{-1}$. We discuss a specific example from a class of similar maps.

**Property I. (Horseshoe property)**

(I-a) $Q(S) \cap S = V_1 \cup V_2$, *where*

$$V_1 = \left\{ (x, y) \mid \frac{1}{8} \le x \le \frac{3}{8},\ 0 \le y \le 1 \right\},$$

$$V_2 = \left\{ (x, y) \mid \frac{5}{8} \le x \le \frac{7}{8},\ 0 \le y \le 1 \right\}.$$

(I-b) $Q^{-1}(V_i) = H_i$, *for* $i = 1, 2$, *where*

$$H_1 = \left\{ (x, y) \mid 0 \le x \le 1,\ \frac{1}{8} \le y \le \frac{3}{8} \right\},$$

$$H_2 = \left\{ (x, y) \mid 0 \le x \le 1, \frac{5}{8} \le y \le \frac{7}{8} \right\}.$$

[3]In dynamical systems, one often considers this case as "flow on a torus", with "$t$" adjoined as an additional independent variable. In that sense, this is not really a separate concept of Poincaré map.

**Property II. (Hyperbolicity)**

(II-a) *On the set* $H_1$, $Q$ *is an affine mapping* $\begin{pmatrix} x \\ y \end{pmatrix} \to A \begin{pmatrix} x \\ y \end{pmatrix} + \begin{pmatrix} b_1 \\ b_2 \end{pmatrix}$ *with constant Jacobian* $A = \begin{pmatrix} \lambda & 0 \\ 0 & \mu \end{pmatrix}$, *where* $0 < \lambda < 1$ *and* $\mu > 1$.

(II-b) *On* $H_2$, $Q$ *is also an affine mapping, now with Jacobian* $-A$.

Thus, on $H_1$, $Q$ expands vertically and contracts horizontally, so that the image of the horizontal strip $H_1$ is the vertical strip $V_1$. $Q$ expands similarly on $H_2$.[4] The minus signs have the effect of mapping the bottom side of the rectangle $H_2$ onto the top side of the rectangle $V_2$. This is illustrated by Figure 15.1, which is well known.

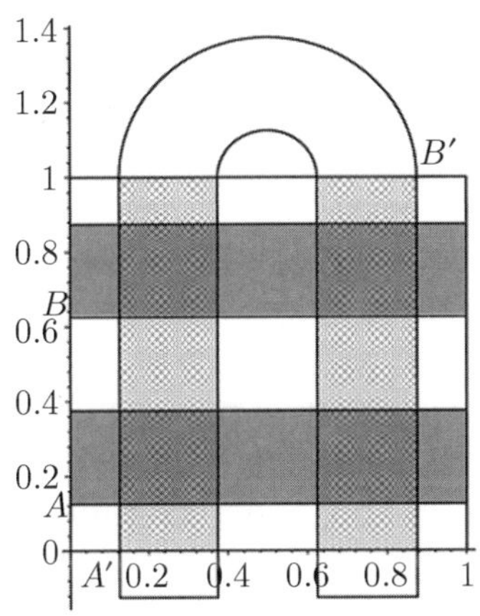


**Figure 15.1.** $Q$ maps the horizontal grey rectangles onto the vertical patterned rectangles, with $A \to A'$, $B \to B'$.

The consequences of these simple properties are surprising. We mention four which taken together describe a form of "chaos".

(1) There is an uncountable set $\Omega$ of points $p$ in $S$ such that $Q^j(p)$ is defined for $-\infty < j < \infty$. Since the domain of $Q$ is $S$, all of these iterates must be in $H_1 \cup H_2$ and in $V_1 \cup V_2$.

(2) For every positive integer $n$ there is a periodic point with least period $n$.[5]

---

[4]From this information we can calculate the restrictions of $Q$ to $H_1$ and $H_2$, and of $Q^{-1}$ to $V_1$ and $V_2$ exactly. For example, $\lambda = \frac{1}{4}$, $\mu = 4$. But these particular sets and numbers aren't important, and so we state the properties more generally. See Exercise 15.3 .

[5]$Q^n(p) = p$ and $Q^j(p) \neq p$ if $1 \leq j \leq n-1$.

(3) There are points $p$ in $S$ such that $Q^j(p) \in S$ for $-\infty < j < \infty$ and yet the trajectory is not periodic. In fact, there are $p$ such that $\{Q^j(p) \mid -\infty < j < \infty\}$ is dense in $\Omega$.

(4) The map exhibits "sensitive dependence on initial conditions". That is, a $\delta > 0$ exists such that for every point in $p \in \Omega$ and every neighborhood $N$ of $p$ in $\Omega$, there is a point $q \in N \cap \Omega$ and a positive integer $n$ such that $|Q^n(p) - Q^n(q)| > \delta$.[6]

Considerably more can be said. A method for studying horseshoe maps called "symbolic dynamics" yields much more detail about the relation between different orbits, but we do not have space to discuss this here [**81**]. Also, horseshoe maps are not limited to the specific setting above. In Section 15.3 we will outline how Melnikov's method can be used to show the existence of a Poincaré map equivalent to a horseshoe for certain second order ode's, thus yielding properties like (1)–(4) above for these differential equations. Later in this chapter we will begin a discussion of how shooting methods can establish similar results. However more precise analysis is required to give property (4), and this will not be addressed with classical methods until Chapter 16.

**15.2.4. Finding horseshoes more generally.** Consider a smooth 1:1 map $P$ from $R^2$ to $R^2$ with a fixed point at $(0,0)$. Suppose that the Jacobian matrix $DP(0,0)$ has no eigenvalue with absolute value equal to 1. Such a fixed point is called "hyperbolic". We wish to examine the behavior of the map $P$ in a neighborhood of $(0,0)$ more carefully.

Just as for ode's, there is a stable manifold theorem for maps (Chapter IX of [**83**]). Suppose that in the case being considered, $DP(0,0)$ has real eigenvalues $\lambda_1, \lambda_2$, with $|\lambda_1| < 1, |\lambda_2| > 1$. The stable manifold theorem for maps tells us that there are smooth curves $\mathcal{S}$ and $\mathcal{U}$ which are stable and unstable manifolds for $P$ at $(0,0)$. For example, define the stable manifold $\mathcal{S}$ as the set of $x_0 \in R^2$ such that $\lim_{j\to\infty} P^j(x_0) = (0,0)$. Clearly $\mathcal{S}$ is an invariant set for $P$. Similarly, if $x_0 \in \mathcal{U}$, then $P^j(x_0) \in \mathcal{U}$ for all $j$, and $\lim_{j\to\infty} P^{-j}(x_0) = (0,0)$.

Here is an example. Let

$$P(x,y) = \left(\frac{1}{2}x, -2y\right).$$

The eigenvalues of $DP(0,0)$ are $\frac{1}{2}$ and $-2$. It is easy to see that the stable manifold is the $x$-axis and the unstable manifold is the $y$-axis. Suppose that the initial point is $x_0 = (1,1)$, which lies on neither $\mathcal{S}$ nor $\mathcal{U}$. Considering

[6] Sensitive dependence on inital conditions has been identified as one of the defining properties of chaos. See [**51**].

forward and backward iterates, we see that

$$x_1 = \left(\frac{1}{2}, -2\right),\ x_2 = \left(\frac{1}{4}, 4\right), \ldots,$$
$$x_{-1} = \left(2, -\frac{1}{2}\right),\ x_{-2} = \left(4, \frac{1}{4}\right),$$

etc. Both forward and backward iterates are unbounded. Forward iterates approach $\mathcal{U}$, while backward iterates approach $\mathcal{S}$. Note also that forward or backward iterates jump back and forth across $\mathcal{S}$.

An interesting situation arises if, as may happen for a nonlinear map, $\mathcal{S}$ and $\mathcal{U}$ intersect at some point $x_0 \neq (0,0)$, and if, moreover, this intersection is nontangential.[7] Then every iterate $x_n = P^n(x_0)$, for $n$ positive or negative, lies on $\mathcal{S} \cap \mathcal{U}$. Also, $\lim_{n\to\infty} x_n = \lim_{n\to-\infty} x_n = (0,0)$. However, points on $\mathcal{S}$ between $x_n$ and $x_{n+1}$ are not necessarily on $\mathcal{U}$, and vice versa.

To clarify this, we can distinguish between the local stable manifold and the global stable manifold. $\mathcal{S}$ is a smooth curve through $(0,0)$. It has a smooth parametrization $t \to \gamma(t)$ with $\gamma(0) = (0,0)$. We consider a small $\mu$ and let $\mathcal{S}_\mu = \{\gamma(t), -\mu < t < \mu\}$. We call this a "local" stable manifold at 0. Similarly, there is a local unstable manifold $\mathcal{U}_\mu$. We can choose $\delta > 0$ such that if $\mathcal{S}_\mu$ and $\mathcal{U}_\mu$ are both contained in $\|x\| < \delta$, then $\mathcal{S}_\mu$ and $\mathcal{U}_\mu$ do not intersect except at 0. Also, if $\|x_0\| < \delta$ and $x_0 \notin \mathcal{S}_\mu$, then $\left\|P^j(x_0)\right\| > \delta$ for some $j > 0$. This is not to say that $x_0 \notin \mathcal{S}$, for the points $x_n$ may leave $\|x\| < \delta$ for small $n$, but return, and lie on $\mathcal{S}_\mu$ for large $n$.

Continuing to assume that $\mathcal{S}$ and $\mathcal{U}$ intersect nontangentially at $x_0$, we consider a point $y$ on $\mathcal{U}_\mu$ which is between $x_N$ and $x_{N+1}$ for a large positive $N$, so large that $x_n \in \mathcal{S}_\mu$ for $n \geq N$. This point $y$ will not necessarily lie on $\mathcal{S}$. It definitely does not lie on $\mathcal{S}_\mu$. Hence there is an $m \geq 0$ such that $\|P^m(y)\| > \delta$.

Now consider the arc $\mathcal{U}_{N,N+1}$ of $\mathcal{U}$ lying between $x_N$ and $x_{N+1}$. Since $x_n \in \mathcal{S}_\mu$ for each $n \geq N$, the points $x_n$ converge to $(0,0)$ as $n \to \infty$, but the segments $P^{n-N}(\mathcal{U}_{N,N+1})$ of $\mathcal{U}$ between $x_n$ and $x_{n+1}$ do not. As the endpoints of these segments get closer and closer to 0, their interiors are stretched along one of the two branches of $\mathcal{U}$ which are extensions of $\mathcal{U}_\mu$ outside the neighborhood $\|x\| < \delta$. In other words, because the map $P$ is continuous, the parts of $\mathcal{U}$ between intersections with $\mathcal{S}_\mu$ tend to follow, alternately, extensions of the two branches of $\mathcal{U}_\mu$ farther and farther out away from 0. This produces the Figure 15.2, called a "homoclinic tangle".

[7]The stable and unstable manifolds of an equilibrium point for an ode cannot intersect nontangentially at a nonequilibrium point. This is because these manifolds are composed of curves which are orbits of solutions, and, assuming that uniqueness holds for initial value problems, two orbits cannot intersect without being identical.

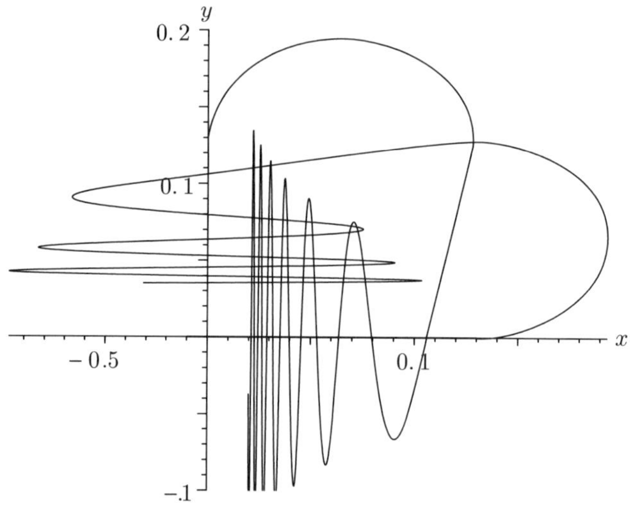


**Figure 15.2.** A "homoclinic tangle". The local stable and unstable manifolds are on the horizontal and vertical axes, respectively.

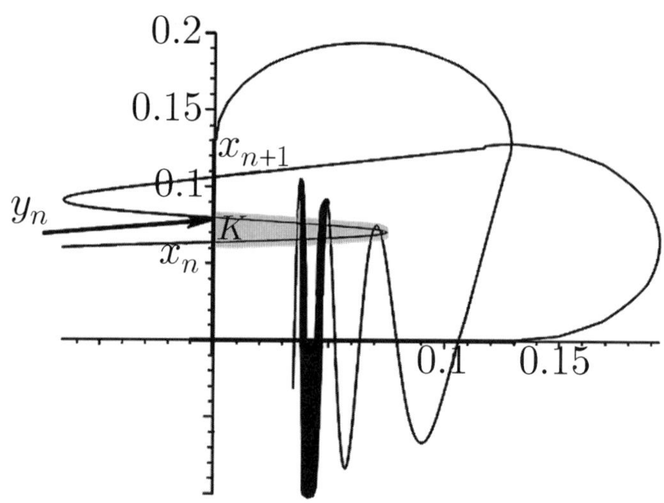


**Figure 15.3.** The mapping $P^4$ takes the grey area $K$ onto the black area.

Similarly, since $x_0 \in \mathcal{U}$, $x_n = P^n(x_0) \to 0$ as $n \to -\infty$ and $x_n \in \mathcal{U}_\mu$ for large negative $n$. However segments of $\mathcal{S}$ between intersections with $\mathcal{U}_\mu$ get stretched along extensions of $\mathcal{S}_\mu$ as $-n$ increases. We consider this particular situation in more detail.

The curve $\mathcal{S}$ may intersect $U_\mu$ at points other than $x_n$. In Figure 15.3 one possibility is shown, in which $\mathcal{S}$ intersects $U_\mu$ at one point, $y_n$, between

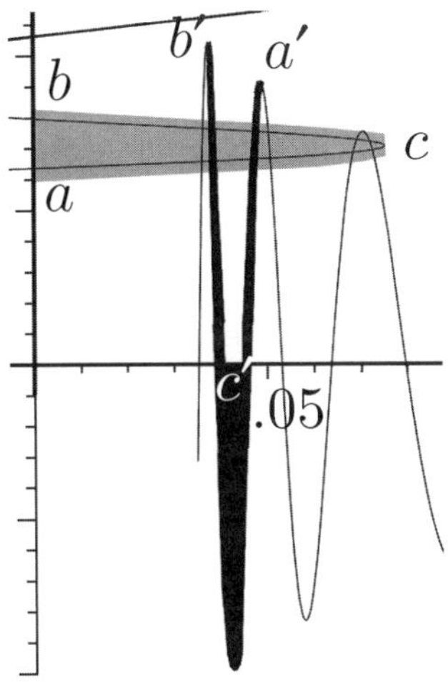


**Figure 15.4.** The points $a, b$, and $c$ are mapped onto $a'$, $b'$, and $c'$, respectively. Compare with Figure 15.1.

$x_{n+1}$ and $x_n$. In this case, let $K_0$ denote the region contained inside the closed curve formed by $\mathcal{U}_\mu$ and $\mathcal{S}$ between $x_n$ and $y_n$. One edge of $K_0$ lies on $\mathcal{U}_\mu$, which in this example is a subset of the vertical axis. Expand $K_0$ slightly to the grey polygonal set $K$ shown, which still has one edge along $\mathcal{U}_\mu$ and which touches $\mathcal{S}$ on an opposite edge. We are interested in the image of $K$ under the mapping $P$. This is illustrated in Figures 15.3 and 15.4, where the region $K$ is mapped by $P$, $P^2$, and so forth until for some $n$ ($n = 4$), $P^n(K) \cap K$ has two components as shown, and so $Q = P^n$ resembles a horseshoe map. The technical challenge is to prove hyperbolicity (Property II above) for such maps. See [**219**], [**187**], [**81**], [**259**].

## 15.3. Melnikov's method

Melnikov's method is one of the few analytic tools for detecting the transverse intersection of stable and unstable manifolds of a Poincaré map. It is often considered a technique of bifurcation theory, and a concise presentation of the theory, without proofs, is given in [**133**]. In this subsection we consider one of the best known examples of this technique. However, our derivation of the so-called "Melnikov function" is not standard, because instead of considering a distance function describing the effect of a small perturbation on stable and unstable manifolds of a Poincaré map geometrically, we base our discussion on energy. We do not give a proof of the Melnikov theorem using this approach but hope that it helps motivate this theorem, at least for the case of perturbed Hamiltonian systems, which constitute most of the known examples. In the following section we use energy to give a simple and complete elementary proof of a form of chaos for the forced pendulum model (15.1). The Melnikov function arises naturally in this proof.

As mentioned in the previous subsection, there are other possibilities in the scenario described above. For example, $\mathcal{S}$ may intersect $\mathcal{U}_\mu$ at two or more points between $x_n$ and $x_{n+1}$. But suppose that $P$ is a Poincaré map for some system of ode's. It can be of either of the two types discussed earlier, for autonomous systems or for time-periodic nonautonomous systems. It turns out that this places some restrictions on what can happen. In particular, the image $P(K)$ must preserve the orientation of $K$. In Figure 15.4 the three points $a, b, c$ are arranged clockwise around $K$. It can be shown that for a Poincaré map $P$, the image points $P(a), P(b), P(c)$ are also arranged clockwise around $P(K)$. Obviously the same is true for $P^n$.[8] This follows because if $q$ is a nonequilibrium point for the system, then $\det D(P)(q) > 0$, where $D(P)(q)$ is the Jacobian of $P$ at $q$. (See Exercise 15.2. There are further figures related to this on page 317 of [**259**] and pages 409–410 of [**191**].)

The result of this is that the situation depicted in Figure 15.3 is what happens. (We are considering $Q = P^4$.) The only essentially different alternative is that the image $Q(K)$ is on the other side of $U_\mu$ from $K$, but then it can be seen that the orientation of $K$ would be reversed. If it can be shown that $P$ has the structure described above, where the stable and unstable manifolds of a hyperbolic fixed point intersect nontangentially, then the solutions of the ode form a complicated structure. For example, if there are infinitely many distinct periodic orbits of $Q$, then there are infinitely many distinct periodic solutions of the ode. All this was first proved by Smale in [**219**].

### 15.3.1. A forced Duffing equation.

Consider the following form of forced Duffing equation:

$$x'' + \varepsilon\delta x' + x^3 - x = \varepsilon\gamma\cos t. \tag{15.7}$$

For $\varepsilon = 0$ the equation is $x'' + x^3 - x = 0$, which is easily analyzed using energy. Its standard system form,

$$\begin{aligned} x' &= y, \\ y' &= x - x^3, \end{aligned} \tag{15.8}$$

is found to have a saddle point at $(0,0)$ and two other equilibrium points, $(\pm 1, 0)$, which are centers. The natural energy function to consider is

$$E(x,y) = \frac{1}{2}y^2 + \frac{x^4}{4} - \frac{x^2}{2}. \tag{15.9}$$

[8] It is easier to see this in the figure by folowing the cyclic order $b', c', a'$.

The curves $E = 0$ give two homoclinic orbits, one with $x > 0$ and the other with $x < 0$.

Now consider small positive $\varepsilon$. ($\delta$ is fixed. We will see the reason for introducing it as a parameter below.) Our system is now

$$\begin{aligned} x' &= y, \\ y' &= x - x^3 + \varepsilon\left(\gamma \cos t - \delta y\right). \end{aligned} \tag{15.10}$$

For $\varepsilon > 0$ this is nonautonomous, and there are no constant solutions. Instead, we will consider Poincaré maps.

An important added feature here is that since (15.10) is nonautonomous, there is a family of Poincaré maps, parametrized by the initial value of $t$. That is, for each $T \in [0, 2\pi)$ we consider the Poincaré map $P_\varepsilon^T$ defined by

$$P_\varepsilon^T\left(x_0, y_0\right) = \left(x_\varepsilon\left(T + 2\pi, x_0, y_0, T\right), y_\varepsilon\left(T + 2\pi, x_0, y_0, T\right)\right), \tag{15.11}$$

where $\left(x_\varepsilon\left(t, x_0, y_0, T\right), y_\varepsilon\left(t, x_0, y_0, T\right)\right)$ is the unique solution of (15.10) with $x\left(T\right) = x_0, y\left(T\right) = y_0$. The formula (15.11) is used to extend the definition of $P_\varepsilon^T$ to include $\varepsilon = 0$. Note that $P_0^T$ does not depend on $T$. We look for fixed points of $P_\varepsilon^T$.

We can find such fixed points by finding fixed points of $P_\varepsilon^0$. If $(x_0, y_0)$ is a fixed point of $P_\varepsilon^0$, then $\left(x_\varepsilon\left(t, x_0, y_0, 0\right), y_\varepsilon\left(t, x_0, y_0, 0\right)\right)$ is a periodic solution of (15.10), and $\left(x_\varepsilon\left(T, x_0, y_0, 0\right), y_\varepsilon\left(T, x_0, y_0, 0\right)\right)$ is a fixed point for $P_\varepsilon^T$.

There is a fixed point of $P_0^0$ at $(0, 0)$. This is not unique, because $(1, 0)$ and $(-1, 0)$ are also fixed points, and the system (15.8) has many other periodic orbits, some of which may also have period $2\pi$. But we can try to apply the implicit function theorem to solve the equation

$$P_\varepsilon^0\left(x_0, y_0\right) = \left(x_0, y_0\right) \tag{15.12}$$

for $(x_0, y_0)$ as functions of $\varepsilon$ in some neighborhood of $(x_0, y_0, \varepsilon) = (0, 0, 0)$. This can be done if the Jacobian of $P_0^0 - I$ at $(0, 0)$ with respect to $(x_0, y_0)$ is nonsingular. This matrix is singular if $\lambda = 1$ is an eigenvalue of $\Phi\left(2\pi\right)$, where $\Phi$ is a fundamental matrix for the linearized system

$$\begin{aligned} u' &= v, \\ v' &= u. \end{aligned} \tag{15.13}$$

It is easily seen that $\Phi\left(2\pi\right) - I$ is nonsingular, and hence for $\varepsilon$ sufficiently small the equation (15.12) has a solution $\left(x_0\left(\varepsilon\right), y_0\left(\varepsilon\right)\right)$ which is unique within some small neighborhood of $(0, 0)$. Further, $\lim_{\varepsilon \to 0}\left(x_0\left(\varepsilon\right), y_0\left(\varepsilon\right)\right) = (0, 0)$.

Now denote a solution $\left(x_\varepsilon\left(t\right), y_\varepsilon\left(t\right)\right)$ of (15.10) by $\phi_\varepsilon\left(t\right)$. The solution such that $\phi\left(T\right) = \hat{\phi}$ will be denoted by $\phi_\varepsilon\left(\cdot, \hat{\phi}, T\right)$. If $\phi_\varepsilon\left(\cdot, \hat{\phi}, T\right)$ has period $2\pi$, then $\hat{\phi}$ is a fixed point for the Poincaré map $P_\varepsilon^T$.

Let $\psi_0(\varepsilon) = (x_0(\varepsilon), y_0(\varepsilon))$. This is a fixed point for $P_\varepsilon^0$. If $\varepsilon$ is small, then $\|\psi_0(\varepsilon)\|$ is small and the $2\pi$-periodic solution $\phi_\varepsilon(t, \psi_0(\varepsilon), 0)$ does not change much in the interval $[0, 2\pi]$. Therefore this solution is always close to $(0,0)$.

Consider solutions of (15.10) with initial conditions near $\psi_0(\varepsilon)$. For small $\varepsilon$ the structure is determined by the eigenvalues $\lambda_1$ and $\lambda_2$ of $DP_0^0(0,0)$, and it is seen from (15.13) that $-1 < \lambda_1 < 0$, $\lambda_2 > 1$. Hence, for small $\varepsilon$, $\psi_0(\varepsilon)$ is a saddle point for $P_\varepsilon^0$.

Keep in mind that for each $T$, $\phi_\varepsilon(T, \psi_0(\varepsilon), 0)$ is a fixed point for $P_\varepsilon^T$. There are then one-dimensional stable and unstable manifolds $\mathcal{S}_\varepsilon^T$ and $\mathcal{U}_\varepsilon^T$ for $P_\varepsilon^T$ at $\phi_\varepsilon(T, \psi_0(\varepsilon), 0)$. These vary continuously with $T$. In the next several paragraphs we discuss $T = 0$, but the same remarks apply to every $T$.

Suppose that $\psi_1$ is on the stable manifold of $P_\varepsilon^0$ at $\psi_0(\varepsilon)$. Then iterates $(P_\varepsilon^0)^n(\psi_1)$ approach $\psi_0(\varepsilon)$ as $n \to \infty$. This implies that the solution to the differential equation (15.10) with $\phi(0) = \psi_1$ tends to the periodic solution through $\psi_0(\varepsilon)$ as $t \to \infty$.[9]

In the previous subsection we discussed local stable and unstable manifolds for a map. For the map $P_\varepsilon^0$ we will denote these by $\mathcal{S}_{\varepsilon,l}^0$ and $\mathcal{U}_{\varepsilon,l}^0$. To define these uniquely, we can choose them for each small $\varepsilon \geq 0$ as maximal connected segments of $\mathcal{S}_\varepsilon^0$ and $\mathcal{U}_\varepsilon^0$ which contain $\psi_0(\varepsilon)$ and lie in a fixed disk $\|\phi\| < \delta$ that is independent of $\varepsilon$ and small enough to ensure that $\mathcal{U}_{\varepsilon,l}^0 \cap \mathcal{S}_{\varepsilon,l}^0 = \psi_0(\varepsilon)$. It can be shown that $\mathcal{S}_{\varepsilon,l}^0$ and $\mathcal{U}_{\varepsilon,l}^0$ depend continuously on $\varepsilon$. When $\varepsilon = 0$, the differential equation is autonomous, and $\psi_0(0) = (0,0)$. This point is a fixed point for the map $P_0^0$ and also an equilibrium point for (15.8). As an equilibrium point for (15.8) it has stable and unstable manifolds, and we have seen that these coincide, because they form two homoclinic orbits.

We claim that the (global) stable and unstable manifolds for $P_0^0$ at $(0,0)$ are the same as the stable and unstable manifolds for (15.8). To show this, consider a point $\hat{\phi}$ on one of the branches of the stable manifold of (15.8) (one of the homoclinic orbits of this system). Then $\phi_0\left(2n\pi, \hat{\phi}, 0\right) = \left(P_0^0\right)^n\left(\hat{\phi}\right)$, which also lies on the homoclinic orbit, and $\lim_{n\to\infty}\phi_0\left(2n\pi, \hat{\phi}, 0\right) = (0,0)$. Hence $\hat{\phi}$ is on the stable manifold for $P_0^0$. Conversely, if $\hat{\phi}$ is on the stable manifold of $P_0^0$, then the sequence of points $\phi_0\left(2n\pi, \hat{\phi}, 0\right)$ tends to $(0,0)$ as $n \to \infty$, which implies that $\lim_{t\to\infty}\phi_0\left(t, \hat{\phi}, 0\right) = (0,0)$, and so $\hat{\phi}$ is on the stable manifold for (15.8) at the origin. The rest of the argument is similar.

[9] More precisely, $\lim_{n\to\infty} \|\phi_\varepsilon(2n\pi + t, \psi_1, 0) - \phi_\varepsilon(t, \psi_0(\varepsilon), 0)\| = 0$ uniformly for $0 \leq t \leq 2\pi$.

Therefore, the stable and unstable manifolds of $P_0^0$ at $(0,0)$ intersect, but not transversely, since they are identical. We do not have the conditions given in the previous subsection for existence of a horseshoe map for $P_0^0$. Indeed, the Poincaré-Bendixson theorem tells us that second order autonomous systems do not exhibit any sort of chaos.

But for small $\varepsilon > 0$, the stable and unstable manifolds of $P_\varepsilon^0$ are perturbed in some way. We do not expect them to be perturbed in exactly the same way, so that they might intersect transversely, or else not intersect at all.

To examine this further, we turn to the energy function, letting $W(t) = E(x(t), y(t)) = \frac{1}{2}y(t)^2 - \frac{1}{2}x(t)^2 + \frac{1}{4}x(t)^4$. We easily see that $W'(t) = \varepsilon x'(t)(\gamma \cos t - \delta x'(t))$.

We now assume that for small $\varepsilon \geq 0$ the manifolds $\mathcal{U}_\varepsilon^0$ and $\mathcal{S}_\varepsilon^0$ intersect transversely at some point $\hat{\phi}(\varepsilon)$ with $\left\|\hat{\phi}(\varepsilon)\right\| > \delta$. Thus, $\hat{\phi}(\varepsilon)$ is not on $\mathcal{U}_{\varepsilon,l}^0 \cup \mathcal{S}_{\varepsilon,l}^0$. As we have seen, $\mathcal{U}_\varepsilon^0$ and $\mathcal{S}_\varepsilon^0$ then intersect transversely at infinitely many points, including points on $\mathcal{U}_{\varepsilon,l}^0$ and $\mathcal{S}_{\varepsilon,l}^0$, but we concentrate on the particular point of intersection $\hat{\phi}(\varepsilon)$, which we assume varies continuously with $\varepsilon$ for $\varepsilon \geq 0$. Our assumptions imply that $\hat{\phi}(0)$ is a nonzero point on a homoclinic orbit $\Gamma$ of (15.8).

Now consider, for each small $\varepsilon > 0$, the segment $\Gamma_\varepsilon^U$ of $\mathcal{U}_\varepsilon^0$ between $\psi_0(\varepsilon)$ and $\hat{\phi}(\varepsilon)$. This segment contains $\mathcal{U}_{\varepsilon,l}^0$. Also, let $\Gamma_\varepsilon^S$ denote the segment of $\mathcal{S}_\varepsilon^0$ between $\hat{\phi}(\varepsilon)$ and $\psi_0(\varepsilon)$. This segment contains $\mathcal{S}_{\varepsilon,l}^0$. Let $\Gamma_\varepsilon = \Gamma_\varepsilon^U \cup \Gamma_\varepsilon^S$. Then $\Gamma_\varepsilon \to \Gamma$ as $\varepsilon \to 0$.

Next we discuss the change in $E$ along the solution $\phi_\varepsilon\left(t, \hat{\phi}(\varepsilon), 0\right)$ of (15.7). Since $\hat{\phi}(\varepsilon) \in \mathcal{S}_\varepsilon^0 \cap \mathcal{U}_\varepsilon^0$, $\lim_{n\to\pm\infty} \phi_\varepsilon\left(2n\pi, \hat{\phi}(\varepsilon), 0\right) = \psi_0(\varepsilon)$, and so $\lim_{n\to\pm\infty} E\left(\phi_\varepsilon\left(2n\pi, \hat{\phi}(\varepsilon), 0\right)\right) = E(\psi_0(\varepsilon))$. Hence, the total change in $E$, as $t$ varies from $-2n\pi$ to $2n\pi$, tends to 0 as $n \to \infty$. Therefore, if $\phi_\varepsilon\left(t, \hat{\phi}(\varepsilon), 0\right) = (x_\varepsilon(t), y_\varepsilon(t))$ then

$$\begin{aligned} (15.14) \quad & \lim_{n\to\infty} \int_{-2n\pi}^{2n\pi} W'(t)\, dt \\ &= \lim_{n\to\infty} \int_{-2n\pi}^{2n\pi} x_\varepsilon'\left(t, \hat{\phi}(\varepsilon), 0\right)\left(\gamma \cos t - \delta x_\varepsilon'\left(t, \hat{\phi}(\varepsilon), 0\right)\right) dt = 0. \end{aligned}$$

Note that a solution of an initial value problem for (15.10) depends continuously on $\varepsilon$ and so approaches a solution of (15.8) as $\varepsilon \to 0$. We assumed earlier that as $\varepsilon \to 0$, $\hat{\phi}(\varepsilon)$ converges to some nonzero point on this homoclinic orbit. Hence we can let $\varepsilon \to 0$ inside the integral in (15.14) and get for each $n$ an integral which is independent of $\varepsilon$. Under our assumptions the center manifold theorem for maps [**28**] implies that $\phi_\varepsilon\left(2n\pi, \hat{\phi}(\varepsilon), 0\right) \to \psi_0(\varepsilon)$ at an exponential rate as $n \to \pm\infty$, independent of $\varepsilon$. Also, $\psi_0(\varepsilon) \to (0,0)$ as $\varepsilon \to 0$. In Exercise 15.6 you are asked to show that the $\varepsilon \to 0$ limit can be interchanged with letting $n \to \infty$, which implies that

$$\int_{-\infty}^{\infty} x_0'\left(t, \hat{\phi}(0), 0\right)\left(\gamma \cos t - \delta x_0'\left(t, \hat{\phi}(0), 0\right)\right) dt = 0,$$

where we note that $\phi_0\left(t, \hat{\phi}(0), 0\right) = \left(x_0\left(t, \hat{\phi}(0), 0\right), x_0'\left(t, \hat{\phi}(0), 0\right)\right)$ is a solution of (15.8) with trajectory on $\Gamma$.

We have been concentrating on the map $P_\varepsilon^0$, but all of the considerations above apply to all of the maps $P_\varepsilon^T$. It may be that $\mathcal{U}_\varepsilon^0$ and $\mathcal{S}_\varepsilon^0$ don't intersect, but $\mathcal{U}_\varepsilon^T$ and $\mathcal{S}_\varepsilon^T$ do, for some $T = T(\varepsilon) \in [0, 2\pi)$. In this case, replace $t$ with $t + T$ in (15.7). If $\lim_{\varepsilon \to 0} T(\varepsilon) = \hat{T}$ then we obtain

$$\int_{-\infty}^{\infty} x_0'\left(t, \hat{\phi}(0), \hat{T}\right)\left(\gamma \cos(t + \hat{T}) - \delta x_0'\left(t, \hat{\phi}(0), \hat{T}\right)\right) dt = 0.$$

So we let

$$I_0(T) = \int_{-\infty}^{\infty} x'(t)\left(\gamma \cos(t + T) - \delta x'(t)\right) dt,$$

where $(x(t), x'(t)) = \phi_0\left(t, \hat{\phi}(0), 0\right)$. This integral exists for every $T$. We have seen that if the stable and unstable manifolds $\mathcal{U}_\varepsilon^{T(\varepsilon)}$ and $\mathcal{S}_\varepsilon^{T(\varepsilon)}$ intersect transversely for $\varepsilon \neq 0$ and $\lim_{\varepsilon \to 0} T(\varepsilon) = T$, then $I_0(T) = 0$.

We hope that this discussion enhances for the reader the plausibility of the following result, which is an application of Melnikov's theory to equation (15.7). See [**81**, Theorem 4.5.3], [**191**], or [**187**] for theorems in a general setting. See also [**191**] for further discussion.

**Theorem 15.1.** *Let $(x(t), y(t))$ be the solution of the system (15.8) which tends to $(0,0)$ at $\pm\infty$ and has $y(0) = 0$, $x(0) > 0$. (This defines a unique solution to this autonomous system). Let*

$$I(\gamma, \delta, T) = \int_{-\infty}^{\infty} x'(t)\left(\gamma \cos(t + T) - \delta x'(t)\right) dt.$$

*If $I(\gamma, \delta, T) \neq 0$ for some $T$ then for small $\varepsilon$, $\mathcal{U}_\varepsilon^T$ and $\mathcal{S}_\varepsilon^T$ do not intersect in a neighborhood of $\mathcal{U}_0^T\ (= \mathcal{S}_0^T)$. If $I(\gamma, \delta, \cdot)$ has a simple zero at $t = T$, then $\mathcal{U}_\varepsilon^T$ and $\mathcal{S}_\varepsilon^T$ intersect nontangentially.*

**Corollary 15.2.** *If $I(\gamma, \delta, \cdot)$ has a simple zero at $T$, then for sufficiently small $\varepsilon$, and some $n$, $\left(P_\varepsilon^T\right)^n$ is equivalent to a hyperbolic horseshoe map on some subset of $R^2$.*

So, which of the two alternatives in Theorem 15.1 holds for the Duffing example? To apply this theorem we first have to find a solution on the homoclinic orbit of the homogeneous equation, which is $x'' + x^3 - x = 0$. It turns out that we can do this, starting with the fact that on this orbit,

$$x'^2 + \frac{1}{2}x^4 - x^2 = 0.$$

Therefore, on the homoclinic orbit in the right half of the $(x, x')$-plane,

$$\frac{dx}{dt} = \pm\sqrt{x^2 - \frac{1}{2}x^4}.$$

This can be solved exactly, and the solution with $x'(0) = 0$ is $x(t) = \sqrt{2}\text{sech } t$.

Since $\frac{d}{dt}\text{sech } t = -\text{ sech } t \tanh t$, we easily derive that

$$I(\gamma, \delta, T) = -\int_{-\infty}^{\infty} \text{sech } t \tanh t \left(\sqrt{2}\gamma \cos(t+T) + 2\delta \text{ sech } t \tanh t\right) dt.$$

This integral can be evaluated using the theory of residues [**81**], and the final answer is

$$I(\gamma, \delta, T) = -\frac{4\delta}{3} - \sqrt{2}\gamma\pi \text{ sech } \left(\frac{\pi}{2}\right) \sin T.$$

We therefore see that if $\frac{4\delta}{3} < \sqrt{2}\gamma\pi \text{ sech } \frac{\pi}{2}$ then $I(T)$ has simple zeros. On the other hand, if the reverse inequality holds, then $I(T)$ is never zero, and the stable and unstable manifolds of $P_\varepsilon$ do not intersect at all, for small $\varepsilon \neq 0$. Again see [**81**] and [**191**] for further discussion.

Melnikov's method is one of the few general tools available to analyze some important ode's which exhibit one form of "chaos" [**171**], [**81**].[10] We have described this method in the context of a particular ode (a forced Duffing equation), referring the reader to several other expositions for the general theory and a variety of applications. Next we will discuss an attractive theorem of Wiggins [**258**], which was originally proved using Melnikov's method, and show how a similar result can be obtained by elementary shooting. We also extend our result to include nonperiodic forcing terms.

## 15.4. Application to a forced pendulum

We first discuss a simple pendulum, consisting of a mass $m$ at the end of a massless rod of length $l$, which pivots on a support that forces the pendulum to move in a vertical plane. Letting $x(t)$ denote the angle that the pendulum

[10] Another general tool is due to Shilnikov [**214**], [**81**]. This was used, for example, in [**86**] to obtain the results mentioned briefly at the end of Chapter 6.

makes with the vertical, a standard derivation leads to an equation of the form

$$x'' + kx' + \sin x = 0, \tag{15.15}$$

where $k$ represents damping due to friction and to air resistance. We will assume that $x' > 0$ represents a counterclockwise movement of the pendulum.

When $k = 0$, the energy $W(t) = \frac{x'(t)^2}{2} - \cos x(t)$ is constant. The orbits of solutions with $W(t) < 1$ are closed smooth curves and represent periodic solutions. These are solutions with low energy. The phase plane obtained by plotting $x'(t)$ against $x(t)$ is well known and can be found in many introductory texts.

If the initial velocity is large, so that $W(t)$ is large, then the trajectory in phase space is unbounded and represents a pendulum which continues to rotate in the same direction, making repeated complete rotations without loss of energy. For $W = 1$ there are two heteroclinic orbits in a "period strip", such as $-\pi \le x \le \pi$, where one heteroclinic orbit tends to $(-\pi, 0)$ in phase space as $t \to -\infty$ and to $(\pi, 0)$ as $t \to \infty$, while the other tends to $(\pi, 0)$ at $-\infty$ and to $(-\pi, 0)$ at $+\infty$.

Corresponding to the trajectory connecting $(-\pi, 0)$ to $(\pi, 0)$ as $t$ increases, there is a unique solution $x_*$ of (15.15) with $k = 0$ such that $x_*(0) = 0$ and

$$\lim_{t\to\infty} x_*(t) = \pi, \qquad \lim_{t\to-\infty} x_*(t) = -\pi. \tag{15.16}$$

Physically, this solution is approximated by a pendulum which starts from rest close to the upright vertical position and makes almost a complete rotation, coming again close to the vertical position.

There is no chaotic motion possible in this model. To obtain more erratic behavior we must add some sort of forcing term. This can take various forms, but the one we study here results from assuming that the support of the pendulum is subjected to a vertical motion which is sinusoidal. This adds a force proportional to $\sin \varepsilon t$ to the gravitational force. We make the assumption that the force on the support varies slowly with time and also that the damping force is small. This results in the equation

$$x'' + \varepsilon\delta x' + (1 + \gamma \sin \varepsilon t) \sin x = 0, \tag{15.17}$$

where $\varepsilon$ is positive but small. To avoid the delicate case where the coefficient of $\sin x$ can be zero, we require that $0 < |\gamma| < 1$. Note that replacing $t$ by $t + \frac{\pi}{\varepsilon}$ is the same as changing the sign of $\gamma$, an unimportant translation if one is interested (as we are) in complete rotations of the pendulum. Therefore the sign of $\gamma$ is not important. Equation (15.17) is one of the equations studied by Wiggins [**258**].

To describe Wiggins' result, suppose as before that $x > 0$ represents displacement from rest in a counterclockwise direction and a full rotation occurs each time $x(t)$ crosses an odd multiple of $\pi$. He then shows that there is a $\delta(\gamma) > 0$ such that irregular behavior can occur if $0 \leq \delta < \delta(\gamma)$ and $\varepsilon$ is sufficiently small. We will describe the function $\delta(\cdot)$ shortly, but first let us specify the nature of this irregular behavior. Our measure of irregularity is that the pendulum makes a sequence of full rotations, alternating between clockwise and counterclockwise rotations in an erratic manner. More precisely (without, however, yet specifying $\delta(\gamma)$), we state this as a theorem.

**Theorem 15.3** (Wiggins [**258**]). *Suppose that $0 < \gamma < 1$ and $0 \leq \delta < \delta(\gamma)$. Then there is an $\varepsilon_1 > 0$ such that for every $\varepsilon$ with $0 < \varepsilon < \varepsilon_1$, and every infinite sequence $\{m_j\}_{j=1,2,3,\ldots}$ of positive integers, there is a solution of* (15.17) *such that the corresponding motion consists of exactly $m_1$ full counterclockwise rotations, followed by exactly $m_2$ full clockwise rotations, $m_3$ full counterclockwise rotations, and so forth.*

It is common to refer to the solutions corresponding to infinite nonrepeating sequences as "chaotic", though as we said, many researchers prefer a stricter interpretation of this term. Wiggins obtains this striking result by applying the technique of Melnikov to the equation (15.17). This involves an extension of earlier applications of Melnikov's method, because previously the forcing term was required to be small in amplitude, whereas here the small parameter measures the frequency of the oscillation, not its amplitude. The necessary extension was also given by Palmer[**188**].

To define the function $\delta(\cdot)$, Wiggins derives the appropriate Melnikov function for (15.17). As we will see, this is similar to the Melnikov function discussed in Section 15.3, but because the nature of the forcing is different, appropriate modifications are needed. Again assume that $x_*$ satisfies (15.15) with $k = 0$, $x_*(0) = 0$, and (15.16). Then the function $\delta(\gamma)$ in Theorem 15.3 is given by

$$\delta(\gamma) = |\gamma| \frac{\int_{-\infty}^{\infty} s x_*'(s) \sin x_*(s) ds}{\int_{-\infty}^{\infty} x_*'(s)^2 ds}.$$

It is therefore seen that Theorem 15.3 implies the existence of chaotic behavior if

$$\int_{-\infty}^{\infty} J_{x*}(s)\, ds > 0, \tag{15.18}$$

where $J_x(s) = -\delta x'(s)^2 + |\gamma| s x'(s) \sin x(s)$. To explain this further, write (15.17) in the autonomous form

$$\begin{aligned} x' &= y, \\ y' &= -\varepsilon \delta y - (1 + \gamma \sin z) \sin x, \\ z' &= \varepsilon \end{aligned}$$

and observe that this is in a different form from (15.10), because the small parameter $\varepsilon$ affects the period of the forcing term, rather than its amplitude. The theory developed in [**258**] or [**188**] adapts Melnikov's technique to this situation and leads to a slightly different Melnikov function. In [**258**] this is given by

$$M(z) = \int_{-\infty}^{\infty} \left( -\delta x_*'(s)^2 + |\gamma| \, s \cos z \; x_*'(s) \sin x_*(s) \, ds \right).$$

It is shown that the Poincaré map for this system has a horseshoe structure if $M(z)$ has simple zeros as a function of $z$. This leads to the criterion of the theorem, namely that $0 \le \delta < \delta(\gamma)$. In fact, since $x_*$ satisfies (15.17) with $\varepsilon = 0$, we can set $\sin x_*(s) = -x_*''(s)$ in the formula for $\delta(\gamma)$, integrate by parts, and use the boundary conditions to obtain that $\delta(\gamma) = \frac{1}{2}|\gamma|$.

We now outline the rest of this chapter, which extends work in [**91**].

(1) We will use a shooting technique to prove Theorem 15.3. In the case where $\delta = 0$ there is a remarkably simple proof, requiring few estimates. This proof makes use of symmetries of the solution when $\delta = 0$. For $\delta > 0$ the proof involves more technicalities but is still conceptually simple.

(2) We will extend Wiggins' result to a wide class of nonperiodic forcing terms, for example, $\sin \varepsilon t + \sin \sqrt{2} \varepsilon t$ or $\sin\left(\varepsilon t + \sin \varepsilon^2 t^2\right)$.

(3) We try to demonstrate that shooting methods give an alternative ode approach to chaos for nonlinear oscillations, by working directly with the differential equation and making no reference to Poincaré maps. We will show that a form of chaos can be found with a very restricted set of initial conditions.

## 15.5. Proof of Theorem 15.3 when $\delta = 0$

We consider solutions of

$$x'' + (1 + \gamma \sin \epsilon t) \sin x = 0 \tag{15.19}$$

with initial conditions $x(0) = \alpha, x'(0) = 0$. When necessary, we will denote the unique solution $x$ by $x(t, \alpha)$. The goal is to obtain the desired solutions by adjusting $\alpha$.

The solutions have certain symmetries, around points $\frac{T_n}{\varepsilon}$ where $n$ is an integer and $T_n = (2n+1)\frac{\pi}{2}$. These are as follows.

(i) If $x\left(\frac{T_n}{\varepsilon}\right) = k\pi$ for some integer $k$, then

$$x\left(\frac{T_n}{\varepsilon} + s\right) - k\pi = k\pi - x\left(\frac{T_n}{\varepsilon} - s\right)$$

for all $s$.

(ii) If $x'\left(\frac{T_n}{\varepsilon}\right) = 0$, then for all $s$, $x\left(\frac{T_n}{\varepsilon} + s\right) = x\left(\frac{T_n}{\varepsilon} - s\right)$.

(iii) If $x\left(t_0\right) = k\pi$ and $x'\left(t_0\right) = 0$ for some $t_0$, then $x\left(t\right) = k\pi$ for all $t$.

These statements are proved by using the uniqueness of solutions to initial value problems for (15.19). For example, to prove (i), consider the functions $u(s) = x(\frac{T_n}{\varepsilon} + s) - k\pi$ and $v(s) = k\pi - x(\frac{T_n}{\varepsilon} - s)$.

The only detailed analysis required in proving Theorem 15.3 when $\delta = 0$ is used to obtain the following lemma.

**Lemma 15.4.** *If $\delta = 0$ and $\varepsilon$ is positive and sufficiently small, then there is an $\overline{\alpha}$ in $(-\pi, 0)$ such that $x\left(t, \bar{\alpha}\right)$ increases monotonically on some interval $[0, t_1]$ and $x\left(t_1, \bar{\alpha}\right) = \pi$.*

We will prove Lemma 15.4 below, in the course of proving a stronger result, Lemma 15.6, needed in later sections. Here we will show how Lemma 15.4, in conjunction with a shooting technique, enables us to prove Theorem 15.3 when $\delta = 0$.

It is convenient to assume that the first complete rotation of the pendulum is in the direction of positive $x$; i.e. as $t$ increases from 0 we want $x\left(t\right)$ to cross $x = \pi$ before crossing $x = -\pi$. We begin by choosing an $\overline{\alpha}$ as in Lemma 15.4. The corresponding solution $x\left(t, \bar{\alpha}\right)$ crosses $x = \pi$ at a point $t = s_1$, with $x' > 0$, and since $x'' > 0$ when $\pi < x < 2\pi$, there must be a $t_1 > s_1$ such that $x' > 0$ on $(0, t_1]$ and $x(t_1) = 2\pi$. By (iii) above, $x'(t_1) > 0$, and so the implicit function theorem implies that the equation $x(t_1(\alpha), \alpha) = 2\pi$ defines a smooth function $t_1(\alpha)$ in a neighborhood of $\overline{\alpha}$. This function is continuous in some maximal interval of the form $(\underline{\alpha}, \overline{\alpha}]$, where $\underline{\alpha} \geq -\pi$ since $x\left(t, -\pi\right) \equiv -\pi$. Because $t_1(\cdot)$ can be extended continuously to an open neighborhood of each point where it is defined,

$$\lim_{\alpha \to \underline{\alpha}^+} t_1(\alpha) = \infty. \tag{15.20}$$

To illustrate the idea of the proof, suppose that $m_1 = 2$, so that we want $x$ to cross $3\pi$ before recrossing $\pi$. This is accomplished by lowering $\alpha$ from $\overline{\alpha}$. Since $t_1(\cdot)$ is continuous, it follows from (15.20) that there is an $\alpha_1 \in (\underline{\alpha}, \overline{\alpha})$ such that $t_1(\alpha_1) = \frac{T_{n_1}}{\varepsilon}$, for some integer $n_1$. The symmetry statement (i) implies that $x\left(t, \alpha_1\right)$ then increases monotonically until it crosses $3\pi$ and

$4\pi$, after which it has a maximum at $t_2 = \frac{2T_{n_1}}{\varepsilon}$ with $x(t_2) < 5\pi$. We have then accomplished the first step of achieving exactly two counterclockwise rotations, and we next wish to obtain a clockwise rotation, since we are assuming that $m_2 \geq 1$.

In a neighborhood of $\alpha_1$ let $t_2(\alpha)$ be the first $t > 0$ where $x(t, \alpha)$ has a local maximum. As we have noted, $4\pi < x(t_2(\alpha_1), \alpha_1) < 5\pi$, and so $x''(t_2(\alpha_1), \alpha_1) \neq 0$. Also, $x'(t, \alpha_1) > 0$ on $(0, t_2(\alpha_1))$. Then $t_2(\cdot)$ can be extended as a continuous function in some neighborhood of $\alpha_1$, as a solution of the equation $x'(t, \alpha) = 0$. As long as $t_2(\cdot)$ is continuous, $x(t_2(\alpha), \alpha)$ is in the interval $(4\pi, 5\pi)$, again since no nonconstant solution can be tangent to a line $x = k\pi$. Further, we claim that as $\alpha$ is lowered from $\alpha_1$, no new zero of $x'(t, \alpha)$ appears in the interval $(0, t_2(\alpha))$ as long as $t_2(\alpha)$ is defined and continuous.

To see this, suppose that there is a largest $\hat{\alpha} < \alpha_1$ such that $t_2(\cdot)$ is continuous in $[\hat{\alpha}, \alpha_1]$ and $x'(\hat{t}, \hat{\alpha}) = 0$ for some $\hat{t} \in (0, t_2(\hat{\alpha}))$. Then necessarily $x''(\hat{t}, \hat{\alpha}) = 0$, but this could only occur if $x(\hat{t}, \alpha) = k\pi$ for $k = 0, 1, 2, 3$, or $4$, and then $x$ is constant, a contradiction.

Moreover, $t_2(\cdot)$ can be extended continuously to a maximal interval of the form $(\underline{\alpha}_1, \alpha_1]$, where $\underline{\alpha} \leq \underline{\alpha}_1 < \alpha_1$ and $\lim_{\alpha \to \underline{\alpha}_1^+} t_2(\alpha) = \infty$. Therefore, we can find $\alpha_2 \in (\underline{\alpha}_1, \alpha_1)$ such that $t_2(\alpha_2) = \frac{T_{n_2}}{\varepsilon}$, for some integer $n_2$. It is important to note that $t_1(\alpha_2)$ is not necessarily of the form $\frac{T_n}{\varepsilon}$, but it is still defined as the first point where $x(t, \alpha_2) = 2\pi$. The solution $x(t, \alpha_2)$ decreases from its maximum at $t_2(\alpha_2)$ to recross $3\pi$ and $\pi$. If $m_2 = 2$ then this completes the second step of the induction process, since with the choice we have made of $\alpha_2$, the next extremum of $x(t, \alpha_2)$ is a minimum where $x = x(0) \in (-\pi, 0)$. If $m_2 \neq 2$, we further adjust $\alpha$ as described in the induction step below.

We complete the proof of Theorem 15.3 (aside from Lemma 15.4) by giving the induction step which we use to choose a decreasing sequence $\{\alpha_j\}$ tending to some $\alpha_\infty > \underline{\alpha}$ such that $x(t, \alpha_\infty)$ is the desired solution as described in the theorem. Suppose that $\alpha_1, \ldots, \alpha_n$ have been chosen. The function $x(t, \alpha_n)$ has its first $n$ local extrema with $x$ in prescribed intervals $((2M_i - 1)\pi, (2M_i + 1)\pi)$ as determined by the sequence $\{m_i\}$. Thus, $M_0 = 0$ and $M_{i+1} = M_i + (-1)^i m_{i+1}$. Suppose for definiteness that $n$ is odd, so that the $n^{\text{th}}$ critical point is a maximum. Let this maximum be at $t_k(\alpha_n)$.

As before, $t_k(\cdot)$ is defined continuously as a zero of $x'(t, \alpha)$ in a neighborhood of $\alpha_n$, and we can find an $\hat{a}_n \in (\underline{\alpha}, \alpha_n)$ such that $t_k(\hat{a}_n) = \frac{T_p}{\varepsilon}$ for some integer $p$. By (ii), the solution $x(t, \hat{\alpha}_n)$ is symmetric around $\frac{T_p}{\varepsilon}$. Since $m_n \geq 1$, $x(t, \hat{\alpha}_n)$ must decrease from its value at $t_k(\hat{a}_n)$, past an odd multiple of $\pi$ and at least to $(2M_n - 2)\pi$.

There are then two cases. If $m_{n+1} = 1$, then we do not want $x$ to decrease past further odd multiples of $\pi$. In that case, we decrease $\alpha$ until the crossing of $(2M_n - 1)\pi$ occurs at some $\frac{T_n}{\varepsilon}$. Then the antisymmetry of the solution around this crossing point ensures that there is a local minimum before $x$ crosses $(2M_n - 3)\pi$.

In the second case, $m_{n+1} \geq 2$, and we want to be sure $x$ crosses $(2M_n - 3)\pi$. To achieve this, we lower $\alpha$ until the crossing of $(2M_n - 2)\pi$ occurs as some $\frac{T_n}{\varepsilon}$. Then the antisymmetry around this crossing ensures that $x$ does decrease below $(2M_n - 3)\pi$. Similarly, if still further clockwise rotations are required in the $n + 1$ stage, we further reduce $\alpha$ to obtain antisymmetry around the point where the solution crosses $(2M_n - 3)\pi$. Continuing in this way we may find that there are too many clockwise rotations. The final step in the induction process is then to reduce $\alpha$ further so as to obtain anti-symmetry around the desired mid-point between the high and low points of $x$ in the $(n+1)^{\text{st}}$ step; that is, the point where $x = (2M_n + (-1)^n m_{n+1})\pi$.

## 15.6. Damped pendulum with nonperiodic forcing

The proof above makes strong use of symmetry. However, this property is lost when $\delta > 0$. In fact, it is not even necessary for the forcing term to be periodic. It is only required to oscillate in a rather general fashion. Here we see an advantage of the shooting method over Poincaré maps, for it is difficult to define such a map for a nonperiodic forced oscillator. The proof of Theorem 15.5 below will also complete our proof of Theorem 15.3.

We now consider the equation

$$x'' + \varepsilon\delta x' + p(\varepsilon t)\sin x = 0 \tag{15.21}$$

where $\varepsilon > 0, \delta \geq 0$, and $p \in C^1[0,\infty)$. We make the following additional assumptions on $p$.

(a) There are numbers $q$ and $Q$ such that $0 < q \leq p(t) \leq Q$ for all $t \geq 0$, and $p'$ is bounded on $[0,\infty)$.

(b) There are positive numbers $\lambda$ and $\eta$ and two infinite sequences of intervals, $I_n = [a_n - \eta, a_n + \eta]$ and $J_n = [b_n - \eta, b_n + \eta]$, with $\{a_n\}$ and $\{b_n\}$ increasing, such that $p' \geq \lambda$ on each $I_n$ and $p' \leq -\lambda$ on each $J_n$. (Obviously $\bigcup_{n=1}^{\infty} I_n$ and $\bigcup_{n=1}^{\infty} J_n$ are disjoint.)

Also, we will modify the description of erratic behavior slightly to allow for intervening swings of the pendulum of less than 360°. Suppose that $\sigma = \{m_j\}_{j=1,2,3,\dots}$ is an infinite sequence of nonnegative integers. We will say that a solution $x$ of (15.21) has the sequence $\sigma$ of left and right rotations if $x$ has successive minima and maxima $0 = t_0 < t_1 < t_2 < t_3 < \cdots$, and in $(t_{j-1}, t_j)$, $x$ crosses exactly $m_j$ odd multiples of $\pi$. Equivalently,

$x(t_j) \in [(2M_j - 1)\pi, (2M_j + 1)\pi]$, where $M_0 = 0, M_j = M_{j-1} + (-1)^{j-1} m_j$, and $x$ is monotonic in $(t_{j-1}, t_j)$.

For each $\alpha$, let $x(t, \alpha)$ denote the solution of (15.21) such that

$$x(0) = \alpha, \qquad x'(0) = 0. \tag{15.22}$$

We can now state our result.

**Theorem 15.5.** *Suppose that the continuous function $p$ satisfies* (a) *and* (b). *Suppose also that for some given $\delta > 0$ and $\lambda > 0$,*

$$\int_{-\infty}^{\infty} \left( -\delta \overline{x}'(t)^2 + \lambda t \overline{x}'(t) \sin(\overline{x}(t)) \right) dt = \mu > 0, \tag{15.23}$$

*where $\overline{x}$ is the unique solution to*

$$x'' + Q \sin x = 0,$$
$$x(0) = 0, \quad x(\infty) = \pi.$$

*Then there is an $\varepsilon_1 > 0$ such that for each $\varepsilon \in (0, \varepsilon_1)$ and each sequence $\{\sigma_j\}$ of nonnegative integers, there is an $\alpha$ such that $x(\tau, \alpha)$ has the sequence $\{\sigma_j\}$ of left and right rotations.*

**Remark:** Since $Q \sin \overline{x}(t) = -\overline{x}''(t)$, (15.23) becomes $\lambda > 2Q\delta$.

**15.6.1. Outline of proof.** A simple rescaling, letting $t' = t/\eta$, allows us to assume in condition (b) that $\eta = 1$. The proof depends on the following two lemmas. Note that Lemma 15.6 is a generalization of Lemma 15.4, as we explain at the end of this section.

**Lemma 15.6.** *There is an $\varepsilon_1 > 0$ such that if $0 < \varepsilon < \varepsilon_1$ and $p' \leq -\lambda$ on some interval $[c-1, c+1]$ and if $x$ is a solution of (15.21) such that*

$$x\left(\frac{c}{\varepsilon}\right) = 2m\pi \quad \text{for some } m \tag{15.24}$$

*and*

$$x' > 0, \qquad x \geq (2m-1)\pi \tag{15.25}$$

*on $\left[\frac{c-1}{\varepsilon}, \frac{c}{\varepsilon}\right]$, then $x$ crosses $(2m+1)\pi$ in $\left(\frac{c}{\varepsilon}, \frac{c+1}{\varepsilon}\right)$ and $x' > 0$ in this interval up to this crossing point.*

**Lemma 15.7.** *In addition, $\varepsilon_1$ can be chosen so that if $0 < \varepsilon < \varepsilon_1$, $p' \geq \lambda$ in $[c-1, c+1]$, and the solution $x$ satisfies (15.24) and (15.25), then $x$ has a maximum in $\left(\frac{c}{\varepsilon}, \frac{c+1}{\varepsilon}\right)$, before any crossing of $(2m+1)\pi$ in this interval.*

We will prove these lemmas in the next subsection. In using them to prove Theorem 15.5 we will again assume that the first full rotation is counterclockwise ($m_1 > 0$). This is obviously unimportant, but convenient for definiteness. As before, we will successively adjust $\alpha$ to obtain, first, $m_1$

upward crossings of odd multiples of $\pi$, then $m_2$ downward crossings, etc. We use Lemmas 15.6 and 15.7, the continuity of the solution with respect to $\alpha$, and the fact (iii) above that if, for some $t \geq 0$ and some integer $k$, $x(t) = k\pi$ and $x'(0) = 0$, then $x(t) \equiv k\pi$. This is still valid for (15.21). We will again choose a nonincreasing sequence $\{\alpha_j\}$ such that the corresponding solutions $x(t, \alpha_j)$ of (15.21) and (15.22) carry out more and more of the turns required by the given sequence $\{m_1, m_2, \dots\}$.

Since $m_1 > 0$, we first choose $\overline{\alpha}$ in $(-\pi, 0)$ so that $x(t, \bar{\alpha})$ increases monotonically until it crosses $\pi$. To achieve this, we observe that for each $\alpha \in (-\pi, 0)$, $x'(t, \alpha) > 0$ as long as $x(t, \alpha) < 0$. Furthermore, for $\varepsilon$ sufficiently small, $x(t, \alpha)$ must cross 0 at some first point $t_0(\overline{\alpha})$. (This is seen, for example, by linearizing the equation $x'' + \epsilon\delta x' + q \sin x = 0$ around $x = 0$.)

While the method is essentially the same as for the case $\delta = 0$ with periodic forcing, there are some technical alterations because $\delta$ may be positive and we allow a broader class of forcing functions. The first zero of $x(t, \alpha)$, $t_0(\alpha)$, is a continuous function of $\alpha$ for $-\pi < \alpha < 0$. However $t_0(-\pi)$ is not defined. Hence $\lim_{\alpha \to -\pi^+} t_0(\alpha) = \infty$. Therefore we can choose $\alpha_1$ so that $t_0(\alpha_1) = \frac{b_j}{\varepsilon}$, for some $j$ with $b_j > 1$. Then $x(t, \alpha_1)$ satisfies the hypotheses of Lemma 15.6, with $c = b_j$, and so $x$ increases monotonically until it crosses $\pi$, at some $s_1(\alpha) \in [\frac{b_j}{\varepsilon}, \frac{b_j+1}{\varepsilon}]$. Also, we can find an $\underline{\alpha}$ in $(-\pi, \alpha_1)$ such that $t_0(\underline{\alpha}) = \frac{a_k}{\varepsilon}$, for some $a_k > 1$. Then Lemma 15.7 implies that $x(t, \underline{\alpha})$ has a maximum before reaching $\pi$. From now on, all values of $\alpha$ will be chosen between $\underline{\alpha}$ and $\alpha_1$.

Suppose now that $m_1 \geq 2$, so that we want $x$ to cross $3\pi$ before its first maximum. Once $x$ crosses $\pi$, $x' > 0$ as long as $x \leq 2\pi$, so that $x$ must cross $2\pi$, at some $t_1 = t_1(\alpha)$, for small $\varepsilon$ and for $\alpha$ sufficiently close to $\alpha_1$. The functions $s_1(\alpha)$ and $t_1(\alpha)$ can be extended continuously to the same maximal interval of the form $(\hat{\alpha}, \alpha_1]$, where $\underline{\alpha} < \hat{\alpha} < \alpha_1$ and $\lim_{\alpha \to \hat{\alpha}^+} s_1(\alpha) = \lim_{\alpha \to \hat{\alpha}^+} t_1(\alpha) = \infty$.

We wish to apply Lemma 15.6 again. It is clear that we can find an infinite sequence of $\alpha$'s tending to $\hat{\alpha}$ such that $t_1(\alpha) = \frac{b_j}{\varepsilon}$ for some $j$. However we must show that for one of these values of $\alpha$, $x \geq \pi$ on $[t_1(\alpha) - \frac{1}{\varepsilon}, t_1(\alpha)]$. In other words, we must show that $s_1(\alpha) \leq t_1(\alpha) - \frac{1}{\varepsilon}$. This is implied by the following result.

**Lemma 15.8.** *As $\alpha \to \hat{\alpha}^+$, $x'(s_1(\alpha), \alpha) \to 0$.*

**Proof.** If not, then there is a sequence $\{\alpha_j\} \to \hat{\alpha}^+$ and a $\rho > 0$ such that $x'(s_1(\alpha_j), \alpha_j) \geq \rho$. But $x'' < 0$ on $[t_0, s_1]$, and so it follows by going backward from $s_1(\alpha)$ that $s_1(\alpha) - t_0(\alpha)$ remains bounded as $\alpha \to \hat{\alpha}^+$ along the sequence $\{\alpha_j\}$. However, $t_0(\cdot)$ is continuous on the compact interval

$[\hat{\alpha}, \alpha_1]$, and hence bounded, while $s_1(\alpha) \to \infty$ as $\alpha \to \hat{\alpha}^+$. This contradiction proves Lemma 15.8. □

Now use the fact that solutions of (15.21) are continuous with respect to initial conditions. If $x(s_1) = \pi$ and $x'(s_1)$ is small, then $x$ is close to $\pi$ over a long interval, which implies that $t_1$, where $x$ crosses $2\pi$, is a long way from $s_1$. Therefore $s_1(\alpha) < t_1(\alpha) - \frac{1}{\varepsilon}$.[11] The solution is monotonic on $\left(t_1(\alpha) - \frac{1}{\varepsilon}, t_1(\alpha)\right)$, and we can lower $\alpha$ so that $t_1(\alpha) = \frac{b_k}{\varepsilon}$ for some (large) $k$, allowing us to apply Lemma 15.6. This shows that for some $\alpha_2 \in (\hat{\alpha}, \alpha_1)$, $x(t, \alpha_2)$ is monotonic at least until it crosses $3\pi$.

Now suppose that $m_1 = 2$. Then we wish to choose $\alpha_3 \in (\hat{\alpha}, \alpha_2)$ such that $x(t, \alpha_3)$ increases monotonically until a maximum where $x \in (4\pi, 5\pi)$. For $\alpha$ close to $\alpha_2$, let $s_2(\alpha)$ be the first point where $x(t, \alpha)$ crosses $3\pi$, and $t_2(\alpha)$ the first point where $x(t, \alpha)$ crosses $4\pi$. As $\alpha$ decreases, $t_1(\alpha) \to \infty$, so that there is a largest $\breve{\alpha} < \alpha_2$ such that $t_1(\breve{\alpha}) = \frac{a_k}{\varepsilon}$ for some $k$. Then Lemma 15.7 implies that $x$ has a maximum before reaching $3\pi$, so that $s_2$ and $t_2$ became unbounded at an earlier point. So we can choose $\alpha_3 \in (\breve{\alpha}, \alpha_2)$ such that $t_2(\alpha_3) = \frac{a_k}{\varepsilon}$ for some $k$. Since $t_1$ is bounded on $[\breve{\alpha}, \alpha_2]$, we can further choose $\alpha_3$ so that $s_2(\alpha_3) < t_2(\alpha_3) - \frac{1}{\varepsilon}$ and apply Lemma 15.7 to obtain the desired solution with a maximum at $\tau_1(\alpha_3)$ with $x(\tau_1(\alpha_3), \alpha_3) \in (4\pi, 5\pi)$. Furthermore, we can choose $k$ as large as we like, and this leads to the next result.

**Lemma 15.9.** *As $k \to \infty$, $x(\tau_1(\alpha_3), \alpha_3) \to 5\pi$.*

(The proof is similar to the proof of Lemma 15.8.) This lemma implies that the distance between the maximum at $\tau_1(\alpha_3)$ and the next downward crossing of $4\pi$ can be made large, allowing us to control whether or not the solution continues in the same direction or turns around again. (At this stage we are applying Lemmas 15.6 and 15.7 to $-x$.)

We now describe the induction step. Having done extra work above relative to the symmetric case, the induction is easier. Suppose that the first $n$ local extrema have been achieved according to the desired pattern $\sigma_1, \ldots, \sigma_n$ and the final extremum is a maximum at $\tau_n(\alpha_n)$. We wish to decrease $\alpha$ from $\alpha_n$, if necessary, so that from $\tau_n(\alpha)$ the solution decreases past exactly $\sigma_{n+1}$ odd multiples of $\pi$. (This corresponds to $\sigma_{n+1}$ clockwise rotations of the pendulum, since we took $x' > 0$ to mean a counterclockwise movement.) After $\tau_n(\alpha_n)$, the solution crosses at least one even multiple of $\pi$. By lowering $\alpha$ we can force the first such crossing to be at either $\frac{b_k}{\varepsilon}$, if $\sigma_{n+1} > 0$, or $\frac{a_k}{\varepsilon}$ if $\sigma_{n+1} = 0$, for some $k$. Further, using Lemma 15.9, we can choose $k$ as large as we like, so that this crossing is a long way (more than $\frac{1}{\varepsilon}$)

[11] Note that $\varepsilon$ is kept fixed all through the argument, once the lemmas have been proved.

beyond the maximum at $\tau_n(\alpha)$, and then we can apply either Lemma 15.6 or Lemma 15.7, as appropriate. If $\sigma_{n+1} > 0$ then a finite induction from 1 to $\sigma_{n+1}$ completes the next step, achieving exactly $\sigma_{n+1}$ clockwise turns of the pendulum and completing the proof of Theorem 15.5 except for the proofs of Lemmas 15.6 and 15.7.

**15.6.2. Proofs of Lemmas 15.6 and 15.7.** As the reader may suspect, energy is the main tool. One new feature is that the energy used to prove each of these lemmas is a function of the variable $c$ used in their statements. Let

$$E_c(x,y) = \frac{1}{2}y^2 - p(c)\cos x.$$

Then for a solution of (15.21), with $W_c(t) = E_c(x(t), y(t))$,

$$W_c'(t) = -\varepsilon\delta x'^2 + x'(t)(p(c) - p(\varepsilon t))\sin x(t). \tag{15.26}$$

To prove Lemma 15.6, note in (15.26) that since $p'(c) < 0$, both $\sin x(t)$ and $(p(c) - p(\varepsilon t))$ change sign at $t = \frac{c}{\varepsilon}$. Hence, the second term of $W_c'$ in (15.26) is positive on $[\frac{c-1}{\varepsilon}, \frac{c}{\varepsilon})$, and on $(\frac{c}{\varepsilon}, \frac{c+1}{\varepsilon})$ as long as $x' > 0$ and $x < (2m+1)\pi$. We must show that the net change in $W_c$ from $\frac{c-1}{\varepsilon}$ to some $T \in (\frac{c}{\varepsilon}, \frac{c+1}{\varepsilon})$ is positive and enough to push $x$ beyond $(2m+1)\pi$.

We first give the proof of Lemma 15.6 (for $\delta \geq 0$) and then make a few comments about Lemma 15.7. The periodicity of sine enables us to consider the case $m = 0$. In the following lemma, $x_{\varepsilon,c}$ denotes the unique solution of

$$\begin{gathered} x'' + p(c)\sin x = 0, \\ x(\pm\infty) = \pm\pi, \quad x\left(\tfrac{c}{\varepsilon}\right) = 0. \end{gathered} \tag{15.27}$$

**Lemma 15.10.** *For each $\varepsilon > 0$ and each $c$, let*

$$\mathcal{Y}_{\varepsilon,c} = \Big\{ \textit{solutions } x \textit{ of } (15.21) \mid x\left(\frac{c}{\varepsilon}\right) = 0 \quad \textit{and on } \left[\frac{c-1}{\varepsilon}, \frac{c}{\varepsilon}\right],\ x' > 0 \textit{ and } -\pi < x < 0 \Big\}.$$

*Then for any $A > 0$,*

$$\lim_{\varepsilon \to 0^+} \left\{ \sup_{c \in R^1,\ x \in \mathcal{Y}_{\varepsilon,c},\ \frac{c}{\varepsilon} - A \leq t \leq \frac{c}{\varepsilon} + A} |x(t) - x_{\varepsilon,c}(t)| \right\} = 0,$$

$$\lim_{\varepsilon \to 0^+} \left\{ \sup_{c \in R^1,\ x \in \mathcal{Y}_{\varepsilon,c},\ \frac{c}{\varepsilon} - A \leq t \leq \frac{c}{\varepsilon} + A} \left|x'(t) - x_{\varepsilon,c}'(t)\right| \right\} = 0.$$

**Proof.** This is where we use the condition that $p'$ is bounded. This implies that for any $A > 0$, $\lim_{\varepsilon \to 0} p(\varepsilon t) = p(c)$ uniformly for $t \in \left[\frac{c}{\varepsilon} - A, \frac{c}{\varepsilon} + A\right]$ and all $c$. We leave the rest of the proof to the reader as Exercise 15.6. ☐

Since $p(c) \le Q$, it follows from (15.23) that $\lambda > 2p(c)\delta$ and hence that

$$\int_{-\infty}^{\infty} I_{x_{\varepsilon,c}}(t)\,dt \ge \mu_1,$$

where

$$I_x(s) = -\delta x'(s)^2 + \lambda s x'(s)\sin x(s)$$

and $\mu_1 = \inf_{c\in R^1,\varepsilon>0}\left(\frac{\lambda}{2p(c)} - \delta\right)\int_{-\infty}^{\infty}\left(x'_{\varepsilon,c}\right)^2 dt > 0.$

We first pick $A$ so large (independent of $c$ and $\varepsilon$) that

$$\int_{\frac{c}{\varepsilon}-A}^{\frac{c}{\varepsilon}+A} I_{x_{\varepsilon,c}}(t)\,dt > \frac{\mu_1}{2},$$

and therefore, for sufficiently small $\varepsilon$ (independent of $c$) and $x \in \mathcal{Y}_{\varepsilon,c}$,

$$\int_{\frac{c}{\varepsilon}-A}^{\frac{c}{\varepsilon}+A} I_x(t)\,dt > \frac{\mu_1}{2}. \tag{15.28}$$

**Lemma 15.11.** *The number $A > 0$ can also be chosen so large (independent of $c$) that for sufficiently small $\varepsilon$ and for each $x \in \mathcal{Y}_{\varepsilon,c}$ and $T \in \left[\frac{c-1}{\varepsilon}, \frac{c}{\varepsilon} - A\right]$,*

$$\left|\int_{T}^{\frac{c}{\varepsilon}-A} I_x(t)\,dt\right| < \frac{\mu_1}{4}. \tag{15.29}$$

**Proof.** This follows by establishing appropriate uniform bounds for $x+\pi$ and $x'$ in $\left[\frac{c-1}{\varepsilon}, \frac{c}{\varepsilon} - A\right]$. To do this, observe from Lemma 15.10 that for each $\nu > 0$ we can choose $A$ and $\varepsilon_A > 0$ such that if $0 < \varepsilon < \varepsilon_A$ and $x \in \mathcal{Y}_{\varepsilon,c}$ for some $c$, then

$$\begin{gathered} 0 < \pi + x(\tfrac{c}{\varepsilon} - A) < \nu, \\ 0 < x'(\tfrac{c}{\varepsilon} - A) < \nu. \end{gathered} \tag{15.30}$$

Lemma 15.10 and the uniform boundedness of $x''$ ensure that this can be done. Now use the fact that $\sin x < -\frac{2}{\pi}(x+\pi)$ if $-\pi < x < -\frac{\pi}{2}$. Let $r^2 = \frac{2}{\pi}q$. For each $t_0 < \frac{c}{\varepsilon} - A$ we wish to compare $x_\varepsilon + \pi$ with the solution $u$ of

$$\begin{gathered} u'' + \varepsilon\delta u' - r^2 u = 0, \\ u(t_0) = x(t_0) + \pi, \quad u'(t_0) = x'(t_0). \end{gathered}$$

Since $x'' + \varepsilon\delta x' - r^2(x_\varepsilon + \pi) > 0$, it is readily seen that $x' > u'$ and $x + \pi > u$ on $(t_0, \frac{c}{\varepsilon} - A]$.

But we can find $u$ exactly, and it does not depend on $c$. We can then show that for sufficiently small $\varepsilon$ and some positive constants $C$ and $D$ depending only on $r$,

$$x_\varepsilon(t) + \pi > u(t) > C(x'_\varepsilon(t_0)(t - t_0) + x_\varepsilon(t_0) + \pi)e^{D(t-t_0)}$$

for $t_0 < t \leq \frac{c}{\varepsilon} - A$. We now set $t = \frac{c}{\varepsilon} - A$ and use (15.30) to obtain that

$$x(t_0) + \pi + x'(t_0)(-t_0 + \frac{c}{\varepsilon} - A) < \frac{\nu}{C} e^{D(A+t_0-\frac{c}{\varepsilon})} \tag{15.31}$$

for $\frac{c-1}{\varepsilon} \leq t_0 \leq \frac{c}{\varepsilon} - A$. But $x_\varepsilon + \pi$ and $x'_\varepsilon$ are both positive in this interval, so that this exponential bound implies (15.29), for some $A$ and sufficiently small $\varepsilon$, proving Lemma 15.11. □

Our goal now is to show that $x(t) > \pi$ for some $t$ in $(\frac{c}{\varepsilon}, \frac{c+1}{\varepsilon})$. Suppose therefore that $x$ does not cross $\pi$ before $x' = 0$. We can further restrict $A$ and $\varepsilon$, if necessary, independent of $c$, to ensure that

$$\begin{aligned} &\textstyle\int_{\frac{c-1}{\varepsilon}}^{\frac{c}{\varepsilon}+A} I_x(t)dt > \frac{\mu_1}{4}, \\ &x(\tfrac{c}{\varepsilon} + A) > \pi - \tfrac{\mu_1}{8}, \\ &0 < x'(\tfrac{c}{\varepsilon} + A) < \min\left\{\tfrac{1}{8}, \tfrac{1}{\delta}\right\}. \end{aligned} \tag{15.32}$$

At this point we consider $A$ as fixed, while lowering $\varepsilon$ further if necessary.

**Lemma 15.12.** *For sufficiently small $\varepsilon$, $x'(t) > 0$ for $t > \frac{c}{\varepsilon} + A$ as long as $t \leq \frac{c+1}{\varepsilon}$, $x < \pi$, and*

$$\int_{\frac{c-1}{\varepsilon}}^{t} I_x(s)\, ds > \frac{\mu_1}{8}. \tag{15.33}$$

**Proof.** Integrate (15.26) from $\frac{c-1}{\varepsilon}$ to $t$. Also, use the inequalities

$$p(\varepsilon t) \leq p(c) - \lambda(\varepsilon t - c) \quad \text{for } \frac{c}{\varepsilon} \leq t \leq \frac{c+1}{\varepsilon}$$

and

$$p(\varepsilon t) \geq p(c) - \lambda(\varepsilon t - c) \quad \text{for } \frac{c-1}{\varepsilon} \leq t \leq \frac{c}{\varepsilon}$$

to obtain from (15.33) that

$$\frac{1}{2}(x'(t)^2 - x'(\frac{c-1}{\varepsilon})^2) + (p(c) + \lambda c)(\cos x(\frac{c-1}{\varepsilon}) - \cos x(t)) > \frac{\varepsilon \mu_1}{8}. \tag{15.34}$$

If $x(t) < -x(\frac{c-1}{\varepsilon})$ then the second term on the left is negative. If $x(t) \geq -x(\frac{c-1}{\varepsilon})$ then

$$0 \leq \cos x\left(\frac{c-1}{\varepsilon}\right) - \cos x\,(t) \leq \cos x\left(\frac{c-1}{\varepsilon}\right) + 1.$$

Setting $t_0 = \frac{c-1}{\varepsilon}$ in (15.31), with $A$ fixed, shows that $\cos x\left(\frac{c-1}{\varepsilon}\right) + 1 = o(\varepsilon)$ as $\varepsilon \to 0$, uniformly for $x$ in $\mathcal{Y}_{\varepsilon,c}$. In either case,

$$x'(t)^2 > \frac{\varepsilon \mu_1}{4} + o(\varepsilon), \tag{15.35}$$

so that $x'$ cannot vanish. This proves Lemma 15.12. □

The next result is unnecessary when $\delta = 0$, since in that case, $I_x > 0$ on $[\frac{c}{\varepsilon} + A, \frac{c+1}{\varepsilon}]$ as long as $x' > 0$ and $x < \pi$.

**Lemma 15.13.** *If* $\int_{\frac{c}{\varepsilon}+A}^{t_1} I_x(s)\,ds = -\frac{\mu_1}{8}$ *for some* $t_1$ *in* $\left(\frac{c}{\varepsilon} + A, \frac{c+1}{\varepsilon}\right)$, *then* $x(t) > \pi$ *somewhere in* $\left[\frac{c}{\varepsilon} + A, t_1\right]$.

**Proof.** This is only possible if $\delta > 0$. From (15.32) it follows that (15.33) holds on $[\frac{c}{\varepsilon} + A, t_1]$. Hence on $\left[\frac{c}{\varepsilon} + A, t_1\right]$, $x' > 0$ and $x'' \leq 0$ as long as $x \leq \pi$, and so again by (15.32), $0 < x' < \frac{1}{8}$. From (15.32) it is also seen that $\delta x'^2 < x'$, and using Lemma 15.12, and (15.32) once more, we get

$$\int_{\frac{c}{\varepsilon}+A}^{t_1} x'(s)ds > \int_{\frac{c}{\varepsilon}+A}^{t_1} \delta x'(s)^2 ds > \frac{\mu_1}{8}.$$

Still another look at (15.32) reveals that $x(t_1) > \pi$. This proves Lemma 15.13.

But now we see from the first relation in (15.32) that if $x$ does not cross $\pi$ in $[\frac{c}{\varepsilon} + A, \frac{c+1}{\varepsilon}]$, then (15.33) holds on this interval. This implies (15.35), so that $x' > \sqrt{\frac{\epsilon\mu_1}{4}} + o(\sqrt{\varepsilon})$. Hence, for small enough $\varepsilon$, independent of $c$, $x$ crosses $\pi$ before $t = \frac{c+1}{\varepsilon}$. This contradiction completes the proof of Lemma 15.6. □

**Proof.** Lemma 15.4 is a special case. To see this, take $\gamma < 0$ in (15.17). (We remarked earlier that the sign of $\gamma$ is not important in proving the theorem.) Also, choose some $\kappa \in (0,1)$ so that (15.18) holds with $\gamma$ replaced by $\kappa\gamma$. Further, let $\eta = \arccos \kappa$. Then $\left|\frac{d}{ds}[\gamma \sin s]\right| = |\gamma \cos s| \geq |\gamma| \cos \eta$ on $-\eta \leq s \leq \eta$. Therefore in Theorem 15.5 take $\lambda = -\gamma \cos \eta$. We define a solution $x$ by $x(-\frac{1}{\varepsilon}) = \hat{\alpha}$, $x'\left(-\frac{1}{\varepsilon}\right) = 0$ where $\hat{\alpha} \in (-\pi, 0)$ is chosen so that $x(0) = 0$. Such a choice of $\hat{\alpha}$ is possible, since if $\alpha$ is close to $-\pi$, then the first zero of $x(t)$ is in $t > 0$, while if $\alpha$ is close to 0 then the first zero of $x$ is in $t < 0$. Then Lemma 15.6 assures us that Lemma 15.4 is valid (with $\overline{\alpha} = \hat{\alpha}$). □

The proof of Lemma 15.7 is similar, making use of the inequalities

$$p(\varepsilon t) \geq p(c) + \lambda(\varepsilon t - c) \quad \text{for } \frac{c}{\varepsilon} \leq t \leq \frac{c+1}{\varepsilon},$$

$$p(\varepsilon t) \leq p(c) + \lambda(\varepsilon t - c) \quad \text{for } \frac{c-1}{\varepsilon} \leq t \leq \frac{c}{\varepsilon}.$$

We omit the details. This completes the proof of Theorem 15.5. □

## 15.7. Final remarks

Theorems 15.3 and 15.5 by no means tell the whole story, even for (15.15) with $\delta = 0$. One should not conclude that most solutions behave chaotically.

While we show that there are uncountably many erratic solutions, it appears from numerical calculations that for small $\varepsilon$ most solutions eventually settle down into small oscillations around an even multiple of $\pi$, that is, the downward vertical position of the pendulum.

However for $\varepsilon$ larger, the downward vertical position can become unstable. This is seen by considering the linearization of (15.17) around $(0,0)$, which is

$$x'' + \varepsilon\delta x' + (1 + \gamma \sin \varepsilon t)\, x = 0,$$

a Mathieu equation (damped if $\delta > 0$). See [**41**] for the undamped case, and there are extensions to the damped case, such as [**238**]. In certain parameter ranges, such as $\varepsilon$ around 2, solutions with low amplitude initially can gain energy and escape the "potential well". One recent study of this for (15.1) is in [**20**].

We have stated elsewhere that one advantage of the shooting method is that sometimes results can be given for larger values of $\varepsilon$. In the papers [**153**] and [**154**] C. Lu showed that this was the case. In [**153**] he modified the proof above to show that for each $\varepsilon > 0$ there is a $\gamma(\varepsilon)$ such that the conclusion of Theorem 15.3 holds if $0 < \gamma < \gamma(\varepsilon)$. In [**154**] he gives specific estimates on $\gamma$ and $\varepsilon$ which ensure that the same conclusions can be proved. The proofs involve only classical analysis.

The type of chaos obtained in this section is weaker in one sense than that obtained using Melnikov's method. As remarked earlier, one hallmark of chaos (e.g. [**51**]) is sensitivity to initial conditions, which was defined above in connection with the horseshoe map. But for the pendulum model we proved only a limited form of this sensitivity. We constructed a mapping from the set of all sequences of zeros and ones to a set of solutions of (15.17), but this map is not 1:1.[12] From our construction it is possible that an open set of initial conditions all correspond to the same infinite sequence of rotations. In Chapters 16 and 18 we discuss a stronger type of sensitivity to initial conditions, corresponding to that found in a hyperbolic horseshoe map. We give an outline of how to establish this using shooting in Chapter 16, but for details the reader is referred to [**90**] and [**4**].

Even a hyperbolic horseshoe mapping does not automatically qualify as "chaotic" in the strongest sense, which requires that the set of complicated solutions is an attractor for the flow. Then, almost all solutions have complicated and unpredictable behavior. The proof of this result, as far as we know, has been carried out only for the well-known equations of Lorenz. This proof, by Tucker in [**243**], required a sophisticated rigorous numerical

[12] Such a mapping has been called a "topological horseshoe", as opposed to the "hyperbolic" horseshoe map described earlier in this chapter.

approach based on the concept of a "geometric attractor", a theory developed by Guckenheimer [80] and found also in [81].

## 15.8. Exercises

**Exercise 15.1.** Does the dynamical system defined by $x' = x$ display sensitive dependence on initial conditions? If so, why is there no "chaos" for this equation? What added feature do the equations in this chapter have?

**Exercise 15.2.** Consider an autonomous system $x' = f(x)$ in $R^3$, where all partial derivatives $\frac{\partial f_i}{\partial x_j}$ are continuous. Suppose (for simplicity) that $f_1(x) \geq 1$ for all $x$. Suppose also (to avoid blowup ) that $f$ is bounded. For each $\alpha \in R^3$, let $x(t, \alpha)$ denote the unique solution such that $x(0) = \alpha$. With $\alpha = (0, \alpha_2, \alpha_3)$, consider the set $K$ of points $\hat{\alpha} = (\alpha_2, \alpha_3)$ in $R^2$ such that $|\alpha_i| \leq 1$ for $i = 2, 3$.

(a) Show that for each such $\alpha$ there is a unique $t(\alpha) > 0$ such that $x_1(t(\alpha), \alpha) = 1$.

(b) Let $P : K \to R^2$ be defined by $P(\alpha) = (x_2(t(\alpha), \alpha), x_3(t(\alpha), \alpha))$. Show that $\det D(P)|_{\hat{\alpha}} > 0$ for each $\hat{\alpha} \in K$. Hint: Consider a fundamental solution of the linear system $y' = D(f)(x(t))\, y$.

(c) Show that $P$ preserves the orientation of every small square subset of $K$. (Hence, it preserves the orientation of each orientable subset of $K$.)

**Exercise 15.3.** Give the exact formulas for $Q|_{H_i}$ and $Q^{-1}|_{V_i}$ for $i = 1, 2$, where $Q$ is the horseshoe map defined in Section 15.2.3.

**Exercise 15.4.** (a) Apply a shooting method to show that for each $\varepsilon > 0$ the Duffing equation

$$\varepsilon x'' = x^3 - x + \cos t \tag{15.36}$$

has at least three solutions with period $2\pi$, one which is positive, one which is negative, and one which changes sign. Hint: Consider solutions with $x'(0) = 0$. Try some computer experiments.

(b) Still using shooting, show that for sufficiently small $\varepsilon > 0$, (15.36) has a solution of least period $2n\pi$, for each $n$. Show that there are bounded solutions which are not periodic. Prove a theorem similar to Theorem 15.3 for this equation, where the unpredictability is in the gaps between sign changes of the solution. (See [5] for more results on this equation, including a result like part (b) for a specific range of values of $\varepsilon$.)

**Exercise 15.5.** (Challenging) Prove that the two limits discussed after (15.14) can be interchanged. Hint: For $\varepsilon > 0$ consider

$\int_0^{2\pi} x'_\varepsilon(t, \psi_0(\varepsilon), 0)(\gamma \cos t - \delta x'_\varepsilon(t, \psi_0(\varepsilon), 0))\, dt.$

**Exercise 15.6.** Prove Lemma 15.10. Show that it is sufficient to prove that

$$\lim_{\varepsilon \to 0} \left( x' \left( \frac{c}{\varepsilon} \right) - x'_{\varepsilon,c} \left( \frac{c}{\varepsilon} \right) \right) = 0$$

uniformly for all $c$ and all $x \in \mathcal{Y}_{\varepsilon,c}$. Use proof by contradiction and the fact that on $\left[\frac{c-1}{\varepsilon}, \frac{c}{\varepsilon}\right]$, $x' > 0$ and $0 \geq x \geq -\pi$. Observe that $|x'|$ is uniformly bounded on $\left[-\frac{1}{\varepsilon}, 0\right]$ for $x \in Y_{\varepsilon,c}$ and the bound can be chosen independent of $c$ and $\varepsilon$ if $\varepsilon$ is small. Also, for any $c$, $x_{c,\varepsilon}\left(\frac{c}{\varepsilon} + t\right)$ depends only on $t$ and $p(c)$ and not explicitly on $c$ or $\varepsilon$. The range of $p(c)$ is restricted to $[q, Q]$.

*Chapter 16*

# Layers and spikes in reaction-diffusion equations, I

## 16.1. Introduction

The theory of pattern formation in physical media is an important source of challenging boundary value problems for both pde's and ode's. For example, the equations in Chapter 8, such as the needle crystal model for freezing (8.8), are of this type. Spatial patterns for which differential equation models exist include rotating spirals in certain chemical reactions [**68**], pattern formation during embryo development and many other biological examples, [**156**], and phase transitions in freezing or melting, to give a small sample. See [**71**] for a number of other examples.

A recent book review [**120**] contains more recent references. However many of these are aimed at broad scientific audiences and do not emphasize mathematical rigor. One that does is the monograph [**190**] by Peletier and Troy, which discusses an important class of fourth order boundary value problems of importance in several areas of physics, such as Rayleigh-Bénard convection, the description of water waves driven by gravity and capillarity, and layering phenomena in second order materials. Shooting methods are employed frequently in this book, which gives a clear exposition of their application to these higher order problems. Variational techniques are also employed, and there is a discussion of one of the unsolved problems in our final chapter, the problem of the suspension bridge.

We will consider reaction-diffusion equations, which are typically of the form

$$u_t = \Delta u + f(u)$$

in a spatially homogeneous setting, or more generally,

$$u_t = \Delta u + f(x, u) \tag{16.1}$$

if spatial inhomogeneities are present. Our focus for most of the next three chapters will be on time-independent solutions, obtained by setting $u_t = 0$. (In Chapter 18 we also prove the stability of some of these solutions.) If the spatial variable $x$ is one dimensional, we get a second order ode

$$u'' + f(x, u) = 0. \tag{16.2}$$

This equation will be studied mostly on a finite interval, say $[0, 1]$, with Neumann boundary conditions

$$u'(0) = u'(1) = 0. \tag{16.3}$$

One may think of $u$ as representing the distribution of a chemical, or other state variable, and, for example, zeros of $u$ could denote points of transition between one state and another. The "pattern" is the way in which these transition points are distributed along the $x$-axis. Neumann boundary conditions indicate that there is no flux of material across the boundaries of the region to be studied. Of course, in higher dimensions, there are many more possibilities, but analysis in such cases is usually much more difficult.[1]

Many problems in pattern formation have been approached using the calculus of variations, because stable physical patterns are often the minima of energy functionals. Some of the equations considered in Chapter 18 have been studied in this manner. We discussed variational approaches earlier, for example in Chapter 11, though not in much detail. Similarly, we will outline the application of this method to a problem in Chapter 18. Of course, we cite references which contain more details. A second important method is to use what are called subsolutions and supersolutions, or upper and lower solutions. Like the calculus of variations, this technique is important for both ode and pde problems. Once again, Chapter 18 presents an example of the use of this method. For a more general exposition, see [**46**] and references therein.

The study of patterns using ordinary differential equation is inherently limited. There is no way that the shape of a snowflake, in all its complexity and beauty, could be completely explained using ode's. Our patterns either are one dimensional or exhibit radial symmetry in two or three dimensions. They do not evolve in time, as they are either steady-state solutions, as

[1] Spatial symmetry can simplify a higher dimensional problem, often leading to an ode. Still, the problem is likely to be more difficult than those we discuss in these three chapters.

described above, or result from a traveling wave substitution, as in the FitzHugh-Nagumo equation studied in Chapter 6. It is therefore important to study the stability of the solution we obtain, with respect to (16.1). Because of space limitations, we study only one stability problem, and that is in Chapter 18.

As in many of our chapters, the equations involve a small parameter $\varepsilon$, and so steady-state solutions change on two spatial scales, with regions of slight variation separated by small intervals of large transitions or oscillations. The model to be discussed in this chapter comes from the theory of shallow water waves, where it has been described as representing features of the patterns resulting from "sloshing" in a bathtub [**182**]. In this ode model, which is quadratic in the unknown function $y$, and nonautonomous, the pattern is seen in steady-state solutions with sharp "spikes", or large deviations from the basic periodic structure over a short interval. The results in this section are a second instance in this book where we describe a kind of "chaos", in this case spatial, and manifested in the sensitivity of the spike pattern to the value of $u(0)$. We discussed chaos in the last section of Chapter 15, where we pointed out the difference between a "topological horseshoe map" and a "hyperbolic horseshoe map", especially as regards sensitivity with respect to initial conditions. For the sloshing model one can prove the type of sensitivity seen in the standard (hyperbolic) Smale horseshoe (see [**90**]), but here we only have space for an outline of this proof. We will also describe briefly some other approaches to the same problem.

## 16.2. A model of shallow water sloshing

We study the equations for two-dimensional irrotational motion induced in a rectangular tank of water which is forced to oscillate horizontally in a periodic fashion. The equations are derived by H. Ockendon, J. Ockendon, and Johnson in [**182**], or see [**90**] for an outline of the derivation. The equation which results is

$$\varepsilon^2 y'' = y^2 - 1 - c - \cos x. \tag{16.4}$$

Here, $y$ is a function defined in terms of the velocity potential of the motion. The parameter $c$ is positive, and the physical interest is in $2\pi$-periodic solutions. After discussing these, we will consider solutions with larger periods, and even solutions which are bounded on $(-\infty, \infty)$ but aperiodic.

We will look for periodic solutions of (16.4) such that

$$y'(0) = 0, \tag{16.5}$$

and from the symmetry of the cosine term in (16.4) it is apparent that such solutions are even functions of $x$. A numerical study is not difficult, and in Figure 16.1 we see what appear to be three periodic solutions. Actually,

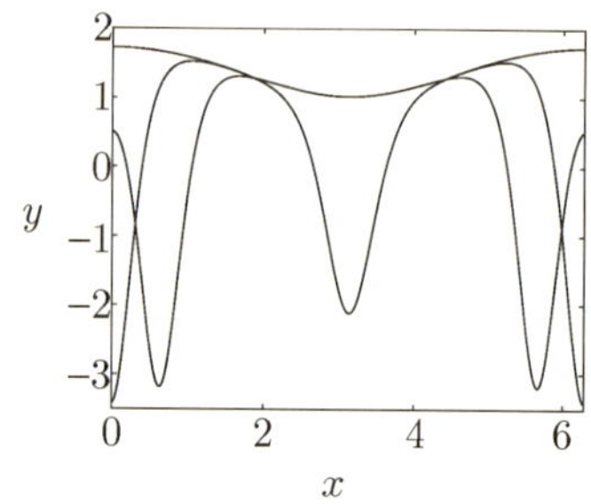


**Figure 16.1.** Six solutions.

however, the plots of six $2\pi$-periodic solutions are given, several with initial conditions too close together to distinguish on this graph. They all satisfy $y'(\pi) = 0$, and so are symmetric around $x = \pi$. The three pairs are denoted at $x = 0$ by $A$, $B$, and $C$ on the graph, and within each pair they are distinguished by whether or not they have a spike at $\pi$.

We will outline the proof that these solutions exist. The first lemma is obvious from the inequality

$$\varepsilon^2 y'' \leq y^2 - c.$$

**Lemma 16.1.** *A solution of* (16.4) *does not have an interior local minimum where* $-\sqrt{c} < y < \sqrt{c}$.

Hence, local minima can be characterized as either positive or negative. Using this lemma, it is not difficult to obtain two of the solutions depicted in Figure 16.1. To describe these two, we let $y(x, \alpha)$ denote the unique solution to (16.4)–(16.5) with $y(0) = \alpha$.

**Theorem 16.2.** *There is a unique* $\alpha_0$ *(depending on* $\varepsilon$*) such that the solution* $y(x, \alpha_0)$ *of* (16.4)–(16.5) *with* $y(0) = \alpha_0$ *is positive and periodic, of period* $2\pi$. *There is an* $\alpha_1 < \alpha_0$ *such that* $y(x, \alpha_1)$ *has a single minimum in* $[0, 2\pi]$, *this minimum being at* $x = \pi$ *with* $y(\pi) < -\sqrt{c}$. *For sufficiently small* $\varepsilon > 0$, $y(0, \alpha_1) > 0$ *and* $\alpha_1$ *is also unique.*

**Remark 16.3.** In **[90]** it is shown that there are no $2\pi$-periodic solutions with a negative minimum close to but not at $\pi$.

The proof of this theorem, and the next, will be discussed in the next section. To obtain a spike, we let $\varepsilon \to 0^+$ and consider the limiting forms of $y(x, \alpha_0)$ and $y(x, \alpha_1)$ as follows. Since we need the asymptotic behavior in order to prove uniqueness of the second solution (though not the first), we will not assume uniqueness in stating the next result.

**Theorem 16.4.** *For each $\varepsilon > 0$, let $y_0^\varepsilon$ be some positive periodic solution of (16.4), and let $y_1^\varepsilon$ be some periodic solution with a single negative minimum in $[0, 2\pi]$, this minimum being at $x = \pi$. Then*

$$\lim_{\varepsilon \to 0^+} y_0^\varepsilon (x) = \sqrt{1 + c + \cos x}$$

*uniformly in $[0, 2\pi]$. Also*

$$\lim_{\varepsilon \to 0^+} y_1^\varepsilon (x) = \sqrt{1 + c + \cos x}$$

*uniformly on each set of the form $[0, \pi - \delta] \cup [\pi + \delta, 2\pi]$, where $0 < \delta < \pi$, while*

$$\lim_{\varepsilon \to 0^+} y_1^\varepsilon (\pi) = -2\sqrt{c}.$$

Previously in this chapter $\varepsilon$ was not necessarily small, but now we are considering a problem in singular perturbation. More detailed information about these solutions, including uniform asymptotic expansions, will be given in Chapter 17.

The solutions $y_1^\varepsilon$ exhibit spikes for sufficiently small $\varepsilon$, while the solutions $y_0^\varepsilon$ do not. We will say that the "spike" in $[0, 2\pi]$ is the section of the graph of $y_1^\varepsilon$ over this interval which is below the $x$-axis. The spike of $y_1^\varepsilon$ is contained in an interval $\pi - \delta \leq x \leq \pi + \delta$ for some $\delta (\varepsilon)$ which tends to zero with $\varepsilon$. In the notation of Theorem 16.2, $y (x, \alpha_1)$ has spikes at odd multiples of $\pi$, while the positive solution $y (x, \alpha_0)$ has no spikes.

Five of the six solutions depicted in Figure 16.1 have spikes. The two solutions mentioned in Theorem 16.4 are labeled $A$. The proof that these exist is given next. The existence of the others shown in Figure 16.1 is proved in [**90**].

## 16.3. Proofs

### 16.3.1. Proof of existence (Theorem 16.2).

**Proof.** As usual, consider the initial value problem

$$\begin{aligned} \varepsilon^2 y'' &= y^2 - 1 - c - \cos x, \\ y (0) &= \alpha, \quad y' (0) = 0. \end{aligned} \tag{16.6}$$

Let $\kappa = \sqrt{2 + c}$, and observe that if $\alpha = \kappa$ then $y'' (0) = y''' (0) = 0$, and $\varepsilon^2 y^{(iv)} (0) = 1$. Hence, $y$ initially increases, and since $y'' > 0$ if $y > \kappa$, $y$ is monotonic increasing and tends to infinity (in finite "time"). This solution is depicted in Figure 16.2, labeled as curve B.

Also, if $\alpha = \kappa - \delta$ for some small $\delta > 0$, then $y'' (0) < 0$, so that $y$ initially decreases, but by continuity with respect to $\alpha$, $y$ turns around,

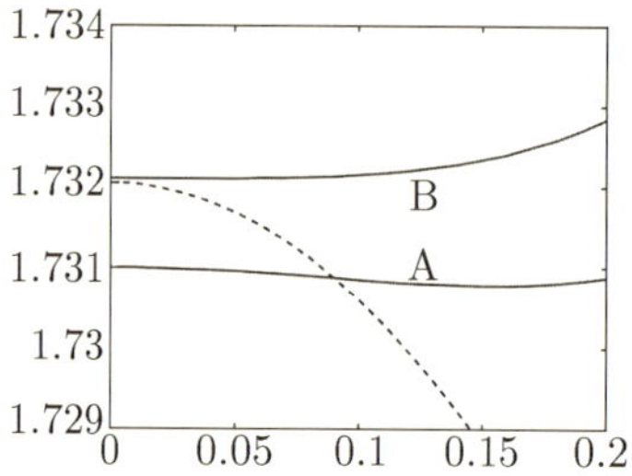


**Figure 16.2.** Two solutions (solid curves) and the upper curve of zeros of $u''$ (dotted). (See text.)

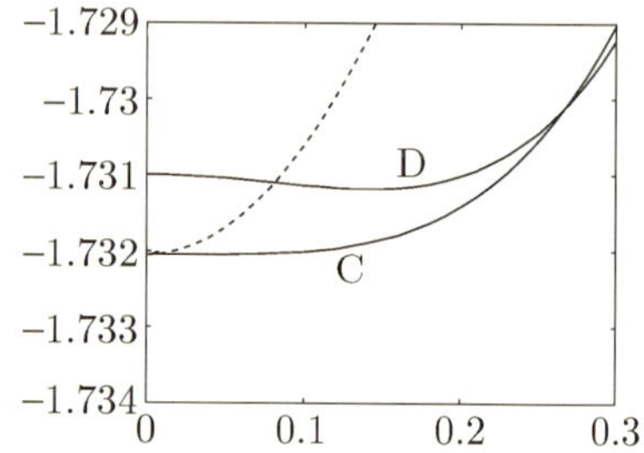


**Figure 16.3.** Two more solutions and the lower curve of zeros of $u''$.

increases beyond $\kappa$, and again blows up. Further, the minimum value of $y$ is close to $\kappa$, and therefore positive. Curve A in Figure 16.2 is such a solution.

Hence, if $\alpha < \kappa$ and $\kappa - \alpha$ is small, then initially $y$ decreases, but $y'$ has a first positive zero, which we denote by $\xi_1(\alpha)$, and $\xi_1(\alpha) \to 0$ as $\alpha \to \kappa^-$. We claim that if $\xi_1 = \xi_1(\alpha) < \pi$ then $y''(\xi_1) > 0$. If not, and $y''(\xi_1) = 0$, then $y'''(\xi_1) = \sin \xi_1 > 0$. But this implies that just to the left of $\xi_1$, $y'' < 0$ and $y' > 0$, and so $y'$ has a previous zero in $(0, \xi_1)$, contradicting the definition of $\xi_1$ as the first zero of $y'$.

It also follows that for small positive $\kappa - \alpha$, $\xi_1(\alpha)$ is the first, and only, minimum of $y$ and that $y$ then tends to infinity.

Because $y''(\xi_1(\alpha)) > 0$, $\xi_1(\cdot)$ can be continued as a continuous function of $\alpha$, as $\alpha$ decreases, at least until a value is reached where $\xi_1(\alpha) = \pi$, at which point the argument that $y''(\xi_1(\alpha)) > 0$ fails. Further, since $y$ has no minimum in $(-\sqrt{c}, \sqrt{c})$, $y(\xi_1(\alpha)) > 0$. However, if $y(0) = 0$, this is obviously false, because $y''(0) < 0$ and so the first minimum of $y$ must be negative. Hence, there is an $\alpha_0 \in (0, \kappa)$ with $\xi_1(\alpha_0) = \pi$. The symmetry of the cosine function implies that $y(x, \alpha_0)$ is symmetric around $\pi$, and therefore must satisfy $y'(2\pi) = 0$, giving us the desired positive $2\pi$-periodic solution of (16.4).

For the second solution, note that if $y(0) = -\kappa$ then at $x = 0$, $y' = y'' = y''' = 0, y^{(iv)} > 0$. This solution is labeled C in Figure 16.3. Therefore, if $\alpha = -\kappa + \mu$ for a sufficiently small $\mu > 0$, then $y$ initially decreases, but $y'$ again has a zero, say at $\xi_1(\alpha)$ as before, with $y''(\xi_1(\alpha)) > 0$. Such a solution is shown in curve D of Figure 16.3. Now the minimum of $y$ is negative, and by increasing $\alpha$ we find a solution $y(x, \alpha_1)$ such that this negative minimum occurs at $\pi$, giving the second solution of Theorem 16.2. We leave further details of the proof to the reader, in Exercise 16.1. □

**16.3.2. Proof of asymptotic behavior (Theorem 16.4).** We concentrate on the second solution, $y_1^{\varepsilon}$. If the result is false then there is a sequence $(\varepsilon_n)$ such that either (i) $y_1^{\varepsilon_n}(\pi)$ converges to some $\mu \neq -2\sqrt{c}$ or else (ii) there is a sequence $x_n$ converging to $\rho \in [0,\pi)$ such that $y_1^{\varepsilon_n}(x_n)$ converges to some $\mu \neq \sqrt{1+c+\cos\rho}$. In case (i) we rescale (16.4) by letting $x = \pi + \varepsilon s$, $y(x) = u(s)$. Then

$$u'' = u^2 - 1 - c - \cos(\pi + \varepsilon s), \tag{16.7}$$
$$u(0) = y(\pi), \quad u'(0) = 0.$$

Consider some compact interval $[-S, S]$ and the solution $U$ to

$$U'' = U^2 - c, \tag{16.8}$$
$$U(0) = \mu. \tag{16.9}$$

The $(U, U')$ phase plane for (16.8) is easily drawn. There is a homoclinic orbit based at $(\sqrt{c}, 0)$, with its minimum at $-2\sqrt{c}$. If $-2\sqrt{c} < \mu < 0$ then the solution of (16.8)–(16.9) is periodic, and so there is an $S > 0$ such that this solution has more than one negative minimum in $[-S, S]$. If $\mu < -2\sqrt{c}$ then this solution blows up in finite time, and $S$ can be chosen so that the solution exists on $[-S, S]$ but rises above $2\kappa$ in this interval. In either case, a contradiction is obtained in case (i) above, as $\varepsilon_n \to 0$, since the solutions $u(s) = y_1^{\varepsilon}(\pi + \varepsilon s)$ of (16.7) have neither of these behaviors.

Exercise 16.2 is to show that a similar argument works for (ii).

**16.3.3. Proofs of uniqueness.**

*The easy case.* The proof that the positive $2\pi$-periodic solution is unique can be done by either of the methods used to prove Proposition 12.1. Suppose, for example, that there are two positive $2\pi$-periodic solutions, $y_1$ and $y_2$. Suppose that $y_1(0) > y_2(0)$, and let $z = y_1 - y_2$. Then $z(0) > 0$, $z'(0) = 0$, and

$$z'' = (y_1 + y_2)\, z.$$

Positivity of $y_1$ and $y_2$ then implies that $z' > 0$ on $[0, 1]$, so that $y_1'(2\pi) > y_2'(2\pi)$, a contradiction.

*The difficult case.* Uniqueness in this and similar problems is often much harder to prove than existence, and we will only outline the argument, referring the reader to [**90**] for details. This result is important not only for establishing uniqueness but also for showing the complicated behavior advertised earlier.

Suppose that there are two solutions, say $y(x, \alpha_1)$ and $y(x, \alpha_2)$ of (16.4)–(16.5) with a negative minimum in $[0, 2\pi]$, this minimum occurring at $x = \pi$. Suppose that $\alpha_1 < \alpha_2$. From the asymptotic behavior found above, we can assume that $y(x, \alpha_1)$ is close to $\sqrt{1+c+\cos x}$ at least on $\left[0, \frac{\pi}{2}\right]$, and the comparison technique used in the "easy case" shows that if $\alpha > \alpha_1$ then $y(x,\alpha) > y(x,\alpha_1)$ on $\left[0, \frac{\pi}{2}\right]$, unless $y(x,\alpha)$ blows up before $x = \frac{\pi}{2}$. Further, there is no blowup before $x = \frac{\pi}{2}$ if $\alpha \in [\alpha_1, \alpha_2]$, since this does not occur at $\alpha_2$. Finally, the same comparison technique shows that if $x_1(\alpha)$ is the first zero of $y(x,\alpha)$ then $x_1$ is an increasing function of $\alpha$ on $[\alpha_1, \alpha_2]$. For small $\varepsilon$, use the rescaling $x = x_1(\alpha) + \varepsilon s$, to get (16.7) and prove the next result.

**Lemma 16.5.** *For sufficiently small* $\varepsilon > 0$, *every solution* $y(x,\alpha)$ *of* (16.4)–(16.5) *with* $\alpha \in [\alpha_1, \alpha_2]$ *has a negative minimum in* $\left[\frac{\pi}{2}, \frac{3\pi}{2}\right]$, *followed by a second zero* $x_2(\alpha) < \frac{3\pi}{2}$.

We also need some quantitative estimates on the solutions $y = y(x,\alpha)$ in the region of a spike. In this lemma, we let $x_i = x_i(\alpha)$ for $i = 1, 2$, and $\xi$ (which also depends on $\alpha$) is the unique zero of $y'$ in $(x_1, x_2)$.

**Lemma 16.6.** *There are constants* $M_1$ *and* $M_2$ *independent of* $\varepsilon$, *and a* $\delta > 0$ *which depends on* $\varepsilon$, *such that for* $\alpha \in [\alpha_1, \alpha_2]$,

$$\begin{aligned} |y'| &\geq \frac{M_1}{\varepsilon} && \textit{on } [x_1, \xi - \delta] \cup [\xi + \delta, x_2], \\ y'' &\geq \frac{M_1}{\varepsilon^2} && \textit{on } [\xi - \delta, \xi + \delta], \\ |y'| &\leq \frac{M_2}{\varepsilon} && \textit{on } [x_1, x_2], \\ |x_2 - x_1| &\leq M_2 \varepsilon. \end{aligned}$$

The proof of this lemma is a bit laborious, but not hard. See [**90**].

The proof of uniqueness in Theorem 16.2 now hinges on the equation of first variation, as introduced in Chapter 12:

$$\varepsilon^2 V'' = 2yV. \tag{16.10}$$

One solution is $v = \frac{\partial y(x,\alpha)}{\partial \alpha}$, which also satisfies the initial conditions

$$\begin{aligned} v(0) &= 1, \\ v'(0) &= 0. \end{aligned} \tag{16.11}$$

In this, as in many uniqueness proofs, the analysis of $v$ is crucial. The case of (16.10) is typical of singularly perturbed uniqueness (and stability) problems, because there is a short interval of rapid change of $y$ during which the behavior of $v$ is hard to determine.

**Lemma 16.7.** *For sufficiently small* $\varepsilon$ *and* $\alpha_1 \leq \alpha \leq \alpha_2$, *the solution* $v$ *of* (16.10)–(16.11) *has exactly one zero in* $[0, 2\pi]$, *and this zero occurs in the interval* $(x_1(\alpha), x_2(\alpha))$. *Further,* $v'(x_2(\alpha)) < 0$.

**Proof.** (Brief outline) The main idea in the proof is to compare the solution $v$ of (16.10)–(16.11) with $w = -y'(x, \alpha)$, which we know a lot about from Lemma 16.6. The reason this can be done is that $v$ satisfies (16.10) while $-y'$ satisfies

$$\varepsilon^2 w'' = 2yw - \sin x,$$

and in most of the interval $[x_1(\alpha), x_2(\alpha)]$ we expect $2yw$ to be much larger than $\sin x$. Note that $w$ has the behavior claimed for $v$ in the lemma.

The initial conditions of $v$ and $w$ are different:

$$v(0) = 1, \quad v'(0) = 0,$$
$$w(0) = 0, \quad w'(0) = -y''(0).$$

It is shown in [**90**] that this does not make much difference.[2] □

This lemma implies that $v(x_2(\alpha))$ is negative. Since $y'(x_2(\alpha)) > 0$, the implicit function theorem shows that $x_2$ is a continuous and increasing function of $\alpha$ in the interval $[\alpha_1, \alpha_2]$. The same was seen to be true for $x_1$. Since $y(x, \alpha_1)$ is symmetric around $\pi$, $\pi - x_1(\alpha_1) = x_2(\alpha_1) - \pi$. But then, $\pi - x_1(\alpha_2) < x_2(\alpha_2) - \pi$, and $y(x, \alpha_2)$ could not be symmetric around $\pi$, which proves the uniqueness of the solution $y_2^\varepsilon$ found in Theorem 16.2 if $\varepsilon$ is sufficiently small.

At this stage we have not used the last conclusion of Lemma 16.7, that $v'(x_2(\alpha)) < 0$. This inequality can be used to strengthen the uniqueness result by ruling out $2\pi$-periodic spike solutions where the spike is not exactly at $\pi$, but it also plays an important role in the next section.

## 16.4. Complicated solutions ("chaos")

Our interest in this section is to consider solutions with longer periods, or which are bounded but possibly aperiodic. We will describe a family of solutions which are close to the solutions $y(x, \alpha_0)$ and $y(x, \alpha_1)$ in Theorem 16.2 most of the time but have a complicated pattern of spikes and nonspikes at successive odd multiples of $\pi$. We are able to analyze these solutions in more detail than was possible for the complicated solutions found in Chapter 15. Our justification for the term chaos is stronger than it was in Chapter 15, because it is possible in this sloshing model to prove "hyperbolicity", giving a greater sensitivity to initial conditions. This result depends on extensions of Lemma 16.7. But the details are again more than we can give in this book. We will state the theorem and discuss a few key ideas of the proof,

[2] In [**90**] there is a misprint in the statement of this lemma, which is Lemma 9 of that paper. The last sentence should say "$u'(t_4) < 0$". Also, in Chapter 18 we give complete details of the proof of a similar result.

showing how they relate to the uniqueness proof outlined in the previous subsection.

**Theorem 16.8.** *There is an* $\varepsilon_1 > 0$ *such that if* $0 < \varepsilon < \varepsilon_1$ *and* $\sigma = (\sigma_1, \sigma_2, \ldots)$ *is a sequence of* 0*'s and* 1*'s, then there is a unique solution* $y_\sigma$ *of* (16.4)–(16.5) *such that the following hold:*

(a) $y_\sigma$ *exists and is bounded on* $[0, \infty)$.

(b) *If* $\sigma_n = 0$ *then* $y > \sqrt{c}$ *in* $I_n = [(2n-2)\pi, 2n\pi]$, *while if* $\sigma_n = 1$ *then the solution has a single negative minimum in this interval, at which* $y < -\sqrt{c}$.

(c)

$$\lim_{\varepsilon \to 0^+} y_\sigma(x) = \sqrt{1 + c + \cos x}$$

*if* $x \neq (2n-1)\pi$. *Further, if* $\sigma_n = 0$ *then this limit holds uniformly on* $I_n$, *while if* $\sigma_n = 1$ *then* $\lim_{\varepsilon \to 0^+} y_\sigma((2n-1)\pi) = -2\sqrt{c}$.

We could not make any uniqueness statements in Chapter 15.

**Idea of proof.** The proof, as in Chapter 15, is by induction, in which we construct a sequence of periodic solutions with longer and longer periods. For example, suppose that the sequence $\{\sigma_n\}$ starts with $\{1, 0, 1, 1, \ldots\}$. Then the construction starts with the solution $y(x, \alpha_1)$ of Theorem 16.2, which has period $2\pi$ and a spike at $\pi$. We will show below that by changing $\alpha$ a small amount from $\alpha_1$, we can find a $4\pi$-periodic solution which has a spike close to (but not exactly at) $\pi$ and is positive in $[2\pi, 4\pi]$. From periodicity it is seen that this solution also has a spike near $5\pi$, and a further small adjustment of $\alpha$ gives a nearby solution which has period $6\pi$ and spikes near $\pi$ and $5\pi$. Again using periodicity, we see that this solution already has the desired spike near $7\pi$ and, with further small adjustment, will have period $8\pi$.

We will give more detail about the first step, in which we change $\alpha$ from $\alpha_1$ just enough to remove the spike near $3\pi$, without affecting the spike near $\pi$. For this we need some additional observations about the solutions of the equation of first variation, (16.10). Suppose that $y > \frac{\sqrt{c}}{2}$ on some interval $[a, b]$ and that $v$ is a solution of (16.10) with $v(a) > 0, v'(a) > 0$. Then $v$ and $v'$ increase at an exponential rate over $[a, b]$, so that $\frac{v(b)}{v(a)} \geq e^{\sigma \frac{b-a}{\varepsilon}}$ and $\frac{v'(b)}{v'(a)} \geq e^{\sigma \frac{b-a}{\varepsilon}}$, for some $\sigma > 0$ which is independent of $\varepsilon$ and of the particular solution $y$ of (16.4). We will indicate this by saying that

$$v(b) >>> v(a),$$
$$v'(b) >>> v'(a).$$

In [**90**] it is shown that for sufficiently small $\varepsilon$ there is a $\delta(\varepsilon) > 0$ such that if $y$ is a solution of (16.4) with exactly one minimum in $[0, 2\pi]$, and this minimum is negative and occurs at or near $\pi$, then

(16.12)
$$v(\pi - \delta) >>> v(0) = 1, \qquad v'(\pi - \delta) >>> 1,$$
$$-\frac{v(\pi-\delta)}{v(\pi+\delta)} \text{ and } -\frac{v'(\pi-\delta)}{v'(\pi+\delta)} \text{ are positive and bounded independent of } \varepsilon,$$
$$-v(2\pi) >>> -v(\pi + \delta), \qquad -v'(2\pi) >>> -v'(\pi + \delta).$$

Recall that $y(x, \alpha_1)$ is periodic and so has a single spike in each of the intervals $[0, 2\pi]$ and $[2\pi, 4\pi]$. This solution corresponds to a sequence which starts off with $\sigma_1 = 1, \sigma_2 = 1$. Lemma 16.7 implies that $\frac{\partial y(2\pi,\alpha)}{\partial\alpha}|_{\alpha=\alpha_1}$ and $\frac{\partial y'(2\pi,\alpha)}{\partial\alpha}|_{\alpha=\alpha_1}$ are both negative. The stronger results in (16.12) allow us to show that an exponentially small decrease in $\alpha$ from the value $\alpha_1$ raises $y(2\pi)$ enough to change the solution significantly in $[2\pi, 4\pi]$, removing its spike at $3\pi$, without changing it much in $[0, 2\pi]$. It is shown in [**90**] that one can construct a unique solution with period $4\pi$ which has a spike in $[0, 2\pi]$, and no spike in $[2\pi, 4\pi]$. This solution corresponds to a sequence beginning with $\sigma_1 = 1, \sigma_2 = 0$.

To continue the induction it is important that for each $n$,

$$|v(2n\pi)| >>> |v(2(n-1)\pi)|, \qquad |v'(2n\pi)| >>> |v'(2(n-1)\pi)|.$$

These imply (as shown in [**90**]) that the adjustment of $\alpha$ needed to give a spike, or no spike, in $[2n\pi, 2(n+1)\pi]$ does not affect the spike pattern in $[0, 2n\pi]$. Periodic solutions are constructed of period $2n\pi$, for every $n$. In accordance with Theorem 16.8, aperiodic solutions with arbitrary aperiodic spike patterns are found as well. Knowledge about the sign of $v(2n\pi)$ for each $n$ also allows us to determine the ordering of initial conditions $\alpha$ which give different spike sequences $\{\sigma_n\}$. All this is spelled out in [**90**]. □

## 16.5. Other approaches

Two other methods have been suggested or applied to this problem. In [**116**], an approach using Melnikov's method (discussed in Chapter 15) was outlined. The relation between the two methods is similar to that discussed in Chapter 15. In addition, a number of problems are suggested where the shooting method may give additional insight complementary to that obtained from dynamical system methods. We are unaware, however, of further work exploring the suggestions made in [**116**].

In [**75**] the problem is viewed as a slowly varying perturbation of a Hamiltonian system. A general result is obtained for such systems, which are in the form

$$\frac{d\mathbf{u}}{dt} = J\nabla H(\mathbf{u}, \varepsilon t) + \varepsilon^2 h(t, \varepsilon t, \varepsilon), \tag{16.13}$$

where $\mathbf{u} \in R^2$, $J = \begin{pmatrix} 0 & 1 \\ -1 & 0 \end{pmatrix}$, and the $C^2$ function $H$ is the so-called Hamiltonian of the system.[3] Several of the systems studied in this book are Hamiltonian systems. However our techniques do not always exploit the special properties which make such systems important, and we do not discuss general Hamiltonian systems.

The equation (16.4) is equivalent to a perturbed Hamiltonian system, with $h = 0$ and

$$H(u, v, \lambda) = \frac{1}{2}v^2 - \frac{1}{3}u^3 + (1 + c + \cos\lambda)u. \tag{16.14}$$

In [**75**] systems of the form (16.13) are studied under hypotheses which are satisfied by (16.14). Among these is the assumption that for each $\lambda$ there are two equilibrium points, one of which has a saddle point structure including a homoclinic orbit which surrounds the second equilibrium. A second assumption is that the area $S(\lambda)$ surrounded by the homoclinic orbit oscillates as $\lambda$ increases, so that in some intervals $S'(\lambda) > 0$ and in others $S'(\lambda) < 0$. For the precise conditions on $H$, see [**75**].

The main tool in the analysis is what is known as the Conley index. This important concept was developed by Conley [**44**] and has been extended and applied by quite a few authors. An outline of the theory is given in [**75**]. Unfortunately we do not have space to consider it here.

Using Conley index the authors of [**75**] obtain the existence of complicated solutions, similar to those found in Theorem 16.2. Thus, they extend the results in [**90**] to more general Hamiltonian systems. However, the technique does not give any information about uniqueness.

## 16.6. Exercises

**Exercise 16.1.** (Routine) Complete the proof of Theorem 16.2.

**Exercise 16.2.** (Routine) Complete the proof of Theorem 16.4.

[3] We revert to standard notation for Hamiltonian systems, in which the independent variable is denoted by $t$.

Chapter 17

# Uniform expansions for a class of second order problems

## 17.1. Introduction

We now consider a class of singularly perturbed second order equations which includes a transformed version of the equation in the last chapter. However the results we will describe are of a different nature. Instead of discussing the existence of solutions with spike behavior, we give a method due to X. Chen and Sadhu [**32**] for obtaining uniform asymptotic expansions of such solutions. Their existence, which can be proved by shooting techniques, will be assumed. Uniform approximations of solutions to singularly perturbed boundary value problems were also obtained in Chapter 7, by a very different technique. At this writing, it appears that the methods are complementary, treating different classes of problems.

The equations treated in [**32**] are of the form

$$\varepsilon^2 u'' = u\,(q\,(x,\varepsilon) - u). \tag{17.1}$$

In the main result of this section we generalize this to

$$\varepsilon^2 u'' = u\,(q\,(x,\varepsilon) - f\,(u)), \tag{17.2}$$

where $f''$ is continuous and

$$\begin{gathered} f\,(0) = 0, \quad f' > 0 \text{ and } f'' \geq 0 \text{ on } (0,\infty), \\ \lim_{u\to\infty} f\,(u) = \infty. \end{gathered} \tag{17.3}$$

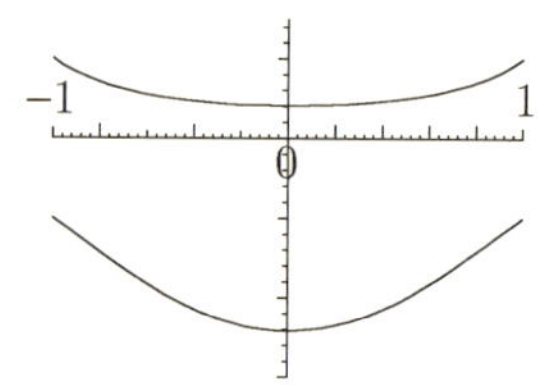


**Figure 17.1**

For simplicity we will prove some important estimates only for $f(u) = u$. It is crucial in this chapter that the right side of (17.2) has exactly two zeros in $0 \leq u < \infty$, in contrast to the equations to be studied in Chapter 18, where the right side is a cubic function of $u$ and there are three zeros. Hence the behavior exhibited by solutions of (17.2) is simpler than that to be considered in Chapter 18. In the terminology of that chapter there are spikes, but no layers except at the boundary. It appears that the method used here does not apply to the case where the right side has three roots.

## 17.2. Motivation

**17.2.1. Carrier's problem.** The example which prompted this work is due to Carrier [**29**]. This problem has been much studied over the years, and we believe that the methods in [**32**] add significantly to the known rigorous results. A formal asymptotic analysis (done originally by Carrier) appears in the well-known monograph by Bender and Orszag [**18**]. For more recent rigorous qualitative results, see [**3**], [**155**], [**152**], and [**256**].

The original form of the Carrier equation is

$$\varepsilon^2 y'' + 2\left(1 - x^2\right) y + y^2 = 1. \tag{17.4}$$

Equation (17.4) is studied on the interval $-1 \leq x \leq 1$, with the Dirichlet boundary conditions

$$y(-1) = y(1) = 0. \tag{17.5}$$

The outer solution $y_o$ is obtained by formally setting $\varepsilon = 0$ in (17.4), after which we solve for $y_o$.[1] The resulting quadratic has two solutions,

$$y_o^{\pm}(x) = x^2 - 1 \pm \sqrt{1 + (1 - x^2)^2}.$$

We plot $y_o^+$ and $y_o^-$ in Figure 17.1. Neither satisfies either of the boundary conditions. Therefore, if solutions to (17.4)–(17.5) exist then we expect boundary layers at $x = \pm 1$. There is a formal inner solution at each $x_0 \in [-1, 1]$, found by setting $\xi = \frac{x - x_0}{\varepsilon}$ and letting $y(x) = Y(\xi)$. Anticipating that we are looking for a boundary layer, we set $x_0 = -1$, for example.

[1] See Chapter 7 for "outer" and "inner" solutions

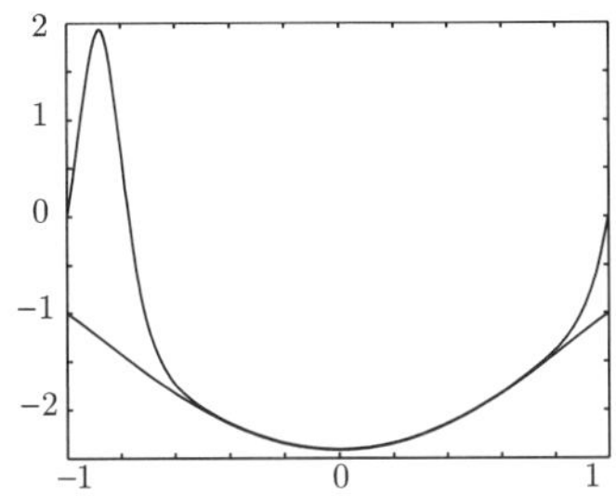


**Figure 17.2**

Making this substitution and then setting $\varepsilon = 0$ gives $Y'' + Y^2 = 1$. A simple phase plane argument shows that there are two solutions such that $Y(0) = 0$ and $\lim_{\xi\to\infty} Y(\xi) = -1$, which is the value of the outer solution $y_o^-$ at the boundary $x = -1$. Each of these solutions lies on a homoclinic orbit based at $(-1, 0)$ in the $(Y, Y')$-plane. One starts at $Y = 0$ on the upper half of this orbit, with $Y$ increasing to a maximum of 2 and then decreasing and tending towards $-1$. The other starts at $Y = 0$ on the lower half of this orbit, with $Y$ decreasing monotonically towards $-1$. In Figure 17.2 we show a solution of (17.4) which exhibits the first of these behaviors at the left boundary and the mirror image of the other at the right boundary. We can use a shooting method to prove that these solutions exist. See Chapter 18 for such proofs in the cubic case.

Our interest here is in uniform asymptotic formulas for these solutions. Such formulas are derived formally, with the method of matched asymptotic expansions, in [**18**], which gives the expansions

$$
\begin{aligned}
(17.6)\quad y = x^2 - 1 - \sqrt{1 + (1 - x^2)^2} &+ 3\,\mathrm{sech}^2\left(\pm\frac{1-x}{\sqrt{2\varepsilon}} + \tanh^{-1}\sqrt{\frac{2}{3}}\right)\\
&+ 3\left(\mathrm{sech}^2\left(\pm\frac{1+x}{\sqrt{2\varepsilon}} + \tanh^{-1}\left(\sqrt{\frac{2}{3}}\right)\right)\right).
\end{aligned}
$$

This represents four different solutions, according to the plus-minus choices. For example the choice of a minus sign in the first $\mathrm{sech}^2$ term and a plus sign in the second $\mathrm{sech}^2$ term gives the solution in Figure 17.2. While all of these types of solutions are analyzed in [**32**], we only have space to give a general theorem, which can be used to obtain the formulas above and also can be applied to a far more general class of equations. We also discuss briefly a second example.

**17.2.2. Shallow water sloshing.** The problem studied in Chapter 16 can also be transformed into the form discussed in this chapter. Thus, we obtain asymptotic expansions of some of the solutions we described earlier. For example, in Theorem 16.2 it was shown that there is a solution, denoted there by $y(x,\alpha_1)$, which is $2\pi$-periodic and has a spike (local minimum) at $x=\pi$. This solution is also a solution to the boundary problem consisting of (16.4) and the Neumann boundary conditions $y'(0)=y'(2\pi)=0$. It is a consequence of the analysis later in this chapter that

$$y(x,\alpha_1)=\sqrt{1+c+\cos x}-3\sqrt{c}\,\mathrm{sech}^2\left(\frac{\sqrt{c}\,(x-\pi)}{\pi\varepsilon}\right)+O(\varepsilon) \tag{17.7}$$

as $\varepsilon\to 0$, uniformly on $[0,2\pi]$.[2] In the language of Chapter 7 the first term can be considered the outer solution and the second is the correction around the spike obtained from the inner solution. It is easy to see that (17.7) implies the results about this solution contained in Theorem 16.4 of Chapter 16 but is more precise. It is not clear, however, that this more precise information shortens the uniqueness proof for this solution, which is important in showing the sensitivity to initial conditions which we described for that problem.

## 17.3. Asymptotic expansion

Following [**32**] we can transform either of the problems from the last section to the standard form (17.1). To transform Carrier's equation, (17.4), for example, we first determine that this equation has a solution $y_g$ which is uniformly within $O(\varepsilon^2)$ of the lower outer solution $y_o^-$, seen in Figure 17.1. (The solution $y_g$ does not satisfy (17.5).) Let $u=y-y_g$. Then $u$ satisfies (17.1) with $q=q_{car}=2\sqrt{1+(1-x^2)^2}+O(\varepsilon^2)$ as $\varepsilon\to 0$.

The main result in [**32**] is about (17.1) for a wide class of functions $q$ which includes $q_{car}$. We will extend this to include solutions to (17.2), with assumptions (17.3) on $f$, though for simplicity we don't give the entire proof in this generality. The following assumptions are made on $q$, where $I=[-1,1]$ and $\varepsilon_0>0$:

$$\begin{gathered}0<q_*:=\min_{I\times(0,\varepsilon_0)}q(x,\varepsilon)\le q^*:=\max_{I\times(0,\varepsilon_0)}q(x,\varepsilon),\\ q_1:=\max_{I\times(0,\varepsilon_0)}\left|q'(x,\varepsilon)\right|<\infty,\qquad q_2=\max_{I\times(0,\varepsilon_0)}\left|q''(x,\varepsilon)\right|<\infty.\end{gathered}\tag{17.8}$$

The result applies to any solution $u$ to (17.2) which exists on an interval $[x_m,x_M]$ and satisfies

$$u'(x_m)=0,\quad u'>0 \text{ on } (x_m,x_M),\ \text{and } u'(x_M)=0. \tag{17.9}$$

There is no assumption on the size of $x_M-x_m$, which may be small or not small. In the statement of the theorem and subsequently we often suppress

[2]See [**32**] for further details.

the dependence of $q$ on $\varepsilon$. Also, we let $u(x_m) = m$ and $u(x_M) = M$. It is not hard to show that there is an $\bar{M} > 0$ depending only on $f$ and $q_*$ such that no solution satisfying (17.9) exists if $M > \bar{M}$. This is because for very large $M$, not depending on $x_M$, if $u(x_M) = M$ and $u'(x_M) = 0$ then the term $uf(u)$ in the equation forces $u$ to become negative as $x$ decreases, before $u' = 0$.

**Remark 17.1.** In the hypotheses of Theorem 17.2 below it is not assumed that $\varepsilon$ or $m$ is small, but the result is of little use without these conditions. Because there are two small parameters, careful consideration must be given to the interpretation of the $O$ symbol, which appears in the statement of the theorem and in its proof. While this interpretation is standard, it is easy to misunderstand what the results are saying. To try to clarify this, we state explicitly:

- There is a constant $K > 0$ depending only on $q_*$, $q^*$, $q_1$, and $q_2$, and not on $M$, $m$, $x_m$, $x_M$, or $\varepsilon$, or the particular solution $u$ being studied, such that at each use, $O(1)$ is a function $\psi$ on $[x_m, x_M]$ (different from one use to the next) such that $|\psi(x)| \leq K$ on this interval. Further, for any positive number $\alpha$, $O(\alpha) = \alpha O(1)$.

**Theorem 17.2.** *Suppose that $u$ is a solution to* (17.2) *for some $\varepsilon > 0$ which satisfies* (17.9) *for some $x_m$ and $x_M$ in $[-1,1]$ with $x_m < x_M$. Suppose that $f$ satisfies conditions* (17.3) *and $q$ satisfies conditions* (17.8)*. Then:*

(I) *For every $x \in (x_m, x_M)$,*

$$\frac{\varepsilon u'(x)}{\rho(u(x), m, M)} = \sqrt{q(x) - \delta_1(x) - \delta_2(x)} \tag{17.10}$$

*where, if $F(u) = \int_0^u sf(s)\,ds$ and*

$$\begin{aligned} &H(u, M, m) \\ &\quad = u^2 (F(M) - F(m)) + m^2 (F(u) - F(M)) + M^2 (F(m) - F(u)), \end{aligned} \tag{17.11}$$

*then*

$$\rho(u, m, M) = \sqrt{\frac{H(u, M, m)}{F(M) - F(m)}}, \tag{17.12}$$

$$\delta_1(x) = \frac{F(M) - F(u)}{H(u, M, m)} \int_{x_m}^{x} \left(u(y)^2 - m^2\right) q'(y)\,dy,$$

$$\delta_2(x) = \frac{F(u) - F(m)}{H(u, M, m)} \int_{x}^{x_M} \left(M^2 - u(y)^2\right) q'(y)\,dy.$$

*(See below for more transparent formulas in the case $f(u) = u$.)*

(II) *Further, on* $[x_m, x_M]$,

$$\delta_1 = O(\varepsilon), \quad \delta_2 = O(\varepsilon), \quad \int_{x_m}^{x_M} \delta_2(x)\,dx = O(\varepsilon^2),$$

$$\int_{x_m}^{x} \frac{\delta_1(y)}{\sqrt{q(y)}}dy = \frac{1}{2}\varepsilon \log\left(\frac{q(x)}{q(x_m)}\right) + O(\varepsilon^2). \tag{17.13}$$

**Corollary 17.3.** *If* $f(u) = u$, *then*

(17.14)

$$\rho(v, m, M) = \sqrt{\frac{(v-m)(M-v)(Mv+mv+mM)}{M^2+Mm+m^2}},$$

$$\delta_1(x) = \frac{M^2+Mu(x)+u(x)^2}{(u(x)-m)\left(u(x)+\frac{Mm}{M+m}\right)} \int_{x_m}^{x} \frac{u(y)^2-m^2}{M^2-m^2} q'(y)\,dy, \tag{17.15}$$

$$\delta_2(x) = \frac{m^2+mu(x)+u(x)^2}{(M-u(x))\left(u(x)+\frac{Mm}{M+m}\right)} \int_{x}^{x_M} \frac{M^2-u(y)^2}{M^2-m^2} q'(y)\,dy. \tag{17.16}$$

**Remark 17.4.** In the case $f(u) = u$ the formula (17.10) can be integrated to give an approximation for the solution $u$ which takes the form $\tilde{u} = \mathcal{E}(z)$, given implicitly by

$$\int_{m}^{\tilde{u}} \frac{dv}{\rho(v, m, M)} = z.$$

Thus, $\tilde{u}$ is an elliptic function.

**Remark 17.5.** Theorem 17.2 is in effect a so-called WKBJ approximation, so useful in the study of the one-dimensional Schrödinger equation

$$u'' = q(x)\,u.$$

Introduced by the physicists Wentzel, Kramers, and Brillouin and the applied mathematician Jeffreys, it was later rigorized by analysts such as Langer and Titchmarsh and is applicable for a remarkably wide range of functions $q$. It gives an approximation to the solution $u$ in the form

$$u \sim q^{-\frac{1}{4}} e^{\int q^{1/2}dx}, \tag{17.17}$$

which leads either to a real exponential or a trigonometric function, depending on the sign of $q$. It appears in the present context because, as we shall see, our solution $u$ to (17.1) is small for "most" of the range of the independent variable, so that (17.1) becomes essentially linear. On the other hand, the appearance of the elliptic function $\mathcal{E}$ rather than the exponential reflects the quadratic nature of (17.1). Finally, the fact that $\int \delta_1/\sqrt{q}$ is a multiple of $\log q$ corresponds to the factor $q^{-\frac{1}{4}}$ in (17.17).

**Proof of Theorem 17.2.** (Part (I)) It is first necessary to verify that the function $H$ defined in (17.11) is nonzero in the interval $(m, M)$. We leave this to Exercse 17.1. Multiply (17.2) by $2u'$ and integrate over $[x_m, x]$ to get

$$\varepsilon^2 u'(x)^2 = \int_{x_m}^{x} 2u'(y)\, u(y)\,(q(y) - f(u(y)))\, dy$$

$$= -2(F(u) - F(m)) + (u^2 - m^2)\, q - \int_{x_m}^{x} (u^2 - m^2)\, q' dy \tag{17.18}$$

where $u = u(x)$ and $q = q(x)$. In the same way, integrating over $[x, x_M]$, we obtain

$$\varepsilon^2 u'(x)^2 = 2(F(M) - F(u)) - (M^2 - u^2)\, q - \int_{x}^{x_M} (M^2 - u^2)\, q' dy. \tag{17.19}$$

Multiply (17.18) by $F(M) - F(u)$ and (17.19) by $F(u) - F(m)$ and add. This yields (17.10), in squared form.

(Part (II)) We will give the proof of (II) only for the case $f(u) = u$, in which case we use equations (17.14)–(17.16). (This is only for notational convenience.) We start with a lemma which assumes that $m < q_*/4$ and, as is clear from (17.1) and (17.9), $M > q_*$. (See (17.8).)

**Lemma 17.6.** *Let $X \in (x_m, x_M)$ be defined by $u(X) = \frac{1}{4}q_*$. Then*

$$x_M - X = O(\varepsilon), \tag{17.20}$$

*and if $x_m \le x_1 < x_2 \le X$, then*

$$x_2 - x_1 = O\left(\varepsilon\left[1 + \log\left(\frac{u(x_2)}{u(x_1)}\right)\right]\right). \tag{17.21}$$

*In particular, for $x \in [x_m, X]$,*

$$X - x = O\left(\varepsilon\left[1 + \log\left(\frac{1}{u(x)}\right)\right]\right). \tag{17.22}$$

**Proof.** Since $\varepsilon^2 u'' = u(q - u) \ge u(q_* - u)$,

$$\frac{1}{2}\varepsilon^2 u'^2|_{x_m}^{x} \ge \frac{1}{2}(u^2 - m^2)\, q_* - \frac{1}{3}(u^3 - m^3)$$

$$= (u - m)\left\{\frac{1}{2}q_*(u + m) - \frac{1}{3}(u^2 + mu + m^2)\right\}.$$

When $x \le X$, $\frac{1}{3}(u^2 + um + m^2) \le u^2 \le \frac{1}{4}q_* u$, and so

$$\frac{1}{2}\varepsilon^2 u'^2 \ge \frac{1}{4}q_*(u^2 - m^2) \tag{17.23}$$

and

$$\varepsilon \int_{u_1/m}^{u_2/m} \frac{dt}{\sqrt{t^2-1}} \geq \sqrt{\frac{1}{2}q_*}\,(x_2 - x_1),$$

where $u_i = u(x_i)$. This immediately leads to (17.21).

To prove (17.20) we rescale the problem by setting $\varepsilon t = x - X$, $U(t) = u(x)$, to get

$$U'' = U\left(q(x,\varepsilon) - U\right), \qquad U(0) = \frac{1}{4}q_*.$$

From (17.23), $U'(0) \geq \sqrt{\frac{1}{2}q_*}\sqrt{\left(\frac{1}{4}q_*\right)^2 - m^2}$, and since $q'$ is bounded, the equation may be written as

$$U'' = U\left(q(X) - U\right) + O(\varepsilon t)$$

on any bounded interval in $t$. Since $M$ is bounded by $\bar{M}$, there is a $T > 0$ depending only on $q_*$ such that $U = M$ somewhere in $[0, T]$, from which it follows that for sufficiently small $\varepsilon$, $u = M$ before $x = \varepsilon T$. This proves (17.20) and Lemma 17.6. □

Using the lemma,

$$\begin{aligned}\delta_2 &= O\left(u\left|x_M - x\right|\right) = O\left(u\left[x_M - X + X - x\right]\right)\\ &= O(\varepsilon) + O\left(\varepsilon u\left[1 + \log\frac{1}{u}\right]\right) = O(\varepsilon).\end{aligned}$$

Then, in the same way,

$$\int_{x_m}^{x_M} \delta_2 dx = O\left(\varepsilon^2\right) + O(\varepsilon)\int_{x_m}^{X} u\left(1 + \log\left(\frac{1}{u}\right)\right) dx.$$

Substituting for $dx$ using (17.23) gives

$$\int_{x_m}^{x_M} \delta_2 dx = O\left(\varepsilon^2\right) + O\left(\varepsilon^2 \int_m^{u(X)} \frac{u\left(1 + \log\left(\frac{1}{u}\right)\right)}{\sqrt{u^2 - m^2}} du\right), \tag{17.24}$$

where the integral on the right is convergent. Further, this integral is bounded independent of $m \in (0, \frac{q_*}{4}]$, so that

$$\int_{x_m}^{x_M} \delta_2 dx = O\left(\varepsilon^2\right), \tag{17.25}$$

as desired. (Note the importance of the boundedness of the integral independent of $m$ for this conclusion, because the constant $K$ in $O\left(\varepsilon^2\right)$ cannot depend on $m$.)

Also,

$$\delta_1 = O\left(\frac{1}{u(u-m)}\int_{x_m}^{x}\left(u^2 - m^2\right) dy\right),$$

where we may assume that $x < X$, since the contribution from $\int_X^{x_M}$ is $O(\varepsilon)$. Again using (17.23) we get

$$\delta_1 = \left( \frac{\varepsilon}{u(u-m)} \int_m^u \sqrt{v^2 - m^2 dv} \right) = O\left( \frac{\varepsilon\sqrt{u^2-m^2}}{u(u-m)}(u-m) \right) = O(\varepsilon).$$

Finally, we consider $\int_{x_m}^{x_M} \frac{\delta_1}{\sqrt{q}} dx$. The key here is to use (17.10) and the bounds on $\delta_1$ and $\delta_2$ which we have just obtained. From these it follows that on $[x_m, x_M]$

$$\frac{dx}{du} = \frac{\varepsilon}{\rho(u, m, M)\sqrt{q(x)}}(1 + O(\varepsilon)). \tag{17.26}$$

We see once again that the contribution from $\int_X^{x_M}$ is $O(\varepsilon^2)$, and thus

$$\begin{aligned} &\int_{x_m}^{x_M} \frac{\delta_1}{\sqrt{q}} dx = O(\varepsilon^2) \\ &+ \int_{x_m}^{X} \frac{M^2 + Mu(x) + u(x)^2}{(u(x) - m)\left(u(x) + \frac{Mm}{M+m}\right)} \left( \int_{x_m}^{x} \frac{u(y)^2 - m^2}{M^2 - m^2} q'(y)\, dy \right) \frac{1}{\sqrt{q(x)}} dx. \end{aligned} \tag{17.27}$$

The integrand of the double integral is bounded, because

$$\begin{aligned} (u(x) - u(m))\left(u(x) + \frac{Mm}{M+m}\right) &\geq \frac{M}{M+m}\left(u(x)^2 - m^2\right) \\ &\geq \frac{M}{M+m}\left(u(y)^2 - m^2\right). \end{aligned}$$

Also, each range of integration is $O\left(\varepsilon \log \frac{1}{m}\right)$, from (17.22). Hence the double integral is $O\left(\varepsilon^2 \log^2 \frac{1}{m}\right)$.

To obtain (17.13) we make three substitutions in (17.27). The first of these is

$$\frac{1}{M^2 - m^2} = \frac{1}{M^2} + O(m^2). \tag{17.28}$$

Note that the error made is $O(m^2)$ times the double integral and hence is $O\left(\varepsilon^2 m^2 \log^2 \frac{1}{m}\right) = O(\varepsilon^2)$. This can then be lumped into the initial $O(\varepsilon^2)$ term. In the same way we can replace $u(x) + \frac{Mm}{M+m}$ by $u(x) + m$, again making an error of $O(m^2)$ which can be ignored. For the third step we consider the numerator in the outer integral of (17.27) and write

$$\begin{aligned} M^2 + Mu + u^2 &= M^2 + Mm + m^2 + O(u - m) \\ &= M^2 + O(m) + O(u - m), \end{aligned} \tag{17.29}$$

since $M$ is bounded. The $O(m)$ term can be dropped as before. The $O(u - m)$ term cancels the $u(x) - m$ in the denominator. This gives an error term of the form

$$O\left(\int_{x_m}^{X} \frac{1}{u(x)+m} \int_{x_m}^{x} \left(u(y)^2 - m^2\right) dy dx\right).$$

We now wish to change variables, first in the inner integral, letting $v = u(y)$. Using (17.23) the error term is then seen to be

$$O\left(\varepsilon \int_{x_m}^{X} \frac{1}{u(x)+m} \int_{m}^{u(x)} \sqrt{v^2 - m^2} dv dx\right).$$

The inner integral is now $O\left(u\sqrt{u^2 - m^2}\right)$, and since $\frac{u}{u+m}$ is bounded we obtain

$$O\left(\varepsilon \int_{x_m}^{X} \sqrt{u(x)^2 - m^2} dx\right).$$

Changing variables in this integral, and using (17.23) again, gives $O\left(\varepsilon^2\right)$ for the third error term as well.

We are left with

$$\int_{x_m}^{x_M} \frac{\delta_1}{\sqrt{q}} dx \tag{17.30}$$
$$= O\left(\varepsilon^2\right) + \int_{x_m}^{X} \frac{1}{u(x)^2 - m^2} \frac{1}{\sqrt{q(x)}} \left(\int_{x_m}^{x} \left(u(y)^2 - m^2\right) q'(y)\ dy\right) dx.$$

We again change variables in the inner integral, letting $v = u(y)$. Since $|\delta_1| + |\delta_2| = O(\varepsilon)$, we see from (17.10) that

$$\frac{dy}{dv} = \frac{\varepsilon}{\rho\sqrt{q}} (1 + O(\varepsilon)).$$

From (17.14) we see that

$$\begin{aligned}
\rho(v, m, M) &= \sqrt{v-m}\sqrt{\frac{M-v}{M^2 + O(m)}}\sqrt{M(v+m) + vm} \\
&= \sqrt{v^2 - m^2}\sqrt{\frac{M-v}{M + O(m)}}\sqrt{1 + O(m)} \\
&= \sqrt{v^2 - m^2}\,(1 + O(m) + O(v-m)),
\end{aligned} \tag{17.31}$$

where we have written $M - v = M - m - (v - m)$. Substituting this into (17.30) and dropping the terms involving $O(m)$ and $O(v - m)$ again results in a total error of $O\left(\varepsilon^2\right)$. We obtain

$$\int_{x_m}^{x_M} \frac{\delta_1}{\sqrt{q}} dx$$
$$= \varepsilon \int_{x_m}^{X} \frac{1}{u(x)^2 - m^2} \frac{1}{\sqrt{q(x)}} \left( \int_m^u \sqrt{v^2 - m^2} \frac{q'(y)}{\sqrt{q(y)}} dv \right) dx \left\{ 1 + O(\varepsilon) \right\}$$
$$+ O\left(\varepsilon^2\right),$$

with $u(y) = v$. The integral converges because the inner integral is $O(u - m)$.

Since $q''$ is bounded, we can write $\frac{q'(y)}{\sqrt{q(y)}} = \frac{q'(x)}{\sqrt{q(x)}} + O(y - x)$, giving

(17.32)
$$\int_{x_m}^{x_M} \frac{\delta_1}{\sqrt{q}} dx = O\left(\varepsilon^2\right)$$
$$+ \varepsilon \int_{x_m}^{X} \frac{1}{u(x)^2 - m^2} \frac{q'(x)}{q(x)} \left( \int_m^u \sqrt{v^2 - m^2} dv \right) dx \left\{ 1 + O(\varepsilon) \right\}$$
$$+ O\left(\varepsilon^2\right) \int_{x_m}^{X} \frac{1}{u^2 - m^2} \int_m^u \sqrt{v^2 - m^2} \left( \int_v^u \frac{dw}{\sqrt{w^2 - m^2}} \right) dv dx.$$

Since $w^2 - m^2 \geq v^2 - m^2$, the last term is

$$O\left( \varepsilon^2 \int_{x_m}^{X} \frac{1}{u^2 - m^2} \left( \int_m^u (u - v) \, dv \right) dx \right) = O\left(\varepsilon^2\right).$$

In the second term on the right of (17.32) we replace $\sqrt{v^2 - m^2}$ by $v$, again making an error of $O(m)$ in the integrand and so an error of $O\left(\varepsilon^2\right)$ overall. This term becomes

$$\frac{1}{2}\varepsilon \int_{x_m}^{X} \frac{q'}{q} dx + O\left(\varepsilon^2\right).$$

Finally, for $x \in (x_m, x_M)$ the same steps show that

$$\frac{1}{2}\varepsilon \int_{x_m}^{x} \frac{\delta_1}{\sqrt{q}} dx = \frac{1}{2}\varepsilon \int_{x_m}^{x} \frac{q'}{q} dx + O\left(\varepsilon^2\right)$$

which implies (17.13) in the case $f(u) = u$. □

Theorem 17.2 enables us to approximate solutions to (17.2) with solutions of the first order equation

$$\frac{\varepsilon u'}{\sqrt{\rho(u, m, M)}} = \sqrt{q(x)}.$$

By integrating this equation and using the estimates in part (II) of the theorem in the case $f(u) = u$, Chen and Sadhu obtain the following result.

**Corollary 17.7.** *Suppose that $q \in C^2(I)$ and that it satisfies* (17.8). *Then there is a $K > 0$ (depending only on $q_*$, $q^*$, $q_1$, and $q_2$) such that if $0 < \varepsilon < \varepsilon_0$, $-1 \leq x_m < x_M \leq 1$, and $u$ is a solution of* (17.1) *satisfying* (17.9), *and $u(x_m) = m$, $u(x_M) = M$, then*

$$M = \frac{3}{2}q(x_M) - \frac{m^2}{M+m} - \frac{|2+O(m)|\,\varepsilon q(x_M)}{\sqrt{q(x_M)}} + O(\varepsilon^2). \tag{17.33}$$

*Also, for $x \in [x_m, x_M]$,*

$$\begin{aligned}\log\frac{1+\sqrt{1-u(x)/M}}{1-\sqrt{1-u(x)/M}} &+ \log\frac{2}{1+\sqrt{1-m^2/u(x)^2}} \\ &= \log\frac{8M}{m} - \int_{x_m}^{x}\left(\frac{\sqrt{q(y)}}{\varepsilon} - \frac{q'(y)}{4q(y)}\right)dy + O(\varepsilon+m).\end{aligned} \tag{17.34}$$

We refer the reader to [**32**] for the proof. Note that equation (17.34) is essentially an implicit approximation for $u$. This result then allows us to approximate solutions to (17.2) over longer intervals. Suppose for example that we wish to approximate the solution to (17.4)–(17.5) which is shown in Figure 17.2. We first transform the equation to the form (17.1). The transformed solution still has a spike just to the right of $x = -1$ and a boundary layer ending at $x = 1$. Extend the equation to the right of $[-1, 1]$, say by setting $q(x) = q(1)$ for $x > 1$. Extend the solution to the right as well, until its next maximum, at $x = x_M > 1$. This solution, say $u$, has a unique minimum in $(-1, 1)$ at $x_m$, where $|x_m|$ is small. Apply Corollary 17.7 on $[x_m, x_M]$. An approximation is obtained, in particular, on $[x_m, 1]$.

By reversing direction from $x_m$, an approximation is also obtained on $\left[x_M^-, x_m\right]$, where $x_M^-$ is the location of the maximum of the boundary spike near $-1$. Finally, we consider the equation on $\left[-1, x_M^-\right]$. Near the maximum of the spike (whose height is approximated implicitly by (17.33)), regular perturbation methods are valid (and even more accurate than the approximation obtained from Corollary 17.7), and these enable us to extend the approximation back to $x = -1$. In this way the appropriate version of formula (17.6) is rigorously justified. In [**32**] it is shown how to handle even more complicated solutions, including all those approximated by (17.6).

The results of this section are not of interest unless $m$ and $\varepsilon$ are both small. If $m$ is not small and $\varepsilon$ is, then regular perturbation theory gives better approximations than we achieve above. Further, the interesting ranges of $\varepsilon$ and $m$ are related. For a given $\varepsilon$, if $m$ is too small then there will be no solution satisfying the hypotheses of the theorem with $[x_m, x_M] \subset [-1, 1]$. On the other hand, for a given $m$, if $\varepsilon$ is too small then the hypotheses cannot be satisfied unless $x_M - x_m = O(\varepsilon)$, in which case both terms on the right of (17.13) are $O(\varepsilon^2)$.

## 17.4. Exercise

**Exercise 17.1.** Use the properties of $f$ in (17.3) to verify that $H$, defined in (17.11), is positive for $m < u < M$.

*Chapter 18*

# Layers and spikes in reaction-diffusion equations, II

In this chapter we continue the study of singularly perturbed reaction-diffusion equations started in Chapter 16. We will discuss a model of gene spreading in a population which, as far as we know, was first analyzed mathematically in the generality considered here by Kurland, in [**132**], and then by Angenent, Mallet-Paret, and Peletier in [**11**].

Again the differential equation is nonautonomous, but now it is cubic in the independent variable, and the particular cubic form turns out to give a large number of steady-state solutions, including single spikes, multiple spikes, and single and multiple layers, in which the solution jumps from a neighborhood of one equilibrium point to a neighborhood of another. Two sorts of results are discussed. One type describes the behavior when the number of oscillations in a solution is kept constant, by adjusting the initial conditions, as the small parameter is decreased. The other, which we have space to describe only briefly, allows the number of layers to increase as $\varepsilon \to 0^+$, and the emphasis is on a sort of averaged equation which describes the larger scale evolution of the envelope of solutions.

The equation to be considered is

$$\varepsilon^2 u'' + u(1-u)(u-\phi(x)) = 0, \tag{18.1}$$

where

$$0 < \phi(x) < 1. \tag{18.2}$$

For convenience we will usually assume that $\phi'$ is continuous. We look for solutions satisfying the Neumann boundary conditions

$$u'(0) = u'(1) = 0. \tag{18.3}$$

In their paper [**11**], Angenent, Mallet-Paret, and Peletier found all of the solutions which are stable steady states of the corresponding reaction-diffusion pde (16.1). These solutions can have single layers (defined below) near the points in $[0,1]$ where $\phi = \frac{1}{2}$. However the set of all solutions (stable or not) is considerably more complicated and may (depending on $\phi$) include solutions with multiple layers clustered near some of the points where $\phi = \frac{1}{2}$, and also single or multiple spikes near critical points of $\phi$.

We will contrast our shooting approach not only with the methods of [**11**] but also with those of Dancer and Yan in [**49**], who used techniques from the calculus of variations. Variational methods are important for many kinds of boundary value problems and especially useful for obtaining stable solutions of reaction-diffusion systems. This approach was also used by Nakashima [**177**], to study equations of the form

$$\varepsilon^2 u'' + \phi(x)\, u\left(1 - u^2\right) = 0.$$

An important reason for discussing variational methods is that they sometimes extend to higher dimensional (pde) problems, as in [**49**] and [**50**]. The same is true for the method of "subsolutions" and "supersolutions" used in [**11**]. So while we believe that shooting gives the most complete set of existence results for the particular ode problem (18.1)–(18.3), it is far from the only technique which should be considered. We call attention also to Section 18.7, where a different class of solutions is described briefly.

As usual in this text, we treat the simplest cases that illustrate the techniques involved. For instance, in our results about layers we will usually assume that $\phi - \frac{1}{2}$ has a single zero in $(0,1)$. When discussing spikes we will assume that $\phi - \frac{1}{2} \neq 0$ and that $\phi'$ has a single zero in $(0,1)$. To see the complications which arise in more general cases, consult the paper [**4**], written with S.-B. Ai and X. Chen. This paper also has a full discussion of the "Morse index" of the solutions, a topic we only touch on here.

## 18.1. A basic existence result

The problem (18.1)–(18.3) has two trivial solutions, $u = 0$ and $u = 1$. A simple shooting argument shows that for every $\varepsilon > 0$ there is at least one other solution. This is true very generally.

**Proposition 18.1.** *If $\phi$ is continuous on $[0,1]$ and satisfies (18.2), then there is a solution $u$ of (18.1)–(18.3) with $0 < u < 1$ on $[0,1]$.*

**Proof.** Consider solutions of (18.1) with $u'(0) = 0$ and $0 < u(0) < 1$. If $u(0)$ is sufficiently small then $0 < u < \phi$ on $[0,1]$, so that $u'' > 0$ and hence $u'(1) > 0$. If $1 - u(0)$ is sufficiently small then $u'' < 0$ on $[0,1]$, and so $u'(1) < 0$. Let

$$f(x,u) = u(1-u)(u - \phi(x)),$$

and consider the modified function

$$\hat{f}(x,u) = \begin{cases} 0 & \text{if } u \geq 1, \\ f(x,u) & \text{if } 0 < u < 1, \\ 0 & \text{if } u \leq 0 \end{cases}$$

instead of $f$ in (18.1). Then, all solutions exist on $[0,1]$, and $u'(1)$ is a continuous function of $u(0)$. Hence there is a value of $u(0) \in (0,1)$ such that $u'(1) = 0$, and it is easy to show that this solution must satisfy $0 < u < 1$ and so be a solution of (18.1) with the original $f$. □

## 18.2. Variational approach to layers

We will now add conditions on $\phi$ and require that $\varepsilon$ be small, in order to obtain what we call a layered solution. Roughly, this is a solution which is close to 0 over some interval or intervals of $x$, close to 1 over other intervals, and the transitions between 0 and 1 are rapid. A look at the variational approach will provide good intuition about why stable layered solutions should be present in this model. For background on calculus of variations and for terminology, see [**62**] and also our Chapter 11.

Again, let

$$f(x,u) = u(1-u)(u - \phi(x)),$$

where $\phi$ is smooth and satisfies (18.2). Then let

$$F(x,u) = \int_0^u f(x,s)\,ds,$$

and consider the functional

$$I_\varepsilon(u) = \int_0^1 \left( \frac{1}{2}\varepsilon^2 u'(x)^2 - F(x,u(x)) \right) dx, \tag{18.4}$$

defined for functions $u \in H^1(0,1)$.[1] Then the Euler-Lagrange equations for the problem of minimizing $I_\varepsilon$ are (18.1)–(18.3). In particular, although functions in $H^1$ satisfy no particular boundary conditions, minimizers can be shown to satisfy so-called "natural" boundary conditions, which in this case turn out to be (18.3).

[1]See Chapter 10 for the definition (in a different context) of $H^1$. Here we say only that functions in this space have "weak" first derivatives which are in $L^2$. Weak derivatives also are defined in Chapter 10.

It is easily shown that if $\phi(x) > \frac{1}{2}$ then

$$\min_{-\infty<u<\infty} [-F(x,u)] = -F(x,0) = 0, \tag{18.5}$$

while if $\phi(x) < \frac{1}{2}$ then this minimum is at $u = 1$. Hence, we expect that a global minimizer for $I_\varepsilon$ will be close, in some sense, to the function

$$u_0(x) = \begin{cases} 0 & \text{if } \phi(x) > \frac{1}{2}, \\ \frac{1}{2} & \text{if } \phi(x) = \frac{1}{2}, \\ 1 & \text{if } \phi(x) < \frac{1}{2}. \end{cases} \tag{18.6}$$

To get layers, we will need $\phi - \frac{1}{2}$ to change sign at least once in $(0,1)$, and then we expect that for small $\varepsilon$ the solution minimizing $I_\varepsilon(u)$ has rapid jumps near the zeros of $\phi - \frac{1}{2}$. Other local minimizers will have jumps near some, but not all, of these points. However, more complicated solutions, such as multiple layers, or solutions with spikes, are not energy minimizers and are not stable for the corresponding reaction-diffusion problem. We will discuss these also, after describing various approaches to the simplest case of a single stable layer.

## 18.3. Three different existence proofs for a single layer in a simple case

We assume that $\phi'$ is continuous on $[0,1]$ and that

$$\begin{aligned} \tfrac{1}{2} < \phi(x) < 1 \quad & \text{if } 0 \le x < x_0, \\ 0 < \phi(x) < \tfrac{1}{2} \quad & \text{if } x_0 < x \le 1. \end{aligned} \tag{18.7}$$

Further, it is assumed that

$$\phi'(x_0) < 0. \tag{18.8}$$

**Theorem 18.2.** *Assume that* (18.7) *and* (18.8) *hold. Then for each sufficiently small* $\varepsilon > 0$ *there is an increasing solution* $u^\varepsilon$ *of* (18.1)–(18.3), *and this family of solutions can be chosen so that*

$$\lim_{\varepsilon\to 0} u^\varepsilon = \begin{cases} 0 & \textit{uniformly on every compact subset of } [0,x_0), \\ 1 & \textit{uniformly on every compact subset of } (x_0,1]. \end{cases} \tag{18.9}$$

In Figure 18.1 we show an example of the type of solution found in this theorem, together with the function $\phi$.

### 18.3.1. Existence using subsolutions and supersolutions.

**Proof.** We give essentially the proof in [**11**]. Some details are adapted from [**210**] and [**46**]. The latter is an excellent general introduction to the method as it applies to ode's.

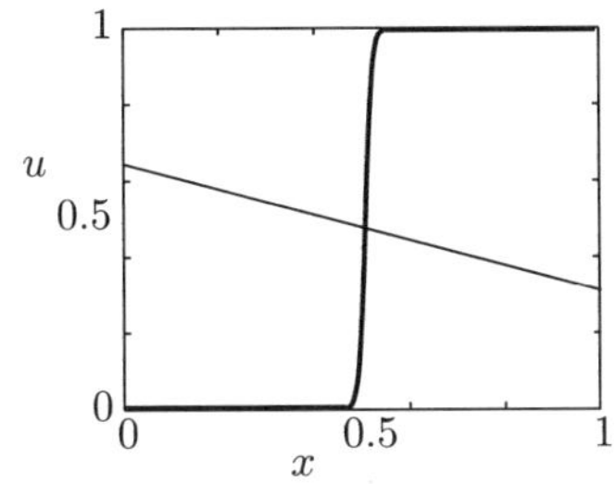


**Figure 18.1.** A single layer and the function $\phi$.

We will not give a general definition of subsolutions or supersolutions (also called lower and upper solutions). For that, see [**46**]. Instead, we will present only what is necessary for the problem at hand. We first introduce what we call "$C^2$ strong subsolutions and supersolutions".

**Definition 18.3.** A $C^2$ strong subsolution of (18.1)–(18.3) is a function $u_1 \in C^2([0,1])$ such that

(i) $u_1$ satisfies (18.3),

(ii) at each $x \in [0,1]$,

$$\varepsilon^2 u_1''(x) + f(x, u_1(x)) > 0. \tag{18.10}$$

A $C^2$ strong supersolution $u^1$ is defined by reversing the inequality in (18.10).[2]

**Lemma 18.4.** *Suppose that $\delta > 0$ with $[x_0 - \delta, x_0 + \delta] \subset (0,1)$. Suppose also that $\rho \in \left(0, \frac{1}{2}\right)$, $f(x, \rho) < 0$, and $f(x, 1-\rho) > 0$ on $[0,1]$. Then there is an $\varepsilon_1 > 0$ such that if $0 < \varepsilon < \varepsilon_1$ then (18.1)–(18.3) has $C^2$ strong subsolutions and supersolutions $u_1$ and $u^1$ which are nondecreasing on $[0,1]$ and satisfy*

$$\begin{aligned} 0 < u_1(x) < u^1(x) < 1 \quad & on\ [0,1], \\ 0 < u^1(x) < \rho \quad & on\ [0, x_0 - \delta], \\ 1 - \rho < u_1(x) < 1 \quad & on\ [x_0 + \delta, 1]. \end{aligned}$$

**Proof.** We start by constructing functions $\underline{u}$ and $\bar{u}$ in $C([0,1])$ which we will use to define $u_1$ and $u^1$. Let

$$\underline{u}(x) = 1 - \rho \quad \text{on } [x_0 + \delta, 1].$$

[2]Usually, subsolutions and supersolutions are defined with weak inequalities. They are not necessarily $C^2$ functions, and the boundary conditions may involve inequalities as well. This leads to the need to invoke the "strong maximum principle" for ode's, as in [**196**]. It turns out that we don't need these technicalities here.

Then consider the initial value problem

$$\varepsilon^2 p'' + f(x,p) = 0, \tag{18.11}$$

$$p(x_0+\delta) = 1-\rho, \quad p'(x_0+\delta) = 0. \tag{18.12}$$

Let $s = \frac{x_0+\delta-x}{\varepsilon}$ and

$$p(x) = q(s).$$

Then

$$q'' + f(x_0+\delta-\varepsilon s, q) = 0, \tag{18.13}$$

$$q(0) = 1-\rho, \quad q'(0) = 0. \tag{18.14}$$

We compare $q$ with the solution to

$$W'' + f(x_0+\delta, W) = 0, \tag{18.15}$$

$$W(0) = 1-\rho, \quad W'(0) = 0. \tag{18.16}$$

Because $\phi(x_0+\delta) < \frac{1}{2}$,

$$\int_0^1 u(1-u)(u-\phi(x_0+\delta))\,du > 0.$$

We then consider the $(W, W')$ phase plane for (18.15). Linearization and a standard energy argument (Exercise 18.1) show that $(1,0)$ is a saddle point for this system, and solutions on one branch of the unstable manifold tend to $-\infty$ in both variables. Hence, if $\rho$ is sufficiently small then the unique solution of (18.15)–(18.16) decreases monotonically to below $W = 0$. We now choose $\varepsilon$ so small that the unique solution $q$ to (18.13)–(18.14) decreases to zero at some $\hat{s} < \frac{\delta}{\varepsilon}$. Let $\hat{x} = x_0+\delta-\varepsilon\hat{s}$, notice that $\hat{x} > x_0$, and set

$$\underline{u}(x) = \begin{cases} 0 & \text{for } 0 \le x < \hat{x}, \\ q\left(\frac{x_0+\delta-x}{\varepsilon}\right) & \text{for } \hat{x} \le x \le x_0+\delta, \\ 1-\rho & \text{for } x_0+\delta < x \le 1. \end{cases}$$

The function $\bar{u}$ is constructed similarly.[3] It satisfies (18.1) on an interval of the form $[x_0-\delta, \tilde{x}]$ where $\tilde{x} < x_0$, and

$$\bar{u}(x) = \rho \quad \text{on } [0, x_0-\delta],$$
$$\bar{u}(x) = 1 \quad \text{on } [\tilde{x}, 1].$$

See Figure 18.2 for the functions $\bar{u}$ and $\underline{u}$.

Now we define $u_1$. First choose $\lambda > 0$ and so large that if $(x,u) \in [0,1]\times[0,1]$ then

$$f_u(x,u) + \lambda > 0. \tag{18.17}$$

[3] The functions $\underline{u}$ and $\bar{u}$ are lower and upper solutions, respectively, according to the definition in [**46**].

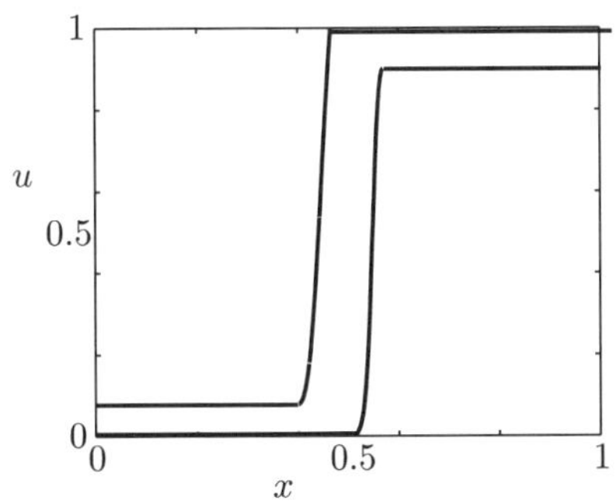


**Figure 18.2.** The functions $\bar{u}$ and $\underline{u}$.

Suppose that $v \in C[0,1]$. We then define an operator $T : C([0,1]) \rightarrow C^2([0,1])$ by considering the boundary value problem

$$\varepsilon^2 u'' - \lambda u = -f(x,v) - \lambda v, \tag{18.18}$$

$$u'(0) = u'(1) = 0. \tag{18.19}$$

The homogeneous problem

$$\varepsilon^2 u'' - \lambda u = 0,$$
$$u'(0) = u'(1) = 0$$

has only the zero solution. Hence, if $v \in C([0,1])$ then (18.18)–(18.19) has a unique solution, $u = Tv$. (This is the Fredholm alternative in the selfadjoint case, [**82**, p. 146].) We then set

$$u_1 = T\underline{u}.$$

**Lemma 18.5.** *$u_1$ is a $C^2$ strong subsolution of* (18.1)–(18.3), *and $\underline{u} < u_1 < 1$ on $[0,1]$.*

**Proof.** Since

$$\varepsilon^2 u_1'' + f(x,u_1) = (f(x,u_1) + \lambda u_1) - (f(x,\underline{u}) + \lambda\underline{u}),$$

(18.17) implies that if $\underline{u} < u_1 < 1$ on $[0,1]$ then $u_1$ is a $C^2$ strong subsolution of (18.1)–(18.3). It also follows from (18.17) that $f(x,\underline{u}) + \lambda\underline{u} \geq 0$ and that this quantity is positive on $(\hat{x},1]$. Suppose that $u_1(x) \leq 0$ for some $x \in [0,1]$. Then

$$\varepsilon^2 u_1''(x) = \lambda u_1(x) - f(x,\underline{u}(x)) - \lambda\underline{u}(x) \leq 0. \tag{18.20}$$

If, in particular, $u_1(x_1) \leq 0$ for some $x_1 > \hat{x}$, then $f(x_1,\underline{u}(x_1)) + \lambda\underline{u}(x_1) > 0$, and so the inequality in (18.20) becomes $\varepsilon^2 u_1''(x_1) < 0$. If $u_1'(x_1) \leq 0$ then $u_1'$ and $u_1$ decrease to the right of $x_1$, implying that $u_1'(1) <$

0, contradicting (18.19). If $u_1'(x_1) > 0$ then (18.20) continues to hold to the left of $x_1$, giving $u_1'(0) > 0$, again a contradiction.

If $u_1(x_1) \leq 0$ for some $x_1 \in [0, \hat{x}]$, then the same contradiction results unless $u_1(x_1) = u_1'(x_1) = 0$. In this case, $u_1 = 0$ on $[0, \hat{x}]$, but the right third derivative of $u_1$ at $\hat{x}$ is negative, again leading to a contradiction at $x = 1$. Hence, $u_1 > 0$ on $[0, 1]$.

Suppose next that $u_1 - \underline{u}$ has a negative minimum at $x_1 \in [0, 1]$, so that

$$(u_1 - \underline{u})''(x_1) \geq 0.$$

Then $x_1 \in (\hat{x}, 1]$, since $\underline{u} = 0$ on $[0, \hat{x}]$. But if $x_1 \in (\hat{x}, x_0 + \delta]$ then

$$\varepsilon^2 (u_1 - \underline{u})''(x_1) = \lambda (u_1 - \underline{u})(x_1) < 0,$$

while if $x_1 \in (x_0 + \delta, 1]$ then

$$\varepsilon^2 (u_1 - \underline{u})''(x_1) = \lambda (u_1 - \underline{u})(x_1) - f(x_1, 1 - \rho) < 0.$$

Finally, suppose that the graphs of $u_1$ and $\underline{u}$ are tangent at $x_1$. If $x_1 = 1$ then $(u_1 - \underline{u})'' < 0$, and so $u_1 - \underline{u}$ has a negative minimum in $[0, 1)$, which was already ruled out. If $x_1 \in [0, \hat{x}]$ then $u_1(x_1) = 0$, also shown to be impossible. Hence $x_1 \in (\hat{x}, 1)$. On $(\hat{x}, x_0 + \delta)$, both $u_1$ and $\underline{u}$ satisfy the equation

$$\varepsilon^2 u'' - \lambda u = -f(x, \underline{u}) - \lambda \underline{u},$$

so that if $\hat{x} < x_1 \leq x_0 + \delta$ then $u_1 = u$ throughout this interval. But we saw that $\underline{u}(\hat{x}) = 0$ and $u_1(\hat{x}) > 0$. If the point of tangency is in $(x_0 + \delta, 1]$, then $f(x_1, \underline{u}(x_1)) > 0$, leading again to the contradictory conclusion that $u_1'(1) < 0$. □

To complete the proof of Lemma 18.4, let $u^1 = T\bar{u}$. A similar analysis shows that $\underline{u} < u_1 < u^1 < \bar{u}$. □

We now show that there is a solution to (18.1)–(18.3). We use an iteration argument, letting $u_{j+1} = Tu_j$ and $u^{j+1} = Tu^j$ for $j = 1, 2, \ldots$. Observe that if $0 \leq u < v \leq 1$ on $[0, 1]$, then (18.17) implies that

$$\varepsilon^2 (Tv - Tu)'' - \lambda (Tv - Tu) = -f(x, v) - \lambda v + f(x, u) + \lambda u < 0.$$

Also, $Tu$ and $Tv$ satisfy (18.3), and the previous maximal principle argument shows that $Tv > Tu$ on $[0, 1]$. Thus, we obtain

$$u_1 < u_j < u_{j+1} < u^{j+1} < u^j < u^1 \tag{18.21}$$

on $[0, 1]$, and it follows that the sequences $\{u_j\}$ and $\{u^j\}$ converge to solutions of (18.1)–(18.3).[4] Condition (18.9) follows from the construction of $\underline{u}$ and $\bar{u}$. Further details of this proof are left to the reader. □

[4] A uniqueness result to be given later shows that they converge to the same solution.

The results in [**11**] go far beyond Theorem 18.2. As stated earlier, all stable solutions are found in this paper. It is proved that the solution constructed in the proof of Theorem 18.2 is locally unique for sufficiently small $\varepsilon$ and that the eigenvalue problem relevant to stability has only eigenvalues with negative real parts.[5] Suppose more generally that $\phi - \frac{1}{2}$ is not zero at the endpoints 0 or 1 and that it has a total of $N$ zeros, with $\phi' \neq 0$ at each of these zeros. Then for every choice $0 < \xi_1 < \cdots < \xi_M < 1$ of zeros of $\phi - \frac{1}{2}$ with the property that numbers $\left(-1^i\right) \phi'\left(\xi_i\right)$ for $i = 1, \ldots, M$ all have the same sign, there is a solution $u$ with a monotonic transition layer near each $\xi_i$, and nowhere else, and this solution satisfies $u'\left(\xi_i\right) \phi'\left(\xi_i\right) < 0$. These are shown to be all of the stable solutions, and this implies the beautiful result that there are exactly $\mathcal{F}_{N+3}$ stable solutions, where $\mathcal{F}_N$ is the $N^{\text{th}}$ Fibonacci number.[6]

The existence proof for these solutions is an extension of the argument given above. Uniqueness and stability follow by a different method. We will discuss the uniqueness and stability of a single layer solution below, after considering its existence by two other methods.

**18.3.2. Existence by a variational method.** In [**49**] Dancer and Yan treat the elliptic Dirichlet problem

$$-\varepsilon^2 \Delta u = u\left(u - a\left(x\right)\right)\left(1 - u\right) \quad \text{in } \Omega, \tag{18.22}$$
$$u = 0 \quad \text{on } \partial\Omega,$$

where $0 < a\left(x\right) < 1$ in $\Omega$. Here $\Omega$ is a bounded domain in $R^N$ with smooth boundary. They point out that their technique also extends to Neumann boundary conditions, of the form

$$D_\nu u = 0 \quad \text{on } \partial\Omega,$$

where $D_\nu u$ is the directional derivative of $u$ at a point on $\partial\Omega$ in the direction of the normal to the boundary at that point. When $N = 1$, therefore, they include the boundary conditions (18.3). Here we will only discuss $N = 1$, letting $\Omega = (0, 1)$.

When $N = 1$ the functional to be minimized is $I_\varepsilon\left(u\right)$, as defined in (18.4). In this case, the minimizer is in $C^2\left((0,1)\right) \cap C^1\left([0,1]\right)$ and solves (18.1)–(18.3).[7] To prove Theorem 18.2, we must prove (18.9). We do this by the method in [**49**], noting that we are dealing with a very simple example of their problem, since they consider the pde case, and in one dimension allow $\phi - \frac{1}{2}$ to have any finite number of zeros in $(0, 1)$.

[5] We give a relatively simple classical proof of uniqueness and stability in Section 18.4. The proof in [**11**] uses dynamical systems methods.

[6] $\mathcal{F}_0 = 0$, $\mathcal{F}_1 = 1$, $\mathcal{F}_{i+2} = \mathcal{F}_i + \mathcal{F}_{i+1}$.

[7] As we point out in Chapter 1, since the one-dimensional case of (18.22) is a smooth ode, a representative of the minimizer is in $C^2\left([0,1]\right)$.

First, though, we briefly discuss the existence of a smooth minimum. Since $-F(x,u) \geq \min\{-F(x,0), -F(x,1)\}$, the functional $I_\varepsilon$ is clearly bounded below. Our first thought might be that the piecewise constant function $u_0$ defined in (18.6) is the minimizer. However, further consideration makes it clear that $u_0$ is not in $H^1$, since if we have a family of smooth functions $v_n$ tending to $u_0$, then for each small $\delta > 0$,

$$\lim_{n\to\infty} \int_{x_0-\delta}^{x_0+\delta} v_n' dx = 1.$$

From the Cauchy-Schwarz inequality,

$$\left(\int_{x_0-\delta}^{x_0+\delta} v_n' dx\right)^2 \leq 2\delta \int_{x_0-\delta}^{x_0+\delta} (v_n')^2 \, dx$$

and it follows that the $H^1$ norm of $v_n$ is not bounded as $n \to \infty$.

Thus, it is reasonable to expect that $u_0$ is smooth and satisfies (18.1). The treatment of the boundary conditions is also straightforward.

The limiting result (18.9) is also not difficult. Suppose that it is not true and in particular that

$$\lim_{\varepsilon\to 0} \max_{0\leq x\leq x_0-\delta} u_0^\varepsilon (x) \geq \mu > 0.$$

Then there exists a sequence $\varepsilon_j \to 0$, with global minimizing function $u_0^j$, such that

$$\max_{0\leq x\leq x_0-\delta} u_0^j \geq \mu.$$

If the maxima occur at $x_j$, then taking a convergent subsequence of $\{x_j\}$ leads to a contradiction, using (18.5) and another Cauchy-Schwarz argument.

We thus see that the variational proof of existence in Theorem 18.2 in this special case is not difficult, once the functional analytic background has been established. But it is more difficult to obtain stable solutions other than the global minimizer (which may exist if $\phi - \frac{1}{2}$ has several zeros in $(0,1)$), because more sophisticated spaces must be considered. (See [**49**].) Variational methods have not often been used to obtain unstable solutions, but see [**54**] for an exception.

**18.3.3. Existence using shooting.** The shooting proof we will give for Theorem 18.2, while simple in outline, is slightly complicated in its details. The reason for using shooting is that, as we will describe below, this method gives a wider variety of solutions than has been obtained by the previous methods. The subsolution and supersolution method and the variational method are mainly of use in finding stable solutions.

We will simplify the shooting proof a bit by assuming that

$$\phi'(x) < 0 \quad \text{for } 0 \le x \le 1, \tag{18.23}$$

instead of (18.8). The changes necessary using (18.8) are slightly tedious but not difficult.

Suppose that $x^* \in [0,1]$, and rescale (18.1) by letting $x = x^* + \varepsilon s$ and $u(x) = y(s)$. Then

$$y'' + f(x^* + \varepsilon s, y) = 0. \tag{18.24}$$

We then consider the equation when $\varepsilon = 0$, which is

$$Y'' + f(x^*, Y) = 0, \tag{18.25}$$

and the corresponding system

$$\begin{aligned} Y' &= Z, \\ Z' &= -f(x^*, Y). \end{aligned} \tag{18.26}$$

This autonomous system is studied by consideration of the energy functional

$$E(x, x^*) = \frac{1}{2} Z(x)^2 + F(x^*, Y(x)),$$

where

$$F(x^*, Y) = \int_0^Y f(x^*, s)\, ds = \int_0^Y s(1-s)(s - \phi(x^*))\, ds,$$

noting that $E(x, x^*)$ is constant along solutions of (18.26).

The system (18.26) has equilibrium points at $(0,0)$, $(\phi(x^*), 0)$, and $(1,0)$. The point $(\phi(x^*), 0)$ is a center while the other two points are saddles. There are three cases to consider: (i) $0 \le x^* < x_0$, (ii) $x^* = x_0$, and (iii) $x_0 < x^* \le 1$, where $x_0$ is defined in (18.7). The three phase planes are shown in Figures 18.3, 18.4, and 18.5, and we leave it to the reader to verify these pictures. We observe that in case (i) there is a homoclinic orbit based at $(1,0)$, while in case (iii) there is a homoclinic orbit based at $(0,0)$. In case (ii) there are two heteroclinic orbits connecting $(0,0)$ and $(1,0)$, and these surround a family of periodic orbits. It turns out that layers follow trajectories close to these periodic orbits, inside the two heteroclinic orbits.

From Figures 18.3, 18.4, and 18.5 we then have the following result (proof left to the reader). In this lemma, we are most interested in small values of $\delta$ and $\gamma$.

**Lemma 18.6.** *Suppose that* $0 < \gamma < \frac{1}{2}$. *Then for each sufficiently small* $\delta > 0$ *there are* $\bar{\alpha} \in (0, \gamma)$, $\bar{\beta} > 0$, *and* $\varepsilon_{\delta,\gamma} > 0$ *such that if* $0 < \varepsilon < \varepsilon_{\delta,\gamma}$ *then*

(a) *if* $0 \le x_1 \le x_0 - \delta$ *and* $u$ *is a solution of* (18.1) *with* $u(x_1) = \bar{\alpha}$ *and* $u'(x_1) \ge 0$, *then* $u' > 0$ *on an interval* $[x_1, x_1 + \eta]$ *and* $u(x_1 + \eta) > 1$,

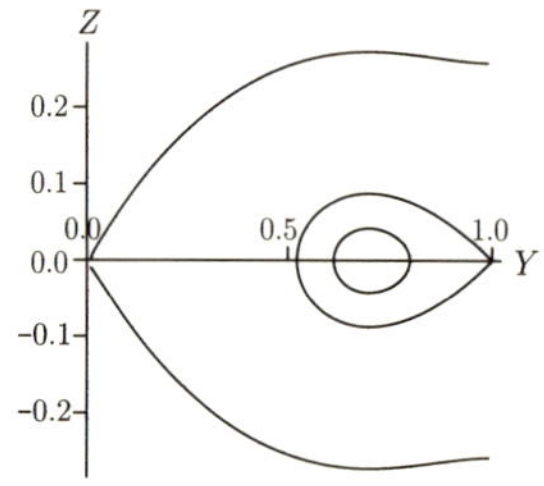


**Figure 18.3.** $0 \leq x^* < x_0$.

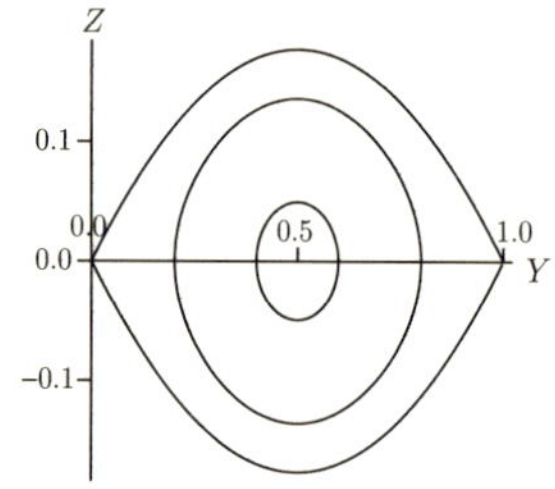


**Figure 18.4.** $x^* = x_0$.

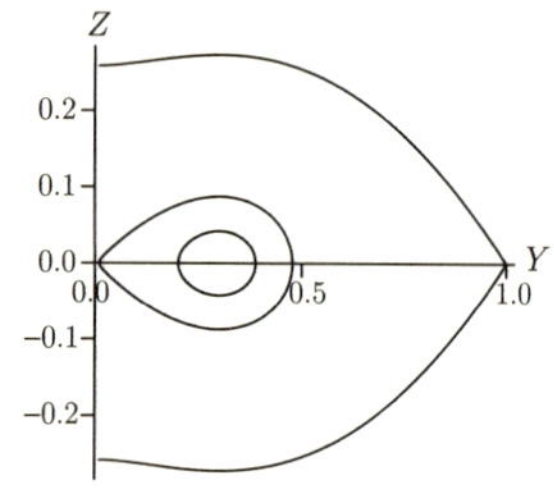


**Figure 18.5.** $x^* = x_0 + \delta$.

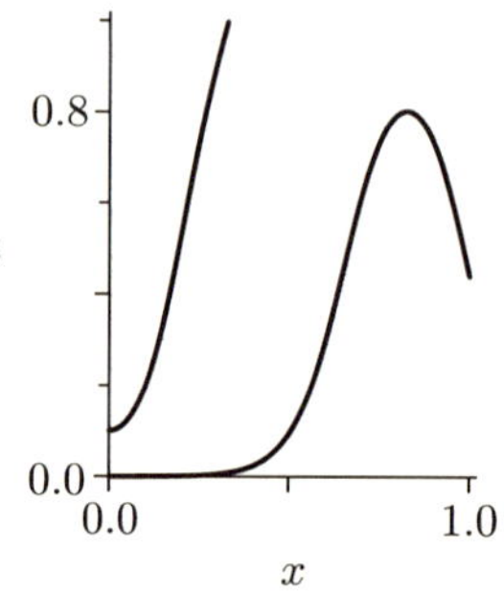


**Figure 18.6.** Solutions with $u(0) = \bar{\alpha}$ and with $u(x_0 + \delta) = \bar{\alpha}$.

(b) *if* $u$ *is a solution of* (18.1) *with* $u(x_0 + \delta) = \bar{\alpha}$ *and* $0 \leq u'(x_0 + \delta) \leq \frac{\bar{\beta}}{\varepsilon}$, *then* $u$ *has a maximum in* $(x_0 + \delta, 1)$ *(see Figure* 18.6*)*,

(c) *if* $u$ *is a solution of* (18.1) *and for some* $x_1 \in (x_0 - \delta, x_0 + \delta)$, $u(x_1) = \bar{\alpha}$ *and* $u'(x_1) \geq 0$, *then* $u$ *increases monotonically to a value above* $1 - \gamma$, *which it reaches before* $x = x_0 + 2\delta$.

*Furthermore,* $\bar{\alpha}$ *can be chosen so that*

$$\sqrt{2\bar{\alpha}M} \leq \bar{\beta}, \tag{18.27}$$

*where* $M = \max_{0 \leq Y \leq 1, 0 \leq x \leq 1} |f(x, Y)|$.

We will assume henceforth that $\delta$, $\gamma$, $\varepsilon_{\delta,\gamma}$, $\bar{\alpha}$, and $\bar{\beta}$ satisfy the conditions given in this lemma. We will also assume that

$$\bar{\alpha} < \phi(1). \tag{18.28}$$

For $0 < \varepsilon < \varepsilon_{\delta,\gamma}$ and $0 < \alpha \leq \bar{\alpha}$, let $u_\alpha$ denote the solution of (18.1) which satisfies the initial condition

$$u(0) = \alpha, \quad u'(0) = 0. \tag{18.29}$$

From (18.28) it follows that $f(x, u) < 0$ for $0 < u \leq \bar{\alpha}$ and $0 \leq x \leq 1$. We therefore see that $u_\alpha'' > 0$ and $u_\alpha' > 0$ on some interval $[0, x_1(\alpha)]$

such that either $x_1(\alpha) = 1$ and $u_\alpha(x_1(\alpha)) < \bar{\alpha}$ or else $u_\alpha(x_1(\alpha)) = \bar{\alpha}$. Clearly, $x_1(\bar{\alpha}) = 0$, and since $0 < \varepsilon < \varepsilon_{\delta,\gamma}$, Lemma 18.6 implies that $u'_{\bar{\alpha}} > 0$ on an interval $[0, x_2(\bar{\alpha})]$ with $u_{\bar{\alpha}}(x_2(\bar{\alpha})) = 1$. As $\alpha$ decreases from $\bar{\alpha}$, $x_2(\alpha)$ continues to be defined as a continuous function of $\alpha$ by the equation $u_\alpha(x) = 1$, at least as long as $x_1(\alpha) \le x_0 - \delta$. However, for sufficiently small $\alpha$, $x_1(\alpha) \ge x_0 + \delta$.

Further, we claim that as long as $x_2(\alpha)$ is continuous (for decreasing $\alpha$) and $u'(x_2(\alpha)) > 0$, $u'$ is positive on $(0, x_2(\alpha)]$. This is trivially true for $\alpha = \bar{\alpha}$, and if it fails be true at some largest $\hat{\alpha} < \bar{\alpha}$, then there is an $\hat{x} < 1$ such that $u_{\hat{\alpha}}(\hat{x}) < 1$, $u'_{\hat{\alpha}} > 0$ on $(0, \hat{x})$, and $u'_{\hat{\alpha}}(\hat{x}) = u''_{\hat{\alpha}}(\hat{x}) = 0$. In this case, $u_{\hat{\alpha}}(\hat{x}) = \phi(\hat{x})$. From (18.23) we see that $u'''_{\hat{\alpha}}(\hat{x}) = -f_x(\hat{x}, \phi(\hat{x})) = \phi'(\hat{x})\phi(\hat{x})(1 - \phi(\hat{x})) < 0$, which implies that $u_{\hat{\alpha}}$ is decreasing in an interval around $\hat{x}$. This is only possible if $u_{\hat{\alpha}} - \phi$ has a previous zero in $(0, \hat{x})$, allowing $u''$ to become negative, which does not occur because $u'_{\hat{\alpha}} > 0$ as long as $\phi' < 0$ in $(0, \hat{x})$.[8]

**Proposition 18.7.** *If $\varepsilon$ is sufficiently small and $x_1(\alpha) = x_0 + \delta$, then $u'$ has a zero before $u = 1$.*

**Proof.** From Lemma 18.6, we must show that $u'(x_1(\alpha)) \le \frac{\beta}{\varepsilon}$. On $[0, x_1(\alpha)]$, $0 < u \le \bar{\alpha}$ and $\varepsilon^2 u' u'' \le u' M$. Integrating this and using (18.27) gives the result. □

Extend the definition of $x_2(\alpha)$ to lower values of $\alpha$ in $(0, \bar{\alpha})$, where possibly $u_\alpha$ does not reach 1, by letting $x_2(\alpha) = 1$ if $u'_\alpha > 0$ on $(0, 1]$ and $u_\alpha(1) < 1$. Proposition 18.7 implies that $x_2(\alpha)$ is not defined if $x_1(\alpha) \ge x_0 + \delta$. Suppose that $x_2$ is continuous on some maximal interval $(\alpha_1, \bar{\alpha}]$. If $u'_{\alpha_1}(1) > 0$ then $x_2$ could be extended continuously to lower values of $\alpha$, and so $u'_{\alpha_1}(1) = 0$ as desired. Since $x_0 - \delta < x_1(\alpha_1) < x_0 + \delta$, part (c) of Lemma 18.6 implies that for sufficiently small $\varepsilon$ the region of steep increase of $u_{\alpha_1}$ is in a narrow vertical strip close to $x_0$. The limit statement in Theorem 18.2 follows from Lemma 18.6, by choosing smaller values of $\delta$ and $\gamma$ and correspondingly smaller values of $\varepsilon_{\delta,\gamma}$. This completes the proof of the theorem. □

## 18.4. Uniqueness and stability of a single layer

In this section we continue to use the notation of Section 18.3.3, so that $u_{\alpha_1}$ is the solution $u^\varepsilon$ of Theorem 18.2. We assume that $\phi' < 0$ on $[0, 1]$. We will prove a local uniqueness result, namely that $u_{\alpha_1}$ is the unique solution of (18.1)–(18.3) among all nearby solutions of (18.1). To state this precisely,

[8] As mentioned earlier, the weaker assumption $\phi'(x_0) < 0$ is sufficient, but then the proof is more complicated.

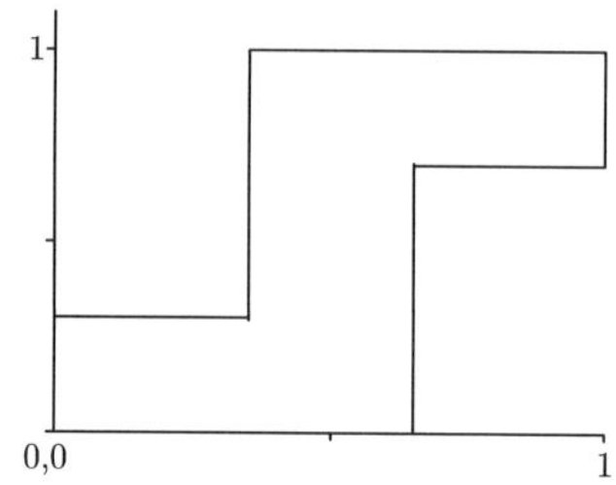


**Figure 18.7.** The set $S_{\gamma,\delta}$.

consider positive values of $\delta$ and $\gamma$ satisfying the conditions of Lemma 18.6, and let $\bar{\alpha}$ be as in that lemma. Also, let

$$S_{\gamma,\delta} = [0, x_0 - \delta] \times [0, \gamma] \cup ([x_0 - \delta, x_0 + \delta] \times [0, 1]) \cup [x_0 + \delta, 1] \times [1 - \gamma, 1].$$

See Figure 18.7.

**Theorem 18.8.** *If $\varepsilon$ is sufficiently small and for some $\alpha_1 \in (0, \bar{\alpha})$, $u'_{\alpha_1} > 0$ on $(0,1)$, $u'_{\alpha_1}(1) = 0$ and $(x, u_{\alpha_1}(x)) \in S_{\gamma,\delta}$ for $0 \le x \le 1$, then there is a $\delta_1 > 0$ such that if $0 < |\alpha - \alpha_1| < \delta_1$ then $u'_{\alpha}(1) \ne 0$.*

**Corollary 18.9.** *The solution $u_{\alpha_1}$ found in Theorem 18.2 is locally unique.*

**Proof.** The hypotheses imply that $u_{\alpha_1}$ is increasing and also that for small $\varepsilon$ and $\gamma$, $u_{\alpha_1}$ is small to the left of $x_0 - \delta$ and $1 - u_{\alpha_1}$ is small to the right of $x_0 + \delta$. Thus, $u_{\alpha_1}$ does not have a "half-spike" near $x = 0$ or $x = 1$, though it is not hard to show that increasing solutions with such half-spikes do exist. (See Exercise 18.3. )

Let $v$ be the unique solution to the linearized equation

$$\varepsilon^2 v'' + f_u(x, u_{\alpha_1}) v = 0 \tag{18.30}$$

with

$$v(0) = 1, \qquad v'(0) = 0. \tag{18.31}$$

Thus, $v(x) = \frac{\partial u_\alpha(x)}{\partial \alpha}|_{\alpha=\alpha_1}$.

**Lemma 18.10.** *For sufficiently small $\varepsilon$, $v > 0$ on $[0,1]$, and further, $v'(1) > 0$.*

**Proof.** Let $w = u'_{\alpha_1}$, so that

$$\varepsilon^2 w'' + f_u(x, u_{\alpha_1}) w = -f_x(x, u_{\alpha_1}), \tag{18.32}$$

$$w(0) = 0, \quad w'(0) = u''_{\alpha_1}(0). \tag{18.33}$$

Exercise 18.2 implies that $u''_{\alpha_1}(0)$ is positive and exponentially small as a function of $\varepsilon$ as $\varepsilon \to 0$. More precisely, there is a $\sigma > 0$ independent of $\varepsilon$ such that

$$\varepsilon^2 u''_{\alpha_1}(0) = O\left(e^{-\frac{\sigma}{\varepsilon}}\right) \tag{18.34}$$

as $\varepsilon \to 0$. Suppose that $v(T) = 0$ for some first $T > 0$. Multiply (18.30) by $w$, (18.32) by $v$, subtract, and integrate by parts. With (18.31) and (18.33), this gives

$$\varepsilon^2 \left(w(T) v'(T) + u''_{\alpha_1}(0)\right) = \int_0^T f_x(x, u_{\alpha_1}(x)) v(x)\, dx > 0.$$

But $w(T) > 0$, $v'(T) \leq 0$, and so

$$\int_0^T f_x(x, u_{\alpha_1}(x)) v(x)\, dx = O\left(e^{-\frac{\sigma}{\varepsilon}}\right) \tag{18.35}$$

as $\varepsilon \to 0$.

To get a contradiction, we note that as $\varepsilon \to 0$, $u'' = O\left(\frac{1}{\varepsilon^2}\right)$ uniformly for $x$ and $u$ in $[0,1]$. Hence,

$$u' = O\left(\frac{1}{\varepsilon^2}\right). \tag{18.36}$$

Further, $f_u(x,0) < 0$, so that there is a $u_0 \in (0,1)$ such that if $0 \leq x \leq 1$ and $0 \leq u \leq u_0$, then

$$f_u(x,u) \leq 0.$$

Hence, from (18.30), $v \geq 1$ as long as $u \leq u_0$.

Also, there is a $\mu > 0$ such that $f_x(x,u) \geq \mu$ as long as $\frac{u_0}{2} \leq u \leq u_0$ and $0 \leq x \leq 1$. The limit statement in Theorem 18.2 shows that for sufficiently small $\varepsilon$, $u(0) < \frac{u_0}{2}$. Hence, (18.36) implies that there is an interval $I_{\varepsilon,\alpha_1}$ of length $K\varepsilon^2$ ($K$ positive and independent of $\varepsilon$), in which $u_{\alpha_1}(x) \in \left[\frac{u_0}{2}, u_0\right]$, and in this interval,

$$f_x(x, u_{\alpha_1}(x)) v(x) \geq \mu.$$

Since $f_x(x,u) > 0$ for all $x \in [0,1]$ and $u \in (0,1)$ and since $v \geq 0$ on $[0,T]$, this gives

$$\int_0^T f_x v\, dx \geq \int_{I_{\varepsilon,\alpha_1}} \mu\, dx \geq K\varepsilon^2 \mu. \tag{18.37}$$

Therefore (18.35) does not hold, showing that $v > 0$ on $[0,1]$.

We have left to prove that $v'(1) > 0$.

We thank S.-B. Ai for this part of the proof. Choose some $x_1 \in (x_0, 1)$. Using (18.30) and (18.32) we obtain

$$\varepsilon^2 (v'u' - u''v)\,|_{x=1} = -\varepsilon^2 u''(0) + \int_0^1 f_x(x, u(x))\, v(x)\, dx$$

$$= \left(-\varepsilon^2 u''(0) + \int_0^{x_1} f_x\; v\; dx\right) + \int_{x_1}^1 f_x\; v\; dx. \tag{18.38}$$

The choice of $x_1$ and Theorem 18.2 imply that $u(x_1) > u_0$ and so, as in (18.37), $\int_0^{x_1} f_x\; v\; dx > K\varepsilon^2\mu$. From (18.34) it follows that the first term in (18.38) is positive for sufficiently small $\varepsilon$. Since $u'(1) = 0$ we obtain

$$v(1)\, u(1)\,(1 - u(1))\,(u(1) - \phi(1)) > \int_{x_1}^1 f_x\; v\; dx.$$

Because $f_x = -u(1-u)\,\phi'(x)$ and $u(1-u)$ is decreasing in $x$ on $[x_1, 1]$,

$$v(1)\, u(1)\,(1 - u(1))\,(u(1) - \phi(1)) > u(1)\,(1 - u(1))\,\eta \int_{x_1}^1 v\; dx, \tag{18.39}$$

where $\eta = \min_{0 \le x \le 1} |\phi'(x)|$.

Note that $f_u(x, 1) < 0$ for $x \in [0, 1]$. From Theorem 18.2 we see that for sufficiently small $\varepsilon$, $u_{\alpha_1}$ is close to 1 on $[x_1, 1]$. Since $f_u(x, 1) = \phi(x) - 1 < -\frac{1}{2}$ on $[x_1, 1]$, there is an $r > 0$ such that if $\varepsilon$ is sufficiently small, then $f_u(x, u_{\alpha_1}(x)) \le -r^2$ in $[x_1, 1]$. Hence,

$$\varepsilon^2 v'' \ge r^2 v \tag{18.40}$$

on $[x_1, 1]$.

Now suppose that $v'(1) \le 0$. Since $v > 0$ on $[0, 1]$, it follows that $v'' > 0$ on $[x_1, 1]$, and so $v' < 0$ on $[x_1, 1)$. From (18.40),

$$\varepsilon^2 v'' v' \le r^2 v v'$$

on $[x_1, 1]$. Integrating this from $x$ to 1 where $x \in [x_1, 1)$ gives

$$\varepsilon^2 \left(v'(1)^2 - v'(x)^2\right) \le r^2 \left(v(1)^2 - v(x)^2\right).$$

Hence,

$$\varepsilon v'(x) \le -r\sqrt{v(x)^2 - v(1)^2}.$$

Integrating this from $x$ to 1 gives

$$v(x) + \sqrt{v(x)^2 - v(1)^2} \ge v(1)\, e^{\frac{r}{\varepsilon}(1-x)},$$

so that

$$v(x) \ge \frac{1}{2} v(1)\, e^{\frac{r}{\varepsilon}(1-x)}.$$

Substituting this into the integral on the right of (18.39) and cancelling $v(1)\, u(1)\,(1 - u(1))$ leads to a contradiction which proves Lemma 18.10. □

It follows that for $|\alpha - \alpha_1|$ sufficiently small, $u'_\alpha(1) \neq 0$ if $\alpha \neq \alpha_1$. This proves Theorem 18.8. □

**Remark 18.11.** Using methods in [**5**] it can be shown that for sufficiently small $\varepsilon$, there is no other $\alpha \in (0, \bar{\alpha})$ such that $(x, u_\alpha(x)) \in S_{\gamma,\delta}$ for all $x \in [0,1]$ and $u'_\alpha(1) = 0$. This is another example where classical methods give a more global result than is usually found with more abstract approaches.

**Remark 18.12.** In [**4**], considerably more delicate analysis is used to determine the Morse index of all layered and spike solutions. For example under the hypotheses of Theorem 18.8 this analysis shows that

$$\frac{\varepsilon v'(1)}{v(1)} = \sqrt{-f_u(1,1)} + o(1) \quad \text{as } \varepsilon \to 0. \tag{18.41}$$

### 18.4.1. Stability.

Stability analysis of steady-state solutions to pde's is almost always limited to a consideration of "linearized" stability, which in many cases can be shown to imply a form of "local" asymptotic stability. A result of this kind for (18.1)–(18.3) is found in [**54**]. See also [**22**]. Hence, in one dimension, we consider the linearization of

$$u_t = u_{xx} + f(x,u) \tag{18.42}$$

with homogeneous Neumann boundary conditions around the steady-state solution $u_\alpha$ satisfying (18.3), namely,

$$\begin{gathered} U_t = U_{xx} + f_u(x, u_\alpha) U, \\ U_x(0,t) = U_x(1,t) = 0. \end{gathered} \tag{18.43}$$

We assume that a solution of (18.43) is written in the form

$$U(x,t) = e^{\lambda t} V(x), \tag{18.44}$$

where $V$ satisfies the boundary conditions (18.3). Then

$$\begin{gathered} V'' + f_u(x, u_\alpha) V = \lambda V, \\ V'(0) = V'(1) = 0. \end{gathered} \tag{18.45}$$

We prove linearized stability by showing that there are no nonnegative eigenvalues for this selfadjoint problem, which implies that every solution of the form (18.44) tends to 0 as $t \to \infty$.

In fact, all the work for this proof has been done. We use the Sturm comparison theorem, as for example in [**41**, Chapter 8, Theorem 1.2]. Write (18.45) in the form

$$\begin{gathered} V'' + (f_u(x, u_\alpha) - \lambda) V = 0, \\ V'(0) = V'(1) = 0 \end{gathered} \tag{18.46}$$

and compare this to (18.30), with the same boundary conditions. Lemma 18.10 implies that if $\lambda \geq 0$ then there is no nontrivial solution to (18.46).

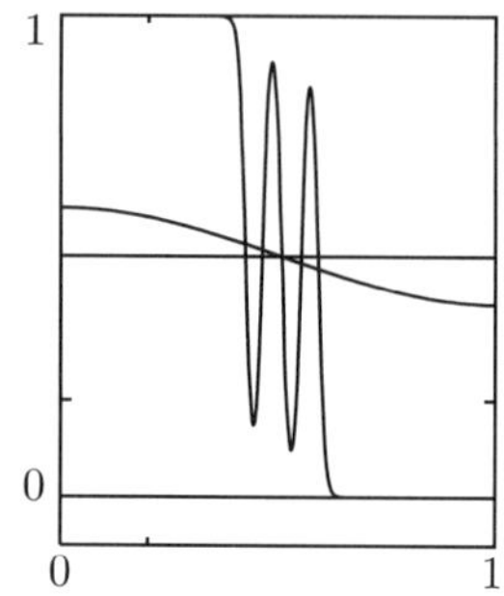


**Figure 18.8.** Multiple layers.

Hence all eigenvalues of (18.45) are negative, proving the linearized stability of the steady-state solution $u_\alpha$ to (18.42).

The stability of the constant solutions $u = 0$ and $u = 1$ is much easier, for if $u$ is one of these solutions, then $f_u(x, u(x)) < 0$ for $0 \leq x \leq 1$. The argument above becomes trivial.

## 18.5. Further stable and unstable solutions, including multiple layers

We continue to examine the simple situation in which $\phi - \frac{1}{2}$ has a single zero in $[0, 1]$, at $x_0 \in (0, 1)$, with $\phi' < 0$. We noted above that there are two trivial stable solutions to (18.1)–(18.3), namely, $u = 0$ and $u = 1$. It was shown in [**11**] that these two, plus the single layered solution of Theorem 18.2, are the only stable solutions in this case.

Hence the solutions considered in the rest of this chapter are unstable. Among these are solutions with multiple layers, as shown in Figure 18.8

It turns out that the solution shown in Figure 18.8 starts with $u(0)$ exponentially close to 1 and ends with $u(1)$ exponentially close to 0. A natural question is whether there are multiple layers going in the other direction, with $u(0)$ small and $u(1)$ close to one. Theorem 18.20 below will show that this is not possible, if $\phi' < 0$. Of course, if $\phi' > 0$ then the result changes so that a single stable layer starts high and ends low, while multiple layers must start low and end high.

To simplify the proof that solutions with multiple layers exist, we will add an assumption which implies that $\phi(0) - \phi(1)$ is not too large. To explain this assumption we recall that for every $x^* \in (0, x_0)$ there is a homoclinic orbit $\Gamma_{x^*}$ of (18.26) which is based at $(1, 0)$. Let $\kappa(x^*)$ denote the minimum value of $Y$ on this orbit. Because $\phi' < 0$, $\kappa$ is a decreasing function of $x^*$ for $0 \leq x^* < x_0$. Also, $\kappa(x^*) \to 0$ as $x^* \to x_0$.

We will assume that $\kappa(0) < \phi(1)$. The conclusions of the theorem hold without this condition, but the proof is slightly more complicated. Much more general results can be found in [**4**].

**Theorem 18.13.** *Suppose that $\phi' < 0$ on $[0,1]$ and that $0 < \phi(1) < \frac{1}{2} < \phi(0) < 1$. Suppose also that $\kappa(0) < \phi(1)$. Then for each integer $n \geq 0$ there is an $\varepsilon_n > 0$ such that if $0 < \varepsilon < \varepsilon_n$ then (18.1)–(18.3) has a solution $u$ with exactly $n$ relative minima and $n$ relative maxima in $(0,1)$, and with $u(0) > \phi(0)$ and $u(1) < \phi(1)$. Further, for each $\delta > 0$, $\varepsilon_n$ can be chosen so that all of the $2n+1$ points where $u = \phi$ are in the interval $(x_0 - \delta, x_0 + \delta)$.*

**Proof.** We start with the following lemma, which is of independent interest [**4**].

**Lemma 18.14.** *Suppose that $n$ is a nonnegative integer. For every $\eta > 0$, there is an $\varepsilon_\eta > 0$ such that if $0 < \varepsilon < \varepsilon_\eta$ and $u$ is a solution of (18.1)–(18.3) on $[0,1]$ with no more than $n$ minima in $(0,1)$, then $|Q(u(x), u'(x), x)| < \eta$ on $[0,1]$, where*

$$Q(u, u', x) = \left[\varepsilon^2 u'^2 + 2F_1(x,u)\right] u(1-u)$$

*and*

$$F_1(x,u) = \begin{cases} \int_1^u f(x,s)\,ds & \text{if } 0 \leq x < x_0, \\ \int_0^u f(x,s)\,ds & \text{if } x_0 \leq x \leq 1. \end{cases}$$

**Remark 18.15.** Consider, for example, an $x \in [0, x_0)$. This lemma says that if $\varepsilon$ is sufficiently small and $u$ is a solution with no more than $n$ minima in $(0,1)$, then either $u(x)$, $1-u(x)$, or $\varepsilon^2 u'^2 + 2F_1(x,u)$ is small. Therefore, if neither $u(x)$ nor $1-u(x)$ is small then $(u(x), \varepsilon u'(x))$ is close to $\Gamma_x$.

**Proof.** (The proof resembles part of the proof of Theorem 16.4.) Suppose that the assertion is false. Then there exist $\eta > 0$ and a sequence $\{\varepsilon_i, u_i, \xi_i\}_{i=1}^\infty$, with $\varepsilon_i > 0$ and $\lim_{i\to\infty} \varepsilon_i = 0$, such that for each $i \geq 1$, $u_i$ is a solution of (18.1)–(18.3) with no more than $n$ minima in $(0,1)$, and

$$Q\left(u_i(\xi_i), u_i'(\xi_i), \xi_i\right) \geq \eta. \tag{18.47}$$

By taking a subsequence if necessary, we can assume that $\lim_{i\to\infty} \xi_i = x_* \in [0,1]$ and that $\lim_{i\to\infty} u_i(\xi_i) = u^* \in [0,1]$. Inequality (18.47) implies that $u^* \in (0,1)$. We consider the case where $x_* \in (0,1)$, with small changes being necessary if $x_* \in \{0,1\}$. Choose $\delta > 0$ so that for large $i$, $0 \leq \xi_i - \delta < \xi_i + \delta \leq 1$.

Let $U_i(s) = u_i(\xi_i + \varepsilon s)$ for $s \in [-\delta/\varepsilon_i, \delta/\varepsilon_i]$. Then $U_i'' = -f(\xi_i + \varepsilon s, U_i)$. Hence $\{U_i''\}_{i=1}^\infty$ is uniformly bounded, and, by taking a subsequence if necessary, there exists $U \in C^2(R^1)$ such that $\lim_{i\to\infty}(U_i, U_i') = (U, U')$ uniformly in each compact subset of $R^1$. In addition $U'' = -f(x_*, U)$ so that $U'^2 = -2F_1(x_*, U) + C$ for some $C$. One notices that if $C > 0$ then the solution is unbounded on $R^1$, and if $C < 0$ then the solution is periodic. But both are impossible, since $U_i$ is bounded and has at most $n$ minima in $[-\delta/\varepsilon_i, \delta/\varepsilon_i]$. Hence, $C = 0$. But this implies that $\lim_{i\to\infty} |\varepsilon_i^2 u_i'^2(\xi_i) + 2F_1(\xi_i, u_i(\xi_i))| = \lim_{i\to\infty} |U_i'^2(0) + 2F_1(\xi_i, U_i(0))| = 0$ and completes the proof of Lemma 18.14. □

We now continue with the proof of Theorem 18.13, which we do in two parts.

(a) Existence of multiple layers:

For $\phi(0) < \alpha < 1$, consider the solution $(Y_\alpha, Z_\alpha)$ of the autonomous system (18.26) with $x^* = 0$ such that $Y(0) = \alpha$, $Z(0) = 0$. Such solutions are periodic, with orbits inside the homoclinic loop $\Gamma_0$. Since $\kappa(0) < \phi(1)$, we can choose $\alpha_0 \in (\phi(0), 1)$ so that if $\alpha_0 \le \alpha < 1$ then for sufficiently small $\varepsilon$ the minimum value of $Y_\alpha$ is less than $\phi(1)$.

Suppose now that $u_{\alpha_0}$ has fewer than $n+1$ minima in $[0,1]$. Then Lemma 18.14 implies that its orbit is close to that of one of the periodic solutions in Figure 18.3 (inside the homoclinic loop), giving the contradictory conclusion that for sufficiently small $\varepsilon$ it has at least $n+1$ minima in $[0,1]$. It follows that for small $\varepsilon$, $u_{\alpha_0}$ has at least $n+1$ local minima in $(0,1)$. On the other hand, for $1-\alpha$ sufficiently small (depending on $\varepsilon$), $u_\alpha$ is close to 1 on $[0,1]$, with $u_\alpha' < 0$. For each small $\varepsilon$, if $\alpha_0 \le \alpha < 1$ then either $u_\alpha$ has $n$ or fewer local minima in $[0,1]$, and so Lemma 18.14 applies to $u_\alpha$, or else $u_\alpha$ has $n+1$ or more local minima. (Note that the choice of $\eta$ (the "degree of closeness") in the lemma does not depend on the particular solution $u_\alpha$.)

The monotonicity of $\phi$ implies a similar monotonicity of successive maxima and minima, which we just saw exist at least for $\alpha = \alpha_0$. The following lemma is similar to one in Chapter 15 and will be used more than once below.[9]

**Lemma 18.16.** *Let $u = u_\varepsilon$ be a solution of (18.1). If $a < b$ are successive local maxima (or successive minima) of $u$ in an interval where $\phi' < 0$, then $u(b) < u(a)$. Furthermore,*

(i) *if $a$ and $b$ are successive local maxima, then*

$$\begin{aligned}
&\int_{u(b)}^{u(a)} v(1-v)\left(\phi\left(\xi_-(v)\right) - v\right) dv \\
&\qquad = \int_{u(c)}^{u(b)} v(1-v)\left(\phi\left(\xi_+(v)\right) - \phi\left(\xi_-(v)\right)\right) dv
\end{aligned} \tag{18.48}$$

*where $c \in (a,b)$ is the intervening local minimum of $u$ and $\xi_-$ and $\xi_+$ are the inverse functions of $u$ on $[a,c]$ and $[c,b]$, respectively;*

[9] We thank Professor S.-B. Ai for much of this lemma and its proof and for several other helpful proofs in the rest of this section where this lemma is used.

(ii) *if $a$ and $b$ are successive local minima, then*

$$\begin{aligned}(18.49)\qquad &\int_{u(b)}^{u(a)} v(1-v)\left(\phi\left(\xi_+(v)\right)-v\right)\,dv \\ &\quad = \int_{u(a)}^{u(c)} v(1-v)\left(\phi\left(\xi_-(v)\right)-\phi\left(\xi_+(v)\right)\right)\,dv\end{aligned}$$

*where $c \in (a,b)$ is the intervening local maximum of $u$ and $\xi_-$ and $\xi_+$ are defined as in* (i).

**Proof.** Suppose that $a$ and $b$ are successive local maxima, and let $c \in (a,b)$ be the intervening local minimum. Consider a pair of points $x_- \in [a,c]$ and $x_+ \in [c,b]$ with $u(x_-) \geq u(x_+)$. Then $u' < 0$ on $(x_-, c)$ and $u' > 0$ on $(c, x^+)$. We obtain

$$\begin{aligned}\frac{\varepsilon^2}{2}\left(u'(x_+)^2 - u'(x_-)^2\right) &= \left(\int_{x_-}^{c} + \int_{c}^{x_+} u(1-u)\left(\phi(x)-u\right)u'dx\right) \\ (18.50)\qquad &= \int_{u(c)}^{u(x_+)} u(1-u)\,\phi\left(\xi_+(u)\right)du + \int_{u(x_+)}^{u(x_-)} u^2(1-u)\,du \\ &\quad - \int_{u(c)}^{u(x_-)} u(1-u)\,\phi\left(\xi_-(u)\right)du,\end{aligned}$$

where $\xi_-$ and $\xi_+$ are the inverse functions of $u$ on $[x_-, c]$ and $[c, x_+]$, respectively. If $u(b) \geq u(a)$, then let $x_- = a$ and choose $x_+$ to be the unique point in $[c,b]$ where $u(x_+) = u(a)$. Then the second term on the right of (18.50) is zero. Also, $u'(x_-) = u'(a) = 0$ so that the left side of this equation is nonnegative. However since $\phi$ is decreasing and $\xi_-(u) < \xi_+(u)$ on $(u(c), u(a))$, the right side is negative. Hence, $u(b) < u(a)$.

Further, letting $x_- = a$ and $x_+ = b$ in (18.50) gives (18.48). Similar arguments give the remaining statements in the lemma. □

The following result will also be useful.

**Lemma 18.17.** *Let $u = u_\varepsilon$ be a solution of* (18.1) *with successive local minimum and maximum $c < b$, respectively, and let $z \in (c,b)$ be such that $u(z) = \phi(z)$. Then*

$$(18.51)\qquad \text{(i) } z - c \geq \varepsilon \log\frac{u(z)}{u(c)} \quad \textit{and} \quad \text{(ii) } b - z \geq \varepsilon \log\frac{1-u(z)}{1-u(b)}.$$

**Proof.** Note that $u' > 0$ on $(c,b)$. On $[c,z]$ we have $\varepsilon^2 u'' \leq u$. Multiplying by $u'$ and integrating from $c$ to $x$ gives $\varepsilon^2 u'^2 \leq u^2$ in $[c,z]$, so that $\varepsilon\frac{u'}{u} \leq 1$ in this interval. A further integration over $[c,z]$ gives (18.51)(i).

Similarly, on $[z, b]$, $0 \geq \varepsilon^2 u'' \geq u - 1$. Multiplying by $u'$ and integrating from $x$ to $b$ gives $\varepsilon^2 u'^2 \leq (1-u)^2$ in $[z, b]$, or $\varepsilon \frac{u'}{1-u} \leq 1$. A final integration, from $z$ to $b$, gives (18.51)(ii) and completes the proof of Lemma 18.17. □

We continue to assume that $\alpha_0$ is chosen so that for sufficiently small $\varepsilon$, $u_{\alpha_0}$ has at least $n+1$ minima in $[0,1]$. To prove the existence of multiple layers we now consider solutions $u_\alpha$ with $\alpha_0 < \alpha < 1$. Either $u_\alpha > \alpha_0$ on $[0,1]$ or $u_\alpha$ decreases on some interval $[0, x^*]$ with $u_\alpha(x^*) = \alpha_0$ and $u'_\alpha(x^*) < 0$. Fix a small $\delta > 0$. We can choose $\varepsilon$ so that the behavior of $u_\alpha$ is different according to whether $0 \leq x^* \leq x_0 - \delta$ or $x^* \geq x_0 + \delta$, with a transition from one behavior to another somewhere in $(x_0 - \delta, x_0 + \delta)$.

In particular, we can choose $\varepsilon$ small enough to ensure that if $x^* \in [0, x_0 - \delta]$ then $u_\alpha$ has at least one minimum in $[0,1]$, while if $x^* \geq x_0 - \delta$, then either $u_\alpha$ has a minimum in $(x^*, x_0 + \delta)$, with this minimum below $\phi(1)$, or else $u_\alpha$ is decreasing in $[0,1]$.[10] If $u_\alpha$ has a minimum then Lemma 18.16 implies that all subsequent minima, if any exist, lie below $\phi(1)$. Hence, $u''_\alpha > 0$ at each of these minima, and so they are all continuous functions of $\alpha$ as $\alpha$ increases. All maxima lie strictly above the graph of $\phi$,[11] and so maxima and minima do not merge as $\varepsilon \to 0$. But as stated earlier, for fixed $\varepsilon$ and $1-\alpha$ sufficiently small, $u_\alpha$ has no minima in $(0,1)$. Hence, holding $\varepsilon$ sufficiently small and fixed and increasing $\alpha$ from $\alpha_0$, the minima and intervening maxima cross $x = 1$ one by one, giving in particular the solution whose existence is asserted in the theorem, with $u(1) < \phi(1)$.

(b) Location of layers:

Having established the existence of this solution, we complete the proof of Theorem 18.13 by showing that the layers all tend to $x_0$ as $\varepsilon \to 0$. Choose a positive number $\gamma$ such that $1 - \gamma > \phi(0)$. In the following lemma, $u$ is a solution of (18.1)–(18.3) with exactly $n$ local minima in $(0,1)$ and with $u(0) > \phi(0)$ and $u(1) < \phi(1)$. The limit claimed, and other limits in the rest of the proof, are uniform over all such solutions $u$, for a fixed $n$.

**Lemma 18.18.** *Let $z_1 = z_1(\varepsilon)$ be the first point where $u = 1 - \gamma$. Then*

$$\lim_{\varepsilon \to 0} z_1 = x_0.$$

**Proof.** Step 1. Assume that the result is false. Then there exists a sequence $\varepsilon_k$ such that $\varepsilon_k \to 0$ and $z_1(\varepsilon_k) \to \bar{x}_0 \in [0,1] \backslash \{x_0\}$ as $k \to \infty$. For notational simplicity, we assume that $z_1 \to \bar{x}_0$ as $\varepsilon \to 0$. A phase plane argument as used in the proof of Lemma 18.14 shows that $\bar{x}_0 \in [0, x_0)$ and

$$\lim_{\varepsilon \to 0} (u(z_1 + \varepsilon\xi), \varepsilon u'(z_1 + \varepsilon\xi)) = (Y(0), Y'(0))$$

---

[10] By truncating $f$ in $u < 0$ and $u > 1$ it can be assumed that all solutions exist on $[0,1]$.

[11] See the paragraph just before Proposition 18.7.

uniformly on every compact interval $|\xi| \le T$ $(0 < T < \infty)$, where $Y$ is the homoclinic solution of

$$Y''(\xi) = f(\bar{x}_0, Y), \quad Y(0) = 1 - \gamma, \quad Y'(0) < 0.$$

We conclude that there exists a $\hat{z}_1 > z_1$ such that as $\varepsilon \to 0$, $\hat{z}_1 = z_1 + O(\varepsilon)$, $u(\hat{z}_1) = 1 - \gamma$, and $\varepsilon u'(\hat{z}_1) = Y'(T) + o(1) > 0$, where $T > 0$ is defined by $Y(T) = 1 - \gamma$.

Step 2. We will assume that $\bar{x}_0 > 0$, with slight changes necessary if $\bar{x}_0 = 0$. Let $x_1 = x_1(\varepsilon) > \hat{z}_1$ be the first local maximum of $u$ in $(0, 1)$. We claim that

$$\lim_{\varepsilon \to 0} x_1 = \bar{x}_0.$$

If not, then assume that there exists a $\zeta > 0$ such that $x_1 > \bar{x}_0 + \zeta$ for all sufficiently small $\varepsilon$ (otherwise replace $\varepsilon$ by a sequence $\varepsilon_k \to 0$ as $k \to \infty$). Then on $[\hat{z}_1, x_1)$, $u' > 0$ and $\varepsilon^2 u'' = u(u-1)(u - \phi(x)) \le \omega^2 (u - 1)$, where $\omega^2 = (1 - \gamma)(1 - \gamma - \phi(0))$. Hence $\varepsilon^2 u'' u' \le \omega^2 (u-1) u'$, and integrating leads to

$$\varepsilon u'(x) \ge \omega \sqrt{(1 - u(x))^2 - (1 - u(x_1))^2}.$$

A further integration yields

$$\varepsilon \log \frac{1 - u(x_1)}{1 - u(x) - \sqrt{(1 - u(x))^2 - (1 - u(x_1))^2}} \ge \omega(x_1 - x),$$

or

$$\varepsilon \log \frac{1 - u(x) + \sqrt{(1 - u(x))^2 - (1 - u(x_1))^2}}{1 - u(x_1)} \ge \omega(x_1 - x). \tag{18.52}$$

Thus,

$$1 - u(x_1) \le 2[1 - u(x)] e^{-\frac{\omega}{\varepsilon}(x_1 - x)}.$$

Setting $x = \hat{z}_1$,

$$1 - u(x_1) \le 2\gamma e^{-\frac{\omega}{\varepsilon}(x_1 - \hat{z}_1)} \le 2\gamma e^{-\frac{\omega}{\varepsilon}(\zeta + o(1))} \tag{18.53}$$

as $\varepsilon \to 0$.

But suppose that $\hat{x}_1$ is the local minimum point of $u$ in $(0, x_1)$ and apply Lemma 18.16 with $a = 0$, $b = x_1$, and $c = \hat{x}_1$. Equation (18.48) becomes

$$\begin{aligned} &\int_{u(x_1)}^{u(0)} v(1 - v)(v - \phi(\xi_-(v)))\, dv \\ &\qquad = \int_{u(\hat{x}_1)}^{u(x_1)} v(1 - v)(\phi(\xi_-(v) - \phi(\xi_+(v))))\, dv. \end{aligned} \tag{18.54}$$

Since $1 - u\,(0) < 1 - u\,(x_1)$, (18.53) implies that the left side of (18.54) is exponentially small as $\varepsilon \to 0$. To estimate the right side we observe that since $\phi' < 0$,

$$\begin{aligned}\phi(\xi_-(v)) - \phi(\xi_+(v)) &\geq K(\xi_+(v) - \xi_-(v)) \geq K(\xi_+(v) - \hat{x}_1) \\ &\geq K(\zeta - \hat{x}_1)\end{aligned} \tag{18.55}$$

for $\phi(\zeta) \leq v \leq u(x_1)$, where $\zeta$ is the point in $(\hat{x}_1, x_1)$ such that $u(\zeta) = \phi(\zeta)$ and where as usual, $K$ may change from one usage to the next. Hence from (18.51)(i),

$$\phi(\xi_-(v)) - \phi(\xi_+(v)) \geq \varepsilon \log \frac{u(\zeta)}{u(\hat{x}_1)} = \varepsilon \log \frac{\phi(\zeta)}{u(\hat{x}_1)}$$

on $[\phi(\zeta), u(x_1)]$. Lemma 18.14 implies that as $\varepsilon \to 0$, $u(\hat{x}_1) - \kappa(\hat{x}_1) \to 0$, and since $\kappa(\hat{x}_1) < \phi(1)$, we obtain a lower bound of the form $K\varepsilon$ for the right side of (18.54), giving a contradiction which completes Step 2.

Step 3. Continuing to vary $K$ when necessary, we will show that

$$x_1 - \hat{z}_1 = O(\varepsilon \log \varepsilon) \text{ as } \varepsilon \to 0 \text{ and } 1 - u(x_1) \geq \sqrt{K\varepsilon + (1 - u(0))^2}.$$

(Note that $1 - u(0) = O(e^{-1/\varepsilon})$ if $\bar{x}_0 > 0$, or else $1 - u(0) = o(1)$.) To see this, observe that $v$ and $v - \phi(\xi^-(v))$ in the integrand on the left of (18.54) are bounded away from zero as $\varepsilon \to 0$, and so we obtain from the argument in Step 2 that

$$\begin{aligned}&(1 - u(x_1))^2 - (1 - u(0))^2 \\ &\qquad \geq K \int_{u(\hat{x}_1)}^{u(x_1)} u(1-u)(\phi(\xi^-(u)) - \phi(\xi^+(u)))\, du \geq K\varepsilon.\end{aligned}$$

Hence

$$1 - u(x_1) \geq \sqrt{K\varepsilon + (1 - u(0))^2}. \tag{18.56}$$

Applying (18.52) with $x = \hat{z}_1$ yields

$$\omega(x_1 - \hat{z}_1) \leq \varepsilon \log \frac{\gamma + \sqrt{\gamma^2 - (1 - u(x_1))^2}}{1 - u(x_1)} = O(\varepsilon \log \varepsilon)$$

as $\varepsilon \to 0$. Furthermore,

$$x_1 = z_1 + (\hat{z}_1 - z_1) + (x_1 - \hat{z}_1) = z_1 + O(\varepsilon \log \varepsilon),$$

which proves Lemma 18.18. □

Now let $z_2 > x_1$ be such that $u(z_2) = 1 - \gamma$ and $u' < 0$ on $(x_1, z_2]$.[12] We will show that as $\varepsilon \to 0$,

$$z_2 = x_1 + O(\varepsilon \log \varepsilon).$$

[12] It follows from the argument below that such a $z_2$ exists if $\varepsilon$ is sufficiently small.

To see this, we note that on $(x_1, z_2]$, $\varepsilon^2 u'' = u(u-1)(u-\phi(x)) < \omega^2(u-1)$ where $\omega$ is the same as in Step 2, and $\varepsilon^2 u'' u' > \omega^2(u-1)u'$. Integration on $[x_1, x] \subset [x_1, z_2]$ gives $\varepsilon^2 u'(x)^2 > \omega^2[(1-u(x))^2 - (1-u(x_1))^2]$ and so $\varepsilon u'(x) < -\omega\sqrt{(1-u(x))^2 - (1-u(x_1))^2}$. From this, as in the derivation of (18.52), we obtain that

$$-\frac{\omega}{\varepsilon}(x - x_1) > \log \frac{1-u(x_1)}{1-u(x) + \sqrt{(1-u(x))^2 - (1-u(x_1))^2}}.$$

With $x = z_2$ and $1 - u(x_1) \geq \sqrt{\varepsilon}$, and using (18.56), we get

$$z_2 - x_1 \leq \frac{\varepsilon}{\omega} \log \frac{\gamma + \sqrt{\gamma^2 - (1-u(x_1))^2}}{1-u(x_1)} = O(\varepsilon \log \varepsilon)$$

as $\varepsilon \to 0$, which is what we asserted above.

Note further that $(u(z_2 + \varepsilon\xi), \varepsilon u'(z_2 + \varepsilon\xi))$ approaches $(Y(\xi), Y'(\xi))$ as $\varepsilon \to 0$ uniformly on each compact interval $|\xi| \leq T$, where $Y$ is the same homoclinic solution as in Step 1. We repeat the argument above inductively to conclude that the local maxima $0 < x_1 < x_2 < \cdots < x_n < 1$ of $u$ satisfy

$$1 - u(x_j) > K\sqrt{\varepsilon} \quad \text{and} \quad x_j = z_1 + O(\varepsilon \log \varepsilon) \quad \text{as } \varepsilon \to 0.$$

From here it is straightforward to show that if $\bar{x}_0 < x_0$ then for sufficiently small $\varepsilon$ the solution has more than $n$ maxima. This completes the proof of Theorem 18.13. □

**Remark 18.19.** In [4] more accurate estimates are obtained. For example it is shown that the separation between the layers obtained in Theorem 18.13 is given by $\Delta x = \varepsilon \log \varepsilon \left(\frac{1}{\alpha_i} + o(1)\right)$ as $\varepsilon \to 0$, where $\alpha_i$ is precisely defined in terms of $f$ and the location of the layer.

### 18.5.1. Orientation of layers.

We now state and prove the result mentioned earlier that with $\phi' < 0$, multi-layered solutions must start high and end low. In this result, $\bar{\alpha}$ is the number found in Lemma 18.6. Hence, if $\varepsilon$ is sufficiently small and $u(0) = \bar{\alpha}$, then $u' > 0$ and $u$ crosses 1.

**Theorem 18.20.** *Suppose that the hypotheses of Theorem* 18.2 *hold and* $n > 1$. *Choose* $\bar{\alpha}$ *as in Lemma* 18.6. *Then for sufficiently small* $\varepsilon$, *there is no solution to* (18.1)–(18.3) *with* $0 < u(0) < \bar{\alpha}$ *and having between two and* $n$ *minima in* $[0, 1]$.

**Corollary 18.21.** *For each* $n > 1$, *a solution with between two and* $n$ *layers near a single root of* $\phi - \frac{1}{2}$ *must have* $u(0)$ *close to* 1, *if* $\varepsilon$ *is small.*

**Proof of Corollary 18.21.** This result follows from Theorem 18.20 and Lemma 18.14, which imply that for each $\delta > 0$, nonmonotonic solutions $u_\alpha$ of (18.1)–(18.3) with $\alpha \in [\bar{\alpha}, 1-\delta]$ have more than $n$ minima if $\varepsilon$ is sufficiently small. Shooting backward is a helpful way to look at this. □

**Proof of Theorem 18.20.** Suppose that there is such a solution $u$ for each small $\varepsilon$. Its first minimum is at $x = 0$. Let $x_1$ be its first local maximum and $x_2$ the next minimum. It follows from Lemma 18.16 that $u(x_2) < u(0)$. Lemma 18.14 implies that $\liminf_{\varepsilon \to 0} x_1 \geq x_0$. Choosing a subsequence if necessary, we assume that as $\varepsilon \to 0$, $x_1 \to x^* \in [x_0, 1]$. Also, $u(0) \to 0$. From Lemma 18.14, $u(x_1) \to Y^*$, where $Y^*$ is the maximum of the homoclinic orbit of (18.25), or else, in the case $x^* = x_0$, $Y^* = 1$.

Choose the unique $z_1 \in (0, x_1)$ such that $u(z_1) = \phi(z_1)$. We then apply (18.51)(ii) from Lemma 18.17, with $b = x_1$, $z = z_1$, to obtain

$$x_1 - z_1 \geq \varepsilon \log \frac{1-\phi(z_1)}{1-u(x_1)} = \varepsilon \log \frac{1-\phi(z_1)}{1-Y^* + o(1)} \quad \text{as } \varepsilon \to 0. \tag{18.57}$$

Note that $z_1 \to x^*$ as $\varepsilon \to 0$, and so for small $\varepsilon$, $Y^* > \phi(z_1)$. Hence

$$x_1 - z_1 \geq K\varepsilon \tag{18.58}$$

for sufficiently small $\varepsilon$. Now use equation (18.49) from Lemma 18.16, with $a = 0$, $b = x_2$, $c = x_1$. We get

$$\int_{u(0)}^{u(x_2)} f(\xi_+(v), v)\, dv = \int_{u(0)}^{u(x_1)} v(1-v)\left(\phi(\xi_+(v)) - \phi(\xi_-(v))\right) dv. \tag{18.59}$$

In Exercise 18.2 you are asked to show that for some $\sigma > 0$, $u(0) = O\left(e^{-\sigma/\varepsilon}\right)$ as $\varepsilon \to 0$, and since $0 < u(x_2) < u(0)$, we see that the left side of (18.59) is also exponentially small as $\varepsilon \to 0$. On the other hand, (18.58) implies that if $u(0) < v < u(z_1)$ then $\phi(\xi_+(v)) - \phi(\xi_-(v)) \geq K\varepsilon$, and this again leads to a contradiction. This proves Theorem 18.20. □

## 18.6. Single and multiple spikes

Here we discuss solutions such as those in Figures 18.9 and 18.10. In fact, most of the work to show that such solutions exist, and are located as shown, was done in the previous section. Here is one of several possible results.

**Theorem 18.22.** *Suppose that $\frac{1}{2} < \phi < 1$ on $[0,1]$, $\phi'$ is continuous, $\phi$ has exactly one critical point in $[0,1]$, say at $x_0$, with $x_0 \in (0,1)$, and $\phi''(x_0) \neq 0$. Then for each integer $n \geq 1$ there is an $\varepsilon_n > 0$ such that if $0 < \varepsilon < \varepsilon_n$ then* (18.1)–(18.3) *has a solution $u_n$ with exactly $n$ local minima in $(0,1)$. If $x_0$ is a local maximum of $\phi$ then, as $\varepsilon \to 0$, the location of one of the local*

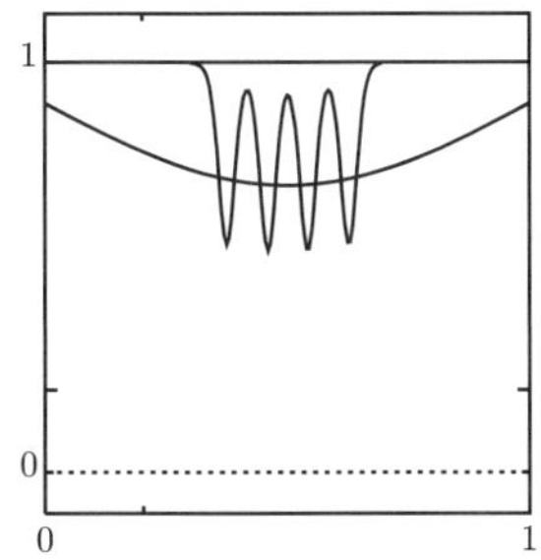

**Figure 18.9.** Multiple central spikes, and $\phi$.

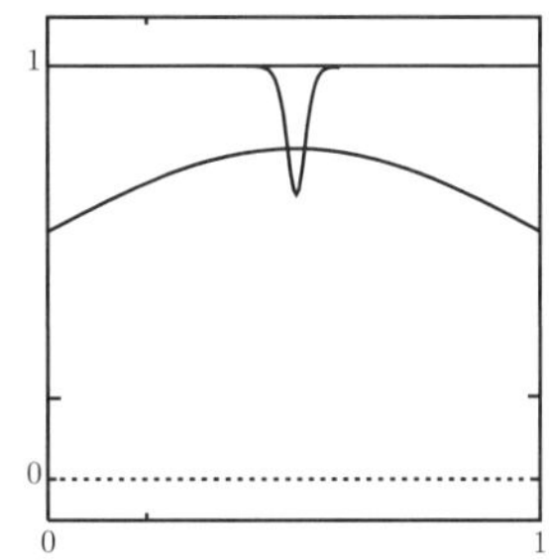

**Figure 18.10.** Single spike, and $\phi$.

*minima tends to* $x_0$*, while all others tend to either* 0 *or* 1*. If* $x_0$ *is a local minimum of* $\phi$ *then all of the minima of* $u_n$ *tend to* $x_0$ *as* $\varepsilon \to 0$.

We refer the reader to [**4**] for the proof, which uses techniques similar to those above.

**18.6.1. Combining layers and spikes.** We have discussed three cases:

(i) If $\phi' < 0$ and $\phi - \frac{1}{2}$ has a single zero, then there can be either a single upward layer, which was shown above to be stable, or a cluster of one or more layers, starting above $\phi(0)$ and ending below $\phi(1)$.

(ii) If $\phi > \frac{1}{2}$ on $[0,1]$ and $\phi'$ has a single zero, at which $\phi'' < 0$, then there can be a single downward spike near the critical point of $\phi$, plus possibly some spikes near the boundary.

(iii) If the same conditions hold except that $\phi'' > 0$ at the critical point of $\phi$, then there can be a cluster of one or more downward spikes, which must be near the critical point of $\phi$.

Similar results hold in related situations, such as $\phi < \frac{1}{2}$, $\phi'' < 0$, where we expect multiple spikes near a critical point of $\phi$.

Now consider a smooth function $\phi$ with the property that there is a partition of $[0,1]$ into subintervals $I_1, \ldots, I_n$ such that in each of these subintervals $\phi$ satisfies one of the above sets of conditions, with $\phi' \neq 0$ and $\phi \neq \frac{1}{2}$ at the endpoints of these intervals. It is shown in [**4**] that all combinations of the above families of solutions, as allowed by the form of $\phi$ in each interval, do exist. (There can be no "boundary" spikes except at 0 and 1.) The proof uses an inductive shooting argument and extends a similar result for the simpler situation of [**5**]. Moreover, detailed stability properties, in the form of the so-called Morse index, are analyzed for all of the constructed solutions. The latter analysis is considerably more delicate than what we have done here.

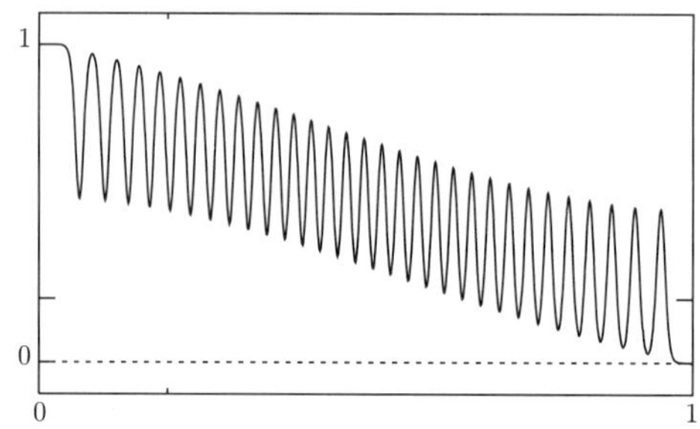

**Figure 18.11.** Many oscillations, defining an envelope.

In addition, the problem can be considered on an infinite interval, $[0, \infty)$ or $(-\infty, \infty)$, and results can be obtained which show sensitivity to initial conditions and chaos. These are similar to the results described for the Smale horseshoe map in Chapter 15 and those for another second order ode which were outlined in Chapter 16. Again see [**4**].

## 18.7. A different type of result for the layer model

In [**63**], Felmer, Martinez, and Tanaka considered a different aspect of the solutions of (18.1) as $\varepsilon \to 0$. This work exploits the fact that for any $x^*$ in $[0, 1]$ the equilibrium $Y = \phi(x^*)$ for the limiting rescaled equation (18.25) is a center, as in Figures 18.3–18.5, so that solutions starting near the point $(\phi(x^*), 0)$ are periodic. Hence solutions to (18.1) near $y = \phi(0)$, $y' = 0$ have oscillations with a frequency of order $\frac{1}{\varepsilon}$. These are very different from the solutions we considered above, where we required that the number of oscillations stays bounded as $\varepsilon \to 0$.

If we allow the number of oscillations to increase as $\varepsilon \to 0$, in the simple case under consideration where $\phi' < 0$ on $[0, 1]$, it is easy to see in numerical examples that solutions lie inside a limiting envelope defined by the successive maxima and minima. See Figure 18.11.

Our theory makes it clear that once oscillations start they continue, until the local minima are close to 0, because the successive maxima and minima of $u$ decrease by a small amount (which could be estimated) from one to the next. It is shown in [**63**] that in an interval where there are oscillations, the set of maxima tend, as $\varepsilon \to 0$, to a smooth curve $e(x)$ (an "envelope") satisfying a certain differential equation

$$e' = H(x, e(x)),$$

where $H$ is a function defined in terms of the homoclinic and heteroclinic orbits of the system (18.26), for various values of $x^*$.

It is further proved in [**63**] that when $\phi$ is not monotonic and $\phi - \frac{1}{2}$ may have more than one zero, there is a class of allowable envelope functions, characterized by a type of energy, and that for each "equal energy" pair of such allowable envelope functions, there is a family of solutions to (18.1)–(18.3) whose maxima and minima tend to the given pair of envelopes as $\varepsilon \to 0$. Their proof of this result uses a "gluing"procedure.

We conjecture that the existence of these solutions can also be obtained by shooting. We have not proved this, but the numerical solution shown above was obtained by a numerical version of the usual shooting argument (set $u(0) = \alpha, u'(0) = 0$ and adjust $\alpha$), and it seems to us that this could be justified rigorously. We believe that this is true not only when $\phi$ is monotonic, but for every smooth function $\phi$ with $0 < \phi < 1$ and with $\phi' \neq 0$ when $\phi = \frac{1}{2}$. We would use an inductive procedure, as in [**4**].

## 18.8. Exercises

**Exercise 18.1.** (Straightforward) Carry out the analysis of the $(W, W')$ phase plane for (18.15).

**Exercise 18.2.** (Straightforward; needed in this chapter) With $\gamma, \delta$, and $\bar{\alpha}$ as in Lemma 18.6, show that there is a $\sigma > 0$ such that if $0 < u(0) < \bar{\alpha}$ and $u$ is bounded, then $0 < u(0) < e^{-\frac{\sigma}{\varepsilon}}$ for sufficiently small $\varepsilon$.

**Exercise 18.3.** Show that under assumption (18.2), if $\varepsilon$ is sufficiently small, then there is an increasing solution to (18.1)–(18.3) which has a half-spike at $x = 0$ and another increasing solution which has a half-spike at $x = 1$. (Interpret "half-spike" in a reasonable way.)

Chapter 19

# Three unsolved problems

## Statements of Problems

There may be readers who turn to this chapter first, being most interested in what is unknown. In this short final chapter we mention three other problems which have occupied a number of researchers in recent decades, and yet have so far failed to reveal their secrets.[1] Space constraints keep us from giving many details about the various approaches which have been tried, but for each problem we cite at least one paper which in turn has a number of citations to earlier work in the field. We would be thrilled if one or more of these problems were to be solved by researchers who first learned of them in this book. We begin by succinctly stating the problems and then give some background for each. The last section contains a new proof, based partly on work of W. C. Troy, of an important piece of the background for the third problem.

### 19.1. Homoclinic orbit for the equation of a suspension bridge

*Prove that if* $0 < q < 2$ then *the equation*

$$u^{(iv)} + qu'' + e^u - 1 = 0 \tag{19.1}$$

*has a nontrivial solution* $u$ *such that* $\lim_{s\to-\infty} u(s) = \lim_{s\to\infty} u(s) = 0.$

[1]The prudent researcher may wish to search recent literature to be sure that these problems haven't been solved since the book went to press!

## 19.2. The nonlinear Schrödinger equation

*Show that for some $a > 0$ the system*

$$
\begin{aligned}
&x'' + \tfrac{2}{\xi}x' + x\left(x^2 + y^2 - 1\right) - ay - a\xi y' = 0, \\
&y'' + \tfrac{2}{\xi}y' + y\left(x^2 + y^2 - 1\right) + ax + a\xi x' = 0
\end{aligned}
\tag{19.2}
$$

*has a solution $(x(\xi), y(\xi))$ on $(0, \infty)$ such that*

$$
\begin{aligned}
\lim_{\xi \to 0^+} y(\xi) = \lim_{\xi \to 0^+} x'(\xi) = \lim_{\xi \to 0^+} y'(\xi) = 0, \\
\lim_{\xi \to \infty} \xi x'(\xi) + x(\xi) - \tfrac{1}{a}y(\xi) = 0, \\
\lim_{\xi \to \infty} \xi y'(\xi) + \tfrac{1}{a}x(\xi) + y(\xi) = 0.
\end{aligned}
\tag{19.3}
$$

## 19.3. Uniqueness of radial solutions for an elliptic problem

*Prove that for each positive integer $k$ the boundary value problem*

$$u'' + \frac{2}{r}u' - u + u^3 = 0, \tag{19.4}$$

$$\lim_{r \to 0^+} u'(r) = 0, \quad \lim_{r \to \infty} u(r) = 0 \tag{19.5}$$

*has at most one solution which is initially positive and has exactly $k$ zeros in $(0, \infty)$.*

# References and some background

## 19.4. Comments on the suspension bridge problem

This is the newest and least studied of these problems, and also the most specialized, but it has challenged a number of researchers for the past two decades. The equations of a suspension bridge were developed largely by McKenna and coworkers in [**146**], [**33**], [**162**] and references therein. To explain the key physical feature, let $u = u(x, t)$ denote the vertical displacement downward of the bridge surface from its rest position at distance $x$ along the bridge and time $t$. When $u > 0$, the cables near that location are stretched and exert an upward restoring force on the bridge. But when $u < 0$, there is no corresponding downward force. Thus, cables are different from springs. The force term $e^u - 1$ is a smooth function showing a version of this asymmetry. It changes sign at $u = 0$, but $|e^u - 1|$ is much larger then $|e^{-u} - 1|$ when $u$ is large and positive. The specific model above is found in [**33**]. It was studied in the monograph [**190**], which is where we found the specific formulation (19.1). A number of interesting solutions are given there but the traveling wave problem is left as a conjecture.

The book [**190**] is also a good source for a number of other references for the problem. We mention particularly the paper [**220**], which gives a variational approach. In this paper it is proved that there is a homoclinic traveling wave solution for almost all values of $q$ in $(0,2)$. Since this result could exclude all rational values of $q$, for example, it seems to us to leave considerable room for improvement. As to the reason that variational techniques do not work, we will say only that their proof involves a certain function $q \to c_q$ which turns out to be decreasing in $q$ but not necessarily differentiable at every $q$. It is known from real analysis that a monotonic function is differentiable at almost all values of $q$, and it was shown in [**220**] that there is a homoclinic orbit at each $q$ where $\frac{dc_q}{dq}$ exists.

## 19.5. Comments on the nonlinear Schrödinger equation

In (19.2)–(19.3) we have reduced the problem to what might be thought to be its simplest terms, though this is perhaps debatable. The nonlinear Schrödinger equation is important in a variety of areas of modern physics. Among these are the propagation of a laser beam through a medium, electromagnetic waves in a plasma, the motion of a vortex filament for the Euler equations of fluid mechanics, and some models of Bose-Einstein condensates. See [**251**] and [**147**] for citations in these areas.

The cubic form of the equation is

$$i\frac{\partial\Phi}{dt} + \Delta\Phi + \Phi\left|\Phi\right|^2 = 0 \tag{19.6}$$

where $\Phi$ is complex. We assume radial symmetry, so that $\Phi$ depends on $t$ and $r$. For the physical background, see [**234**] or, more briefly, [**147**].[2]

The problem (19.2)–(19.3) arises in studying finite time blowup of solutions to (19.6). (Physicists refer to this as "wave collapse".) Radially symmetric solutions of equation (19.6) can be studied in each dimension $d \geq 1$, using the radial Laplacian $\Delta = \frac{\partial^2}{\partial r^2} + \frac{d-1}{r}\frac{\partial}{\partial r}$. In (19.2) we have taken $d = 3$, and this is the case we will concentrate on below. Then, as shown, for example, in Section 7.1 of [**234**], suitable scaling of $t, r$, and $\Phi$ together with an appropriate similarity substitution results in a complex ode

$$Q_{\xi\xi} + \frac{2}{\xi}Q_\xi - Q + ia\left(\xi Q\right)_\xi + Q\left|Q^2\right| = 0, \tag{19.7}$$

where $a$ is a real parameter. As seen in [**234**], the rescalings of both $r$ and $\Phi$ are $t$-dependent, becoming more exaggerated as $t$ tends to the blowup time.

[2]In some physical applications, including the important example of self-focusing in laser propagation, "$t$" is replaced by "$z$", because this variable represents distance along the beam. This was studied, for example, in [**65**], and indeed, we wish to thank a co-author of that paper who is one of the major researchers in the field, G. Papanicolau, for introducing us to this problem.

The function $Q$ is to satisfy boundary conditions

$$\begin{aligned} \lim_{\xi\to 0} Q_\xi(\xi) = 0, \qquad \lim_{\xi\to 0} \operatorname{Im} Q(\xi) = 0, \\ \lim_{\xi\to\infty} |Q(\xi)| = 0. \end{aligned} \tag{19.8}$$

The second of these three conditions is, basically, that $Q(0)$ is real. We note that (19.7) is invariant under a rotation $Q \to Qe^{i\theta}$ in the complex plane, so that this condition fixes the phase of $Q$. There is the extra physical constraint that the Hamiltonian associated with (19.6),

$$H(\psi) = \int_{R^n} |\nabla\psi|^2 - \frac{1}{2}|\psi|^4 \; dx,$$

is finite, and it is shown in Section 7.7.1 of [**234**] that this is equivalent to the condition

$$H(Q) = 0. \tag{19.9}$$

The system (19.2) is obtained from (19.7) by setting $Q = x + iy$. Condition (19.9) is a different sort of constraint from the boundary conditions we have discussed up to now, but it is not unusual to have integral constraints in problems derived from physics. We now give a heuristic argument showing the effect of this on the problem. See the cited literature for more details.

Linearizing the complex equation (19.7) around $Q = 0$ results in the linear equation

$$\tilde{Q}'' + \frac{2}{\xi}\tilde{Q}' - \tilde{Q} + ia\left(\xi\tilde{Q}(\xi)\right)' = 0.$$

It is further shown in Section 7.1.1 of [**234**] that this equation has solutions $\tilde{Q}_1$ and $\tilde{Q}_2$, where as $\xi \to \infty$,

$$\tilde{Q}_1(\xi) \sim \xi^{-1-\frac{i}{a}}, \qquad \tilde{Q}_2(\xi) \sim \xi^{-\left(2-\frac{i}{a}\right)} e^{-ia\frac{\xi^2}{2}}.$$

Examining the singularities of $\tilde{Q}_1$ and $\tilde{Q}_2$ at $\infty$, we see that $\tilde{Q}_1$ is in $L_4\left(R^3\right)$ and $\tilde{Q}_1'$ is in $L_2\left(R^3\right)$. On the other hand, $\tilde{Q}_2'$ is not in $L_2\left(R^3\right)$, because the volume integral introduces an $r^2$ into the integrand, and so $H(Q)$ is finite only for solutions which are asymptotic to a multiple of $\tilde{Q}_1$ as $\xi \to 0$.

The complex exponents in the asymptotic expressions for $\tilde{Q}_1$ and $\tilde{Q}_2$ show that the real and imaginary parts of each are oscillatory, but $\tilde{Q}_1$ oscillates much more slowly than $\tilde{Q}_2$. The challenge is to show for this nonlinear problem that, for some $a$, a solution with the given initial conditions picks out just the slow oscillation in its behavior as $\xi \to \infty$. This is known to follow from (19.3) and to be sufficient to give condition (19.9). See Exercise 19.1.

Using either the complex form of the problem, (19.7)–(19.8), or the real form (19.2)–(19.3), we see that we are dealing with a nonlinear "connection" problem, and in this sense the problem is similar to that considered for

Painlevé II in Chapter 3. A third equivalent version can be found by using polar coordinates for the dependent variable $Q = x + iy$. It appears from numerical computation that the problem has a solution for each $d$ with $2 < d \leq 4$. However the only existence proofs are those by Kopell and Landman [**123**] and by Rottshäfer and Kaper [**205**], each of which covers a range $2 < d < 2 + \delta$ for a small $\delta$.[3]

## 19.6. Comments on the elliptic problem and a new existence proof

Our final problem can be obtained from the nonlinear Schrödinger equation as well as a number of other problems in partial differential equations. Starting with (19.6), we ask if there are standing wave solutions. These are solutions of the form

$$\Phi(x,t) = e^{i\lambda^2 t}\psi(x).$$

Substituting this into (19.6) and looking for real solutions for $\psi$, we get

$$\Delta\psi - \lambda^2\psi + \psi^3 = 0. \tag{19.10}$$

Then, we look for radially symmetric solutions $\psi(x) = u(r)$ where $r = |x|$. Setting $\lambda = 1$ and taking the spatial dimension $n$ to be 3 gives equation (19.4):

$$u'' + \frac{2}{r}u' - u + u^3 = 0.$$

The boundary conditions (19.5), namely

$$u'(0) = 0, \quad u(\infty) = 0,$$

are obtained from the physical problems where these equations arise [**251**].

**19.6.1. Existence and uniqueness of solutions.** The first result we know about was by Nehari [**179**], who proved that the problem (19.4)–(19.5) has at least one positive solution. Positive solutions are the most important physically, since they are solutions of minimum energy and are the most stable [**251**]. In [**42**] Coffman proved that there is only one positive solution, and we repeated his proof in Chapter 12. This solution is called the "ground state", because it is the solution of lowest energy in an appropriate sense.

Solutions which have sign changes, called "bound states", are also important, as localized finite energy solutions [**252**]. In [**207**] it was shown by Ryder that for each $k \geq 1$ there is a solution to (19.4)–(19.5) with exactly $k$ zeroes in $(0,\infty)$. The problem posed above is whether these solutions are unique. There are no results for this problem. At the end of this section, and this book, we give a new proof of the existence of the bound states.

[3] The range of $d$ in [**205**] is slightly larger than that in [**123**].

**19.6.2. Extensions.** Following Coffman's paper, K. McLeod and Serrin [**168**] considered equations of the form

$$u'' + \frac{n-1}{r}u' + |u|^{p-1}u - u = 0 \tag{19.11}$$

with boundary conditions

$$u'(0) = 0, \quad u(\infty) = 0. \tag{19.12}$$

It is known from work of Pohozaev [**193**] that for $n > 2$ there are no solutions if $p \geq \frac{n+2}{n-2}$. Coffman's result was for the case $p = n = 3$. In [**168**] uniqueness of the ground state was shown under a variety of conditions on $p$ and $n$. In particular, the proof covers those $(n, p)$ such that $2 < n \leq 4$ and $1 < p < \frac{n+2}{n-2}$. In [**134**], M.-K. Kwong completely solved the ground state uniqueness problem by removing the restriction on $n$, showing that for every $n > 2$ there is a unique positive solution if $1 < p < \frac{n+2}{n-2}$. His proof uses techniques of Coffman and Sturm oscillation theory. The result was extended, and the proof shortened, by K. McLeod [**167**], and a further shortening, using geometric methods, was given by Clemons and Jones in [**39**]. The latter proof made use of an "Emden-Fowler" transformation, and it would be interesting to see if a shorter classical proof could be found using this transformation.

There are also solutions to problem (19.11)–(19.12) which are not positive. Once again the condition $1 < p < \frac{n+2}{n-2}$ is imposed, and then, for every $k \geq 1$ there is a solution with exactly $k$ zeros. This result was proved in [**111**] by a dynamical systems method and in [**169**] by a classical method.[4] The uniqueness of these solutions is unknown for any $k > 0$. In [**241**] Troy obtained a uniqueness result for $k = 1, 2$ in the case where the nonlinear term is a piecewise linear function mimicking $u - u^3$ [**241**].

**Remark 19.1.** Equation (19.10) is the basis of both our second and third unsolved problems. It is a special case of

$$\Delta u + f(u) = 0$$

where $f(u) \sim |u|^{p-1} u$ as $|u| \to \infty$. This equation has been intensively studied in $R^n$ for arbitrary positive integers $n$, and the properties of solutions depend markedly on the values of $n$ and $p$. The literature is too vast to summarize here, or to list a significant number of references. One way to search for these would be to look for citations of the landmark paper of Pohozaev [**193**]. For papers more closely related to the problem (12.44)–(12.45) and its extensions, look for citations of [**42**].

[4]The classical proof is considerably shorter.

**19.6.3. Existence of bound states for (19.4).** The proofs of Nehari and Ryder referred to in Section 19.6.1 used the calculus of variations. The results in [**111**] and [**169**] discussed above include the existence of bound states for (12.44) as a special case. Here we will give a new proof, a major step of which is based on a transformation of Troy which greatly simplifies part of the problem. We thank Professor Troy for allowing us to use it here. We discuss only the case of (19.4), but the technique extends to the equation (19.11). The theorem also covers existence of the ground state, as was promised in Chapter 12.

**Theorem 19.2.** *For each nonnegative integer $k$ there is a solution of* (19.4)–(19.5) *with exactly $k$ zeros in* $(0,\infty)$.

**Proof.** The first step is to rescale the equation to obtain what appears to be a simpler problem. As usual, for $\alpha>0$ let $u(r,\alpha)$ denote the unique solution of (19.4) such that $u(0)=\alpha$, $u'(0)=0$. (See Exercise 1.3.) Following [**169**], define $v$ by

$$u(r)=\alpha v(\alpha r).$$

Then

$$v''+\frac{2}{s}v'+v^3-\frac{1}{\alpha^2}v=0$$

and

$$v(0)=1,\quad v'(0)=0, \tag{19.13}$$

where $s=\alpha r$. Letting $\delta=\frac{1}{\alpha}$ we get

$$v''+\frac{2}{s}v'+v^3-\delta v=0. \tag{19.14}$$

Denote the solution of (19.14) and (19.13) by $v(s,\delta)$. Let $w(s)=v(s,0)$. Then

$$w''+\frac{2}{s}w'+w^3=0, \tag{19.15}$$
$$w(0)=1,\quad w'(0)=0.$$

A modification of Exercise 1.3 can be used to show that for $s>0$, $v(s,\alpha)$ depends continuously on $\delta$. This continuity is uniform in every interval $0<s\le S$. Hence, if $w$ has $k$ zeros in $(0,S)$ then for sufficiently small $\delta$ (depending on $k$), $v$ has at least $k$ zeros in $(0,S)$ and $u$ has at least $k$ zeros in $(0,\frac{1}{\delta}S)$. (Note that $w'\neq 0$ at each zero of $w$.) On the other hand, we will see that $w$ has infinitely many zeros in $(0,\infty)$, while by linearizing (19.4) around $u=0$ it is found that $u$ has only a finite number of zeros in this interval.

**Lemma 19.3.** *There is no solution of* (19.15) *which is positive on some interval of the form* $(\sigma,\infty)$.

**Proof.** The history of this simple sounding lemma is complicated, because (19.15) and the more general equation

$$w'' + \frac{n-1}{s} w' + |w|^{p-1} w = 0 \tag{19.16}$$

have been widely studied. The result as we have stated it appears in [**30**], with a proof and references to previous work on the same problem. The result is used in [**169**], from where another chain of references can be traced. The proof below, based on Troy's transformation, is considerably simpler than any other we know of, and the method can be extended to (19.16).

Suppose that $w$ is a solution of (19.15) which is positive on some interval $(\sigma, \infty)$. Define the function $\gamma$ on $(\log \sigma, \infty)$ by[5]

$$w(s) = \frac{1}{s} \gamma(\log s).$$

Then

$$\gamma'' - \gamma' + \gamma^3 = 0. \tag{19.17}$$

Since this equation is autonomous, it can be written as

$$\begin{aligned} x' &= y, \\ y' &= y - x^3. \end{aligned}$$

Then $(0,0)$ is the only equilibrium point. It is nonhyperbolic, because the eigenvalues of the linearization around $(0,0)$ are 0 and 1. Since

$$\left(\frac{1}{2} x^4 + y^2\right)' = 2y^2,$$

all nonzero solutions can be shown to be unbounded and, more importantly, to spiral away from the origin. (See our final exercise at the end of this chapter.) Therefore no solution satisfies $x > 0$ on an interval $(T, \infty)$, implying that $w$ cannot be positive on an interval $(e^T, \infty)$. This proves Lemma 19.3.[6] □

The next step is to recall the energy functional $W$ which was introduced in Section 12.6, namely

$$W(t) = \frac{1}{2} u'(t)^2 + F(u(t)), \tag{19.18}$$

where

$$F(u) = \frac{1}{4} u^4 - \frac{1}{2} u^2. \tag{19.19}$$

---

[5]For (19.16), with $p > 1$, the transformation is $w(s) = s^{-\frac{2}{p-1}} \gamma(\log s)$. Note that $n$ is not involved.

[6]In a recent paper [**129**], Krisner and Troy used a similar transformation to determine some other important properties, not previously known, of (19.16).

Recall that

$$W' = -\frac{2}{r}\left(u'\right)^2. \tag{19.20}$$

**Lemma 19.4.** *If $u$ is a solution of* (19.4) *with $W(r_0) < 0$ for some $r_0 > 0$, then $u$ has no zeros in $(r_0, \infty)$ and either* $\lim_{r\to\infty} u(r)$ *does not exist or this limit is nonzero.*

**Proof.** It is easy to show that if $u \to 0$ then $u' \to 0$. The lemma then follows from (19.18), (19.19), and (19.20), since these imply that $W$ is decreasing, $W \geq 0$ if $u = 0$, and $\lim_{r\to\infty} W(r) = 0$ if $u \to 0$. □

Lemma 19.3 shows that for each positive integer $k$ the set of positive numbers $\alpha$ such that $u(r,\alpha)$ has more than $k$ zeros is nonempty. On the other hand, Lemma 19.4 implies that if $0 < \alpha < \sqrt{2}$ then $W(0) < 0$ and $u(r,\alpha)$ has no zeros. Hence, for each $k \geq 0$

$$\alpha_k = \inf\{\alpha > 0 \mid u(r,\alpha) \text{ has more than } k \text{ zeros}\}$$

is well defined, and $\alpha_k \geq \sqrt{2}$. We wish to show that $u(r,\alpha_k)$ has exactly $k$ zeros and that $\lim_{r\to\infty} u(r,\alpha_k) = 0$.

**Remark 19.5.** The situation here is not unique. Recall, for example, that in Chapter 11 we presented a fairly straightforward proof that a certain boundary value problem had solutions with at least $k$ zeros, but we had to work harder to show there were solutions with exactly $k$ zeros. For the problem under consideration here, we can choose $\alpha$ so large that there are more than $k$ zeros. As we lower $\alpha$, zeros disappear. The issue is how many can disappear at once. While the proof below is not phrased in exactly those terms, this is what it amounts to: we show that as $\alpha$ is lowered, zeros only disappear one at a time, and so counting down, we come to a solution with exactly $k$ zeros.

It follows from the continuity of $u$ with respect to $\alpha$ that $u_k = u(r,\alpha_k)$ has no more than $k$ zeros, for otherwise $u(r,\alpha)$ would still have more than $k$ zeros at values of $\alpha$ just below $\alpha_k$. Let

$$W_k = \frac{1}{2}\left(u_k'\right)^2 + F(u_k).$$

Since $F(u_k) \geq -\frac{1}{2}$ and $W_k$ is decreasing, $L = \lim_{r\to\infty} W_k(r)$ exists. We can assume that after its last zero, $u_k > 0$. Suppose that $\lim_{r\to\infty} u_k(r)$ does not exist. Then $u_k$ has an unbounded sequence $\{r_m\}$ of successive maxima and minima. Since $u_k'(r_m) = 0$, $\lim_{m\to\infty} F(u_k(r_m)) = L$. By considering the graph of $F(u)$ for $u > 0$ we see that this is only possible if $-\frac{1}{4} < L < 0$. In this case, solutions $u(r,\alpha)$ sufficiently close to $u_k$ reach the region where $W < \frac{L}{2}$, after which they have no further zeros. But up to this point they must have the same number of zeros as $u_k$, if $|\alpha - \alpha_k|$ is sufficiently small, and thus they have the same number of zeros as $u_k$ in $(0, \infty)$, contradicting the definition of $\alpha_k$. Hence, $\lim_{r\to\infty} u_k(r)$ exists. This limit is either zero or one, and if it is one, then nearby solutions also tend to one, with no more zeros then $u_k$. Again this is a contradiction of the definition of $\alpha_k$. Hence, $\lim_{r\to\infty} u_k(r) = 0$. Lemma 19.4 then implies that $W_k > 0$ on $[0, \infty)$.

We still have to show that if $k > 0$ then $u_k$ has exactly $k$ zeros, and not fewer. Suppose that $u_k$ has exactly $j$ zeros for some $j < k$. We will show that this is impossible by showing that if $\alpha - \alpha_k$ is sufficiently small and positive, then $u_\alpha$ has no more than $j+1$ zeros, contradicting the definition of $\alpha_k$.

To show this we again use the functional $W$ and introduce an identity which was used in [**189**] and [**169**]. From the equations (19.18) and (19.20) it is found that

$$W' + \frac{2}{r}W = -\frac{1}{r}u'^2 + \frac{2}{r}F(u),$$

and using the integrating factor $r^2$ gives

$$(r^2W)' = -ru'^2 + 2rF(u). \tag{19.21}$$

For the next step recall that $u_k(r) \to 0$ as $r \to \infty$.

**Lemma 19.6.** *With* $W_k = \frac{1}{2}(u_k')^2 + F(u_k)$, $\lim_{r\to\infty} r^2 W_k(r) = 0$.

**Proof.** Considering (19.4) for large $r$, we expect that solutions which tend to zero eventually behave like solutions of the equation

$$u'' + \frac{2}{r}u' - u = 0.$$

These are linear combinations of modified Bessel functions, and it is not hard to see that there is a one-parameter family of such solutions, say of the form $\{cP(r)\}$ for a particular solution $P(r)$, which approach zero as $r$ approaches infinity. Further, $P(r) = O(e^{-\mu r})$ as $r \to \infty$ for some $\mu > 0$.

Then a standard integral equation argument, like the one outlined in Section 3.3, shows that if $u$ is a solution of (19.4) which approaches zero at infinity, then both $u$ and $u'$ tend to zero at an exponential rate. This conclusion implies the lemma. □

Suppose now for definiteness that $j$ is odd, and so $u_k(r) < 0$ for large $r$. For each $\delta > 0$ there is a $\rho_k > 1$ so large that if $r \geq \rho_k$ then $-1 < u_k(r) < 0$ and $0 < r^2 W_k(r) < \delta$. From Lemma 19.4 we see that $\rho_k$ can also be chosen so that if $r \geq \rho_k$ then $u_k' > 0$.

Suppose now that for some $\alpha > \alpha_k$, $u = u(\alpha, r)$ has a $(j+1)^{\text{st}}$ zero $r_{j+1}$. As $\alpha \to \alpha_k^+$, $r_{j+1} \to \infty$. Then $u$ is positive in some interval $(r_{j+1}, R)$, and we wish to show that if $\alpha - \alpha_k$ is sufficiently small then $R = \infty$ and there are no further zeros of $u$.

We can choose $\alpha$ so close to $\alpha_k$ that $-1 < u(\rho_k) < 0$, $u'(\rho_k) > 0$, $r_{j+1} > \rho_k$, and

$$\rho_k^2 W(\rho_k) < \delta. \tag{19.22}$$

We see from (19.21) that $r^2 W(r)$ is decreasing as long after $r = \rho_k$ as $|u| < \sqrt{2}$. Recall that if $W$ is ever negative then $u$ cannot have further zeros. It is therefore our goal to show that $W$ becomes negative. Using (19.21) we see that if $|u| \leq 1$ then

$$\left(r^2 W\right)' \leq -\frac{1}{2} r u^2. \tag{19.23}$$

If $u'$ has a zero before $u = \sqrt{2}$ then $W$ is negative at that zero and we are done. Hence we can assume that $u' > 0$ as long as $u \leq 1$, and also that $u$ reaches 1. On the other hand, we know that $W \leq \frac{1}{4}\alpha^4 - \frac{1}{2}\alpha^2$. Choosing $\alpha < \alpha_k + 1$, for example, we see that there is a bound for $|u'|$ on $[0, \infty)$ which is independent of $\alpha$. Hence there is an interval of length $d > 0$, also independent of $\alpha$, in which $\frac{1}{2} \leq u \leq 1$. Recall that $r_{j+1} > \rho_k \geq 1$. Integrating (19.23) and using (19.22) shows that if $\delta$ is sufficiently small then $W$ becomes negative, and $u$ has no further zeros. This proves Theorem 19.2. □

## 19.7. Exercises

**Exercise 19.1.** (Easy) Show that the asymptotic form given for $\tilde{Q}_1(\xi)$ satisfies the boundary conditions at $\infty$ in (19.3).

**Exercise 19.2.** (Straightforward) Complete the proof that every solution of (19.17) has an unbounded set of zeros.

# Bibliography

[1] M. J. Ablowitz and P. A. Clarkson, *Solitons, Nonlinear Evolution Functions and Inverse Scattering,* Cambridge University Press, 1991.

[2] M. Abramowitz and I. Stegun (editors), *Handbook of Mathematical Functions,* National Bureau of Standards, Applied Mathematics Series #55, 1964. See also NIST Digital Library of Mathematical Functions at `http://dlmf.nist.gov`.

[3] S.-B. Ai, Multibump solutions to Carrier's problem, *Jour. Math. Anal. Appl.* **277** (2003), 405-422.

[4] S.-B. Ai, X. Chen, and S. P. Hastings, Layers and spikes in nonhomogeneous bistable reaction-diffusion equations, *Trans. Amer. Math. Soc.* **358** (2006), 3169-3206.

[5] S.-B. Ai and S. P. Hastings, A shooting approach to layers and chaos in a forced Duffing equation, *J. Diff. Eqns.* **185** (2002), 389-436.

[6] G. B. Airy, On the intensity of light in the neighborhood of a caustic, *Transactions of the Cambridge Philosophical Society* **6** (1838), 379-402.

[7] K. T. Alligood, T. Sauer, and J. A. Yorke, *Chaos: An Introduction to Dynamical Systems,* Springer New York, 1996.

[8] D. Allwright, The Hopf bifurcation in some biochemical control loops, *J. Math. Biol.* **11** (1981), 85-93.

[9] C. Amick and J. B. McLeod, A singular perturbation problem in needle crystals, *Arch. Rat. Mech. Anal.* **109** (1990), 139-171.

[10] C. Amick and J. B. McLeod, A singular perturbation problem in water waves, *Stability Appl. Anal. Contin. Media* **1** (1991), 127-148.

[11] S. B. Angenent, J. Mallet-Paret, and L. A. Peletier, Stable transition layers in a semilinear boundary value problem, *J. Diff. Eqns.* **67** (1987), 212-242.

[12] D. G. Aronson, H. F. Weinberger, Nonlinear diffusion in population genetics, combustion and nerve propagation in *Partial Differential Equations and Related Topics,* Springer Lectures Notes in Math **446** (1975), 5-49.

[13] T. Bakri, Y. A. Kuznetsov, F. Verhulst, and E. Doedel, Multiple solutions of a generalized singular perturbed Bratu problem, preprint.

[14] A. P. Bassom, P. A. Clarkson, C. K. Law, and J. B. McLeod, Application of uniform asymptotics to the second Painlevé transcendent, *Arch. Rat. Mech. Anal.* **143** (1988), 241-271.

[15] A. P. Bassom, P. A. Clarkson, A. C. Hicks, and J. B. McLeod, Integral equations and exact solutions for the fourth Painlevé equation, *Proc. Roy. Soc. Lon. Ser. A* **437** (1992), 1-24.

[16] J. Bebernes and D. Eberly, *Mathematical Problems from Combustion Theory,* Springer, 1989.

[17] E. S. Benilov, S. J. Chapman, J. B. McLeod, J. R. Ockendon, and S. Zubkov, On liquid films on an inclined plate, *J. Fluid Mech.* **663** (2010), 53-69.

[18] C. M. Bender and S. A. Orszag, *Advanced Mathematical Methods for Scientists and Engineers*, Springer, 1999.

[19] M. V. Berry and K. E. Mount, Semi-classical approximations in wave mechanics, Reports on Progress in Physics **35** (1972), 315-397.

[20] S. R. Bishop, A. Sofroniou, and P. Shi, Symmetry-breaking in the response of the parametrically excited pendulum model, *Chaos, Solitons and Fractals* **25** (2005), 257-264.

[21] J. Bona and F. Weissler, Traveling fronts of a conservation law with hyper-dissapation, Adv. Diff. Eqns. **16** (2011), 917–935.

[22] W. Borsch-Supan and M. Fiebig-Wittmaack, Stability of stationary solutions of a one-dimensional parabolic equation with homogeneous Neumann boundary conditions, *J. Diff. Eqns.* **94** (1991), 55–66.

[23] E. F. F. Botta, F. J. Hut, and A. E. P. Veldman, The role of periodic solutions in the Falkner Skan equation for $\lambda > 0$, *J. Eng. Math.* **20** (1986), 81-93.

[24] J. S. Bramley and S. C. R. Dennis, The calculation of eigenvalues for the stationary perturbation of Poiseuille flow, *Journal of Computational Physics* **47** (1982), 179-198.

[25] H. Brezis, *Analyse Functionnelle,* Masson, Paris, 1983.

[26] P. Brunovsky, Tracking invariant manifolds without differential forms, *Acta Math. Univ. Comenianai* **LXV** (1996), 23-32.

[27] G. Carpenter, A geometric approach to singular perturbation problems with applications to nerve impulse equations, *J. Diff. Eqns.* **23** (1977), 335-367.

[28] J. Carr, *Applications of Center Manifold Theory,* Springer, 1981.

[29] G. F. Carrier, Singular perturbation theory and geophysics, *SIAM. Rev.* **12** (1970), 175-193.

[30] S. Chen, W. R. Derrick, and J. Cima, Positive and oscillatory radial solutions of semilinear elliptic equations, *J. Appl. Math and Stochastic Anal.* **10** (1997), 95-108.

[31] X. Chen, Existence, uniqueness, and asymptotic stability of traveling waves in non-local evolution equations, *Adv. Diff. Eqns* **2** (1997), 125-160.

[32] X. Chen and S. Sadhu, Asymptotic expansions of solutions of an inhomogeneous equation, preprint.

[33] Y. Chen and P. J. McKenna, Traveling waves in a nonlinear suspended beam: Theoretical results and numerical observations, *J. Diff. Eqns.* **136** (1997), 325-355.

[34] C. Chicone, *Ordinary Differential Equations with Applications,* Springer, 1999.

[35] M. Chipot, S. P. Hastings, and D. Kinderlehrer, Transport in a molecular motor system, *Math. Model. and Numer. Anal.* **38** (2004), 1011-1034.

[36] M. Chipot, D. Kinderlehrer, and M. Kowalczyk, A variational principle for molecular motors, *Meccanica* **38** (2003), 505-518.

[37] D. C. Clark, A variant of the Ljusternik-Schnirelman theory, *Indiana Univ. Math. J.* **22** (1972), 65-74.

[38] P. A. Clarkson, Painlevé equations: Nonlinear special functions, in F. Marcellán and W. Van Asschel (eds.), *Orthogonal Polynomials and Special Functions: Computation and Application,* Lecture Notes in Mathematics **1883** (2006), 331-411.

[39] C. B. Clemons and C. K. R. T. Jones, A geometric proof of the Kwong-McLeod uniqueness result, *SIAM J. Math. Anal.* **24** (1993), 436-443.

[40] M. J. Clifford and S. R. Bishop, Approximating the escape zone for the parametrically excited pendulum, *Journal of Sound and Vibration* **172** (1994), 572-576.

[41] E. A. Coddington and N. Levinson, *Theory of Ordinary Differential Equations,* McGraw-Hill, 1955.

[42] C. Coffman, Uniqueness of the ground state for $\Delta u - u + u^3 = 0$, *Arch. Rat. Mech. Anal.* **46** (1972), 81-95.

[43] D. S. Cohen, A. Fokas, and P. A. Lagerstrom, Proof of some asymptotic results for a model equation for low Reynolds number flow, *SIAM J. Appl. Math.* **35** (1978), 187–207.

[44] C. Conley, *Isolated invariant sets and the Morse index,* CMBS Lecture Notes **38**, Amer. Math. Soc., 1978.

[45] W. A. Coppel, On a differential equation of boundary layer theory, *Philos. Trans. Roy. Soc. London Ser. A* **253** (1960), 101-136.

[46] C. De Coster and P. Habets, An overview of the method of lower and upper solutions for odes, in *Nonlinear Analysis and Its Applications to Differential Equations,* vol. 43 of Progress in Nonlinear Differential Equations and Their Applications, Birkhäuser, 2001, 3–22.

[47] M. G. Crandall and P. H. Rabinowitz, Bifurcation, perturbation of simple eigenvalues and linearized stability, *Arch. Rat. Mech. Anal.* **52** (1973), 161-180.

[48] E. N. Dancer, On the structure of solutions of an equation in catalysis theory when a parameter is large, *J. Diff. Eqns.* **37** (1980), 404-437.

[49] E. N. Dancer and S. Yan, Construction of various types of solutions for an elliptic problem, *Calculus of Variations* **20** (2004), 93-118.

[50] E. N. Dancer and S. Yan, Multi-layer solutions for an elliptic problem, *J. Diff. Eqns.* **194** (2003), 382-405.

[51] R. Devaney, *An Introduction to Chaotic Dynamical Systems* (2nd edition), Westview Press, 2003.

[52] P. G. Drazin and W. H. Reid, *Hydrodynamic Stability* (2nd edition), Cambridge Press, 2004.

[53] S. R. Dunbar, Traveling wave solutions of diffusive Lotka-Volterra equations: A heteroclinic connection in $R^4$, *Trans. Amer. Math. Soc.* **286** (1984), no. 2, 557–594.

[54] M. Del Pino, P. Felmer, and K. Tanaka, An elementary construction of complex patterns in nonlinear Schrödinger equations, *Nonlinearity* **15** (2002), 1653-1671.

[55] C. Doering, G. B. Ermentrout, and G. Oster, Rotary DNA Motors, *Biophysical Journal* **69** (1995), 2256-2267.

[56] Y. Du and Y. Lou, Proof of a conjecture for the perturbed Gelfand equation from combustion theory, *J. Diff. Eqns.* **173** (2001), 213-230.

[57] Y. Du, Bifurcation and related topics in elliptic problems, in *Handbook of Differential Equations, Stationary Partial Differential Equations,* volume **II** (2005), Elsevier, 127-209.

[58] G. B. Ermentrout, *Simulating, Analyzing, and Animating Dynamical Systems, A Guide to XPPAUT for Researchers*, SIAM, 2002.

[59] G. B. Ermentrout and D. Terman, *Mathematical Foundations of Neuroscience*, Springer, 2010.

[60] A. Erdelyi, *Asymptotic Expansions*, Dover, 2010.

[61] G. B. Ermentrout and J. B. McLeod, Existence and uniqueness of travelling waves for a neural network, *Proc. Roy. Soc. Edinburgh* Sect. A **123** (1993), 461-478.

[62] C. Evans, *Partial Differential Equations*, 2nd edition, AMS, Graduate Studies in Mathematics, Vol. 19, 2010.

[63] P. Felmer, S. Martinez, and K. Tanaka, High-frequency chaotic solutions for a slowly varying dynamical system, *Ergodic Theory and Dynamical Systems* **26** (2006), 379-407.

[64] N. Fenichel, Geometric perturbation theory for ordinary differential equations, *J. Diff. Eqns.* **31** (1979), 53-98.

[65] G. Fibich and G. Papanicolau, Self-focusing in the perturbed and unperturbed nonlinear Schrödinger equation in critical dimension, *SIAM J. Appl. Math.* **60** (1999), 183-240.

[66] P. Fife and J. McLeod, The approach of solutions of nonlinear diffusion equations to traveling front solutions, *Arch. Rat. Mech. Anal.* **65** (1977) 335–361.

[67] P. Fife and J. B. McLeod, A phase plane discussion of convergence to travelling fronts for nonlinear diffusion, *Arch. Rat. Mech. Anal.* **75** (1980/81), 281-314.

[68] P. Fife, Understanding the patterns in the BZ reagent, *Jour. Stat. Phys.* **39** (1985), 687-703.

[69] R. FitzHugh, Impulses and physiological states in models of nerve membrane, *Biophysical Journal,* 1 (1961), 445-466.

[70] A. S. Fokas, A. R. Its, A. Kapseev, and V. Yu Novokshenov, *Painlevé Transcendents, the Riemann-Hilbert Approach,* Amer. Math. Soc., 2006.

[71] A. C. Fowler, *Mathematical Models in the Applied Sciences*, Cambridge University Press, 1997.

[72] P. O. Frederickson and A. C. Lazer, Necessary and sufficient damping in a second order oscillator, *J. Diff. Eqns.* **5** (1969), 252-270.

[73] R. E. Gaines and J. Mawhin, *Coincidence Degree and Nonlinear Differential Equations,* Springer, Berlin, 1977.

[74] G. P. Galdi, *An Introduction to the Mathematical Theory of the Navier Stokes Equations,* vols I and II, Springer, 1994, 2001.

[75] T. Gedeon, H. Kokubo, K. Mischaikov, and Hiroe Oka, Chaotic solutions in slowly varying perturbations of Hamiltonian systems with application to shallow water sloshing, *J. Dynamics and Diff. Eqns.* **14** (2002), 63-84.

[76] I. M. Gelfand, Some problems in the theory of quasilinear equations, *Amer. Math. Soc. Transl. (2)* **29** (1963), 295-381.

[77] B. Gidas, W.-M. Ni, and L. Nirenberg, Symmetry and related properties via the maximum principle, *Comm. Math. Physics* **68** (1979), 209-243.

[78] P. Glendenning, *Stability, Instability and Chaos*, Cambridge Univ. Press, 1994.

[79] B. C. Goodwin, Oscillatory behavior in enzymatic control processes, *Adv. Enzyme Regul.* **3** (1965), 425-438.

[80] J. Guckenheimer, A strange, strange attractor, in Marsden and Mc-Cracken (eds.), *The Hopf Bifurcation and Its Applications*, Springer, 1976.

[81] J. Guckenheimer and P. Holmes, *Nonlinear Oscillations, Dynamical Systems, and Bifurcations of Vector Fields,* Springer, 1983.

[82] J. K. Hale, *Ordinary Differential Equations,* Wiley-Interscience, 1969.

[83] P. Hartman, *Ordinary Differential Equations,* John Wiley & Sons, 1964.

[84] J. M. Hammersley and G. Mazzarino, A differential equation connected with the dendritic growth of crystals, *IMA J. Appl. Math.* **42** (1989), 43–75.

[85] S. P. Hastings, On the uniqueness and asymptotic behavior of a similarity solution for a compressible boundary layer, *Arch. Rat. Mech. Anal.* **42** (1971), 128-136.

[86] S. P. Hastings, Single and multiple pulse waves for the FitzHugh-Nagumo equations, *SIAM J. Applied Math* **42** (1982), 247-260.

[87] S. P. Hastings, D. Kinderlehrer, and J. B. McLeod, Diffusion mediated transport in multiple state systems, *SIAM J. Math. Anal.* **39** (2007), 1208-1228.

[88] S. P. Hastings, C. Lu, and Y. H. Wan, A three parameter shooting method as applied to a problem in combustion theory, *Physica D* **19** (1986), 301-306.

[89] S. P. Hastings and J. B. McLeod, A boundary value problem related to the KDV equation, *Arch. Rat. Mech. Anal.* **73** (1980), 31-51.

[90] S. P. Hastings and J. B. McLeod, On the periodic solutions of a forced second-order equation, J. *Nonlinear Science* **1** (1991), 225-245.

[91] S. P. Hastings and J. B. McLeod, Chaotic motion of a pendulum with oscillatory forcing, *Amer. Math. Monthly* **100** (1993), 563-572.

[92] S. P. Hastings and J. B. McLeod, Short proofs of results by Landesman, Lazer, and Leach on problems related to resonance, *Differential and Integral Equations* **24** (2011), 435–441.

[93] S. P. Hastings and J. B. McLeod, An elementary approach to a model problem of Lagerstrom, *SIAM J. Math. Anal.* **40** (2009), 2421-2436.

[94] S. P. Hastings and J. Murray, The existence of oscillatory solutions in the Field-Noyes model for the Belousov-Zhabotinsky reaction, *SIAM J. Appl. Math.* **28** (1975), 678-688.

[95] S. P. Hastings and W. C. Troy, Oscillating solutions of the Falkner-Skan equation for positive $\beta$, *J. Diff. Eqns.* **71** (1988), 123-144.

[96] S. P. Hastings, J. Tyson, and D. Webster, Existence of periodic solutions for negative feedback cellular control systems, *J. Diff. Eqns.* **25** (1977), 39-64.

[97] B. D. Hassard, N. D. Kazarinoff, and Y. H. Wan, *Theory and Applications of Hopf Bifurcation,* London Mathematical Society Lecture Notes **41** (1978), Cambridge University Press.

[98] P. Hess, On a theorem by Landesman and Lazer, *Indiana Univ. Math. J.* **23** (1973/74), 827-829.

[99] E. J. Hinch, *Perturbation Methods*, Cambridge, 1991.

[100] M. W. Hirsch, *Differential Topology,* Springer, 1976.

[101] M. W. Hirsch and H. L. Smith, Monotone dynamical systems, in *Handbook of Differential Equations: Ordinary Differential Equations,* vol. **2** (2005), Elsevier/North-Holland, 239-357.

[102] P. Holmes and D. Spence, On a Painlevé type boundary-value problem, *Quart. J. Mech. Appl. Math.* **37** (1984), 525-538.

[103] A. L. Hodgkin and A. F. Huxley, A quantitative description of membrane current and its application to conduction and excitation in nerve, *Journal of Physiology* (London) **117** (1952), 500–544.

[104] A. P. Hooper and R. Grimshaw, Nonlinear instability at the interface between two viscous fluids, *Phys. Fluids* **28** (1985), 37-45.

[105] K.-C. Hung and S.-H. Wang, A theorem on S-shaped bifurcation curves for a positone problem with convex-concave nonlinearity and its application to the perturbed Gelfand problem, J. Diff. Eqns. **251** (2011), 223–237.

[106] E. L. Ince, *Ordinary Differential Equations*, Dover, 1956.

[107] S. Invernizzi and G. Treu, Quantitative analysis of the Hopf bifurcation in the Goodwin $n$-dimensional metabolic control system, *J. Math. Biol.* **29** (1991), 733-742.

[108] C. K. R. T. Jones, Stability of the traveling wave solutions of the Fitzhugh-Nagumo system, *Trans. Amer. Math. Soc.* **286**(1984), 431-469.

[109] C. K. R. T. Jones, Geometric singular perturbation theory, *Dynamical Systems (Montecatini Terme, 1994),* Lecture Notes in Mathematics, **1609**, Springer Verlag, 1995.

[110] C. K. R. T. Jones, N. Kopell, and R. Langer, Construction of the FitzHugh-Nagumo pulse using differential forms, in *Patterns and Dynamics in Reactive Media,* IMA Volumes in Mathematics and Its Applications **37** (1991), 101-116.

[111] C. K. R. T. Jones and T. Küpper, On the infinitely many solutions of a semilinear elliptic equation, *SIAM J. Math. Anal.* **17** (1986), 803-835.

[112] J. Jones, W. C. Troy, and A. D. MacGillivray, Steady solutions of the Kuramoto-Shivishinsky equation for small wave speed, *J. Diff. Eqns.* **96** (1992) 28-55.

[113] D. Joseph and T. Lundgren, Quasilinear Dirichlet problems driven by positive sources, *Arch. Rat. Mech. Anal.* **49** (1973), 241-269.

[114] N. Joshi and A. V. Kitaev, On Boutroux's tritronquée solutions of the first Painlevé equation, *Studies in Applied Mathematics* **107** (2001), 253-291.

[115] N. Joshi and M. D. Kruskal, The Painlevé connection problem: an asymptotic approach, *Studies in Applied Mathematics* **86** (1992), 315-376.

[116] T. J. Kaper and S. Wiggins, A commentary on "On the periodic solutions of a forced second-order equation", by S. P. Hastings and J. B. McLeod, *J. Nonlinear Science* **1** (1991), 247-253.

[117] A. Kapila, B. Matkowsky, and J. Vega, Reactive-diffusive systems with Arrhenius kinetics: Peculiarities of the spherical geometry, *SIAM J. Appl. Math.* **38** (1980), 382-401.

[118] S. Kaplun and P. A. Lagerstrom, Low Reynolds number flow past a circular cylinder, *J. Math. Mech.* **6** (1957), 595-603.

[119] A. V. Kitaev, C. K. Law, and J. B. McLeod, Rational solutions of the fifth Painlevé equation, *Differential Integral Equations* **7** (1994), 967-1000.

[120] E. Knobloch, Review of *Pattern Formation and Dynamics in Nonlinear Systems* by M. Cross and H. Greenside, *SIAM Review* **52** (2010), 567-572.

[121] A. N. Kolmogorov and S. V. Fomin, *Elements of the Theory of Functions and Functional Analysis*, Dover, 1999.

[122] N. Kopell and L. Howard, Bifurcations and trajectories joining critical points, *Advances in Mathematics* **18** (1975), 306-358.

[123] N. Kopell and M. Landman, Spatial structure of the focusing singularity of the nonlinear Schrödinger equation: A geometrical analysis, *SIAM J. Math. Anal.* **55** (1995), 1297-1323.

[124] P. Korman, Global solution branches and exact multiplicity of solutions for two point boundary value problems, in *Handbook of Differential Equations, volume 3,* Elsevier, 2006.

[125] P. Korman and Y. Li, On the exactness of an $S$-shaped bifurcation curve, *Proc. Amer. Math. Soc.* **127** (1999), 1011-1020.

[126] P. Korman, Y. Li, and T. Ouyang, An exact multiplicity result for a class of semilinear equations, *Comm. Partial Diff. Eqns.* **22** (1997), 661-684.

[127] M. A. Krasnoselski, *Positive Solutions of Operator Equations,* Noordhoff, Groningen, 1964.

[128] M. Krupa, B. Sandstede, and P. Szmolyan, Slow and fast waves in the FitzHugh-Nagumo equation, *J. Diff. Eqns.* **133** (1997), 49-97.

[129] E. Krisner and W. C. Troy, Radially symmetric solutions of $\Delta w + |w|^{P-1}w = 0$, preprint.

[130] M. Kruskal and H. Segur, Asymptotics beyond all orders in a model of crystal growth, *Studies in Appl. Math.* **85** (1991), 129-181.

[131] Y. Kuramoto and T. Tsuzuki, Persistent propagation of concentration waves in dissipative media far from thermal equilibrium, *Progr. Theoret. Phys.* **55** (1976), 356-369.

[132] H. Kurland, Monotone and oscillatory equilibrium solutions of a problem arising in population genetics, in *Contemporary Mathematics* **17,** Amer. Math. Soc., 1983, 323–342.

[133] Y. Kuznetsov, *Elements of Applied Bifurcation Theory,* Springer, 2004.

[134] M.-K. Kwong, Uniqueness of positive radial solutions of $\Delta u - u + u^p = 0$ in $R^n$, *Arch. Rat. Mech. Anal.* **105** (1989), 243-266.

[135] T. Laetsch, The number of solutions of a nonlinear two point boundary value problem, *Indiana Univ. Math. J.* **20** (1970), 1-13.

[136] P. A. Lagerstrom, *Matched Asymptotic Expansion: Ideas and Techniques,* Springer, 1988.

[137] P. A. Lagerstrom and R. G. Casten, Basic concepts underlying singular perturbation theory, *SIAM Review* **14** (1972), 63-120.

[138] L. D. Landau and E. M. Lifschtiz, *Fluid Mechanics,* 2nd edition, Reed Educational and Professional Publishing, 1987.

[139] E. M. Landesman and A. C. Lazer, Nonlinear perturbations of linear elliptic boundary value problems at resonance, *J. Math. Mech. (Indiana J. Math.)* **19** (1970), 609-623.

[140] E. M. Landesman and A. C. Lazer, Linear eigenvalues and a nonlinear boundary value problem, *Pacific J. Math.* **33** (1970), 311-328.

[141] J. L. Langer, Existence of needle crystals in local models of solidification, Phys. Rev. A **33** (1986), 435-441.

[142] P. Lax, *Linear Algebra,* Wiley-Interscience, 1997.

[143] A. C. Lazer, A second look at the first result of Landesman-Lazer type, *Electronic J. Diff. Eqns. Conf.* **5** (2000), 113-119.

[144] A. C. Lazer and D. E. Leach, Bounded perturbations of forced harmonic oscillators at resonance, *Ann. Mat. Pura Appl.* **82** (1969), 49-68.

[145] A. C. Lazer and D. E. Leach, On a nonlinear two point boundary value problem, *J. Math. Anal. Appl.* **26** (1969), 20-27.

[146] A. C. Lazer and P. J. McKenna, Large-amplitude oscillations in suspension bridges: Some new connections with nonlinear analysis, *SIAM. Rev.* **32** (1990), 537-578.

[147] B. LeMesurier and P. Christiansen, Regularization and control of self-focusing in the 2D cubic Schrödinger equation by attractive linear potentials, *Physica D* **184** (2003), 226-236.

[148] M. Levi, Quasi-periodic motions in super-quadratic time-periodic potentials, *Comm. Math. Phys.* **143** (1991), 43-83.

[149] B. Liu, Quasiperiodic solutions of nonlinear Liénard equations, *Disc. Cont. Dyn. Sys.* **12** (2005), 137-160.

[150] C. Lu, Multiple steady states in a biochemical system, *SIAM J. Math. Anal.* **18** (1987), 1771-1783.

[151] C. Lu, Global bifurcation of steady-state solutions on a biochemical system, *SIAM J. Math. Anal.* **21** (1990), no. 1, 76–84.

[152] C. Lu, Asymptotic analysis of the Carrier-Pearson problem, *Elect. J. Diff. Eqns.* Conference **10** (2003), 227-240.

[153] C. Lu, Chaotic motions of a parametrically excited pendulum, *Comm. Nonl. Sci. Num. Sim.* **11** (2006), 861-884.

[154] C. Lu, Chaos of a parametrically excited undamped pendulum, *Comm. Nonl. Sci. Num. Sim.* **12** (2007), 45-57.

[155] A. D. MacGillivray, Perturbation analysis of a problem of Carrier's, *Stud. Appl. Math.* **104** (2000), 293-311.

[156] P. Maini and H. Othmer, eds., Mathematical Models for Biological Pattern Formation, in *IMA volumes in Mathematics and Applications*, **121,** Springer, 2001.

[157] J. L. Massera, The existence of periodic solutions of systems of differential equations, *Duke J. Math.* **17** (1950), 457-475.

[158] W. Massey, *Algebraic Topology: An Introduction*, Springer, 1989.

[159] J. Mawhin, Resonance and nonlinearity, a survey, *Ukrainian Mathematical Journal* **59** (2007), 197-214.

[160] J. Mallet-Paret, The global structure of traveling waves in spatially discrete systems, *J. Dyn. Diff. Eq.* **11** (1999), 49–127.

[161] J. Mallet-Paret and H. L. Smith, The Poincaré-Bendixson theorem for monotone cyclic feedback systems, *J. Dyn. Diff. Eqns.* **2** (1990), 367-421.

[162] P. J. McKenna and W. Walter, Traveling waves in a suspension bridge, *SIAM J. Appl. Math.* **50** (1990), 703-715.

[163] J. B. McLeod, *Swirling Flow,* in *Lecture Notes in Mathematics,* **448,** Springer, 1975.

[164] J. B. McLeod and P. J. Olver, The connection between partial differential equations soluble by inverse scattering and ordinary differential equations of Painlevé type, *SIAM J. Math. Anal.* **14** (1983), 488-506.

[165] J. B. McLeod and S. Sadhu, Existence of solutions and asymptotic analysis of a class of singularly perturbed boundary value problems, preprint.

[166] J. B. McLeod and J. Serrin, The existence of similar solutions for some laminar boundary layer problems, *Arch. Rat. Mech. Anal.* **31** (1968), 288-303.

[167] K. McLeod, Uniqueness of positive radial solutions of $\Delta u + f(u) = 0$ in $R^n$: II, *Trans. Am. Math. Soc.* **339** (1993), 495-505.

[168] K. McLeod and J. Serrin, Uniqueness of positive radial solutions of $\Delta u + f(u) = 0$ in $R^n$, *Arch. Rat. Mech. Anal.* **99** (1987), 115-145.

[169] K. McLeod, W. C. Troy, and F. B. Weissler, Radial solutions of $\Delta u + f(u) = 0$ with prescribed numbers of zeros, *J. Diff. Eqns.* **83** (1990), 368-378.

[170] A. I. Mees and P. E. Rapp, Periodic metabolic systems: Oscillations in multiple-loop negative feedback biochemical control networks, *J. Math. Biol.* **5** (1978), 99-114.

[171] V. K. Melnikov, On the stability of the center for time periodic solutions, *Trans. Moscow Math. Soc.* **12** (1963), 3-52.

[172] K. Mischaikov and M. Mrozek, Conley index, in *Handbook of Dynamical Systems, volume 2*, B. Fiedler, ed., Elsevier, 2002.

[173] J. W. Milnor, *Topology from the Differentiable Viewpoint,* University Press of Virginia, 1965, reprinted by Princeton University Press, 1997.

[174] G. Morris, A case of boundedness in Littlewood's problem on oscillatory differential equations, *Bull. Austral. Math. Soc.* **14** (1976), 71-93.

[175] J. Moser, *Stable and Random Motions in Dynamical Systems, with Special Emphasis on Celestial Mechanics,* Princeton University Press, 1973, reprinted 2001.

[176] J. Nagumo, S. Arimoto, and S. Yoshizawa, An active pulse transmission line simulating nerve axon, *Proc IRE.* **50** (1962), 2061–2070.

[177] K. Nakashima, Multi-layered stationary solutions for a spatially inhomogeneous Allen-Cahn equation, *J. Diff. Eqns.* **191** (2003), 234-276.

[178] Z. Nehari, Characteristic values associated with a class of nonlinear second order differential equations, *Acta. Math.* **105** (1961), 141-175.

[179] Z. Nehari, On a nonlinear differential equation arising in nuclear physics, *Proc. Roy. Irish Academy* **62** (1963), 117-135.

[180] M. H. A. Newman, *Elements of the Topology of Plane Sets of Points*, Cambridge Univ. Press, 1951.

[181] W.-M. Ni and R. Nussbaum, Uniqueness and nonuniqueness for positive radial solutions of $\Delta u + f(u, r) = 0$, *Comm. Pure Appl. Math.* **38** (1985), 67-108.

[182] H. Ockendon, J. R. Ockendon, and A. D. Johnson, Resonant sloshing in shallow water, *J. Fluid Mech.* **167** (1986), 465-479.

[183] R. E. O'Malley Jr., *Singular Perturbation Methods for Ordinary Differential Equations,* Springer, 1991.

[184] R. Ortega, Boundedness in a piecewise linear oscillator, and a variant of the small twist theorem, *Proc. Lon. Math. Soc.* **79** (1999), 381-413.

[185] R. Ortega, Twist maps, invariant curves and periodic differential equations, *Progr. Nonlinear Diff. Eqns. Appl.* **43** (2001), 85-112.

[186] P. Painlevé, Sur les équations différentielles du second ordre et d'ordre supérieur dont l'intégrale générale est uniforme, *Acta Math.* **25** (1902), 1–85.

[187] K. J. Palmer, Exponential dichotomies and transverse homoclinic points, *J. Diff. Eqns.* **55** (1984), 225-256.

[188] K. J. Palmer, Transversal heteroclinic points and Cherry's example of a nonintegrable Hamiltonian system, *J. Diff. Eqns.* **65** (1986), 321-360.

[189] L. A. Peletier and J. Serrin, Uniqueness of positive solutions of semilinear equations in $R^n$, *Arch. Rat. Mech. Anal.* **81** (1983), 181-197.

[190] L. A. Peletier and W. C. Troy, *Spatial Patterns: Higher Order Models in Physics and Mechanics*, Birkhäuser, 2001.

[191] L. Perko, *Differential Equations and Dynamical Systems*, Springer Texts in Applied Mathematics 7, Third Edition, 2001.

[192] É. Picard, Mémoire sur la théorie des équations aux dérivées partielles et la méthode des approximations successives, *J. Math. Pures Appl.* **6** (1890), 145-210.

[193] S. Pohozaev, Eigenfunctions of the equation $\Delta u + \lambda f(u) = 0$, *Soviet Math.* **6** (1965), 1408-1411.

[194] N. Popović and P. Szmolyan, A geometric analysis of the Lagerstrom model problem, *J. Diff Eqns.* **199** (2004), 290-325.

[195] N. Popović and P. Szmolyan, Rigorous asymptotic expansions for Lagerstrom's model equation—a geometric approach, *Nonlinear Analysis* **59** (2004), 531-565.

[196] M. H. Protter and H. F. Weinberger, *Maximum Principles in Differential Equations*, Prentice Hall, 1958, reprinted by Springer, 1984.

[197] C. Pugh, *Real Mathematical Analysis,* Springer, 2002.

[198] O. Purcell, N. J. Savery, C. S. Grierson, and M. di Bernardo. A comparative analysis of synthetic genetic oscillators, *Journal of The Royal Society Interface* **7** (2010), 1503-1524.

[199] P. J. Rabier, Invertibility of the Poiseuille linearization for stationary two-dimensional channel flows: Symmetric case, *J. Math. Fluid. Mech.* **4** (2002), 327-350.

[200] P. J. Rabier, Invertibility of the Poiseuille linearization for stationary two-dimensional channel flows: Nonsymmetric case, *J. Math Fluid Mech.* **4** (2002), 351-373.

[201] A. G. Ramm, A simple proof of the Fredholm alternative and a characterization of the Fredholm operators, *Amer. Math. Monthly* **108** (2001), 855.

[202] M. Renardy and R. Rogers, *An Introduction to Partial Differential Equation*, Springer, 1993.

[203] S. V. Rhagavan, J. B. McLeod, and W. C. Troy, A singular perturbation problem arising from the Kuramoto-Shivishinsky equation, *Differential and Integral Equations* **10** (1997), 1-36.

[204] R. R. Rosales, The similarity solution for the Korteweg-De Vries equation and the related Painlevé transcendent, *Proc. Roy. Soc. Lon.* **361** (1978), 265-275.

[205] V. Rottschäfer and T. Kaper, Blowup in the nonlinear Schrödinger equation near critical dimension, *J. Math. Anal. Appl.* **268** (2002), 517-549.

[206] P. Ruoff et al., The Goodwin model: Simulating the effect of light pulses on the circadian sporulation rhythm of *Neurospora crassa, J. Theor. Biol.* **209** (2001), 29-42.

[207] G. H. Ryder, Boundary value problems for a class of nonlinear equations, *Pac. J. Math.* **22** (1967), 477-503.

[208] S. Sadhu, *Existence and Asymptotic Analysis of Solutions of Singularly Perturbed Boundary Value Problems*, Thesis, University of Pittsburgh, 2011.

[209] B. Sandstede, Stability of traveling waves, in *Handbook of Dynamical Systems II* (B. Fiedler, ed.), North-Holland, 2002, 983-1055.

[210] D. Sattinger, Monotone methods in nonlinear elliptic and parabolic boundary value problems, *Indiana Univ. Math. J.* **21** (1972), 979-1000.

[211] D. Salamon, Connected simple systems and the Conley index of isolated invariant sets, *Trans. Amer. Math. Soc.* **291** (1985), 1-41.

[212] L. Segel, *Mathematical Models in Molecular and Cellular Biology,* Cambridge University Press, 1980.

[213] J. Serrin, Existence theorems for some compressible boundary layer problems, *Studies in Applied Mathematics 5; Advances in Differential and Integral Equations*, SIAM (1970), 35-42.

[214] L. P. Shilnikov, A case of the existence of a denumerable set of periodic motions, *Sov. Math. Dokl.* **6** (1965), 163-166.

[215] M. Shub, *Global Stability of Dynamical Systems,* Springer, 1987.

[216] N Sidorov, B. Loginov, A. Sinitsyn, and M. Falaleev, *Lyapunov-Schmidt Methods in Nonlinear Analysis and Applications,* Kluwer Academic Publishers, 2002.

[217] G. Sivashinsky, Nonlinear analysis of hydrodynamic instability in laminar flames, I. Derivation of basic equations, *Acta Astronaut.* **4** (1977), 1117-1206.

[218] M. Slemrod, Monotone increasing solutions of the Painlevé I equation $y'' = y^2 + x$ and their role in the stability of the plasma-sheath transition, *Eur. J. Appl. Math.* **13** (2002), 663-680.

[219] S. Smale, Diffeomorphisms with many periodic points, in *Differential and Combinatorial Topology,* S. S. Cairns (ed.), Princeton University Press (1965), 63-80.

[220] D. Smets and J. B. van der Berg, Homoclinic solutions for Swift-Hohenberg and suspension bridge type equations, *J. Diff. Eqns.* **184** (2002), 78-96.

[221] H. L. Smith, Periodic orbits of competitive and cooperative systems, *J. Diff. Eqns.* **65** (1986), 361-373.

[222] H. L. Smith, *Monotone Dynamical Systems, Math. Surveys and Monographs*, vol. 41, Amer. Math. Soc., 1995.

[223] J. Smoller, *Shock Waves and Reaction-Diffusion Equations*, 2nd edition, Springer, 1994.

[224] J. Smoller and A. Wasserman, Global bifurcation of steady state solutions, *J. Diff. Eqns.* **39** (1981), 269-290.

[225] R. Srzednicki, Ważewski method and Conley index, in *Handbook of Differential Equations: Ordinary Differential Equations,* vol. **1** (2004), Elsevier/North-Holland, 591-684.

[226] I. Stakgold, Branching of solutions of nonlinear equations, *SIAM Review* **13** (1971), 289-382.

[227] K. Stewartson, Correlated compressible and incompressible boundary layers, *Proc. Roy. Soc.* **A 200**, (1949), 84-100.

[228] M. Struwe, *Variational Methods,* Springer, 1990.

[229] C. A. Stuart, Buckling of a heavy tapered rod, *J. Math. Pures. Appl.* **80** (2001), 281-337.

[230] C. A. Stuart, On the spectral theory of a tapered rod, *Proc. Roy. Soc. Edin.* **132A** (2002), 729-764.

[231] C. A. Stuart and G. Vuillaume, Buckling of a critically tapered rod: Properties of some global branches of solutions, *Proc. Roy. Soc.* Lon. A **460** (2004), 3261-3282.

[232] H. P. F. Swinnerton-Dyer and C. T. Sparrow, The Falkner-Skan equation I. The creation of strange invariant sets, *J. Diff. Eqns.* **119** (1995), 336-394.

[233] H. P. F. Swinnerton-Dyer and C. T. Sparrow, The Falkner-Skan equation II. Dynamics and bifurcation of $P$ and $Q$ orbits, *J. Diff. Eqns.* **183** (2002), 1-55.

[234] C. Sulem and P.-L. Sulem, *The Nonlinear Schrödinger Equation,* vol. **139** in Applied Mathematical Sciences, Springer, 1999.

[235] F. Takens, Singularities of vector fields, *Publ. Math. IHES* **43** (1974), 47-100.

[236] K. Tam, On the Lagerstrom model for flow at low Reynolds numbers, *J. Math. Anal. Appl.* **49**(1975), 286–294.

[237] A. Taylor and D. Lay, *Introduction to Functional Analysis,* 2nd edition, Wiley, 1980.

[238] J. H. Taylor and K. S. Narendra, Stability regions for the damped Mathieu equation, *SIAM Journal on Applied Mathematics* **17** (1969), 343-352.

[239] E. C. Titchmarsh, *Eigenfunction Expansions Associated with Second Order Differential Equations,* 2nd edition, Oxford University Press, 1962.

[240] W. C. Troy, The existence of steady solutions of the Kuramoto-Sivashinsky Equations, *J. Diff. Eqns.* **82**(1989), 269-313.

[241] W. C. Troy, The existence and uniqueness of bound-state solutions of a semi-linear equation, *Proc. Roy. Soc. A* **461** (2005), 2941-2963.

[242] J. F. Toland, Existence and uniqueness of heteroclinic orbits for the equation $\lambda u''' + u' = f(u)$, *Proc. Roy. Soc. Edin.* A **109** (1988), 23-36.

[243] W. Tucker, A rigorous ODE solver and Smale's 14th Problem, *Found. Comp. Math.* **2** (2002), 53–117.

[244] J. Tyson, On the existence of oscillatory solutions in negative feedback cellular control processes, *J. Math. Biol.* **1** (1975), 311-315.

[245] F. Verhulst, *Methods and Applications of Singular Perturbations,* Springer, 2005.

[246] G. Vuillaume, Studying the buckling of a tapered rod with the genus of a set, *SIAM J. Math. Anal.* **34** (2003), 1128-1151.

[247] G. Vuillaume, A singular nonlinear eigenvalue problem: Bifurcation in non-differentiable cases, *Thése No. 2775*, École Polytechnique Fédérale de Lausanne, 2003.

[248] W. Walter, *Ordinary Differential Equations,* Springer, 1998.

[249] S. H. Wang, On S-Shaped Bifurcation curves, *Nonlinear Analysis* **22** (1994), 1475-1485.

[250] T. Ważewski, Sur un principe topologique de l'examen de l'allure asymptotique des intégrales des équations différentielles ordinaires, *Ann. Soc. Polon. Math.* **20** (1947), 279-313.

[251] M. I. Weinstein, Modulational stability of ground states of nonlinear Schrödinger equations, *SIAM J. Math. Anal.* **16** (1985), 472-491.

[252] M. I. Weinstein, The nonlinear Schrödinger equation—singularity formation, stability, and dispersion, *Contemporary Mathematics* **99** (1989), Amer. Math. Soc., 213-232.

[253] H. Weyl, On the differential equations of the simplest boundary-layer problems, *Ann. Math.* **43** (1942), 381-407.

[254] S. D. R. Wilson, The development of Poiseuille flow, *J. Fluid Mech.* **38** (1969), 793-806.

[255] S. D. R. Wilson, The "drag-out" problem in film coating theory, *J. Eng. Math.* **16** (1982), 209-221.

[256] R. Wong and Y. Zhao, On the number of solutions to Carrier's problem, Stud. Appl. Math. **120** (2008), 213-245.

[257] E. Yanagida, Stability of fast traveling pulse solutions of the FitzHugh-Nagumo equation, *Journal of Math. Bio.* **22** (1985), 81-104.

[258] S. Wiggins, On the detection and dynamical consequences of orbits homoclinic to hyperbolic periodic orbits and normally hyperbolic invariant tori in a class of ordinary differential equations, *SIAM J. Applied Math.* **48** (1988), 262-285.

[259] S. Wiggins, *Global Bifurcations and Chaos: Analytic Methods*, Springer, 1988.

# Index